# LOW TEMPERATURES AND COLD MOLECULES

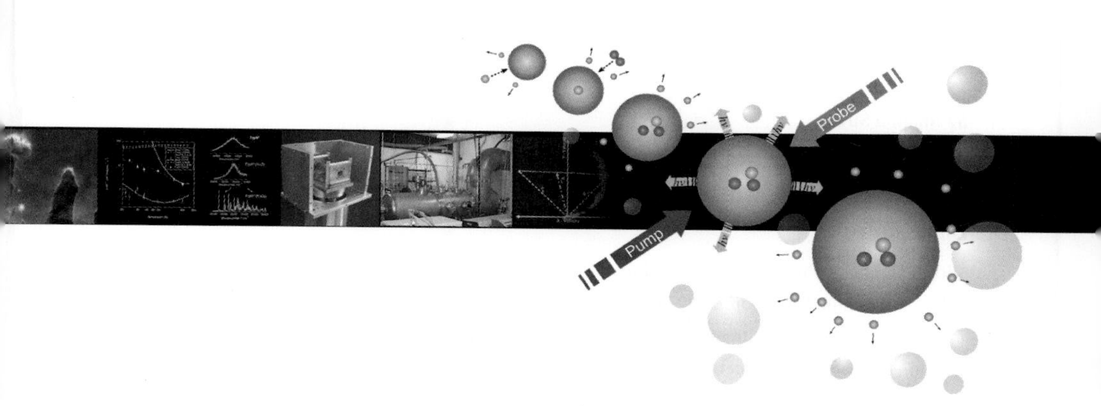


# LOW TEMPERATURES AND COLD MOLECULES

Editor

**Ian W M Smith**

Universities of Birmingham and Cambridge

ICP

Imperial College Press

*Published by*

Imperial College Press
57 Shelton Street
Covent Garden
London WC2H 9HE

*Distributed by*

World Scientific Publishing Co. Pte. Ltd.
5 Toh Tuck Link, Singapore 596224
*USA office:* 27 Warren Street, Suite 401-402, Hackensack, NJ 07601
*UK office:* 57 Shelton Street, Covent Garden, London WC2H 9HE

**British Library Cataloguing-in-Publication Data**
A catalogue record for this book is available from the British Library.

**LOW TEMPERATURES AND COLD MOLECULES**



ISBN-13 978-1-84816-209-9
ISBN-10 1-84816-209-X

Typeset by Stallion Press
Email: enquiries@stallionpress.com

Printed by FuIsland Offset Printing (S) Pte Ltd, Singapore

# Preface

The attainment of extreme conditions and the examination and understanding of how matter behaves under previously unattainable conditions is a constant stimulus to scientists. The concept of an absolute zero of temperature has been with us for many years and the challenge of reaching as close as possible to that limit is nothing new. However, dramatic advances have been made in recent years and this book provides articles describing some of these advances, written by leading experts in these fields. The title of the volume also stresses '*molecules*'; that is, here we are concerned with the production and behaviour of molecules — as distinct from atoms — under cold and ultra-cold conditions. In particular, emphasis is placed on the chemical reactions of species at low temperatures — and on how the study of chemical reactions can be pushed to ever lower temperatures.

It should be appreciated that the lowest temperatures known to science have been generated in earthly laboratories. The distribution of the cosmic background radiation with frequency corresponds to a temperature of 2.728 K. Although there are colder places in the universe, as far as we know at this time, temperatures below 1 K have only been achieved in laboratories. Molecules are observed in regions of space warmer than the cosmic background, but still extremely cold by most standards: the cold cores of dense interstellar clouds have temperatures of *ca.* 10 K. The first chapter in this book summarises our present state of knowledge regarding these huge interstellar regions, where most of the observed interstellar molecules are found. The next three chapters summarise experimental and theoretical efforts to explore chemical reactivity at these *low* temperatures and to provide the necessary information to incorporate into astrochemical models, which are constructed in order to see how well we understand the chemistry that leads to the observed interstellar molecules — and to the abundances of these species relative to that of molecular hydrogen, which is much the most abundant interstellar molecule.

Most of the known interstellar molecules have been identified from their spectra, either rotational spectra observed in emission using ground-based telescopes sensitive in the mmwave region of the electromagnetic spectrum, or *via* infrared spectroscopy conducted from satellites and using stars in the line-of-sight as background sources. In addition, in the laboratory — at least, the laboratories of chemical physicists — spectroscopic observations led the way in the use of gaseous expansions to generate very low temperatures. Chapter 5 describes some of the latest progress in this field, concentrating on the spectra of inherently unstable species, such as radicals and molecular ions and, in some senses, it acts as a bridge to the second half of the book, where the emphasis is on *ultra-low* temperatures and *ultra-cold* molecules.

A common theme in the second half of the book, Chapters 6 to 10, is the attainment of temperatures below, indeed well below, that of the cosmic background, 2.728 K. The holy grail in these experiments — or perhaps one of the holy grails — is to explore molecular collisions under conditions where the wavelength associated with the relative velocity of two species approaches or exceeds the size of the colliding species so that treatments of the collisions based on a classical approach to the dynamics of the relative motion are no longer appropriate.

Chapter 8 describes the examination of molecules in liquid helium droplets at 0.37 K ($^4$He) and 0.15 K ($^3$He). Initially, such experiments were spectroscopically led. They demonstrated the remarkable properties of helium as a 'soft' matrix that, in some cases, scarcely interacts with the molecules embedded in the droplets. Now this medium is being employed to study the dynamics of reactions.

The other chapters in this second half of the book explore approaches to producing ultra-cold charged or neutral molecules, with the ultimate objective of studying collisions at ultra-low relative velocities. The techniques vary from multi-pole traps in the case of molecular ions (Chapter 6), to kinematic methods of bringing molecules to rest (Chapter 8), to the use of rapidly switched electric fields to remove kinetic energy from neutral species (Chapter 9), and finally to methods more akin to those employed to generate atomic Bose-Einstein condensates (Chapter 10). All of these chapters describe work at the cutting-edge of research in molecular physics.

In all cases, the editor has encouraged the contributors both to give a critical survey of their field and to look forward to further developments in it. He believes that they have reacted splendidly to his exhortation. He further believes that this book should be of value to those already immersed in one or other of these areas of research, to newcomers to this field, and to more general readers who wish to know the present state-of-art in this very exciting area of chemical physics.

Ian Smith
Cambridge
May, 2008

# Contents

# Chapter 1

# The Chemistry of Cold Interstellar Cloud Cores

Eric Herbst

*Department of Physics, The Ohio State University,*
*Columbus, OH 43210, USA**

Tom J. Millar

*Astrophysics Research Centre, School of Mathematics and Physics*
*Queen's University Belfast, Belfast BT7 1NN,*
*Northern Ireland*

## Contents



*Also Departments of Astronomy and Chemistry.

We review the chemical processes that occur in cold (10 K) and dense ($n \approx 10^4$ cm$^{-3}$) interstellar cores, which are the coldest objects in larger assemblies of gas and dust known as dense interstellar clouds. We show how these processes produce the wide variety of exotic and normal molecules detected in these portions of the interstellar medium. Although much of the chemistry occurs in the gas phase, a significant portion takes place on the surfaces of the dust particles. Both types of chemistry are discussed. Some emphasis is placed on deuterium isotopic fractionation, which enhances the abundances of deuterium-containing isotopologues by very large factors compared with the low deuterium-to-hydrogen elemental abundance ratio of $\approx 10^{-5}$. The strengths and weaknesses of large time-dependent models of the chemistry are discussed.

## 1.1. Introduction

Baryonic matter in the universe is localised to a great extent in large assemblies of stars and interstellar matter known as galaxies. Galaxies come in a variety of shapes and sizes (e.g. spiral, elliptical, irregular, giant, dwarf); ours, the Milky Way, is a rather typical specimen of the spiral variety and has a baryonic mass of roughly $10^{11}$ solar masses. Except for a halo around the galactic centre, the Milky Way is rather flat, with its spiral arms tracing out high densities of material. Approximately 10% of the matter lies in between the stars. In external galaxies, the amount of interstellar matter can be significantly less. Careful study of the spectra of stellar atmospheres in which the matter is mainly atomic yields the relative abundances of each element. Although all stars are not the same in this regard, average or "cosmic abundances" are well-known for our galaxy and tell us that the Milky Way consists mostly of hydrogen, with the abundance of helium roughly 10% of that of hydrogen by number. Both of these elements were produced in the Big Bang. All heavier elements, which are produced in stellar interiors, have much lower abundances. For example, the biogenic elements carbon, nitrogen, and oxygen have abundances relative to hydrogen of $3 \times 10^{-4}$, $9 \times 10^{-5}$, and $7 \times 10^{-4}$, respectively. The abundance of deuterium is somewhat variable but is roughly 1–2 $\times 10^{-5}$ that of hydrogen throughout much of our galaxy. Elemental abundances need not be the same in other galaxies; for example, the abundances of elements heavier than helium, known collectively as the "metallicity" despite the fact that the most abundant elements are certainly not metals, can be much lower in nearby irregular galaxies.

The so-called interstellar medium is rather heterogeneous in density and temperature.[1,2] Molecules exist in the colder portions of the medium, which consists of "clouds" of varying density that can be labelled as "diffuse" or "dense" and contain matter both in the form of gas and tiny dust particles. Diffuse clouds are rather transparent to the radiation of background stars whereas dense clouds have sufficient material to be opaque to background stars in the visible and ultra-violet regions of the spectrum. The opacity is due mainly to the dust particles, and its wavelength dependence can be used to estimate their size distribution. The typical gas-phase temperature in a diffuse cloud lies between 50–100 K with densities $n$ in the range $10^{2-3}\,\mathrm{cm}^{-3}$. Although a small number of diatomic and triatomic gas-phase molecules have been detected in these regions, they are of low fractional abundance except for molecular hydrogen, which can have a concentration as large as its atomic form. Since dust particles tend to be formed out of heavy refractory elements, these are reduced in abundance compared with stellar cosmic abundances. Typical "depletions" in diffuse clouds range from factors of a few for carbon and oxygen to much greater factors for refractory elements such as silicon, iron, and calcium.

Although small and relatively homogeneous dense clouds, known as globules, can occur by themselves, the larger dense clouds are quite heterogeneous in density, and depending on size are known as "assemblies" or "giant" clouds. In these large objects, the dense matter often exists as "cold cores" in a bath of more diffuse material, which can be as rarefied as in diffuse clouds or somewhat denser. These cold cores, with a size of a light year ($9.46 \times 10^{17}$ cm) or less, a temperature of 10 K, and a density $n = 10^4\,\mathrm{cm}^{-3}$, are presumably formed from their surroundings by gravitational collapse. Their temperature is still significantly above the black-body background temperature of the universe, 2.728 K, which shows that there are additional sources of heat. The background material surrounding the cores is typically somewhat warmer, ranging in temperature up to $\approx$25 K for the more diffuse portions. In the dense gas, there is very little atomic hydrogen; the conversion from atomic to molecular hydrogen is essentially complete. Indeed, large numbers of gas-phase molecules can be detected, mainly through their rotational emission spectra. The two-best known cold cores are labelled TMC-1 and L134N; in the former, more than 50 different gas-phase molecules have been detected. The gas-phase elemental abundances in these objects can be difficult to determine since much of the matter is no longer in atomic form. Nevertheless, models of the chemistry work best with higher depletions for selected heavy elements than observed in diffuse clouds.

Cold cores are not stable indefinitely; some disperse while others collapse further. The collapse, which leads to the formation of solar-type stars, proceeds initially in an isothermal manner, since molecules radiate away the kinetic energy developed in the collapse. Eventually, a central dense condensation develops, which can have a temperature perhaps as low as 5 K and a density of $10^6$ $cm^{-3}$ or more. Such objects are known as "pre-stellar cores." Once the central condensation becomes sufficiently dense to be optically thick to cooling radiation, it can no longer remain cool but starts to heat up, becoming what is known as a "protostar", which emits in the infrared. As the protostar heats up, it warms its environment and this material becomes known as a hot corino.[1] Larger and warmer structures, known as hot cores, are associated with high-mass star formation. With temperatures ranging from 100–300 K depending upon the mass of the protostar, hot cores and corinos show a rich chemistry that is quite different from their colder forebears. Meanwhile, the central protostar becomes hot enough to ignite nuclear reactions and becomes a baby star. Some of the surrounding material collapses into a planar rotating disk of dense gas and dust, known as a "protoplanetary disk". In some instances, the dust in the disk can agglomerate into larger solid objects such as planetary systems. As the star-planet system develops, it can blow off the surrounding material. Before this event occurs, the nascent stars can strongly radiate the region around them and produce so-called photon-dominated regions.

### 1.1.1. *Interstellar Molecules and Their Chemistry*

Table 1.1 lists the gas-phase molecules detected in all phases of the interstellar medium, mainly dense clouds, and in very old stars that have developed a large circumstellar shell of gas and dust similar in physical conditions to cold interstellar cores. The molecules range in size from simple diatomics to a thirteen-atom nitrile, and are overwhelmingly organic in nature.[3] The second most abundant molecule to $H_2$ in dense regions is CO, which has a so-called fractional abundance with respect to $H_2$ of $10^{-4}$. A large percentage of the molecules are exotic by terrestrial standards, including positive and negative molecular ions, radicals, and isomers. Molecular positive ions have fractional abundances of at most $10^{-8}$; these are for the least reactive of such species. The most abundant polyatomic species have fractional abundances of $10^{-6}$. Moreover, most of the organic species found in cold cores are very unsaturated, such as the radical series $C_nH$ and the cyanopolyyne series ($HC_{2n}CN$), while more saturated, terrestrial-type organic species such as methanol, ethanol, methyl formate, propionitrile, etc. are associated with hot cores/corinos either because their abundance is higher in such sources or because they are only found there.[1]

In the cooler portions of large dense clouds, the dust particles can be studied by vibrational absorption spectroscopy using both ground-based

Table 1.1. Gas-phase interstellar and circumstellar molecules

| $H_2$ | PO | $N_2O$ | $CH_4$ | $C_4H_2$ | $H_2C_6$ |
|---|---|---|---|---|---|
| CH | SO | $SO_2$ | $SiH_4$ | $H_2C_4$ | $C_6H_2$ |
| $CH^+$ | $SO^+$ | SiCN | $H_2COH^+$ | $HC_2CHO$ | $C_7H$ |
| NH | FeO | SiNC | $CH_2NH$ | c-$C_3H_2O$ | |
| OH | | AlNC | $H_2C_3$ | $HC_3NH^+$ | $CH_3CHCH_2$ |
| SH | $H_3^+$ | MgCN | c-$C_3H_2$ | $C_5N$ | $CH_3OCH_3$ |
| $C_2$ | $CH_2$ | MgNC | $CH_2CN$ | $HC_4N$ | $CH_3CONH_2$ |
| CN | $NH_2$ | | $NH_2CN$ | | $CH_3C_4H$ |
| CO | $H_2O$ | $CH_3$ | $CH_2CO$ | | $C_8H$ |
| $CO^+$ | $H_2S$ | $NH_3$ | HCOOH | $CH_3CHO$ | $C_8H^-$ |
| $CF^+$ | CCH | $H_3O^+$ | $C_4H$ | $CH_3NH_2$ | $HC_7N$ |
| CP | HCN | $H_2CO$ | $C_4H^-$ | $CH_3CCH$ | |
| CS | HNC | HCCH | $HC_3N$ | $C_2H_3OH$ | $C_2H_5CHO$ |
| HF | $HCO^+$ | $H_2CN$ | $HC_2NC$ | c-$CH_2OCH_2$ | $CH_3COCH_3$ |
| NO | $HOC^+$ | $HCNH^+$ | HNCCC | $C_2H_3CN$ | $HOCH_2CH_2OH$ |
| PN | HCO | $H_2CS$ | $C_5$ | $HC_5N$ | $CH_3C_5N$ |
| NS | $HN_2^+$ | $C_3H$ | $C_4Si$ | $C_6H$ | |
| AlF | HCP | c-$C_3H$ | | $C_6H^-$ | $CH_3C_6H$ |
| AlCl | HNO | HCCN | | | $HC_9N$ |
| NaCl | $HCS^+$ | HNCO | $C_2H_4$ | $C_2H_6$ | |
| KCl | $C_3$ | $HOCO^+$ | $CH_3OH$ | $HCOOCH_3$ | $C_6H_6$ |
| SiC | $C_2O$ | HNCS | $CH_3SH$ | $CH_3COOH$ | |
| SiN | $C_2S$ | $C_3N$ | $CH_3CN$ | $HOCH_2CHO$ | $HC_{11}N$ |
| SiO | c-$C_2Si$ | $C_3O$ | $CH_3NC$ | $C_2H_3CHO$ | |
| SiS | $CO_2$ | $C_3S$ | $CH_2CNH$ | $CH_3C_3N$ | |
| $N_2$? | OCS | c-$SiC_3$ | $NH_2CHO$ | $CH_2CCHCN$ | |

Notes: "c" stands for cyclic species; "?" stands for ambiguous detections; isotopologues excluded. (August 2007)

and space-borne spectrometers with either protostars or background stars as the lamp.[2,4] In this manner, it is found that dust particles are core-mantle objects, with cores mainly of amorphous silicates and mantles of mixed ices containing water, $CO_2$, CO, methanol, and other species with lesser abundances. These mantles evaporate as hot cores are formed, and their molecular inventory becomes part of the gas phase, accounting in part for the chemical differences between cold and hot cores.[1] There is also evidence for small particles of amorphous carbon ranging down in size to large molecules known as PAH's (polycyclic aromatic hydrocarbons), which are associated with the soot formed in combustion on Earth.[2]

Interstellar molecules are formed in the clouds themselves from atoms and dust particles ejected from older stars, either explosively, as in the case of supernovae, or less violently, as in the case of low-mass stars. The stellar ejecta eventually lead to the formation of interstellar clouds through the force of gravity, which also causes the formation of denser smaller structures mediated possibly by turbulence, shock waves, and ambipolar diffusion. During the early stages of cloud formation, the diffuse material is rather inhospitable to molecular development, because the density is low and photons penetrate the regions so as to photodissociate species. Nevertheless, molecular hydrogen is formed efficiently.[2] Under low-density conditions, the only gas-phase process capable of producing a diatomic species from precursor atoms is radiative association, in which the diatomic complex is stabilised by the emission of a photon. This process,

$$\mathrm{H} + \mathrm{H} \longrightarrow \mathrm{H_2} + \mathrm{h}\nu, \tag{1.1}$$

is very inefficient for a variety of reasons and cannot possibly explain the large abundance of molecular hydrogen.[2] Alternative associative processes, which are important in the high-temperature early universe, involve ions; e.g.,

$$\mathrm{H} + \mathrm{e}^- \longrightarrow \mathrm{H}^- + \mathrm{h}\nu, \tag{1.2}$$

$$\mathrm{H}^- + \mathrm{H} \longrightarrow \mathrm{H_2} + \mathrm{e}^-. \tag{1.3}$$

These processes are too slow to produce much $H_2$ under cold interstellar conditions. The only viable alternative is the formation of $H_2$ on dust particles by successive adsorption of two hydrogen atoms. In the last decade, experiments and theory have shown that the process is an efficient one on a variety of amorphous and irregular surfaces over a range of temperatures that include those of diffuse and dense interstellar clouds.[1] Moreover, the exothermicity of reaction can provide enough energy to desorb the newly-formed $H_2$ from the dust particle, often in excited vibrational states.[5] After a significant amount of gaseous $H_2$ is produced, it can shield itself against photodissociation since this process occurs via line radiation to specific states of an excited electronic state followed by emission to the continuum of the ground state. The discrete radiation needed is then more strongly diminished than continuum radiation.[2]

Once there is a sizeable amount of $H_2$ in the gas, gas-phase reactions can occur, and lead to the production of most but not all of the species seen in diffuse clouds and dense cores. At the same time, in the cold dense regions, molecules in the gas slowly adsorb onto dust particles. Some are unreactive on surfaces, whereas others, mainly atoms and radicals, are reactive. The result is that mantles of ice grow with the most abundant species formed by chemical reaction rather than simple deposition from the gas.[2,6]

The gas-phase chemistry in cold cores is strongly constrained by the low density and the low temperature. The former condition rules out three-body processes such as ternary association, whereas the latter rules out ordinary chemical reactions, which possess significant amounts of activation energy. After all, with a standard Arrhenius expression for the rate coefficient, an activation energy of 1 eV or more and a temperature of 10 K lead to a very small rate coefficient and, given the long intervals between collisions, high rate coefficients, near the collisional limit, are needed. The chemical processes that dominate the gas-phase chemistry are thus exothermic reactions without activation energy. These processes comprise both reactions involving ions (e.g., ion-molecule and electron recombination reactions) and those involving one or two radicals. Actually, the role of ion-molecule processes was recognized first, and it is still reasonable to state that such processes do dominate the chemistry of cold interstellar gas.[7] Studies of even colder gas; viz. $\leq 1$ K, as discussed in this volume, will improve our understanding of low-temperature chemistry and allow us to make more accurate predictions of rate coefficients at low interstellar temperatures.

Ions are formed by a variety of processes, but the dominant one is ionisation by cosmic rays, which are high energy (MeV-GeV) bare nuclei travelling at velocities near the speed of light and formed in highly energetic events such as supernovae. Although the energy spectrum of cosmic rays can be measured above the terrestrial atmosphere, the low-energy tail interacts sufficiently with the solar wind that it does not reach the orbit of the Earth. Since cosmic rays reflect cosmic abundances, they are dominated by protons. As energetic protons enter interstellar clouds, they can ionise many atoms and molecules before losing their high kinetic energies. Moreover, the secondary electrons produced in the ionisation can themselves ionise other atoms and molecules. Since the flux of cosmic rays is rather low, the fractional ionisation obtained in dense sources is only on the order of $10^{-7}$. The actual rate of ionisation is parameterized by a first-order rate coefficient $\zeta$ ($s^{-1}$), and is typically estimated to be in the vicinity of 1–5 $\times 10^{-17}$ $s^{-1}$ for dense cold cores and possibly 1–2 orders of magnitude higher in diffuse clouds.[2,8] These values come from estimates based on the measured cosmic ray flux above the Earth and the calculated ionisation cross section or by fitting models of the type discussed below with observed abundances of molecules. The difference between diffuse and dense sources comes from the penetration depth of cosmic rays as a function of energy; only higher energy rays can penetrate through the column of material associated with dense clouds whereas lower energy cosmic rays can penetrate through the lesser columns associated with diffuse clouds. The lower penetration depth is associated with a higher cross section for ionisation. Despite the higher value of $\zeta$ in diffuse clouds, the major source of ionisation in such sources is ultra-violet radiation from background stars, which is sufficient to ionise the gas-phase carbon totally, leading to a fractional ionisation as high as $10^{-4}$. In dense clouds, on the other hand, external photons do not penetrate and the major source of ultra-violet radiation comes from cosmic rays, which excite molecular hydrogen, which in turn fluoresces. The radiation field from this indirect process is large enough that it competes with chemical reactions in the destruction of gas-phase neutral molecules.

The most important ionisation caused by cosmic ray protons (CRP) in dense clouds is that of $H_2$, the dominant species:

$$H_2 + CRP \longrightarrow H_2^+ + e^-, \tag{1.4}$$

in which the major channel is single ionisation without dissociation. The parameter $\zeta$ normally refers to the ionisation of $H_2$ and includes ionisation by secondary electrons. Once $H_2^+$ is formed, it immediately (within a day or so) reacts with $H_2$ via a well-studied ion-molecule reaction to form the well-known triangular ion $H_3^+$:

$$H_2^+ + H_2 \longrightarrow H_3^+ + H. \tag{1.5}$$

Since this ion does not react with ubiquitous $H_2$, it has a higher abundance than $H_2^+$ and has been detected via vibrational transitions in both dense and diffuse clouds.[9] Cosmic ray ionisation of helium is also a key process; the resulting $He^+$ ion does not react rapidly with $H_2$.

In the following section, we consider the different types of gas-phase chemical processes that occur in the cold interstellar gas, both from the point of view of their synthetic and destructive roles, and from what is known about them.[2,7]

## 1.2. Gas-Phase Chemical Processes

Let us start by discussing the combined roles of ion-molecule and dissociative recombination reactions in a simple synthesis, that of water and the hydroxyl radical. The first reaction in the sequence is the ion-molecule reaction between atomic oxygen and $H_3^+$:

$$O + H_3^+ \longrightarrow OH^+ + H_2. \tag{1.6}$$

The hydroxyl ion is depleted "quickly" by a hydrogen atom transfer reaction with $H_2$, as is the reaction product $H_2O^+$:

$$OH^+ + H_2 \longrightarrow H_2O^+ + H, \tag{1.7}$$

$$H_2O^+ + H_2 \longrightarrow H_3O^+ + H. \tag{1.8}$$

The oxonium ion, $H_3O^+$, does not react with $H_2$ and instead reacts with electrons in a process known as dissociative recombination, leading to neutral fragments such as water and the hydroxyl radical. Both of these species are then depleted by gas-phase reactions. This small subset of reactions is a paradigm of ion-molecule syntheses in cold interstellar clouds. Note that similar processes occur in diffuse clouds, but the harsh ultraviolet radiation field penetrating these clouds limits the chemistry by rapid photodissociation of species other than $H_2$ and, to a lesser extent, CO, which are shielded by self-absorption of the distinctive radiation needed.[2]

### 1.2.1. *Ion-molecule and Dissociative Recombination Reactions*

Ion-molecule reactions with $H_2$ as the neutral reactant play an important role in the chemistry of interstellar clouds. Indeed, ion-molecule reactions involving many other neutral species, non-polar and polar, are also critical. A useful compendium of ion-molecule reactions has been put together by Anicich.[10] Reactions with non-polar reactants such as $H_2$ often occur with rate coefficients very close to that given by a simple capture approach known as the Langevin model, in which only the long-range ion-induced dipole potential is used and the polarisability is assumed to be a scalar so that the problem reduces to that of a central force. With the assumption that all trajectories that overcome the centrifugal barrier lead to reaction, the Langevin rate coefficient $k_{\mathrm{L}}$ is given by the expression (in cgs-esu units)[11]

$$k_{\mathrm{L}} = 2\pi e\sqrt{\alpha/\mu}, \tag{1.9}$$

where $\alpha$ is the polarisability, $e$ is the electronic charge, and $\mu$ is the reduced mass. A typical value of $k_{\mathrm{L}}$ is $\approx 10^{-9}\,\mathrm{cm}^3\ \mathrm{s}^{-1}$.

For the case of polar neutral reactants, the problem is no longer one which assumes a central force.[11,12] The simplest treatment of such reactions — the locked-dipole approach — is a capture approach based on the assumption that the dipole stops rotating and locks onto the incoming ion in a linear configuration. The result is a large and temperature-dependent rate coefficient of the form

$$k_{\mathrm{LD}} = [1 + (2/\pi^{1/2})x]k_{\mathrm{L}}, \tag{1.10}$$

where $k_{\mathrm{L}}$ is the Langevin rate coefficient, and $x$ (cgs-esu units) is given by the expression

$$x = \frac{\mu_{\mathrm{D}}}{\sqrt{2\alpha k_{\mathrm{B}} T}}. \tag{1.11}$$

Here $\mu_{\mathrm{D}}$ is the dipole moment of the neutral, and $k_{\mathrm{B}}$ is the Boltzmann constant. The result is normally too large compared with experimental values but it is an asymptotic limit as the temperature approaches zero and the dipolar neutral loses angular momentum. In cold interstellar cores, there is also the possibility that rotation is subthermal at the low densities, so that the neutral species lie preferentially in low rotational states even at higher temperatures. For reactions involving non-linear neutral species, a more reasonable approximation, based on classical trajectory techniques, is known as the "trajectory scaling (ts) approach." Here the rate coefficient is given by the expression

$$k_{\rm ts} = [0.62 + 0.4767x]k_{\rm L}. \quad (1.12)$$

For large $x$, as occurs at low temperature, the temperature dependence of both approaches is $T^{-1/2}$. More detailed theoretical methods exist, ranging from refined capture theories to full quantum scattering treatments.

Of course, the simple models are not always accurate, nor do they consider the role of multiple potential surfaces arising from degenerate or near-degenerate electronic states of reactants, or yield any information on the branching fractions of assorted possible exothermic products. Experimental results, performed in a variety of ways although mainly with SIFT (Selected Ion Flow Tube) and ICR (Ion Cyclotron Resonance) techniques,[13,14] are indeed needed by modellers. Useful compendia of measured ion-molecule rates and products exist, such as those of Anicich[10] and the electronic *UMIST Database for Astrochemistry.*[15] The many unstudied ion-molecule reactions can often be grouped into families by types of reaction, and correct products often guessed at. For example, reactions between an ion and $H_2$ often yield products in which a hydrogen atom is transferred to the ion, as occurs in the reactions leading to $H_3O^+$. Eventually the sequence is terminated either because a reaction does not occur rapidly due to barriers or endothermicity, or because the saturated ion is produced. Thus, the oxonium ion does not react with $H_2$, but undergoes dissociative recombination with an electron. The low abundance of electrons in cold cores make such reactions much slower than ion-$H_2$ reactions, although the rate coefficients are very large, unlike the recombination of atomic ions with electrons.

Dissociative recombination reactions[16] have been studied via a variety of techniques, and rate coefficients with typical values $10^{-(6-7)}(T/300)^{-0.5}\,\mathrm{cm}^3\ \mathrm{s}^{-1}$ measured, but the assorted products proved difficult to determine for many years. The order of magnitude of the rate coefficient and the temperature dependence can be rationalised by a simple model in which reaction occurs when the reactants, which are attracted by a Coulomb long-range potential, come within a critical distance $R_{\rm crit}$ of one another. This assumption leads eventually to an approximate thermal rate coefficient $k_{\rm dr}$ (cm$^3$ s$^{-1}$) given by the relation[17]

$$k_{\rm dr} = e^2 R_{\rm crit}\sqrt{\frac{8\pi}{m_{\rm e} k_{\rm B} T}}, \qquad (1.13)$$

where $m_{\rm e}$ is the rest mass of the electron. The arbitrary assumption of a 1 Å critical distance leads to an overall rate coefficient of $1.9 \times 10^{-6}$ $(T/300)^{-0.5}$. Naturally, more detailed, quantum mechanical treatments have been undertaken, especially for the particularly complex case of $H_3^+$+e. For most systems, two mechanisms compete with one another: a direct mechanism, in which the system crosses over into a repulsive neutral state with a potential surface that intersects that of the ion near its equilibrium geometry, and an indirect one, in which Rydberg states mediate the transfer.

Following a small number of experiments that utilised the flowing afterglow apparatus[18] and spectroscopic detection of neutrals, the products of a significant number of reactions have been studied with storage rings, mainly in Stockholm (CRYRING)[19] and in Aarhus (ASTRID).[20] In these latter experiments, a beam of ions is merged with electrons and, following reaction, the neutral products exit the curved beam and strike detectors, allowing a determination of their mass. For the dissociative recombination of the oxonium ion, the following product channels have been detected: $H_2O$ + H, OH + 2H, OH + $H_2$, and O + H + $H_2$. The product branching fractions have been measured by the flowing afterglow method, and via two storage rings, and the experimental results vary, especially those between the two techniques. Both storage ring experiments show that OH + 2H is the dominant channel, while $H_2O$ + H has a branching fraction of 0.18–0.25. The break-up into atomic oxygen is negligible. On the other hand, the flowing afterglow results show that there is a negligible amount of water product while the channel O + $H_2$ + H has a branching fraction of 0.30. Despite the disagreements, all the experiments are in agreement that three-body product channels are important.

The short reaction sequence starting with O + $H_3^+$ illustrates how ion-molecule/dissociative recombination syntheses produce unreactive ions ($H_3O^+$), normal neutral species ($H_2O$), and radicals (OH). The example is too simple to lead to the production of isomers, but the dissociative recombination involving the four-atom ion $HCNH^+$ is thought to lead to the following three sets of products: HCN+H, HNC+H, and CN+2H.[21] Although experiments have not yet been able to distinguish between HCN and HNC, theoretical treatments indicate that they are produced in equal amount. In general, ion-molecule syntheses account at least semi-quantitatively for much and possibly most of the exotic chemistry detected in cold interstellar cloud cores.

### 1.2.2. *Neutral-Neutral Reactions*

Neutral-neutral reactions are also of importance, however, as long as there is no activation energy barrier.[22] A useful compendium can be found in the *UMIST Database for Astrochemistry*.[15] For example, starting with the hydroxyl radical, neutral-neutral reactions with atoms lead to the diatomics $O_2$, NO, and $N_2$ via well-known processes:

$$OH + O \longrightarrow O_2 + H, \tag{1.14}$$

$$OH + N \longrightarrow NO + H, \tag{1.15}$$

$$NO + N \longrightarrow N_2 + O. \tag{1.16}$$

The reaction leading to $O_2$ has been studied down to 39 K in a CRESU (Cinétique de Réaction en Ecoulement Supersonique Uniforme) apparatus. At this temperature, the measured rate coefficient is $3.5 \times 10^{-11}\,\text{cm}^3\,\text{s}^{-1}$. Although the CRESU result appears to be independent of temperature in the range 39–142 K,[23] there is some possible disagreement with other results at slightly higher temperatures. Several theoretical treatments suggest that the rate coefficient drops, possibly appreciably, at still lower temperatures.[24]

If one starts with the hydrocarbon radicals CH and $CH_2$ (see below), carbon monoxide can be formed by reactions with atomic oxygen:

$$CH + O \longrightarrow CO + H, \tag{1.17}$$

$$CH_2 + O \longrightarrow CO + 2H. \tag{1.18}$$

Indeed, CO is formed by many reactions, both neutral-neutral and ion-molecule, and is not easily destroyed. Not surprisingly, it is a dominant species in the cold core gas.

A particularly reactive atom in neutral-neutral reactions is atomic carbon, since, not only can it react with radicals and "semi-radicals" such as $O_2$, it can also react with a variety of non-radicals such as unsaturated hydrocarbons. Indeed, the radicals CN and CCH have the same ability. Such reactions have been studied with the CRESU apparatus down to temperatures as low as ≈10 K.[25,26] Often the rate coefficients show a weak inverse temperature dependence below 300 K so that at low temperatures the rate coefficients can be almost as large as Langevin values. Progress has recently been made in understanding this temperature dependence in terms of variational transition state theory[27] (see also Georgievskii and Klippenstein, this volume). These reactions will be discussed individually in the organic chemistry section below.

Once the diatomics CO and $N_2$ are formed, they can be protonated by reaction with $H_3^+$ to form the formyl ion $HCO^+$ and $HN_2^+$. Neither of these species reacts with $H_2$ and so they have relatively high abundances for molecular ions (e.g., fractional abundances approaching $10^{-8}$ in cold dense cores), although $HN_2^+$ is destroyed by reaction with CO to form the formyl ion. Other abundant ions in the gas include the atomic ions $C^+$, $H^+$, and $He^+$. In addition, the existence of $N_2$ leads to ammonia via the sequence of reactions

$$\mathrm{He^+ + N_2 \longrightarrow N^+ + N + He,} \tag{1.19}$$

$$\mathrm{N^+ + H_2 \longrightarrow NH^+ + H,} \tag{1.20}$$

$$\mathrm{NH^+ + H_2 \longrightarrow NH_2^+ + H,} \tag{1.21}$$

$$\mathrm{NH_2^+ + H_2 \longrightarrow NH_3^+ + H,} \tag{1.22}$$

$$\mathrm{NH_3^+ + H_2 \longrightarrow NH_4^+ + H,} \tag{1.23}$$

$$\mathrm{NH_4^+ + e \longrightarrow NH_3 + H.} \tag{1.24}$$

The sequence cannot begin with neutral atomic nitrogen, because its protonation reaction with $H_3^+$, unlike reaction (1.6), is endothermic.

### 1.2.3. *Radiative Association*

If the elemental carbon is initially in the form of $C^+$, as is detected in diffuse clouds, then the initial reaction undergone by this ion is with $H_2$ but it is *not* the ion-molecule reaction

$$\mathrm{C^+ + H_2 \longrightarrow CH^+ + H,} \tag{1.25}$$

since this process is endothermic by 0.4 eV; rather it is an unusual, low-density process known as radiative association,[11] in which the collision complex is stabilised by emission of a photon:

$$\mathrm{C^+ + H_2 \rightleftharpoons CH_2^+\dagger \longrightarrow CH_2^+ + h\nu.} \tag{1.26}$$

Although radiative association has been occasionally studied in the laboratory (e.g., in ion traps[28]), most experiments are undertaken at densities high enough that ternary association, in which collision with the background gas stabilises the complex, dominates. A variety of statistical treatments, such as the phase-space theory, have been used to study both radiative and ternary association.[29] These approximate theories are often quite reliable in their estimation of the rate coefficients of association reactions. In the more detailed treatments, microscopic reversibility has been applied to the formation and re-dissociation of the complex.[11] Enough experimental and theoretical studies have been undertaken on radiative association reactions to know that rate coefficients range downward from a collisional value to one lower than $10^{-17}\,\mathrm{cm}^3\,\mathrm{s}^{-1}$ and depend strongly on the lifetime of the complex and the frequency of photon emitted.[11,28,29] The lifetime of the complex is related directly to the density of vibrational states, a key parameter in statistical theories, which is determined by the size of the complex and its bond energy. The type of photon emitted — either between vibrational levels in the ground electronic state of the complex or between a reachable excited electronic state and the ground state — determines in part the stabilization rate. If stabilization occurs via transitions from vibrational levels of the ground electronic state above the dissociation limit to such levels below the limit, its rate can be equated with the average Einstein $A$ coefficient for vibrational emission ($\mathrm{s}^{-1}$) in a system of coupled oscillators, which has been shown to be given approximately by the relation[11]

$$A = (E_{\mathrm{vib}}/s) \sum_{i=1}^{s} A_{1-0}^{(i)}/h\nu_i \tag{1.27}$$

where $E_{\mathrm{vib}}$ is the vibrational energy with respect to the potential minimum, $s$ is the number of oscillators, $A_{1-0}^{(i)}$ is the Einstein emission rate for the fundamental transition of the individual mode $i$ and the sum is over the modes $i$ with frequencies $\nu_i$.

The association of $\mathrm{C}^+$ and $\mathrm{H}_2$ has a calculated rate coefficient, derived from a statistical theory, of $4 \times 10^{-16}(T/300)^{-0.2}\,\mathrm{cm}^3\ \mathrm{s}^{-1}$ in the range 10–300 K.[30] Although the rate coefficient is small, the fact that $\mathrm{H}_2$ is a reactant makes the process competitive. A variety of other radiative association processes are important in interstellar chemistry, perhaps the simplest being the corresponding reaction involving neutral atomic carbon:

$$\mathrm{C} + \mathrm{H}_2 \longrightarrow \mathrm{CH}_2 + \mathrm{h}\nu, \tag{1.28}$$

which, however, has not been studied in great detail. Often, there is laboratory information on the ternary association but not the radiative one.

In this case, one can make a partial use of theory to convert the experimental information into an estimate for the radiative association rate coefficient.[11,29]

Consider a radiative association between an ion $A^+$ and a neutral B that occurs via a complex $AB^+\dagger$:

$$A^+ + B \rightleftharpoons AB^+\dagger \rightarrow AB^+ + h\nu. \tag{1.29}$$

If we label the rate coefficients for formation, dissociation, and radiative stabilization of the complex $k_1$, $k_{-1}$, and $k_r$ respectively, the steady-state approximation yields that the second-order rate coefficient $k_{\rm ra}$ for radiative association is

$$k_{\rm ra} = \frac{k_1 k_r}{k_{-1} + k_r} \approx (k_1/k_{-1})k_r, \tag{1.30}$$

assuming, as is most often the case, that the dissociation rate exceeds the radiative stabilization rate. If a quasi-thermal theory is used, the ratio $k_1/k_{-1}$ can be treated as an equilibrium constant, while if phase-space theory is used, one must compute a value of $k_1$ for each initial state of the reactants and the overall rate coefficient involves a summation over the states of the reactants and the angular momentum of the complex. A state-specific treatment of some type is necessary to obtain the correct density dependence for association over the whole range of densities, especially at high densities, when ternary association gives way to a saturated two-body rate law. A similar steady-state analysis to that used for Eq. (1.30) for ternary association yields the third-order rate coefficient $k_{\rm 3B}$:

$$k_{\rm 3B} \approx (k_1/k_{-1})k_2, \tag{1.31}$$

where $k_2$ is the rate coefficient for collisional stabilization of the complex. Conversion of $k_{\rm 3B}$ into $k_{\rm ra}$ then requires knowledge only of $k_2$ and $k_r$.[29] The collisional stabilization rate coefficient depends on the bath gas, which is often helium for ion-molecule systems. The efficiency is found to be less than unity, and it is typical to approximate this rate coefficient in the range $\approx 10^{-9} - 10^{-10}\,\rm cm^3\ s^{-1}$ for ion-helium relaxation, with the lower limit appropriate for room temperature.[29] The radiative stabilization rate via vibrational photons is given by Eq. (1.27) for the average Einstein $A$ coefficient. For a standard bond energy of 4 eV and typical fundamental intensities, $k_r \approx 10^{2-3}\ \rm s^{-1}$. To convert $k_{\rm ra}$ to a lower temperature, as is often needed, one can use the temperature dependence from the quasi-thermal approach of Bates & Herbst,[29] in which the rate coefficient for radiative association goes as $T^{-r/2}$, where $r$ is the total number of rotational degrees of freedom of the reactants assuming that rotation can be treated classically. For light reactants, especially $H_2$, this approximation is poor and it is better to ignore its two rotational degrees of freedom. As an example, suppose a reaction between $A^+$ and B is measured to have a ternary rate coefficient at 300 K of $10^{-27}\,\mathrm{cm^6\,s^{-1}}$. Using $k_2 = 10^{-10}\,\mathrm{cm^3\,s^{-1}}$ and $k_r = 10^2\,\mathrm{s^{-1}}$, we obtain that $k_{\mathrm{ra}} = 10^{-15}\,\mathrm{cm^3\,s^{-1}}$ at room temperature, with a temperature-dependent rate coefficient of $k_{\mathrm{ra}} = 10^{-15}\,(\mathrm{T}/300)^{-3}\,\mathrm{cm^3\,s^{-1}}$, where it has been assumed that the reactants are non-linear species with three degrees of rotational freedom each. Of course, this approach is a crude one because the individual rate coefficients are highly averaged quantities.[29] An alternative approach is to use a statistical method for the exact temperature dependence, since this can be more complex if rotation cannot be treated classically, if there is atomic fine structure, and for other reasons as well.[30]

### 1.2.4. *Organic Chemistry*

In addition to the basic radiative association between $C^+$ and $H_2$, the reaction

$$C + H_3^+ \longrightarrow CH^+ + H_2 \tag{1.32}$$

also serves to "fix" atomic carbon into a molecular form and thus start the build-up of organic molecules, or, more specifically, hydrocarbons. This proton-transfer reaction has yet to be studied in the laboratory. Once $CH^+$ and $CH_2^+$ are produced, H-atom transfer reactions with $H_2$ lead to the methyl ion, $CH_3^+$. Like $C^+$, the methyl ion does not react exothermically with $H_2$ but does undergo a radiative association to form $CH_5^+$, which has been studied in two different laboratories and via statistical theory.[28] Although the three approaches are not in perfect agreement, an estimated value for $k_{\mathrm{ra}}$ of $1.3\times 10^{-14}(T/300)^{-1.0}\,\mathrm{cm^3\,s^{-1}}$ in the range 10–300 K is used in interstellar model networks. Note that the radiative association rate coefficient is larger for the reaction involving more atoms. Interestingly, at 10 K the radiative association of $CH_3^+$ with $H_2$ is faster than dissociative recombination because of the low abundance of electrons. Once $CH_5^+$ is produced, it can undergo dissociative recombination with electrons to form a variety of single-carbon hydrocarbons, although methane is a minor product. This species is formed more efficiently by the reaction

$$CH_5^+ + CO \longrightarrow CH_4 + HCO^+. \tag{1.33}$$

The formation of more complex hydrocarbons then proceeds via three types of processes — carbon "insertion", condensation, and radiative association of heavy species — which can occur for both ion-molecule and neutral-neutral reactions.[2,7] Many of the ion-molecule reactions important in both synthesis and depletion have been tabulated in either experimental compilations[10] or compilations based on interstellar networks.[15] In the ion-molecule realm, both $C^+$ and neutral atomic carbon can undergo a variety of insertion reactions. Note that by the term "insertion", we do not mean that the attacking carbon atom/ion must insert itself into a chemical bond, but also that it can add itself to a terminal atom. Many reactions involving $C^+$ and hydrocarbons have been studied, and these often show products in which the insertion/addition of the carbon leads to a more complex but less saturated hydrocarbon. Consider, for example, the reaction between $C^+$ and methane:

$$C^+ + CH_4 \longrightarrow C_2H_3^+ + H; C_2H_2^+ + H_2. \tag{1.34}$$

Experimental studies show that the first set of products dominates with a branching fraction of 0.73.[10] Reactions between neutral atomic carbon and hydrocarbon ions are thought to also lead to carbon insertion, but experiments are lacking. Without H-atom transfer or radiative association reactions with $H_2$, carbon-insertion reactions eventually lead to bare clusters ($C_n$ or $C_n^+$). For example, consider the case of $C_2H_3^+$, which undergoes no reaction with $H_2$. Instead, dissociative recombination occurs to form species such as $C_2H$ and $C_2H_2$. Reactions with $C^+$ then lead to ions such as $C_3^+$ and $C_3H^+$. These ions react with $H_2$ to form hydrocarbon ions only as saturated as $C_3H_3^+$, which is partly isomerised to a cyclic form. Dissociative recombination leads to c-$C_3H_2$, c-$C_3H$, as well as the linear and carbene forms CCCH and $H_2CCC$. As the number of carbon atoms increases past four, hydrogenation reactions seem only to saturate the ions to the extent of putting two H atoms on the carbon chain. Dissociative recombination then leads to bare clusters and to radicals of the form $C_nH$. Although the radicals are a salient feature of cold cores, the need to detect carbon clusters via vibrational transitions has so far precluded their detection in these sources. They have been detected in the cold envelopes of old carbon-rich stars.

Condensation reactions also produce more complex species, and these can be somewhat more saturated. Consider, for example, the following well-studied ion-molecule examples:

$$CH_3^+ + CH_4 \longrightarrow C_2H_5^+ + H_2, \tag{1.35}$$

$$C_2H_2^+ + C_2H_2 \longrightarrow C_4H_3^+ + H. \tag{1.36}$$

Dissociative recombination of the ethyl ion can then lead to a species as saturated as $C_2H_4$. Since condensation reactions are generally not as important as carbon insertion reactions, the unsaturated nature of the hydrocarbon chemistry remains a salient prediction of ion-molecule chemistry, which appears to be in good agreement with observations of cold interstellar cores.

There are analogous neutral-neutral processes to the ion-molecule carbon insertion and condensation mechanisms. Reactions between unsaturated hydocarbons and neutral atomic C often result in carbon insertion. The best-analysed reaction of this class is[31,32]

$$C + C_2H_2 \longrightarrow c - C_3H + H, C_3H + H, C_3 + H_2, \tag{1.37}$$

which has been studied via the CRESU technique, a merged-beams apparatus, and a crossed-beams apparatus. Note that both the cyclic and linear isomers of $C_3H$ are formed. Indeed, this reaction appears to be the most important for the formation of the $C_3H$ isomers, although the more abundant cyclic $C_3H_2$ isomer is formed predominantly not by a neutral-neutral reaction but via the ion-molecule synthesis discussed above. For condensation reactions, a well-studied reaction is the following:[25]

$$CCH + C_2H_2 \longrightarrow C_4H_2 + H. \tag{1.38}$$

Let us now broaden our discussion to consider organic species more complex than hydrocarbons. The reaction between $C^+$ and ammonia leads to the $HCNH^+$ ion, which is the precursor of HCN and its isomer HNC, in addition to CN, as discussed above. The radical CN is especially reactive with unsaturated hydrocarbons at low temperatures, as shown in CRESU studies. The reaction with acetylene leads to cyanoacetylene, a well-known interstellar molecule:[25]

$$CN + C_2H_2 \longrightarrow HC_3N + H. \tag{1.39}$$

Presumably the higher members of the cyanopolyyne series detected in cold cores are produced via analogous reactions between CN and $C_4H_2$, $C_6H_2$, etc. There are competitive ion-molecule pathways, starting with reactions involving atomic nitrogen and hydrocarbon ions. Also, the two isomers of $HC_3N$ - HNCCC and HCNCC — detected in the cold core TMC-1, are likely produced via protonation of $HC_3N$ followed by dissociative recombination to form the unusual structures.

In addition, a number of organic species are produced from precursor ions that are the products of radiative association reactions. Although radiative association is most important when $H_2$ is a reactant because of the overwhelming relative abundance of $H_2$, such processes between two heavy species are also valuable on occasion. There are some neutral-neutral systems that have been studied, but the majority of these systems are thought to be ion-molecule ones. For the most part, the systems do not possess competitive exothermic channels, although there are exceptions, either because the collision complex can only dissociate to exothermic products with a significant barrier that does not quite choke off reaction but lengthens the lifetime of the complex, or because the association and regular channels are parallel in the sense that they take quite different pathways. The methyl ion, which is reasonably abundant because it reacts only slowly with $H_2$, seems to be especially fruitful as a reactant in association processes leading to more complex species. A list of such reactions involving $CH_3^+$ follows:

$$CH_3^+ + HCN \longrightarrow CH_3CNH^+ + h\nu, \tag{1.40}$$

$$CH_3^+ + NH_3 \longrightarrow CH_3NH_3^+ + h\nu, \tag{1.41}$$

$$CH_3^+ + CO \longrightarrow CH_3CO^+ + h\nu, \tag{1.42}$$

$$CH_3^+ + H_2O \longrightarrow CH_3OH_2^+ + h\nu. \tag{1.43}$$

The reaction involving ammonia has only been observed as a ternary process. The product ions, after dissociative recombination, should produce the neutrals $CH_3CN$, $CH_3NH_2$, $CH_2CO$, and $CH_3OH$, respectively. Dissociative recombination of $CH_3CNH^+$ may also lead to the recently detected $CH_2CNH$ (ketenimine) and the fact that $CH_3CNH^+$ and $CH_3NCH^+$ can reach a quasi-equilibrium means that $CH_3NC$ can also be produced. Radiative association involving more complex hydrocarbon ions, such as $C_2H_3^+$, also occur with neutral reactants such as CO and HCN. Recently, however, some evidence has begun to accumulate that the radiative association processes may not be as important as previously thought. In the particular case of the formation of methanol, it appears from experimental evidence that (a) the radiative association to form protonated methanol is much slower than previously thought, and (b) the dissociative recombination reaction involving protonated methanol does not yield a significant product channel $CH_3OH + H$.[19] For the case of methyl amine, the molecule is not detected in cold cores, but in warmer ones, which argues against the involvement of radiative association, since this type of reaction gets slower with increasing temperature.

### 1.2.5. *Negative Ion Formation*

Many years ago, one of us[33] suggested that negative molecular ions could be formed in space via the process of radiative attachment. The simplest treatment of this process[33,34] is to assume that it proceeds via a collision complex:

$$A + e \rightleftharpoons A^-\dagger \rightarrow A^- + h\nu, \tag{1.44}$$

where the complex is a negative ion with an energy slightly above the electron affinity of the neutral. Then, the overall rate coefficient for radiative attachment $k_{\rm ratt}$ is given by the same expression used above for radiative association although the individual rate coeffficients for $k_1$ and $k_{-1}$ are not the same. If one assumes that only $s$-wave scattering occurs, both of these rate coefficients refer to a complex with low angular momentum equal to that of the neutral reactant. Simplification of the more general phase-space result with the additional assumption that the complex relaxes quickly to its ground electronic state leads to the simple expression

$$k_{-1} = c/\rho \tag{1.45}$$

where $c$ is the speed of light in cgs units and $\rho$ $(({\rm cm}^{-1})^{-1})$ is the density of vibrational states of the negative ion complex in its ground electronic state at a vibrational energy equal to the electron affinity. The phase space theory result for the rate coefficient of formation of the complex is given by the expression[34]

$$k_1 = \hbar^2 G \sqrt{2\pi/(m_e^3 k_{\rm B} T)} \approx 4.98 \times 10^{-7} G (T/300)^{-1/2}, \tag{1.46}$$

where $G$ is the ratio of the ground-state electronic degeneracy of the anion to the product of those of the reactants. If the density of vibrational states is sufficiently large that $k_{-1} < k_r$, then the overall rate coefficient for radiative attachment is given by $k_1$.

Even more so than for the case of radiative association, very little experimental information exists for radiative attachment, and the reliability of the simple phase-space treatment has not been tested. One possible problem is that the formation of the negative ion complex may require so-called "doorway" states known as dipole-bound states, in a similar manner to the Rydberg mechanism for dissociative recombination of positive molecular ions.[35] Such states have been seen in the spectra of negative ions, although a detailed calculation employing them for radiative attachment has not been attempted.

Radiative attachment is not the only mechanism for the production of anions; their production in laboratory discharges probably occurs via dissociative attachment, in which electrons attach themselves to neutrals while a chemical bond is broken; viz.,

$$\rm e + C_2H_2 \longrightarrow C_2H^- + H. \tag{1.47}$$

For the most part, however, dissociative attachment is an endothermic process because even a single chemical bond is typically greater in energy than the electron affinity of the neutral. Exceptions do exist; exothermic dissociative attachment has been suggested to form the anion $CN^-$ from the parent neutral MgNC, which is found in the envelope surrounding the carbon-rich star IRC+10216.[34]

Use of the simple phase-space approach leads to the prediction that radiative attachment is most efficient ($k_{ratt} \approx k_1$) for neutral species with large electron affinities (3–4 eV) and negative ions with large densities of vibrational states, requiring species, probably radicals, with more than a few atoms in size.[33,36] In such situations, it was predicted that the abundance ratio of the negative ion to the neutral species in cold cores could be as high as 1% or even more, assuming that the major destruction routes are so-called associative detachment reactions with atomic hydrogen:

$$A^- + H \longrightarrow AH + e. \tag{1.48}$$

Given the paucity of experimental studies on the rotational spectra of negative ions, this early prediction was not tested for more than 25 years. Quite recently, a number of anions have been studied in the laboratory by the Thaddeus group,[37] including $CN^-$, $CCH^-$, $C_4H^-$, $C_6H^-$, and $C_8H^-$, the neutrals of which all possess large electron affinities. Of these species, subsequent searches of interstellar and circumstellar sources have so far resulted in the detection of $C_6H^-$ and $C_8H^-$ in the best-studied cold core TMC-1, the detection of $C_4H^-$, $C_6H^-$, and $C_8H^-$ in the cold, carbon-rich circumstellar envelope IRC+10216, and the detection of $C_6H^-$ in the protostellar source L1527. Moreover, the detected abundance of $C_4H^-$ relative to $C_4H$ is well below 1% in IRC+10216, and the upper limit to this ratio in TMC-1 is even smaller. In qualitative agreement,[36,38] use of the phase-space theory (and suitable destruction routes for anions) predicts that the $C_4H^-/C_4H$ abundance ratio will be considerably lower than that for the larger anions of this class, for which radiative attachment is predicted to occur at the collisional limit. However, the predicted ratio for $C_4H^-/C_4H$ is still significantly higher than the observed value in IRC+10216 and the upper limit in TMC-1, so that there is astronomical evidence, at least, that the radiative attachment of electrons to $C_4H$ is slower than predicted. Because the dipole moment of this species is also much lower than those of the neutrals $C_6H$ and $C_8H$, the argument that dipole-bound states play a role in radiative attachment gains some force, since these states only appear for dipoles larger than $\approx$2 Debye. Of course, astronomical evidence need not be definitive, because predicted abundances rely on uncertain knowledge of physical conditions as well as possible additional formation and destructive processes for anions such as exothermic dissociative attachment channels and photo-detachment.

## 1.3. Surface Chemistry on Cold Dust Grains: Basic Considerations

Although the major interstellar reaction that occurs on the surfaces of dust particles is the formation of molecular hydrogen, more complex molecules are also thought to be formed in this manner.

### 1.3.1. *Low-Temperature Surface Chemistry*

In both diffuse and dense interstellar clouds, the dust grains tend to be rather cold, with temperatures in equilibrium with the gas in cold dense cores and temperatures at or below $\approx 25\,\mathrm{K}$ in the warmer diffuse medium. Gas-phase species accreting onto dust particles can bind either by weak long-range (van der Waals) forces or by strong chemical forces; the former is known as physisorption and the latter as chemisorption. Since chemisorption most often occurs with an activation energy barrier, low temperature accretion is dominated by physisorption, with binding (desorption) energies $E_{\mathrm{D}}$ ranging from $\approx 0.03$–$0.5\,\mathrm{eV}$. These desorption energies can be measured by classical techniques or by more modern methods such as temperature-programmed desorption (TPD).[39] Once on the grain and thermalised to the grain temperature, the weakly-bound species can diffuse to other sites on the grain via a random-walk. The potential for such motion depends on the adsorbate, the surface material, and its morphology. For a totally regular surface, one can imagine a potential with regularly spaced wells and barriers; the former are often referred to as binding sites. The barrier heights, which we label $E_{\mathrm{b}}$, are normally smaller than the desorption energies, but are more difficult to measure. From limited measurements augmented by quantum chemistry, $E_{\mathrm{b}}$ is found to range in size from values close to $E_{\mathrm{D}}$ to values near zero for special surfaces such as graphite. For irregular surfaces, with assorted imperfections, the potential for diffusion is certainly irregular as well, and its local character depends on the nature of the imperfection.

There are three well-discussed mechanisms for surface reactions, and they can pertain to both chemisorption and physisorption depending on the temperature range.[39] The best studied mechanism, which involves the diffusion of adsorbates followed by their reaction when in close proximity, is known as the Langmuir-Hinshelwood mechanism. At low temperature, this mechanism proceeds best in the absence of chemical activation energy, but can occur competitively even in the presence of small activation energy barriers because of a competition between reaction and diffusion, which also has barriers. In the absence of low-temperature tunnelling of adsorbates between binding sites, the Langmuir-Hinshelwood mechanism for a given type of binding operates between a lower temperature limit, under which the adsorbates cannot diffuse rapidly to find reaction partners, to a higher temperature limit, where desorption is more rapid than diffusion to find the partners. Even if the surface species cannot move, there are two other mechanisms that can become operative. The first is known as the Eley-Rideal mechanism; here a gas-phase species strikes an adsorbate and the two react. Very similar is the so-called "hot atom" mechanism, in which a hot gas-phase species sticks to the surface but fails to thermalise instantly and can hop around the surface and react before thermalising.

### 1.3.2. *$H_2$ Formation*

Since a significant amount of molecular hydrogen is formed in diffuse clouds, where the dust particles are thought to be mainly bare, experiments and theories directed at $H_2$ formation in such clouds either use some form of silicate or carbon, the two materials associated with bare grains. The first low-temperature experiments on $H_2$ formation with suitable grain analogs were performed by Vidali and co-workers with a TPD apparatus, initially on olivine and amorphous carbon.[40] They found that the process occurs via a Langmuir-Hinshelwood mechanism, but only over a very narrow temperature range which, especially for the case of olivine, is much lower than typical surface temperatures in diffuse clouds. From this experiment, they were able to determine the desorption energies for H and $H_2$, the barrier to diffusion for H, and the fraction of nascent $H_2$ molecules immediately ejected into the gas-phase. Determination of these parameters comes from a kinetic model of the reaction, in which random walk is modelled by a simple treatment involving rate equations (see Sec. 1.6). The initial interpretation that diffusion occurs classically, by thermal hopping, rather than by tunnelling, has been challenged. More recently, Vidali *et al.*[41] found that using a "rough" (highly irregular) silicate surface expands the range of temperatures over which $H_2$ formation occurs efficiently to include diffuse cloud surface temperatures, in agreement with an earlier stochastic theory.[42]

Other studies, on porous amorphous solid water, relevant to dense interstellar clouds where the grains have icy mantles, found a more complex process, in which $H_2$ can be formed at very low temperatures by a rapid process possibly involving tunnelling between pores, but that the desorption of $H_2$ into the gas occurs gradually depending on the binding site.[43] The formation of $H_2$ on graphite has been studied at higher temperatures, such as might pertain to photon-dominated regions. The mechanism appears to be a complex one based on chemisorption, in which the two H atoms first form long-range clusters before combining. Another experiment, undertaken at University College London by Price and co-workers, has as its major goal determination of the vibrational-rotational state distribution of ejected $H_2$ molecules. Current results show considerable vibrational excitation.[5] In the interstellar medium, such excitation might still be detectable for many years after the formation of the hydrogen molecule. A variety of theoretical treatments have been applied to the *dynamics* of $H_2$ formation on surfaces. Clary and co-workers, in particular, have studied both the Eley-Rideal and Langmuir-Hinshelwood mechanisms on a form of graphite (HOPG) using quantum mechanical methods.[44]

### 1.3.3. *Reactions in Ice Mantles*

Although a large number of surface reactions have been put into gas-grain chemical models of cold interstellar cloud cores, the dominant ones for long periods of time involve atomic hydrogen, which still has an appreciable gas-phase abundance and is uniquely reactive on surfaces because of its diffusive capability at low temperatures. In addition to combining with themselves to form $H_2$, H atoms can react with slower-moving heavy atoms and radicals without activation energy to saturate these species. The result is a very different chemistry from that occurring in the cold gas, where unsaturated species are the norm. Some of these reactions have been studied in the laboratory, although without as much detail as applied to the study of $H_2$ formation. Perhaps the most important sequence of reactions involves the conversion of atomic oxygen first to the hydroxyl radical and then to water. Model calculations indicate that this synthesis of water leads to far more of the water ice detected in cold cores than does simple accretion of gas-phase water, which has a rather low abundance. Another hydrogenation sequence starts with CO, which is made copiously in the gas, and leads through the radical HCO to formaldehyde, and subsequently, through the radicals $CH_2OH$ and $CH_3O$, to methanol.[45] Two of these reactions have small chemical activation energies, but, as will be discussed below, this need not prevent their occurrence. The sequence of reactions is an important one, since there is no known gas-phase synthesis of methanol efficient enough to explain its abundance in cold cores, which means that it must be accounted for by some sort of non-thermal desorption of methanol ice.

In addition to the formation of normal isotopologues, deuterated species can also be formed, as will be discussed later. This result can happen in two ways. First, in addition to association reactions with atomic hydrogen, there are association reactions with atomic deuterium. For example, the isotopologue $CH_2DOH$ can be formed by the following sequence of reactions:

$$\mathrm{CO + H \rightarrow HCO;\ \ HCO + H \rightarrow H_2CO;\ \ H_2CO + H \rightarrow CH_2OH}, \tag{1.49}$$

$$\mathrm{CH_2OH + D \rightarrow CH_2DOH}. \tag{1.50}$$

This type of mechanism is a general one and may well occur in regions with high abundances of atomic deuterium such as the centres of pre-stellar cores, discussed in Sec. 1.5. Secondly and more specifically, atomic deuterium can react with methanol to form $CH_2DOH$ although the exact mechanism by which this happens is not known definitively.[46] Surface deuteration followed by desorption may well be an important source of gas-phase fractionation, as discussed in Sec. 1.5 following a detailed discussion of gas-phase fractionation.

## 1.4. Models with Gas-Phase Chemistry

We begin this section by discussing time scales appropriate for the physical and chemical evolution of cold cores, also known as dark interstellar clouds, before discussing various models for describing the gas-phase chemistry and the uncertainty associated with such models. It should be mentioned that kinetic treatments of the chemistry are needed because the time scales to reach chemical equilibrium are far longer than the lifetimes of the clouds, and because the chemistry is powered by an external energy source.

### 1.4.1. *Time Scales*

The time scale for chemical evolution for a particular species can be determined crudely by consideration of the inverse of its destruction rate.[2] In the unshielded interstellar medium, neutral species are destroyed by the interstellar ultraviolet radiation field with photo-rates which are typically $\sim 10^{-10} - 10^{-11}\,s^{-1}$ per event, leading to time-scales of ~300–3000 yr. Inside cold, dark cores, the presence of dust grains causes effective absorption of the UV photons with the result that time-scales are determined by binary reactions and can be written as $[kn(X)]^{-1}$ s, where $k$ is an appropriate rate coefficient ($cm^3\,s^{-1}$) and $n(X)$ is the number density ($cm^{-3}$) of species X. For radicals, X corresponds to reactive atoms such as C and O with $n(X) \sim 10^{-6}n$, where $n$ is the total number density of the cloud, and $k \sim 10^{-10}$–$10^{-11}$ $cm^3\,s^{-1}$ at low temperatures. Hence the time scale is $\sim 3 \times 10^4$ yr for $n \sim 10^4\,cm^{-3}$. On the other hand, the time scale for destruction of stable neutral molecules such as CO or $H_2O$ can be much longer. Such species are destroyed primarily through binary reactions with abundant cations, $X^+$. The rate coefficients can be taken crudely to be the Langevin value, $\sim 10^{-9}$ $cm^3\,s^{-1}$ with $n(X^+) \sim 10^{-8}\,n$, leading to a time scale of $\sim 3 \times 10^5$ yr. These chemical time-scales should be compared with two other important time scales in cold clouds. The first is the time scale taken for accretion, or freezing out, of the gas onto the cold dust grains which populate these cores. For gas-phase species X, this is given by:

$$t_{acc} = (S_X \pi a^2 v_X n_d)^{-1} \tag{1.51}$$

where $S_X$ is the fraction of collisions of X with grains which actually lead to accretion and is close to unity for dust grains at 10 K, $\pi a^2$ is the cross section $\sigma$ of a dust grain assuming spherical, neutral particles of radius $a$, $n_d$ is the number density of dust grains per unit volume, and $v_X$ is the mean thermal velocity of X. A more detailed form of this equation can be written to take into account the fact that grains are charged and have a size and shape distribution, which can be determined from studies of interstellar extinction. Using typical parameters for cold clouds leads to a time scale for accretion of $t_{acc} \sim 3 \times 10^9/n$ yr, where $n$, the *gas* density, is measured per $cm^3$. To derive this formula, we have used $a = 0.1\mu$ and converted the grain density to the gas density by the standard dust-to-gas abundance ratio of $\approx 1 \times 10^{-12}$.

The second important time scale is that associated with the dynamics of the cloud itself. If the core is massive enough to be dominated by its self-gravity, then the free-fall time scale, which is the time taken for gravitational collapse to high density — $t_{ff} \sim 4 \times 10^7/n^{1/2}$ yr — is appropriate. Even if the cores are not massive enough for self-gravity to dominate, the small physical sizes of cold cores implies that the sound-crossing time, $t_{sc} \sim L/c_s$, a measure of the time needed to establish pressure equilibrium, where $L$ is the size, typically $3 \times 10^{17}$ cm, and $c_s$ is the sound speed, is $\sim 10^5$ yr. For $n = 10^4\,cm^{-3}$, one sees that the chemical, accretion and dynamical time scales are all comparable, indicating that the adoption of steady state kinetics is likely to be a poor approximation and that time-dependent solution of the chemical (and physical) evolution of the gas must be undertaken.

### 1.4.2. *Homogeneous Sources*

Although there have been a number of models which incorporate both dynamics and chemistry, we begin by concentrating on "pseudo-time-dependent" models; that is, those in which chemical evolution occurs under fixed physical conditions. Models in which physical conditions are fixed by a single set of parameters are known as "one-point-models" and are most readily applied to clouds which have a homogeneous structure. Although they have limited applicability to real interstellar cloud cores, they have the advantage of allowing one to investigate easily the importance of particular reactions and rate coefficients and to incorporate the effects of accretion on the chemistry. A major weakness of this approach is that, in reality, the physical conditions change as the chemistry takes place, with a more appropriate starting point at much lower densities and diffuse cloud abundances.

The advent of fast computers means that it is now feasible to integrate several hundred ordinary differential equations (ODEs) describing a gas-phase chemistry of up to 10,000 reactions in cold cores. As long as the physical conditions do not alter, the system of ODEs is first-order, although non-linear, and can be integrated using standard packages. The general form of the ODE to be solved is:

$$\frac{dn(X_i)}{dt} = \Sigma_l \Sigma_m k_{lm} n(X_l) n(X_m) - n(X_i) \Sigma_j k_{ij} n(X_j) + \Sigma_p k_{ip} n(X_p) - n(X_i) \Sigma_q k_{iq} + k_i^{des} n(X_i^d) - k_i^{acc} n(X_i) \quad (1.52)$$

where the first two terms on the right-hand side represent the binary reactions which form and destroy $X_i$, the following two terms represent the first-order reactions which form and destroy $X_i$, and the final two terms represent the desorption of $X_i$ from the grain, where it is labelled $X_i^d$, and accretion onto the dust. Further terms need to be added if $X_i$ is injected to the gas from other processes at the grain surface, e.g. dissociative recombination following collisions between gas-phase cations and negatively-charged grains. Grains in cold cores are mainly negatively charged because of the greater thermal speeds of electrons as compared with heavy positive ions.

A similar set of rate equations can be written for the granular species $X_i^d$ and can include chemical reactions, perhaps driven by interaction with UV photons or cosmic ray particles, which alter the composition of the grain mantle. Although it is relatively easy to extend the set of ODEs to include surface species and reactions, it is not clear that such an approach is valid; small number statistics means that the notion of "averages" inherent in the rate equation approach is questionable. This issue and alternative approaches to describe surface chemistry are discussed more fully in Sec. 1.6.

The set of initial conditions which must be supplied in order to integrate the system of ODEs includes the rate coefficients together with the parameters that describe the physical conditions in the cold core — number density, elemental abundances, temperature, UV radiation field, cosmic-ray ionisation rate, and grain properties: size, number density, and opacity. Of these parameters, elemental abundances are the most uncertain due to the unknown fraction of heavy elements — carbon, nitrogen, oxygen, silicon, magnesium, etc. — which are incorporated into the refractory cores or ice mantles of the dust grains. It is common if simplistic to start with atomic gas except for hydrogen, which is at least partially molecular.

A number of groups now make available sets of gas-phase rate coefficients for use in interstellar chemistry. The *UMIST Database for Astrochemistry* (the most recent version of which is also known as RATE06)[15] contains information on some 4500 reactions of which some 35% have been measured experimentally, some at temperatures down to 20 K, and has been updated and released anew every 5 years or so. The UDfA database is aimed at describing both hot and cold chemistries and software is available to select the appropriate rate coefficients for particular uses. The Ohio State University group[22] has an alternative database of approximately the same number of reactants and reactions, but focused more on low temperature chemistry. New versions are released on irregular time scales and designated by month and year (the current relevant version of the OSU program is osu_03_2008). This group also releases particular data sets which have been used in published calculations, as does the Durham/Paris group.[47] The UDfA database assigns an uncertainty and a temperature range of applicability to each rate coefficient; these uncertainties are also available in the recent osu networks. The uncertainties can be translated into predicted uncertainties for model abundances and used in sensitivity calculations. The method discussed by Wakelam *et al.* [48] has been used recently for cold cores, and consists of running the chemical model thousands of times with individual rate coefficients picked randomly from Gaussian distributions determined by their uncertainties. The calculated abundances also show Gaussian distributions in most instances, and these can be compared with observational results. Occasionally, the abundance distributions bifurcate into two peaks, a phenomenon known as bistability; this arises because of the non-linear nature of the coupled ODEs.

Figure 1.1, calculated with the RATE06 network,[15] shows the typical time evolution of the carbon chemistry for a purely gas-phase model of a cold core ($T = 10\,\mathrm{K}$, $n = 10^4\,\mathrm{cm}^{-3}$) in which carbon is initially ionised, hydrogen is in molecular form and the elemental abundance of oxygen is greater than that of carbon ("oxygen-rich abundances"). As one can see, $C^+$ is transformed first to neutral carbon atoms. As discussed in Sec. 1.2, $C^+$ and C are processed into hydrocarbons before the excess of oxygen locks up "all" of the available carbon into the strongly bound CO molecule and hydrocarbon abundances decrease to their steady-state values. A comparison between such models and observations of cold cores such as TMC-1 and L134N shows that best agreement occurs at so-called early time, $\sim 10^5$ yr, and is strengthened by appropriate choice of the elemental abundances. A variety of criteria have been developed to quantify the degree of agreement, the simplest being the fraction of detected molecules for which observation and calculation lead to abundances within an order of magnitude of one another. The more recent approach of Wakelam *et al.*[48] has also been utilised. In calculations done with the osu network, the abundances of approximately 80% of the 40 or so molecules seen in L134N can be reproduced to within an order-of-magnitude at early-time; the agreement for the 60 or so molecules in TMC-1 is worse unless carbon-rich abundances are utilised.[48] The reason for this difference is that TMC-1 is uniquely

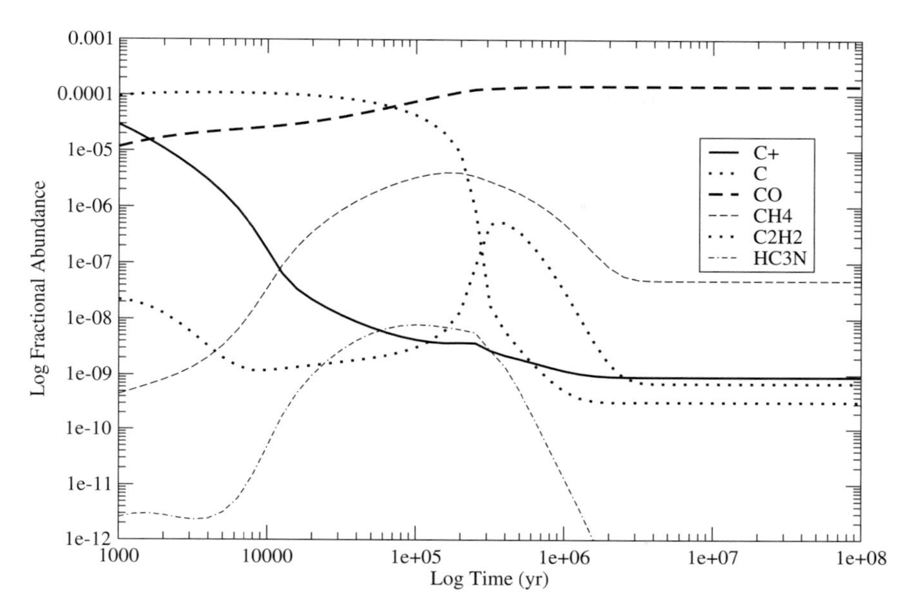


Fig. 1.1. Time evolution of gas-phase carbon chemistry at $T = 10\,\mathrm{K}$ and $n = 10^4\,\mathrm{cm}^{-3}$.

rich in organic species, and these cannot be grown very efficiently under oxygen-rich conditions. Recent work using the RATE06 network with the addition of negative ions shows that the abundance of $C_6H^-$ in TMC-1 is well accounted for.[36]

As mentioned previously, steady-state abundances, which take in excess of $10^6$ yr to be reached, are not appropriate since material will freeze out onto the dust grains, a process not included in Fig. 1.1. If one includes accretion at the standard rate, the results in Fig. 1.2 pertain. One sees that, with the exception of hydrogen and helium and their associated ions, all species are removed from the gas phase in a few million years. Since this time-scale is inversely proportional to density, then higher density cores should be unobservable via transitions in gas-phase CO. In fact, although such higher density cores exist, there is no observational evidence for *completely* depleted cores indicating that either there are mechanisms which return material, albeit inefficiently, to the gas phase or the surface area of dust grains is reduced in dense, cold cores perhaps through grain coagulation processes. Observational evidence in support of this latter conjecture is difficult to obtain using current techniques. A variety of desorption processes, discussed below, have been included in models.

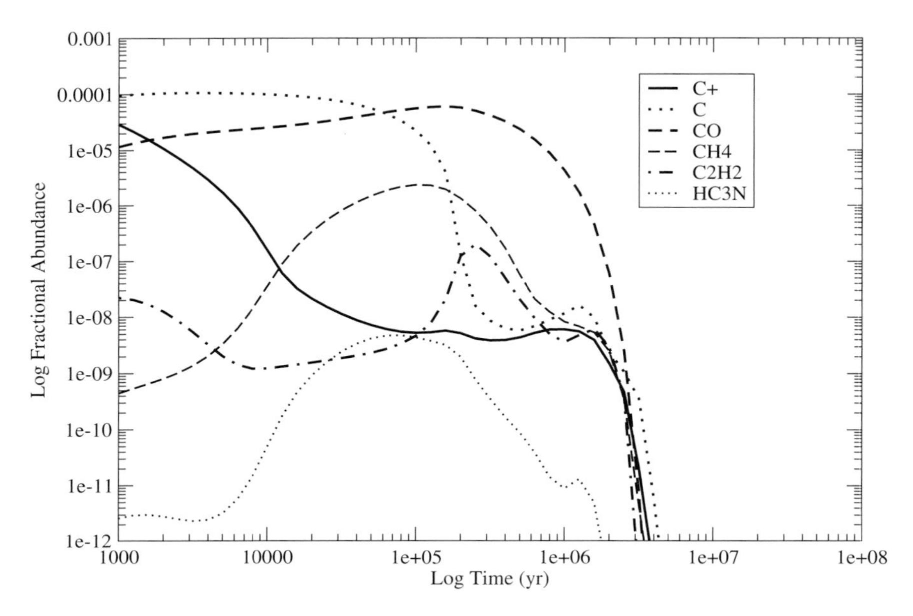


Fig. 1.2. Time evolution of gas-phase carbon chemistry but with accretion at $T = 10\,\mathrm{K}$ and $n = 10^4\,\mathrm{cm}^{-3}$.

*Thermal evaporation* of a grain surface species $X^d$ occurs with a rate coefficient[2]

$$k_{\mathrm{te}} = \nu_0 e^{-E_{\mathrm{D}}/k_{\mathrm{B}}T_d} \tag{1.53}$$

where $\nu_0$ is a frequency, which can be estimated as $(2n_s E_{\mathrm{D}}/\pi^2 m)^{1/2}$.[49] Here $n_s$ is the number of sites per unit surface area, $m$ is the mass of the species $X^d$, $E_{\mathrm{D}}$ is its binding (desorption) energy to the grain, and $T_d$ is the grain temperature, which is $\sim$10 K in cold cores. Note that astronomers often refer to energies with units of temperature so that the Boltzmann constant $k_{\mathrm{B}}$ is often set to unity. Typically the frequency is about $10^{12}\,\mathrm{s}^{-1}$ for physisorbed species. The rate of evaporation is very sensitive to the grain temperature. For example, water on water ice has a binding energy $E_{\mathrm{D}}/k_{\mathrm{B}}$ $\sim$5770 K and a temperature rise of 10 K from $T_d = 95\,\mathrm{K}$ increases the evaporation rate by a factor of 300. For species which are weakly bound, such as CO and $N_2$, an even smaller temperature rise from 30 K gives the same effect. Although grain temperatures can be derived in interstellar clouds from observations of submillimeter continuum emission from dust, it is difficult to argue that the grain temperature is known to better than 10% accuracy given the uncertainties associated with their size, shape, chemical composition and emissivity, adding further uncertainty to gas-grain models. In fact, recent experiments have shown that the situation is more complex. Astronomical infrared observations of the CO fundamental absorption band have shown that CO is trapped both within the polar, $H_2O$-rich, component of interstellar ices as well as in the non-polar, CO- and $CO_2$-rich, component.[2] Laboratory TPD experiments have shown that CO can be trapped at a variety of sites within a porous $H_2O$ ice structure and that up to four binding energies may be appropriate for this and other species.[50] In any case, since binding energies are typically greater than $1000\,k_B$, the time scale for thermal evaporation from a grain at 10 K is longer than the age of the Universe. For thermal evaporation to occur within $10^5$ yr at 10 K, one requires a binding energy less than about 550 K, not satisfied for species other than hydrogen and helium. Thus thermal evaporation is unlikely to play any significant role in returning material to the gas in cold cores.

*Cosmic ray-induced heating* refers to the passage of a high energy cosmic ray through a grain that heats material along its path of interaction and allows evaporation to occur. This process has a slow rate as the cosmic-ray ionisation rate is small and such events relatively rare.[51]

Absorption of external UV photons by ices can lead to *direct photodesorption* particularly at the cloud surface where the flux of UV photons is high.[2] If clouds are homogeneous, then the process is not competitive elsewhere due to efficient extinction of the UV field by the dust particles. The photons can penetrate the cloud more efficiently if either the surface area of the dust is much smaller than anticipated due to coagulation or the cloud is very clumpy on small scales allowing UV photons access to much greater depths from the surface. Typically, models of cold cores have ignored this desorption mechanism due to the lack of photons and the small yield of desorbed molecules per photon absorbed. Recently, it has been shown that the yield of CO desorption from CO ice is about two orders of magnitude more efficient than previously thought, indicating that it may be the dominant process at least in the outer regions of cold clouds.[52]

As discussed in Sec. 1.1, cosmic-ray ionisation of $H_2$ produces secondary electrons which excite $H_2$ and cause it to fluoresce in the Lyman and Werner bands. Since cosmic rays penetrate cold cores easily, these UV photons are generated internally throughout the cloud and are not subject to the levels of extinction suffered by externally-generated photons.[53] In principle these photons can both dissociate gas-phase molecules and photodesorb mantle material; the latter process is known as *cosmic ray-induced photodesorption.*

*Chemical reactions* can also lead to non-thermal desorption.[2,5,6] The formation of the $H_2$ molecule from the recombination of two H atoms on the surface of an interstellar grain releases 4.5 eV of excess energy. Much of this goes into excitation of the $H_2$ molecule but a certain fraction is deposited into the grain and may be sufficiently localised to desorb nearby molecules. More generally, the exothermicity of any surface chemical reaction can be used to eject the products from the surface in analogy with unimolecular chemical reactions. This process will be discussed in more detail in Section 1.6.3 in the context of surface chemistry. Exothermicity-based desorption can follow photodissociation of molecules on surfaces, since the resulting fragments can recombine exothermically.

Laboratory experiments on irradiated ices result in the production of radicals which, at low temperatures, are effectively immobile in the ice. As the temperature of the ice increases, the radicals become mobile and reactions between them release energy which can be used to generate a runaway process resulting in a catastrophic desorption event or *explosion* at a temperature of about 27 K according to the experiments.[2] Such a temperature is too high to be achievable easily in a cold core but may occur through the absorption of a single UV photon if the grain has a very small size.

The influence of some of these desorption processes on cold core abundances has been investigated by Willacy & Williams,[54] Shalabeia & Greenberg,[55] and Willacy & Millar.[56] The latter also included deuterium fractionation, a process known to be efficient at low temperatures and which provides additional observational constraints. Willacy & Millar compared predictions with observations of some 25 species in the cold core TMC-1 and with the D/H abundance ratios of a further 8 species, finding best agreement (to within a factor of 5) in 19 abundances and 5 ratios for the case of $H_2$ driven desorption. The global effect of these desorption processes is to cause a quasi-steady state to develop at times in excess of a million years due to the balance between accretion and desorption. Such a quasi-steady state can also occur in models with surface chemistry, as discussed in Section 1.6.3.

Despite this apparent success, there is as yet no consensus on which, if any, of the above processes should be included in models of cold cores. Most rely on ad-hoc assumptions about efficiency, and few have been studied in the laboratory under realistic physical conditions.

### 1.4.3. *Non-Homogeneous Sources*

It is difficult to extract three-dimensional information on the physical properties of cold cores from astronomical observations. Molecular line emission traces minor components of the mass — molecular hydrogen being homopolar and therefore not possessing allowed dipole transitions — but it does have the advantage of allowing dynamical information (eg. infall, outflow, rotation) to be derived. As mentioned in the previous sub-section, there may also be other selection effects at work, for example, high-density regions may have very weak line emission because of accretion. In such cases, thermal emission from dust grains is a better tracer of the mass distribution in the cloud. Submillimeter observations of dust emission are now used routinely to derive density and temperature profiles of the dust under some simplifying assumptions about geometry and emissivity of the particles. Such observations show that cold cores, particularly those involved in low mass star formation, have power-law density distributions, sometimes with a flatter profile or constant density in the central regions. Quiescent cores; that is, those which show thermal line widths, have density profiles well fit by those of the Bonnor-Ebert (BE) sphere, appropriate for an isothermal object in pressure equilibrium with an external medium. The BE sphere has a uniform density inner region surrounded by an envelope with density $\rho \propto r^{-2}$, approximately.[57]

For quiescent cores, the simplest approach is to model the cloud as a series of concentric shells treating each shell as a one-point model with fixed physical conditions. Such models have been constructed by Roberts *et al.*[58] in their study of deuterium fractionation in the pre-stellar cores L1544 and $\rho$ Oph D. These cores were modelled by at most 6 shells and much better agreement with observation was obtained than through using a single one-point model. The non-homogeneous models, which have high densities in their central regions, show that molecular distributions such as CO will have central "holes" in their distributions due to accretion and that the size of such holes increases with increasing time. Such central holes, although detected in some species such as CS, are not detected in all molecules for reasons which are not yet clear. Thus, while CO is depleted, $N_2H^+$ is not, despite the fact that laboratory studies show that CO and $N_2$ have very similar binding energies on ices.[59] Indeed Tafalla *et al.*[60] find that $N_2H^+$ has a constant abundance and that $NH_3$ actually increases toward the core centre. Clearly, a model in which accretion and evaporation are described by essentially one parameter for each species, namely the binding energy, will be limited in its success in explaining the diversity of results seen in cold cores.

Since stars do actually form in such regions it may also be appropriate to include dynamics, particularly that of infall which is detected at velocities of about 0.1–0.2 km s$^{-1}$ in several well-studied objects. Lee, Bergin & Evans[61] combined chemistry, dynamics, and a calculation of molecular excitation to follow the evolution of molecular rotational line profiles. Their model used a sequence of BE spheres and the so-called "inside-out" collapse to follow the chemistry of a parcel of gas as it falls towards the centre of the source. The results show complex, time-dependent effects in which molecular abundances are affected by the competition between accretion and evaporation. The species which has the most effect on the chemistry is CO, not surprisingly since it is the most abundant interstellar molecule after $H_2$. When CO is abundant in the gas phase, reaction with $He^+$ produces $C^+$ and O which proceed to form C-bearing and O-bearing molecules. When CO is frozen onto dust grains then the abundances of these other molecules decrease appreciably.

Aikawa *et al.*[62] have included deuterium chemistry in a BE sphere model to which infall has been added. Infall is determined by the ratio of gravitational and pressure forces with the low infall velocities observed in starless cores reproduced when the ratio is slightly greater than 1. The model contains accretion, desorption and surface chemistry using the diffusive rate approach (see Sec. 1.6). For a ratio of 1.1, a sphere with an initial central density of $2 \times 10^4\,\mathrm{cm}^{-3}$ takes $1.17 \times 10^6\,\mathrm{yr}$ to reach a central density of $3 \times 10^7\,\mathrm{cm}^{-3}$, with a flat density distribution within 4000 astronomical units ($1\ \mathrm{AU} = 1.5 \times 10^{13}\,\mathrm{cm}$) and a density profile scaling as $r^{-2.3}$ outside this. The infall is rather slow and, since both the central and outer boundary of the cloud are assumed to have zero velocity, the infall velocity has a maximum value at a radius which moves inwards as the collapse proceeds. This model reproduces the observational result that molecules such as CS and CO are depleted significantly inside 5000 AU by accretion onto grains. It is perhaps worth noting at this point that such depletions are difficult to observe, particularly for CO since the size of the depleted region is small compared to typical telescope beam sizes, and emission from CO is dominated by the outer regions of the cloud where it is undepleted rather than by the central high-density core.

These dynamical models of cold cores are typically much more complex than static models and it is worth asking whether such models actually help better constrain the physics and chemistry of star formation. Lee *et al.*[63] have tried to answer this question by comparing dynamical models of collapse with static models having empirical abundance distributions. Such models also seek to reproduce line profiles but do so by modifying abundance distributions, for example by adopting large steps in abundance to represent material either accreted onto or removed from grain surfaces. Such models do not contain any chemistry and are mostly used by observers to make a fit to observations so that evidence that such distributions arise naturally in dynamic models would be powerful support for the observationally-derived picture. Lee *et al.* also considered non-homogeneous, non-dynamic models and showed that differences occur between these and dynamic models even when one adopts the underlying density and temperature for the core from dust continuum observations. As mentioned above, these non-dynamic models allow large central holes to occur in the molecular distributions as material freezes rapidly on to the dust at high density. For example, the CS abundance falls by some five orders of magnitude in the central region. When infall is included the situation is much different since dynamics allow the depleted zone to be replenished by infall of fresh material from the outer region with the result that the CS abundance decreases by only an order of magnitude. When these inner regions eventually become warm due to their proximity to a nascent central star, differences in abundance distributions will be more evident in the higher energy rotational transitions and will be prime targets for the new ALMA (Atacama Large Millimeter Array) interferometer in Chile.

## 1.5. Deuterium Fractionation

Deuterium is formed efficiently only in the Big Bang with an abundance relative to hydrogen which is sensitive to the baryon density, that is normal matter, in the Universe. Since this baryon density determines to a large extent the evolution of the Universe, measuring the deuterium abundance has been a matter of great importance to astrophysicists for almost 50 years. Such measurements are difficult to do because the deuterium equivalent of the hydrogen 21-cm line (the transition between the first-excited and lowest hyperfine split states of the 1s configuration) lies in a very unfavourable part of the radio spectrum while its UV Lyman-alpha line (2p-1s) can only be observed with space telescopes. Furthermore, the efficient extinction of UV radiation by dust means that absorption due to this line can only be studied in nearby regions, perhaps up to 3000 light years, from the Sun, a radius only about one-tenth of that of the Milky Way galaxy. Observations with the *FUSE* (Far Ultraviolet Satellite Explorer) satellite show that there is variation in the D/H ratio by about a factor of two, perhaps an indication that D is preferentially absorbed onto dust grains.[64] The local abundance ratio is in the region of $(1–2)\times10^{-5}$, which we shall call the cosmic ratio, although it is likely to be smaller by a factor of a few than the ratio set in the Big Bang since D is — and has been — destroyed rather easily once incorporated into stars.

Despite the difficulty of observing atomic deuterium, it turns out that it is readily detected in interstellar molecules; 30 isotopologues of the 140 or so interstellar molecules have been observed (see Table 1.2). The initial detections were a surprise to radio astronomers because the intensity ratios implied abundances of deuterated molecules on the order of a few percent of their hydrogenated parents. Subsequently, astronomers detected the triply-deuterated molecules $ND_3$ and $CD_3OH$, which have abundance ratios around $10^{-4}$, an enhancement of some 11 orders of magnitude over the statistical value — large even by astronomical standards. This enhancement, or fractionation, occurs because of small zero-point-energy differences between

Table 1.2. Deuterated interstellar molecules

| | | | | | |
|---|---|---|---|---|---|
| HD | $H_2D^+$ | $D_2H^+$ | $DCO^+$ | $N_2D^+$ | HDO |
| $D_2O$ | DCN | DNC | HDS | $D_2S$ | $C_2D$ |
| HDCO | $D_2CO$ | HDCS | $D_2CS$ | $NH_2D$ | $NHD_2$ |
| $ND_3$ | $C_4D$ | $DC_3N$ | $DC_5N$ | $CH_3OD$ | $CH_2DOH$ |
| $CHD_2OH$ | $CD_3OH$ | $CH_2DCN$ | $CH_3CCD$ | $CH_2DCCD$ | c-$C_3HD$ |

reactants and products containing H and D.[58,65,66] These differences are negligible at room temperature but not at the temperatures of cold cores. In these regions the primary reservoirs of hydrogen and deuterium are $H_2$ and HD, respectively, with $HD/H_2 = 2(D/H)_{cosmic} \sim 3 \times 10^{-5}$.

The most important reactions which extract deuterium from HD involve ion-neutral isotope exchange reactions:

$$H_3^+ + HD \longrightarrow H_2D^+ + H_2 + \Delta E_1, \tag{1.54}$$

$$CH_3^+ + HD \longrightarrow CH_2D^+ + H_2 + \Delta E_2, \tag{1.55}$$

$$C_2H_2^+ + HD \longrightarrow C_2HD^+ + H_2 + \Delta E_3, \tag{1.56}$$

where the reaction exoergicities (defined as positive here), although small — $\Delta E_1/k_B \sim 230\,K$, $\Delta E_2/k_B \sim 375$ K, $\Delta E_3/k_B \sim 550\,K$ — are much larger than the temperatures of cold interstellar clouds. Hence at 10–20 K, the reverse reactions do not occur efficiently despite the large abundance of $H_2$. As a result, the degree of fractionation, particularly in $H_2D^+$, can become large. Once formed, these deuterated ions can pass on their enhanced deuterium content to other species in chemical reactions.

It is instructive to consider the fractionation which arises in $H_2D^+$. In addition to the reverse of reaction (1.54), it can be destroyed by dissociative recombination with electrons and by reaction with neutral molecules, which we represent here by CO and HD. If one assumes steady-state, which is a reasonable approximation since the reactions are fast:

$$\frac{[H_2D^+]}{[H_3^+]} = \frac{k_{1f}}{k_{1r} + k_{1e}[e] + k_{CO}[CO] + k_{HD}[HD]} \frac{[HD]}{[H_2]}. \tag{1.57}$$

where [X] represents the fractional abundance of X relative to $H_2$, $k_{1f}$ and $k_{1r}$ are the forward and reverse rate coefficients of reaction (1.54), $k_{1e}$ is the dissociative recombination rate coefficient of $H_2D^+$, and $k_{CO}$ and $k_{HD}$ are the rate coefficients for the reactions of $H_2D^+$ with CO and HD. The ratio is a function of temperature because the rate coefficients are temperature dependent. For conditions typical of undepleted cold cores, $k_{1r}$ is negligible, as is dissociative recombination, the ion-neutral rate coefficients are roughly equal, and $[CO] \sim 10^{-4} \sim 10[HD]$. Thus, $[H_2D^+]/[H_3^+] \sim 10^4$

$[HD/H_2] \sim 0.1$. This large fractionation in $H_2D^+$ can be transmitted to other species through rapid deuteron transfer reactions, such as

$$H_2D^+ + N_2 \longrightarrow N_2D^+ + H_2. \tag{1.58}$$

If we assume that the transfer of the deuteron occurs statistically, i. e. in one-third of reactions, then this process leads to fractionation ratios of a few percent in $N_2D^+$, as observed in molecular clouds such as TMC-1. The $H_2D^+$ ion is the dominant source of fractionation at 10 K as it is more abundant than either $CH_2D^+$ or $C_2HD^+$. For clouds with temperatures larger than about 20 K, reaction with $H_2$ (the reverse of reaction (1.54) dominates the denominator and the $[H_2D^+]/[H_3^+]$ abundance ratio rapidly falls to its cosmic value. At these intermediate temperatures, fractionation by the hydrocarbon ions becomes important until 40–60 K when they too are destroyed by reaction with $H_2$. The reactivities of the hydrocarbon ions are different from that of $H_2D^+$, in particular the proton affinity of $N_2$ is small so that $N_2D^+$ can only be formed by $H_2D^+$. As such, $N_2D^+$ is a tracer of cold, gas-phase chemistry.

We have seen that in high-density cold cores CO and other abundant molecules can be accreted onto dust grains. Once these molecules are depleted by an order of magnitude or so, reaction with HD now dominates the loss of $H_2D^+$, again assuming that dissociative recombination is unimportant. In this case $D_3^+$ can be formed rapidly via:

$$H_2D^+ + HD \longrightarrow D_2H^+ + H_2 + \Delta E_4, \tag{1.59}$$

$$D_2H^+ + HD \longrightarrow D_3^+ + H_2 + \Delta E_5, \tag{1.60}$$

where $\Delta E_4/k_B \sim 187\,K$ and $\Delta E_5/k_B \sim 159$ K. Since $D_3^+$ is unreactive with HD, then it is possible that $D_3^+$ can be more abundant than $H_3^+$ and thus a more important deuteron donor than $H_2D^+$, not only because it is more abundant but also because it will transfer a deuteron on every reactive collision. This leads to very large fractionation in the gas phase.[67]

The construction of detailed chemical kinetic models describing deuterium chemistry is non-trivial and contains of necessity several approximations which are open to debate. For example, the reaction

$$C^+ + H_2O \longrightarrow HCO^+ + H \tag{1.61}$$

has three deuterated equivalents:

$$C^+ + HDO \longrightarrow HCO^+ + D, \tag{1.62}$$

$$C^+ + HDO \longrightarrow DCO^+ + H, \tag{1.63}$$

$$C^+ + D_2O \longrightarrow DCO^+ + D. \tag{1.64}$$

Typically, one adopts the same total rate coefficient for each of the three reactive systems ($C^+ + H_2O$, HDO, $D_2O$) but applies a statistical branching ratio to the products of the $C^+ + HDO$ reaction, in this case 0.5. For more complex systems, additional approximations, often having to do with the preservation of functional groups within a reaction, are made. The proton transfer reaction:

$$H_3^+ + CH_3OH \longrightarrow CH_3OH_2^+ + H_2, \tag{1.65}$$

has four possible product channels when the neutral is singly-deuterated methanol:

$$H_3^+ + CH_2DOH \longrightarrow CH_2DOH_2^+ + H_2, \tag{1.66}$$

$$H_3^+ + CH_2DOH \longrightarrow CH_3OHD^+ + H_2, \tag{1.67}$$

$$H_3^+ + CH_3OD \longrightarrow CH_2DOH_2^+ + H_2, \tag{1.68}$$

$$H_3^+ + CH_3OD \longrightarrow CH_3OHD^+ + H_2. \tag{1.69}$$

One can include all four reactions in a reaction network or prescribe that the D atom must be retained in its functional group, thereby omitting the middle two reactions.

The number of reactions to be included in models of deuterium chemistry increases rapidly when one considers multiply-deuterated species. The reaction between $H_3^+$ and $CH_3OH$ actually has a second, dissociative, product channel in addition to proton transfer:

$$H_3^+ + CH_3OH \longrightarrow CH_3^+ + H_2O + H_2. \tag{1.70}$$

The $H_3^+$ ion has three deuterated counterparts while methanol has seven. Thus there are now $4 \times 8$ reactive collisions to be considered, each with multiple product channels. Each of the ions $H_2D^+$ and $D_2H^+$ has two possible product channels in each proton transfer reaction since they can transfer either a D or a H. If one applies the functional group rule, there are then 48 reactions involving proton transfer. Reaction (1.70) gives rise to even more possibilities because there are three products each containing hydrogen; even if the functional group rule is used to restrict possible channels there is a total of 64 reactions generated. Thus two reactions in a non-deuterated chemistry must be replaced by 112 in a fully deuterated model. Typically the inclusion of deuterium increases the number of reactions by a factor of four and species by a factor of two. Thus, as an example, a fairly modest interstellar model containing 199 species and 2717 reactions becomes one with 425 species and 13,848 reactions when fully deuterated using the functional group rule. Finally, there is some indication that in the dissociative recombination of deuterated ions, hydrogen atoms are lost preferentially over deuterium atoms so that statistical branching is not appropriate. Since deuteron transfer is one of the major mechanisms by which deuterium is fractionated, such preferential retention of deuterium can have a large effect on the calculated abundances; unfortunately, very few systems have been studied experimentally.

Detailed models of deuterium chemistry have been made using realistic density profiles for static pre-stellar cloud cores.[58] These show that $D_3^+$ becomes more abundant than $H_3^+$ within a radial distance of 6000 AU from the centre of the cloud. At the very centre:

$$\frac{[D_3^+]}{[H_3^+]} \sim \frac{k_{1f}}{k_{3e}[e]}\frac{[HD]}{[H_2]} \sim 20, \tag{1.71}$$

where $k_{3e}$ is the rate coefficient for the dissociative recombination of $D_3^+$ with electrons. Subsequent support for this model has come from the detection of a rotational transition of para-$D_2H^+$ at 690 GHz with a similar intensity to the 372 GHz line of ortho-$H_2D^+$. Observations at these frequencies are difficult due to water vapour absorption in the atmosphere and more accurate abundances in more sources will be a target for future instruments such as ALMA. The models also show that large fractionation occurs in multiply-deuterated species in these central, depleted regions within cold cores, and give $[ND_3]/[NH_3] \sim 0.002$ and $[NHD_2]/[NH_3] \sim 0.14$, seemingly in good agreement with observation. These deuterated species are transient since they too will freeze out onto the dust particles with the implication that very highly fractionated sources will be difficult to detect in species like $NHD_2$ since time scales are short. In essence, the only molecules predicted to be detectable via radio astronomy in the coldest, highest density cores are $H_2D^+$ and $D_2H^+$;[68] high-resolution spectral and spatial observations with ALMA will give unique dynamical and physical information on the late stages of low-mass star formation.

Figure 1.3 shows calculated fractionation ratios (abundance ratios of deuterated to normal species) for $H_2D^+$, $D_3^+$, and $D_2CO$ as functions of time for a homogeneous cloud of molecular hydrogen density $10^6\,cm^{-3}$ and temperature 10 K. Results are shown both for no grain accretion (filled objects) and for grain accretion (unfilled objects). It can be seen that in the case of grain accretion, the fractionation ratios increase dramatically, especially for $D_3^+$, where the result approaches the approximation given in Eq. (1.71). Note also that the high $D_2CO/H_2CO$ ratio seen for the case of grain accretion ends at a time of $1 \times 10^5$ yr, when the species are almost totally accreted onto grain surfaces. At times later than this, the limiting case envisaged by Walmsley *et al.*[68] pertains in the absence of desorption mechanisms.

Although it is clear that deuterated cations observed in interstellar clouds are formed in gas-phase reactions, the situation is not so clear for

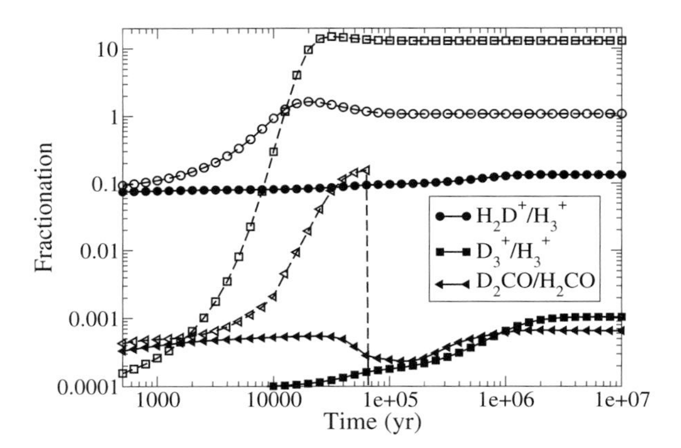


Fig. 1.3. Fractionation ratios plotted against time for a cloud with $H_2$ density $10^6$ cm$^{-3}$ and temperature 10 K. Results with unfilled objects are for the case of grain accretion, whereas those with filled objects are for the case of no grain accretion. Calculation performed by Dr. Helen Roberts.

the neutral species. Consider highly depleted regions for which $[D_3^+]\sim[H_3^+]$. Dissociative recombination of these ions with electrons will lead to a high fractionation in atomic D. Since collisions with grains are rapid at high density, a large gas phase D/H ratio will be reflected in a large D/H ratio in the dust. As discussed in Sec. 1.3, it is believed that surface chemistry may play a substantial role in the synthesis of certain species, in particular $CH_3OH$, which is observed to be abundant in interstellar ices, and $H_2S$, for which no efficient gas-phase synthesis has been found at low temperature. Given the large abundance of CO which freezes out onto the grains it seems reasonable that addition and other reactions with H and D could lead to very large fractionation in $CH_3OH$ and its intermediary $H_2CO$.[45,46] Indeed, some sources have $[D_2CO]/[H_2CO] \sim 0.4$ and have a higher abundance of deuterated methanol, added over all isotopic forms, than normal methanol. As mentioned above, detailed models incorporating multiple deuteration can give rise to very large fractionation in the gas-phase in the innermost regions of the core. In fact, if one integrates the number density through the cloud to obtain column densities, which are derived from the radio telescope observations, a second difficulty arises, namely that the ratio of column densities can be much less than the local abundance ratios. For example, although $[D_2CO]/[H_2CO]$ can be large in the central region of the core and close to that observed, the column density ratio is about an order of magnitude less than observed. This occurs because $H_2CO$, but not $D_2CO$, is formed readily in the outer region of the cloud and "dilutes" the ratio calculated for the inner region.

A further reason for thinking that grain surface chemistry might be responsible for creating large fractionation in some species comes from consideration of the structure of protonated neutrals. Consider formaldehyde as an example. Proton transfer leads to two structures, one, the less stable, protonated at the carbon end, the other protonated at the oxygen end, written $H_2COH^+$, an ion which has actually been detected in interstellar clouds. Thus when $H_2CO$ is deuterated in the gas phase, the deuteron attaches preferentially to the oxygen atom. If the basic structure of the complex formed in the dissociative recombination with electrons is preserved, then the products are $H_2CO$ + D and not HDCO + H. Isomerisation of the $H_2COD$ intermediate may lead to HDCO but the efficiency is likely to be less than that assumed in models which ignore structure.[69] Similar arguments may be applied to the fractionation of the cyanopolyynes $HC_3N$ and $HC_5N$ and to methanol in the gas-phase. The situation for methanol is, in fact, somewhat more critical since the latest laboratory results show that only a few percent of dissociative recombinations of protonated methanol actually produce methanol.[19] The implication is that the methanol detected in cold clouds must have been formed on and released from cold grains. Since surfaces will be enriched in deuterium, large abundances of deuterated methanol are expected. The detailed processes by which a species like CO becomes hydrogenated or deuterated on the surface are discussed in Sec. 1.3.3.

In our discussion of deuteration we have implicitly assumed that all reactants are in their ground state and that rotational excitation does not affect the kinetics of the primary fractionation reactions. In fact this may not be the case. Since the $J = 1$ (ortho) state of $H_2$ lies 170 K above the $J = 0$ (para) state, even a small fraction of $H_2$ in $J = 1$ will result in a faster destruction rate coefficient for $H_2D^+$ than with $J = 0$. In highly depleted gas, destruction of $H_2D^+$ via ortho-$H_2$ is more rapid than reaction with HD for a fractional abundance of ortho-$H_2$ greater than about one percent. Since the formation of $H_2$ on interstellar dust is a highly exothermic process and the ejected molecule is likely to be superthermal internally, the initial ortho-para ratio is likely to be the high-temperature value of 3. This value is reduced through proton exchange reactions with $H^+$ and $H_3^+$ and reaches a steady-state value in the range $10^{-3}$–$10^{-4}$ compared with the equilibrium value of $4 \times 10^{-7}$ at 10 K, on a time scale which may be longer than that of a particular cold core but shorter than that of molecular gas as a whole. Thus fractionation in $H_3^+$ and related species may depend on the thermal and chemical history of the gas. The detailed processes and selection rules for the various processes which interconvert nuclear spin in $H_2$, $H_3^+$, $H_2D^+$, $D_2H^+$ and $D_3^+$ are still a matter of debate[70,71] (see also chapter 3 by Gerlich) but it does appear that the very high fractionation ratios observed in highly depleted cores may not all be the result of gas-phase reactions involving multiply-deuterated $H_3^+$.

## 1.6. Surface Chemistry: Mathematical Details and Gas-Grain Models

### 1.6.1. *Rate Equations*

The *kinetics* of $H_2$ formation (and other surface reactions) via the Langmuir-Hinshelwood (diffusive) mechanism can be treated by rate equations, as in Eq. (1.52), or by stochastic methods.[72] There are two main objections to the former approach: it does not handle random-walk correctly and it fails in the limit of small numbers of reactive species. The latter objection is a far more serious one in the interstellar medium because dust particles are small, and the number of reactive atoms and radicals on their surfaces can be, on average, less than unity. Nevertheless, with rare exceptions, the few large models of interstellar chemistry that include surface processes as well as gas-phase chemistry do so via the rate equation approach, so we discuss it here. In the treatment below, we do not use the ordinary units of surface chemistry — areal concentrations or monolayers[39] — but instead refer to numbers of species on the mantle of an individual but average grain. Numbers can be converted to bulk concentrations, as used in Eq. (1.52), by multiplication by the grain number density $n_d$.

Consider two species, A and B, that can react on a grain surface via diffusion without activation energy. In the following discussion, we ignore the possibility that diffusion can occur by tunnelling. There are two reasons for this neglect: (1) careful experimental work[40] appears to exclude it for H atoms on olivine and amorphous carbon and (2) calculated potential barriers for physisorbed species tend to be very wide compared with chemical barriers (D. Woon 2007, private communication). If we follow the hopping of a single molecule of A from one site to another, the average (ensemble) probability of reaction with B after a hop to an adjacent binding site is given by the product of the hopping rate and the probability that the adjacent site is occupied by B. The hopping rate for species A $k_{\mathrm{hop,A}}$ ($\mathrm{s}^{-1}$) is given by the expression[40]

$$k_{\mathrm{hop,A}} = \nu_0 \exp\left(-E_{\mathrm{b,A}}/T\right) \tag{1.72}$$

where $\nu_0$ is the "attempt" frequency (see also Eq. 1.53), typically taken as $1\text{–}3 \times 10^{12}\,\mathrm{s}^{-1}$ for physisorbed species, and the energy barrier is written, as is common in astronomy, in units of K. The probability that B resides on an adjacent site is simply $N(B)/N$, where $N(B)$ is the number of sites on a grain occupied by B, and $N$ is the total number of sites. With a standard site density of $\approx 10^{15}$ cm$^{-2}$, the number of sites on a grain of radius 0.1 $\mu$m is $\approx 10^6$. Taking account of the fact that species B can also hop from site to site, we arrive at a rate law for the disappearance of A by reaction with B of

$$\frac{dN(A)}{dt} = -(k_{\mathrm{hop,A}} + k_{\mathrm{hop,B}})N(A)N(B)/N. \tag{1.73}$$

Equation (1.73) can be rewritten as

$$\frac{dN(A)}{dt} = -(k_{\mathrm{diff,A}} + k_{\mathrm{diff,B}})N(A)N(B) = -k_{\mathrm{AB}}N(A)N(B), \tag{1.74}$$

where the first-order rate coefficients for diffusion are loosely defined as the hopping rates over a number of sites equivalent to the total number of sites on the dust particle, although, of course, the process of "back diffusion" means that the total grain is not seen by the diffusing particle in only $N$ moves. If atomic hydrogen is a reactant, its uniquely low barrier against diffusion means that the diffusion of a heavier reactant need not be considered. Indeed, for cores at 10 K, the only major reactant that moves on surfaces rapidly is atomic hydrogen. It is important to reiterate that although $H_2$ is the dominant form of hydrogen in dense clouds, there can be sufficient atomic hydrogen to render it an important surface reactant.

If there is chemical activation energy $E_{\mathrm{A}}$ in addition to the barrier to diffusion, two revisions of Eq. (1.74) have been suggested.[73] The simpler is to add either a tunnelling probability or hopping probability (Boltzmann factor), whichever is greater, as a simple factor, labelled $\kappa$, so that the rate coefficient $k_{\mathrm{AB}}$ is multiplied by $\kappa$. In the case of hopping,

$$\kappa = \exp\left(-E_{\mathrm{A}}/T\right), \tag{1.75}$$

whereas in the case of tunnelling,

$$\kappa = P_{\mathrm{tunn}}. \tag{1.76}$$

For a rectangular potential of width $a$,

$$P_{\mathrm{tunn}} = \exp[-2(a/\hbar)\sqrt{(2\mu E_{\mathrm{A}})}], \tag{1.77}$$

where $\mu$ is the reduced mass of the A-B system.

The more complex revision is to consider diffusive hopping from one site to another as competitive with tunnelling through or hopping over the activation energy barrier. If we assume that tunnelling under the activation energy barrier dominates hopping over it, the additional factor $\kappa$ can be written in the steady-state limit as

$$\kappa = \frac{k_{\mathrm{tunn}}}{k_{\mathrm{tunn}} + k_{\mathrm{hop,A}} + k_{\mathrm{hop,B}}}, \tag{1.78}$$

where

$$k_{\rm tunn} = \nu_0 \times P_{\rm tunn}. \tag{1.79}$$

In this case, if the tunnelling rate exceeds the two diffusive hopping rates, the activation barrier does *not* impede the rate of reaction! On the other hand, if the tunnelling rate is much slower than the diffusive hopping rates, the more complex expression for $\kappa$ reduces to

$$\kappa \approx \frac{k_{\rm tunn}}{k_{\rm hop,A} + k_{\rm hop,B}}, \tag{1.80}$$

and the rate coefficient $k_{\rm AB}$ becomes

$$k_{\rm AB} \approx k_{\rm tunn}/N, \tag{1.81}$$

which is formally independent of the rate of diffusion. Nevertheless, the difference between the result of Eq. (1.81) and the simple approach with tunnelling is often minimal. The microscopic stochastic approach discussed below suggests that the competitive mechanism is the correct one,[74] but this view is not universal.[75]

To place species A into a chemical model with both gas-phase and surface processes also requires inclusion of adsorption and desorption. Adsorption onto a single grain follows a simple bimolecular rate law of the type

$$\frac{dN(A)}{dt} = S_{\rm A} v_{\rm A} \sigma n(A), \tag{1.82}$$

where $n(A)$ is the gas-phase density of $A$ and the other parameters have been defined in the discussion following Eq. (1.51). If one wishes to estimate a rate for the bulk adsorption of A, one must also multiply both sides of Eq. (1.82) by the dust grain density.

Thermal desorption (evaporation) can be handled by the so-called first-order Polanyi-Wigner equation, which for an individual grain takes the form[39]

$$\frac{dN(A)}{dt} = -k_{\rm te,A} N(A), \tag{1.83}$$

where $k_{\rm te}$ is defined in Eq. (1.53). Experimentalists often report evaporation as a zeroth-order process if many monolayers are involved and no net change is detected. There is even a second-order Polanyi-Wigner equation, for situations in which desorption follows reaction.[39,40] Finally, in cold interstellar cloud cores, little evaporation takes place for any species more strongly bound that H and $H_2$, so that non-thermal desorption processes must be considered. Mechanisms based on the exothermicity of surface chemical reactions, cosmic ray bombardment, photon bombardment, and grain-grain collisions have been included (see Sec. 1.4.2). A recent experiment shows a measurable rate of photodesorption for the case of pure CO ice bombarded with far-UV photons.[52]

In addition to the problems already mentioned, the rate equation treatment makes no distinctions about where the adsorbates are located; they can be in any monolayer of the mantle (and the number of monolayers in an icy mantle can range up to several hundred) and they can be in any site of a monolayer. In other words, the details of reactions are averaged over vertical and horizontal features. The vertical problem can be partially removed by restricting processes to the topmost monolayer, although in porous systems this restriction is not obviously true.

### 1.6.2. *Stochastic Approaches*

The assumptions and problems inherent in the rate equation approach can be removed by assorted stochastic treatments, although these typically involve far more computer time and are difficult to couple with gas-phase rate equations.[72] The problem of average adsorbate abundances that are less than unity on individual dust particles can be removed by so-called macroscopic stochastic methods, which take both the discrete numbers of adsorbates per grain and their fluctuations into account. The methods involve either a Monte Carlo treatment, in which random numbers are chosen to determine what process occurs during a time interval, or a master equation approach, which contains differential equations as in the rate-equation treatment. Instead of solving for the average number of adsorbates per grain, however, these methods solve for the probability as a function of time that a certain number of adsorbates are present.[76] For example, one replaces the one differential equation for $N(A)$ above, with a series of equations for $P_{\rm n}(A)$, the probability that $n$ atoms of A are present on a grain. Unfortunately, although such a treatment increases the number of equations, it does not represent the full complexity of the method because probabilities of different species are correlated; that is, one must solve for the joint probability of, say, $n$ atoms of H, $m$ atoms of O, $p$ atoms of C, etc. In order to accomplish this goal, it is useful to divide the set of different adsorbates into reactive atoms and radicals, which never build up a large surface abundance, and inert or weakly reactive species, such as CO, which can achieve large abundances of more than 1–10 monolayers. The atoms/radicals require a stochastic treatment whereas the chemistry of the species with large abundances can be treated by rate equations. Nevertheless, there are still convergence problems if average abundances of reactive species exceed unity, and as of the present the macroscopic master equation method has only been used in models with small networks of surface reactions. A new approximation to the master equation treatment may be sufficiently efficient to allow its usage with large models under a wide range of conditions.[75]

A more detailed treatment of surface chemistry has recently been reported; this method, known as the continuous-time random-walk approach, is a Monte Carlo treatment in which random walk is treated exactly and the details of local surface morphology included.[42,72,77] The grain surface is divided into a two-dimensional lattice of sites, corresponding to potential minima. One can start with a flat surface, in which all sites are identical, or a rough surface, in which all sorts of imperfections cause local changes in the potential. Rough surfaces with given percentages of irregularities can themselves be obtained by Monte Carlo methods. To perform the calculation, one considers adsorption, desorption, hopping, and possible tunnelling under activation energy barriers to be Markovian (memory-less) events, in which the time interval between two consecutive events follows a Poisson distribution of the type

$$f(t) = k \exp(-kt), \tag{1.84}$$

where $k$ is a first-order rate coefficient, such as that given for diffusive hopping in Eq. (1.72). Once the calculation commences, random numbers are called to determine the time interval for each process, and this is added to the current time. Then, as the clock moves forward, the events are allowed to occur at the appropriate time, when, in addition, the next interval for the process is determined by the choice of another random number. The hopping of all individual adsorbates can be followed, and activation-less reactions occur if two adsorbates occupy the same site. If there is activation energy, the competition between hopping and tunnelling through an activation energy barrier can also be followed. If, for any adsorbate, the time interval for desorption is shorter than that for hopping, then the species desorbs at the appropriate time.

Although the method has been used to study the formation of $H_2$ on surfaces pertaining to diffuse clouds, and the formation and morphology of ices grown in cold cores,[42,74] it has proven very difficult to use it simultaneously with a gas-phase chemistry that is treated adequately by rate equations. The basic problem appears to be that the subroutines used to solve the coupled differential equations for gas-phase kinetics choose time intervals that are different from the random time intervals chosen from the Poisson distributions. A preliminary paper that reports an attempt to couple the two methods by a linearization procedure has just appeared in the literature. Even with this approximation, only a small number of surface reactions can be treated simultaneously.[77]

### 1.6.3. *Gas-Grain Models of Cold Cores*

We have previously focused attention on purely gas-phase chemical models, gas-phase models with accretion onto grains, and gas-phase models with both accretion and desorption. Gas-grain models, such as that of Aikawa *et al.*[62] briefly mentioned above, differ from these models in that they include surface chemistry. For models with large numbers of surface reactions, either the simple rate equation treatment is used, or the rate coefficients $k_{\mathrm{AB}}$ (see Eq. (1.74)) are modified in a semi-empirical manner to handle fractional average adsorbate abundances to an extent. The so-called "modified rate treatment" has been tested against stochastic methods in small systems of equations.[72]

For a non-collapsing cold interstellar core such as TMC-1, we have already shown that the time scale against striking a grain is $\approx 3 \times 10^5$ yr, which is equivalent to the gas-phase chemical time scale to reach early-time abundances. During this time, the major chemical processes that occur on granular surfaces in a gas-grain chemical model are the production of mantle species from the initial constituents of the gas and from species formed early in the gas. If the initial gas is assumed to be atomic, except for the high abundance of $H_2$, the most important early process is the formation of water ice from oxygen and hydrogen atoms colliding with and sticking to dust particles. As the gas-phase abundance of CO increases towards the end of the first $10^5$ yr of the existence of the core and its accretion rate also increases, the rate of its hydrogenation into surface formaldehyde and methanol proceeds until significant abundances of these surface species exist. Within a decade of time later, however, the gas-phase becomes strongly depleted of heavy species, as in simple accretion models, in the absence of desorption (see Fig. 1.2). Unless non-thermal desorption mechanisms exist, the results of gas-grain models for gas-phase and mantle abundances can only be compared with observation for a short period of time during which a substantial mantle has developed and the gas-phase has only lost a small fraction of its heavy material. If, however, non-thermal desorption mechanisms are included, this period of time can be extended.

Let us briefly consider a very recent model of the gas-grain chemistry of a cold interstellar core, in which several mechanisms for non-thermal desorption are considered, including cosmic-ray-induced desorption, "indirect" photodesorption, and exothermicity-induced desorption.[6] Of these, the dominant process is the third. Most surface reactions in these models are associative in nature, with the grain acting in part as a third body to remove enough energy to stabilise the single product. Since these reaction exothermicities are typically much larger than physisorption binding energies, it is entirely possible that sufficient energy can be directed into the physisorption bond to break it. In the absence of much experimental information on the efficiency of this process, one can utilise unimolecular rate theory to estimate the probability that a surface association reaction can lead to the ejection of the product molecule. With the crude RRK approach, Garrod *et al.*[6] estimated that the fraction $f$ of desorbed products in a competition with loss of energy to the grain is given by the expression

$$f = \frac{aP}{1 + aP}, \tag{1.85}$$

where $a$ is treated as a parameter, and $P$ is the RRK expression for the probability that a greater amount of energy than needed for desorption is present in the adsorbate-surface bond. The standard expression for $P$ is given by

$$P = [1 - E_{\mathrm{D}}/E_{\mathrm{exo}}]^{s-1}, \tag{1.86}$$

where $E_{\mathrm{D}}$ is the energy associated with the bond between the surface and the adsorbed molecule, $E_{\mathrm{exo}}$ is the reaction exothermicity and $s$ is the number of vibrational modes in the molecule-surface bond system. Note that $a = 1$ yields the pure RRK treatment.

Although there is no direct photodesorption in the model, there is photodissociation of surface adsorbates, and such photodissociation produces radical and atomic fragments, which can recombine with other atoms and radicals, leading to the possibility of desorption via exothermic reactions. The final mechanism considers the temperature increase in a grain following attack by a cosmic ray particle, and follows the rate of evaporation as the grain cools.

Use of this gas-grain model seems to lengthen the time of best agreement with gas-phase observation for the cold cores L134N and TMC-1 up to perhaps $2\text{–}3 \times 10^6$ yr, at which time the fraction of molecules with abundances predicted to be within an order-of-magnitude of the observed abundances is 0.80–0.85 for L134N and 0.65–0.70 for TMC-1 (see a similar effect in the accretion-desorption model of Willacy & Millar[56]). The dependence on time is rather flat, however, especially for L134N. Values of the parameter $a$ of 0.01, 0.03, and 0.10 were used and yield results statistically insignificant from each other; a value of $a = 0$ yields worse results. One important gas-phase molecule produced sufficiently in this model but not in purely gas-phase approaches is methanol, which comes off the grains during its exothermic formation.

As regards the agreement with mantle abundances at these times, the cold core sources have not been studied in detail in infrared absorption, so it is best to use the well-studied source Elias 16, which lies near TMC-1, astronomically speaking. The level of agreement can best be described as moderate, with the high abundance of water ice securely reproduced, but the abundances of CO and $CO_2$ ice underproduced and the abundances of $NH_3$ and $CH_4$ ice overproduced. It should be noted that smaller but stochastic models, in which only surface chemistry is considered, with nearly constant fluxes of gas-phase material upon the grains, can achieve much better agreement for the mantle abundances.[77]

What happens as the age of the cold core becomes considerably greater than $10^6$ yr? It is instructive to follow the dominant form of carbon. First, a high flux of gaseous CO is adsorbed onto grains and reacts to form sufficient surface methanol so that it starts to dominate, although by $10^7$ yr the dominant form of carbon shifts to methane ice via a complex sequence of reaction, non-thermal desorption, and accretion. Finally, at very long times of $10^8$ yr or so, the dominant form of carbon is a mixture of mantle hydrocarbons more complex than methane. Meanwhile, the non-thermal desorption processes are sufficiently efficient that sizeable portions of gas-phase material remain. For example, gas-phase CO achieves a nearly steady-state fractional abundance of $10^{-6}$ and ammonia of $10^{-7}$. Of course, such long times may have little physical meaning since collapse leading to star formation through stages such as pre-stellar cores and hot cores will probably have already begun.

A late-time chemistry may have been discovered recently in a different context in the star-forming region L1527, in which a general temperature of $\approx$30 K pertains in the vicinity of the protostar.[78] In this protostellar source, the observed molecules strongly resemble those in TMC-1, and include unsaturated radical hydrocarbons and cyanopolyynes. What chemistry can lead to such species at such a late stage of stellar evolution following the pre-stellar core stage where most heavy molecules deplete onto cold 10 K grains? There are two possibilities. We may be dealing with gas that did not accrete onto grains, perhaps from the outer regions of pre-stellar cores. Or, the TMC-1-like molecules originate from the evaporation of mantle material. Unlike the case of hot cores, where the temperature has already risen to 200–300 K in the vicinity of star formation, here a temperature of only 30 K has been achieved. At this temperature, methane and CO ice desorb efficiently into the gas. Of the two, methane is the likely precursor of the TMC-1 carbon-chain molecules, starting with reactions such as Eq. (1.34). Is the 30 K material gradually heating up to form a hot core? Or is 30 K the asymptotic temperature after the heat-up phase? At present, we do not know the answers to these questions. Indeed, despite more than 35 years of active investigation in astrochemistry, there still remain many unanswered questions in this exciting interdisciplinary area of science.

## Acknowledgement

Astrophysics at QUB is supported by a grant from the Science and Technology Facilities Council (UK). EH acknowledges support for his research program in astrochemistry by the National Science Foundation (USA).

## References


1. Lis D, Blake GA, Herbst E. (eds). (2006) *Astrochemistry: Recent Successes and Current Challenges*, Cambridge University Press, Cambridge.
2. Tielens AGGM. (2005) *The Physics and Chemistry of the Interstellar Medium*, Cambridge University Press, Cambridge.
3. Woon DE. (2006) http://www.astrochymist.org/
4. Whittet DCB. (2002) *Dust in the Galactic Environment*, Institute of Physics, London.
5. Williams D, Brown WA, Price SD, Rawlings JMC, Viti S. (2007) Molecules, ices and astronomy. *Ast. & Geophys.* **48**: 1-25-1.34.
6. Garrod RT, Wakelam V, Herbst E. (2007) Nonthermal desorption from interstellar dust grains via exothermic surface reactions. *Astron. & Astrophys.* **467**: 1103–1115.
7. Herbst E. (2005) Chemistry of Star-Forming Regions. *J. Phys. Chem.* **109**: 4017–4029.
8. Le Petit F, Roueff E, Herbst E. (2004) $H_3^+$ and other species in the diffuse cloud towards $\zeta$ Persei: A new detailed model. *Astron. & Astrophys.* **417**: 993–1002.
9. Oka T. (2006) Introductory remarks. *Phil. Trans. R. Soc. A* **364**: 2847–2854.
10. Anicich VG. (2003) *An index of the literature for bimolecular gas-phase cation-molecule reaction kinetics.* Jet Propulsion Laboratory (JPL-Publ-03-19).
11. Herbst E. (2006) Gas Phase Reactions. In GWF Drake (ed.) *Handbook of Atomic, Molecular, and Optical Physics*, pp. 561–573, Springer Verlag, Leipzig.
12. Herbst E, Leung CM. (1986) Effects of Large Rate Coefficients for Ion-Polar Neutral Reactions on Chemical Models of Dense Interstellar Clouds. *Astrophys. J.* **310**: 378–382.
13. Smith D, Adams NG. (1987) The selected ion flow tube (SIFT): studies of ion-neutral reactions. *Adv. Atom. Mol. Phys.* **24**: 1–49.
14. Huntress WT Jr. (1977) Laboratory studies of bimolecular reactions of positive ions in interstellar clouds, in comets, and in planetary atmospheres of reducing composition. *Astrophys. J. Suppl.* **33**: 495–514.
15. Woodall J, Agúndez M, Markwick-Kemper AJ, Millar TJ. (2007) The UMIST database for astrochemistry 2006. *Astron. & Astrophys.* **466**: 1197–1204. See also http://www.udfa.net.
16. Florescu-Mitchell AI, Mitchell JBA. (2006) Dissociative recombination. *Physics Reports* **430**: 277–374.
17. Weston Jr RE, Schwarz HA. (1972) *Chemical Kinetics.* Prentice-Hall, Englewood Cliffs.
18. Herd CR, Adams NG, Smith D. (1990) OH Production in the dissociative recombination of $H_3O^+$, $HCO_2^+$, and $N_2OH^+$ — Comparison with theory and interstellar implications. *Astrophys. J.* **349**: 388–392.
19. Geppert WD, Hamberg M, Thomas RD, Österdahl F, Hellberg F, Zhaunerchyk V, Ehlerding A, Millar TJ, Roberts H, Semaniak J, af Ugglas M,

Källberg A, Simonsson A, Kaminska M, Larsson M. (2006) Dissociative recombination of protonated methanol. *Faraday Discussions* **133**: 177–190.
20. Jensen MJ, Bilodeau RC, Safvan CP, Seiersen K, Andersen LH, Pedersen HB, Heber O. (2000) Dissociative Recombination of $H_3O^+$, $HD_2O^+$, and $D_3O^+$. *Astrophys. J.* **543**: 764–774.
21. Semaniak J, Minaev BF, Derkatch AM, Hellberg F, Neau A, Rosen S, Thomas R, Larsson M, Danared H, Paäl A, af Ugglas M. (2001) Dissociative Recombination of $HCNH^+$: Absolute Cross Sections and Branching Ratios. *Astrophys. J. Suppl.* **135**: 275–283.
22. Smith IWM, Herbst E, Chang Q. (2004) Rapid neutral-neutral reactions at low temperatures: a new network and first results for TMC-1. *Mon. Not. R. Astron. Soc.* **350**: 323–330. See also www.physics.ohio-state.edu/~eric.
23. Carty D, Goddard A, Köhler SPK, Sims IR, Smith IWM. (2006) Kinetics of the Radical-Radical Reaction, $O(^3P_J) + OH(X\ ^2\Pi_\Omega) \longrightarrow O_2 + H$, at Temperatures Down to 39 K. *J. Phys. Chem. A* **110**: 3101–3109.
24. Xu C, Xie D, Honvault P, Lin SY, Guo H. (2007) Rate constant for $OH(^2\Pi) + O(^3P) \rightarrow H(^2S) + O_2(^3\Sigma_g^-)$ reaction on an improved *ab initio* potential energy surface and implications for the interstellar oxygen problem. *J. Chem. Phys.* **127**: 024304.
25. Sims IR. (2006) Experimental Investigation of Neutral-Neutral Reaactions and Energy Transfer at Low Temperatures. In Lis D, Blake GA, Herbst E. (eds). *Astrochemistry: Recent Successes and Current Challenges*, pp. 97–108. Cambridge University Press, Cambridge.
26. Smith IWM, Sage AM, Donahue NM, Herbst E, Quan D. (2006) The temperature-dependence of rapid low temperature reactions: experiment, understanding and prediction. *Faraday Discuss.* **133**: 137–156.
27. Sabbah H, Bienner L, Sims IR, Georgievskii, Klippenstein SJ, Smith IWM. (2007) Understanding Reactivity at Very Low Temperatures: The Reactions of Oxygen Atoms with Alkenes. *Science* **317**: 102–105.
28. Gerlich D, Horning S. (1992) Experimental Investigations of Radiative Association Processes as Related to Interstellar Chemistry. *Chem. Rev.* **92**: 1509–1539.
29. Bates DR, Herbst E. (1988) Radiative Association. In TJ Millar & DA Williams (eds). *Rate Coefficients in Astrochemistry*, pp. 17–40, Kluwer Academic Publishers, Dordrecht.
30. Herbst E. (1985) An Update of and Suggested Increase in Calculated Radiative Association Rate Coefficients. *Astrophys. J.* **291**: 226–229.
31. Chastaing D, Le Picard SD, Sims IR, Smith IWM. (2001) Rate coefficients for the reactions of $C(^3P_J)$ atoms with $C_2H_2$. $C_2H_4$, $CH_3CCH$ and $H_2CCCH_2$ at temperatures down to 15 K. *Astron. & Astrophys.* **365**: 241–247.
32. Cartechini L, Bergeat A, Capozza G, Casavecchia P, Volpi GG, Geppert WD, Naulin C, Costes M. (2002) Dynamics of the C + $C_2H_2$ reaction from differential and integral cross-section measurements in crossed-beam experiments. *J. Chem. Phys.* **116**: 5603–5611.
33. Herbst E. (1981) Can negative molecular ions be detected in dense interstellar clouds? *Nature* **289**: 656–657.

34. Petrie S, Herbst E. (1997) Some interstellar reactions involving electrons and neutral species: Attachment and isomerization. *Astrophys. J.* **491**: 210–215.
35. Güthe F, Tulej M, Pachkov MV, Maier JP. (2001) Photodetachment Spectrum of $l$-$C_3H_2H^-$: The Role of Dipole Bound States for Electron Attachment in Interstellar Clouds. *Astrophys. J.* **555**: 466–471.
36. Millar TJ, Walsh C, Cordiner MA, Ní Chuimín R, Herbst E. (2007) Hydrocarbon Anions in Interstellar Clouds and Circumstellar Envelopes. *Astrophys. J.* **662**: L87–L90.
37. McCarthy MC, Gottlieb CA, Gupta H, Thaddeus P. (2006) Laboratory and Astronomical Identification of the Negative Molecular Ion $C_6H^-$. *Astrophys. J.* **652**: L141–L144.
38. Remijan AJ, Hollis JM, Lovas FJ, Cordiner MA, Millar TJ, Markwick-Kemper AJ, Jewell PR. (2007) Detection of $C_8H^-$ and Comparison with $C_8H$ toward IRC + 10216. *Astrophys. J.* **664**: L47–L50.
39. Kolasinski K. (2002) *Surface Science. Foundations of Catalysis and Nanoscience*, John Wiley & Sons, Ltd., West Sussex, UK.
40. Katz N, Furman I, Biham O, Pirronello V, Vidali G. (1999) Molecular Hydrogen Formation on Astrophysically Relevant Surfaces. *Astrophys. J.* **522**: 305–312.
41. Peretz HB, Lederhendler A, Biham O, Vidali G, Li L, Swords S, Congiu E, Roser J, Manicó G, Brucato JR, Pirronello V. (2007) Molecular Hydrogen Formation on Amorphous Silicates under Interstellar Conditions *Astrophys. J.* **661**: L163–L166.
42. Cuppen H, Morata O, Herbst E. (2005) Monte Carlo simulations of $H_2$ formation on stochastically heated grains. *Mon. Not. R. Astron. Soc.* **367**: 1757–1765.
43. Hornekaer L, Baurichter A, Petrunin VV, Field D, Luntz AC. (2003) Importance of Surface Morphology in Interstellar $H_2$ Formation. *Science* **302**: 1943–1946.
44. Tiné S, Williams DA, Clary DC, Farebrother AJ, Fisher AJ, Meijer AJHM, Rawlings JMC, David CJ. (2003) Observational Indicators of Formation Excitation of $H_2$. *Astrophys. & Sp. Sci.* **288**: 377–389.
45. Watanabe N, Nagaoka A, Shiraki T, Kouchi A. (2004) Hydrogenation of CO on Pure Solid CO and CO-$H_2O$ Mixed Ice. *Astrophys. J.* **616**: 638–642.
46. Nagaoka A, Watanabe N, Kouchi A. (2005) H-D Substitution in Interstellar Solid Methanol: A Key Route for D Enrichment. *Astrophys. J.* **624**: L29–L32.
47. See the URL massey.dur.ac.uk/drf/protostellar.
48. Wakelam V, Herbst E, Selsis F. (2006) The effect of uncertainties on chemical models of dark clouds. *Astron. & Astrophys.* **451**: 551–562.
49. Landau LD, Lifshitz EM. (1960) *Course of Theoretical Physics. Vol. 1, Mechanics.* Pergamon Press, New York.
50. Collings MP, Anderson MA, Chen R, Dever JW, Viti S, Williams DA, McCoustra MRS. (2004) A laboratory survey of the thermal desorption of astrophysically relevant molecules. *Mon. Not. R. Astron. Soc.* **354**: 1133–1140.

51. Bringa EM, Johnson RE. (2004) A New Model for Cosmic-Ray Ion Erosion of Volatiles from Grains in the Interstellar Medium. *Astrophys. J.* **603**: 159–164.
52. Öberg KI, Fuchs GW, Awad Z, Fraser HJ, Schlemmer S, van Dishoeck EF, Linnartz H. (2007) Photodesorption of CO Ice. *Astrophys. J.* **662**: L23–L26.
53. Gredel R, Lepp S, Dalgarno A, Herbst E. (1989) Cosmic-ray-induced Photodissociation and Photoionization Rates of Interstellar Molecules. *Astrophys. J.* **347**: 289–293.
54. Willacy K, Williams DA. (1993) Desorption processes in molecular clouds — Quasi-steady-state chemistry. *Mon. Not. R. Astron. Soc.* **260**: 635–642.
55. Shalabiea OM, Greenberg JM. (1994) Two key processes in dust/gas chemical modelling: photoprocessing of grain mantles and explosive desorption. *Astron. & Astrophys.* **290**: 266–278.
56. Willacy K, Millar TJ. (1998) Desorption processes and the deuterium fractionation in molecular clouds. *Mon. Not. R. Astron. Soc.* **298**: 562–568.
57. Ward-Thompson D, Motte F, Andre P. (1999) The initial conditions of isolated star formation - III. Millimetre continuum mapping of pre-stellar cores. *Mon. Not. R. Astron. Soc.* **305**: 143–150.
58. Roberts H, Herbst E, Millar TJ. (2004) The chemistry of multiply deuterated species in cold, dense interstellar cores. *Astron. & Astrophys.* **424**: 905–917.
59. Bisschop SE, Fraser HJ, Öberg KI, van Dishoeck ER, Schlemmer S. (2006) Desorption rates and sticking coefficients for CO and $N_2$ interstellar ices. *Astron. & Astrophys.* **449**: 1297–1309.
60. Tafalla M, Myers PC, Caselli P, Walmsley CM, Comito C. (2002) Systematic Molecular Differentiation in Starless Cores. *Astrophys. J.* **569**: 815–835.
61. Lee JE, Bergin EA, Evans II NJ. (2004) Evolution of Chemistry and Molecular Line Profiles during Protostellar Collapse. *Astrophys. J.* **617**: 360–383.
62. Aikawa Y, Herbst E, Roberts H, Caselli P. (2005) Molecular Evolution in Collapsing Prestellar Cores. III. Contraction of a Bonnor-Ebert Sphere. *Astrophys. J.* **620**: 330–346.
63. Lee JE, Evans II NJ, Bergin EA. (2005) Comparisons of an Evolutionary Chemical Model with Other Models. *Astrophys. J.* **631**:351–360.
64. Linsky J + 17 co-authors. (2006) What is the Total Deuterium Abundance in the Local Galactic Disk? *Astrophys. J.* **647**: 1106–1124.
65. Roberts H, Millar TJ. (2000) Modelling of deuterium chemistry and its application to molecular clouds. *Astron. & Astrophys.* **361**: 388–398.
66. Roberts H, Millar TJ. (2006) Deuterated $H_3^+$ as a probe of isotope fractionation in star-forming regions. *Phil. Trans. R. Soc. A* **364**: 3063–3080.
67. Roberts H, Herbst E, Millar TJ. (2003) Enhanced Deuterium Fractionation in Dense Interstellar Cores Resulting From Multiply Deuterated $H_3^+$. *Astrophys. J.* **591**: L41–L44.
68. Walmsley CM, Flower DR, Pineau des Forêts G. (2004) Complete depletion in prestellar cores. *Astron. & Astrophys.* **418**: 1035–1043.
69. Osamura Y, Roberts H, Herbst E. (2005) The Gas-Phase Deuterium Fractionation of Formaldehyde. *Astrophys. J.* **621**: 348–358.

70. Uy D, Cordonnier M, Oka T. (1997) Observation of Ortho-Para $H_3^+$ Selection Rules in Plasma Chemistry. *Phys. Rev. Lett.* **78**: 3844–3847.
71. Gerlich D, Windisch F, Hlavenka P, Plasil R, Glosik J. (2006) Dynamical constraints and nuclear spin caused restrictions in $H_mD_n^+$ collision systems. *Phil. Trans. R. Soc. A* **364**: 3007–3034
72. Herbst E, Chang Q, Cuppen H. (2005) Chemistry on interstellar grains. *J. Phys.: Conf. Ser.* **6**: 18–35.
73. Awad Z, Chigai T, Kimura Y, Shalabiea OM, Yamamoto T. (2005) New Rate Constants of Hydrogenation of CO on $H_2O$-CO Ice Surfaces. *Astrophys. J.* **626**: 262–271.
74. Cuppen HM, Herbst E. (2007) Simulation of the Formation and Morphology of Ice Mantles on Interstellar Grains. *Astrophys. J.* **668**: 294–309.
75. Barzel B, Biham O. (2007) Efficient Simulations of Interstellar Gas-Grain Chemistry Using Moment Equations. *Astrophys. J.* **658**: L37–L40.
76. Biham O, Furman I, Pirronello V, Vidali G. (2001) Master Equation for Hydrogen Recombination on Grain Surfaces. *Astrophys. J.* **553**: 595–603.
77. Chang Q, Cuppen HM, Herbst E. (2007) Gas-grain chemistry in cold interstellar cloud cores with a microscopic Monte Carlo approach to surface chemistry. *Astron. & Astrophys.* **469**: 973–983.
78. Sakai N, Sakai T, Hirota T, Yamamoto S. (2008) Abundant Carbon-Chain Molecules toward Low-Mass Protostar IRAS04368 + 2557 in L1527. *Astrophys. J.* **672**: 371–381.

# Chapter 2

# Gas Phase Reactive Collisions at Very Low Temperature: Recent Experimental Advances and Perspectives

André Canosa[a], Fabien Goulay[b], Ian R. Sims[a] and Bertrand R. Rowe[a]

[a]*Institut de Physique de Rennes, Equipe: "Astrochimie Expérimentale," UMR UR1-CNRS 6251, Bât. 11 C, Université de Rennes 1, Campus de Beaulieu, 35042 Rennes Cedex, France*

[b]*Combustion Research Facility, Mail Stop 9055, Sandia National Laboratories, Livermore, California 94551-0969, USA*

## Contents

## 2.1. Introduction

The study of gas phase reactivity (kinetics and dynamics) at temperatures as low as 10 K is of major importance for the understanding of a wide range of cold natural environments in which a rich chemistry takes place. Beyond our own atmosphere, for which recent observations indicate that the temperature at the mesopause can be lower than 150 K at high latitudes, the development of space missions, ground telescopes and satellites has opened up new vistas and demonstrated the existence of even colder extraterrestrial environments. In the Solar system, the temperature of atmospheres of the Giant Planets or their satellites is only several tens of Kelvin: 70 K at the tropopause of the now well known satellite of Saturn, Titan, and 50 K in the atmosphere of Neptune or Pluto. In the coma of comets, temperatures ranging from 10 K to 200 K have been determined according to the distance to the Sun and the position with respect to the nucleus.

Beyond the solar system, one penetrates into the so-called *interstellar medium* (ISM hereafter), an apparently empty space which separates stars from each other. In this very tenuous environment (average density $\sim 1\,\text{cm}^{-3}$) which is actually quite heterogeneous, a large number of local concentrations of cold (typically 10 K) and dense ($\sim 10^4\,\text{cm}^{-3}$) matter were discovered over the last century. To date, thanks to radio astronomy and satellites such as ISO (Infrared Space Observatory) or orbiting telescopes such as the HST (Hubble Space Telescope), about 150 different molecules (230 including isotopologues) have been identified in the ISM, revealing an amazingly rich and complex chemistry. At the time of writing, the largest molecule currently identified is the cyanopolyyne $HC_{11}N$ (cyanodecapentayne) which was detected in 1997. Most of the observed molecules were detected in the denser parts of the ISM in the so-called molecular clouds and a significant number of these are highly unsaturated. More details can be found in the contribution in this book by Herbst and Millar including a list of known interstellar molecules. The total number of these interstellar molecules continues to grow and the development of the ALMA interferometer in Chile and the future space telescope Herschel will certainly enrich the list considerably. How such a variety of molecules, of which many are organic, can be produced and survive in such harsh environments is a fundamental question that gave birth to a new field in astronomy about 40 years ago: *astrochemistry*.

While only a few positive ions have been identified in the ISM (about 10% of the observed molecules), they play a major role in the chemistry because many ion-molecule reactions proceed with large rate coefficients at room temperature and below ($>10^{-9}\,\text{cm}^3\,\text{molecule}^{-1}\,\text{s}^{-1}$) due to the strong attractive forces between an ion and the permanent or induced dipole on the neutral reagent. Therefore, although very few experimental rate coefficients were determined at low temperatures when the first generation of photochemical models of interstellar clouds (ISCs) appeared in the mid-seventies, these reactions were dominant in the proposed chemical schemes. Initially rate coefficients were considered to be independent of temperature, as often observed by the experiments available at that time. In the early- and mid-1980s, the development of new experiments and collision theories gave modellers the opportunity to refine their simulations by introducing more reliable temperature dependences for ion-molecule reactions.

Until recently no negative ions were known to be present in the ISM, although their existence was assumed. In 2006 and 2007, the detection of the first three anions, $C_4H^-$, $C_6H^-$ and $C_8H^-$, has significantly changed our vision of the ISM since electrons were hitherto expected to be the main carriers of negative charges. Polyyne anions as well as other probable negative charge carriers, such as polycyclic aromatic hydrocarbon anions, might be formed through low-energy (0.05 meV to 10 meV) electron attachment to cold molecules. It is also interesting to note that in Titan's ionosphere, massive negative ions have been detected by the Cassini-Huygens mission and are now believed by some to be a major source of organic matter at the surface of Saturn's main moon. One of the most efficient ways to produce negative ions is by dissociative electron attachment, which consists of the fragmentation of a polyatomic molecule into smaller species by impact of a low-energy electron. This process can depend strongly on temperature, either through the electron kinetic energy or the internal state populations of the neutral molecules. Therefore, in order to predict the abundance of large carbon-containing anions and understand the complex ionic chemistry occurring in the ISM, as well as in Titan's ionosphere, it is essential to know the attachment rate coefficient when the electrons are thermalised with the reactive molecule. Although very few such reactions have been integrated into photochemical models of cold environments, interest in electron attachment is growing, and laboratory data for collisions between very low energy electrons and cold molecules under conditions of true thermodynamic equilibrium are clearly needed. Dissociative recombination of polyatomic cations with electrons, although not discussed in the present chapter, must be mentioned as well, because it is also an important family of processes included in ISC models which was thought to be the main source for the formation of neutral species.

For many years most neutral-neutral reactions were expected to be slow at the very low temperature of molecular clouds as they were thought to have positive activation energies $E_a$. If the temperature dependence of the rate coefficient $k$ is described by the Arrhenius equation ($k(T) = A\exp(-E_a/RT)$) then the rate coefficients at *ca* 10 K will be insignificantly small. Only a few atom-radical reactions such as $O + OH \rightarrow O_2 + H$, $O + CH \rightarrow CO + H$, $N + OH \rightarrow NO + H$ etc... were considered in models because this class of processes was expected to proceed without any barrier. In the early 1990s experimental developments shed new light on radical-neutral reactions and demonstrated that a large number of these processes, even some involving a saturated molecular partner, can be very efficient at temperatures as low as 13 K.[1–4] This discovery was carefully considered by ISC modellers who analysed the impact of these new and revolutionary observations.[5] The number of neutral–neutral processes included in models was significantly increased and rate coefficients were re-evaluated on the basis of the new experimental data. The main consequence of this upgrading was to reduce the rate of formation of complex species as a result of the prompter destruction of small radicals which initiate the chemistry leading to complex molecules. Further details can be found in the chapter by Herbst and Millar.

From a fundamental point of view, theoretical calculations of rate coefficients for molecular collisions have made great progress in recent years (see the contribution by Klippenstein and Georgievskii in this book). However, at the current time and more particularly for neutral-neutral reactions, the predictive power of such calculations has still to reach its full potential, principally owing to the extreme difficulty of calculating the necessary multidimensional potential energy surface (PES) or hypersurface which governs the interaction. Comparison of experimental measurements of rate coefficients with theory is especially important at very low temperatures, as it is here that the rate coefficients are most sensitive to the finer details of the potential energy surfaces, and therefore the comparison is most revealing (remember that 10 K corresponds to less than 1 meV). Several groups[6] have calculated rate coefficients for barrierless reactive processes in the zero temperature limit, employing capture theories. One advantage of such calculations is that the capture rate coefficient in the limit of very low temperature should depend only upon the long range attractive forces between the two species, which are much easier to calculate than the full PES. The calculations of Clary, for example, predict that the rate coefficient for the reaction between two dipolar molecules is zero at temperature $T = 0$, increases as $T^{1/6}$ to a maximum at a very low temperature (typically close to 10 K) and then decreases, eventually having a temperature dependence of $T^{-1/6}$. The combination of such interesting effects, the presence of long range attractive forces, and the possibility of detecting resonances are all fundamental theoretical motivations to perform low temperature experimental studies of molecular collisions, especially at temperatures below 10 K.

In the case of bimolecular ion-molecule reactions the long range interaction potential between reagents is particularly simple and well characterized. There exists a variety of formulations for predicting the capture rate coefficients, the simplest analytical one being the Langevin formula for the reaction of an ion with a non-polar molecule (ion-induced dipole potential). For charge-dipole capture, there has been a very large body of theoretical work, ranging from statistical approaches such as the SACM (Statistical Adiabatic Channel Model)[7] to classical trajectory (CT) calculations.[8] Analytical relationships have been deduced which yield essentially the same results in the temperature range relevant to ISCs, in good agreement with available measurements. Interestingly, it is at the lowest temperatures treated by theory, say in the 1–10 K temperature range, that the calculated results start to deviate from each other.[7,9] In fact, even for the ion induced dipole case, deviations from the Langevin formula are expected.[10] It would therefore be extremely interesting to perform experiments in this temperature range.

The present chapter is organised as follows: in section II we will present the general aspects and problems that have to be tackled to investigate the field of gas phase reactive collisions at very low temperatures. The basic techniques will be briefly reviewed according to the kind of reaction families to be studied. Special attention will be paid to the CRESU technique (a French acronym standing for *Cinétique de Réaction en Ecoulement Supersonique Uniforme* which can be translated as Reaction Kinetics in Uniform Supersonic Flow) in section III. Basic features and its range of applicability will be presented in detail. Section IV will be dedicated to the contribution of this technique to the understanding of the chemistry of charged species, with particular application to ion-molecule reactions and electron attachment. Section V will focus on recent results obtained for radical-neutral reactions by the continuous and pulsed versions of the CRESU technique. Some emphasis will be put on recent breakthroughs: namely, the reaction O + OH, which is the first radical-radical process involving two unstable species to be studied at very low temperature; the chemistry of condensable species at temperatures as low as 60 K; the reactions of atomic oxygen with several alkenes that demonstrated new theoretical capabilities; and finally the dimerisation of benzene which is the first step to further nucleation of molecules. The chapter will be concluded by presenting in section VI a series of potential future developments in the field.

## 2.2. Experimental Techniques for the Measurement of Rate Coefficients at Very Low Temperatures

### 2.2.1. *Specific Aspects of Low and Very Low Temperature Chemistry Measurements*

#### 2.2.1.1. *Nature of the Measurements: Rate Coefficients and Branching Ratios or Cross Sections*

The rate coefficient $k(T)$ of a bimolecular chemical reaction in the gas phase at a given temperature $T$ results from the thermal average of a very large number of bimolecular reactive collisions. These collisions involve reagents in a variety of quantum states, the populations of which follow Boltzmann's law at this temperature, with a Maxwell-Boltzmann distribution of centre of mass kinetic energies $E_T$ (hereafter $KE_{cm}$). Ignoring any dependence of the reaction cross section ($\sigma$) on the internal states of the reagents, the relationship between the rate coefficient and the reaction cross section is given by:

$$k(T) = \left(\frac{8k_{\mathrm{B}}T}{\pi\mu}\right)^{1/2} (k_{\mathrm{B}}T)^{-2} \int_0^\infty \sigma(E_T)E_T \exp(-E_T/k_{\mathrm{B}}T)dE_T \qquad (2.1)$$

where $k_{\mathrm{B}}$ and $\mu$ are the Boltzman constant and the reduced mass respectively. Termolecular association reactions are usually the result of a sequence of bimolecular events involving intermediate species and therefore can be considered in the same way.

From the application point of view most often the thermal rate coefficient and the branching ratios for a reaction at a given temperature constitute the required data. However, conditions of local thermodynamic equilibrium (hereafter LTE) do not always prevail in natural and artificial media especially when they are irradiated by high energy particles (photons, cosmic rays). From a theoretical point of view, the thermal rate coefficient/ branching ratio could be easily modelled by summation of results on single collisions, and be compared to experimental data. However, laboratory data concerning well specified quantum states are also of great interest for comparison to theory.

It is evident from this short discussion that chemical reactivity can, in principle, be studied experimentally using several approaches, leading either to the measurement of thermally averaged quantities such as rate coefficients, or to more microscopic ones such as cross sections. To obtain thermal averages, experiments are most often conducted at a density sufficient to ensure that the large number of collisions maintain LTE. This is not the case for cross section measurements, which are performed with beams in the molecular free regime and with very few collision events. These are also very difficult measurements, since very low temperatures correspond to very low $\mathrm{KE_{cm}}$. In order to obtain a rate coefficient by convolution following Eq. (2.1), such experiments need to be done with a sufficiently high resolution. Very low values of $\mathrm{KE_{cm}}$ ($<1\,\mathrm{meV}$) with high resolution are extremely difficult to achieve. To date the equivalent minimum temperature at which rate coefficients can be extracted from these data remains higher ($T > 30\,\mathrm{K}$ for neutral chemistry) than the lowest temperature achieved in thermal systems ($T = 7\,\mathrm{K}$), such as uniform supersonic expansions.

#### 2.2.1.2. *Nature of the Reagent: Neutral or Charged Species*

Historically the experimental techniques that have been developed in the field of very low temperature chemistry have first concerned charged species. As mentioned in the Introduction, the first reason for this was that ion-molecule reactions were thought to be the main route to interstellar molecule synthesis.[11] The second reason, and maybe the most important one, is that ions can be easily manipulated, stored by electromagnetic fields and observed with high sensitivity, giving rise to a large variety of techniques which are not applicable to neutral chemistry. Most of these techniques were developed in the 1980s and early 1990s and have been reviewed elsewhere.[12]

Collisions involving electrons deserve a special discussion since, due to their very low mass, $KE_{cm}$ is essentially the electron kinetic energy. It is rather easy, especially with laser techniques, to generate electrons of very low energies which do not relax efficiently to the higher neutral gas temperature.[13] It is therefore quite simple to study cross sections at low $KE_{cm}$ for processes such as electron attachment[13] but such studies have been conducted mostly with internal temperature of the molecules close to 300 K.

#### 2.2.1.3. *Nature of the Cooling: Cryogenic Cooling or Supersonic Expansion*

Except for techniques relying on electromagnetic fields (see the chapters by Gerlich and by van der Meerakker *et al.*) or the special case of electrons, two general methods allow one to cool gases to low temperature: ***expansion techniques*** and ***cryogenic cooling***. The latter suffers from the disadvantage that any gas which is present above its saturated vapour pressure will condense on the refrigerated walls of the cooled apparatus. The solution is to keep the reagents at partial pressures below their vapour pressure at the wall temperature. However, the concentration must be high enough to perform kinetic experiments which are often carried out under pseudo-first order conditions and only gases with a significant vapour pressure at the wall temperature are then suitable.

Techniques which are based on the supersonic expansion of a non condensable buffer gas do not suffer from this problem. They have been widely used in experimental science, for example in the first attempt to liquefy gases or, much more recently, as a source of cold molecules for spectroscopy (see the chapter by Davis *et al.*) or of molecular beams for collision dynamics studies. Cooling arises from conservation of energy which implies, that for a gas flow under adiabatic conditions, the sum of specific enthalpy and kinetic energy remains constant during the expansion. For a perfect gas with a constant specific heat capacity $c_p$ and with a well defined temperature, the energy conversion is driven by the following equation:

$$c_p T + \frac{u^2}{2} = c_p T_0 \tag{2.2}$$

where $T_0$ and $T$ are the temperatures of the gas prior and after expansion respectively and $u$ is the supersonic flow velocity. It is clear from this equation that the velocity of the supersonic gas cannot be greater than

$$u_{\max} = \sqrt{2c_p T_0} \tag{2.3}$$

corresponding to a complete conversion of the thermal energy into kinetic energy. Flows for which this condition is nearly fulfilled are known as hypersonic.

The simplicity of Eq. (2.2) could lead one to think that supersonic expansions will be easy to use for the study of chemical reactions at low and very low temperatures. The simplest type is the so-called free jet expansion, in which a gas at high pressure escapes through a small orifice into a high vacuum chamber. Here the cooling can be extreme: down to *ca.* 0.1 K. In this case the medium is inhomogeneous and the thermodynamic conditions are extremely complex. As the expansion evolves, the collision frequency decreases and therefore the different temperature moments (translation, rotation and vibration) cannot reach equilibrium[14] making it difficult to assign a "true" thermodynamic temperature. The strong gradients of temperature and density, the low frequency of collisions and the short useable area of the jet have prevented its generalised use in studies of chemical reactions although several ion-molecule reactions have been investigated.[12] For these processes, the long range electrostatic forces are strong enough to attract together reagents travelling with very similar translational velocity. The insertion of a "skimmer" into the supersonic zone of the free jet generates a molecular beam which, coupled with a second one at small angle, can be used for measuring cross sections at low $KE_{cm}$.

Finally, a low temperature chemical reactor can be obtained using "collimated" uniform supersonic flows generated by the isentropic expansion of a buffer gas through an axisymmetric convergent-divergent Laval nozzle. Uniform supersonic flows were initially developed in a continuous flow version for the study of ion-molecule reactions.[15] Later, the methodology was adapted to the study of reactions between neutral species in a continuous flow version[16–18] and also in a pulsed flow version.[19–21] This technique known as CRESU will be described in detail in Section 2.3.

### 2.2.2. *Brief Overview of Techniques Used to Date*

#### 2.2.2.1. *Ion-Molecule Reactions*

In the 1980s and early 1990s there was a formidable development of techniques allowing the study of ion-molecule reactions down to temperatures close to 10 K. The state of the art was reviewed by M.A. Smith 1994[12] and a more up-to-date review of the methods presently used in this field, excepting supersonic flow reactors, is given in the chapter by Gerlich.

A first group of methods makes use of cryogenic cooling which is not a problem for the ionic partner of the reaction, since ions can be maintained far from the walls using suitable fields, therefore inhibiting their heterogeneous condensation. Ions are then cooled by collisions with a buffer gas or by elastic collisions with the neutral reagent. At the lowest temperatures, these techniques are usually restricted to co-reagents of high vapour pressure, especially hydrogen and its deuterated forms. The very first measurements on ion-molecule reactions at temperatures below 80 K were reported in 1983.[22] Using a drift tube cryogenically cooled by liquid helium, temperatures as low as 20 K were routinely obtained. However, this technique suffered from some difficulties due to the presence of electric fields,[12] which means that ions are not strictly equilibrated at the apparatus temperature and that their velocity distribution is not Maxwell-Boltzmann. This method, as well as the cryogenically cooled Penning trap experiment of Dunn and co-workers,[23] has not been widely used and most of the developments of cryogenically cooled methods have been performed by Gerlich and co-workers using radiofrequency traps with special geometries to minimise heating of the ions due to radiofrequency fields.

The second group of methods relies on supersonic expansion as the source of cooling and the use of supersonic flows as flow reactors. In the early 1980s the group of Rowe, at the rarefied wind tunnel facility, SR3, in the "laboratoire d'Aérothermique du CNRS" in Meudon (France) developed the CRESU technique: an ion-molecule reactor that made use of the uniform supersonic flows generated by a Laval nozzle. The first measurements of rate coefficients down to 20 K were reported in 1984 and soon results for temperatures as low as 8 K were obtained.[24] Note that in the CRESU method the temperature is a "true" temperature, i.e. LTE prevails in the flow. The up-to-date version of this apparatus and the basis of the technique will be described in some detail in the following paragraphs.

Following closely the work of Rowe and co-workers, M.A. Smith and co-workers have developed an ion-molecule reactor based on the free jet expansion.[25,26] To avoid very large pumping capacities and high gas consumption the jet is operated in a pulsed mode at 10 Hz, which is easier in this case due to the small dimension of the nozzle throat, typically ranging between 0.1 and 1 mm. Ions are created by laser multiphoton ionisation of a suitable precursor, ten to thirty nozzle diameters downstream of the nozzle exit plane, where the velocity has reached a value very close to the hypersonic limit (see Eq. (2.3)). From this point the flow continues to expand at all downstream positions and ions eventually react with neutral molecules. Reagent and product ions are monitored versus distance to the laser beams using a time of flight mass spectrometer set at right angles to the flow. To extract the rate coefficient, M.A. Smith and co-workers have to take into account that density continuously drops along the flow axis. Therefore, in the ion reaction zone the neutral-neutral collision rate is negligible, causing strong thermodynamic disequilibrium within the jet. The velocity distribution is no longer Maxwell-Boltzmann but it was shown that it remains possible to assign a well-defined kinetic average temperature for the bulk rate coefficient measured.[12] More problematic is the question of rotational temperature which can diverge quite far from the kinetic temperature[26] and cannot always be modelled due to the lack of data on rotational relaxation in this energy range. Even if the temperature and the thermodynamic conditions cannot be assigned as well as in a CRESU experiment, the free jet reactor has been able to produce data for some reactions at an equivalent kinetic temperature of 0.3 K which is still an ultimate performance in this field.[12] However it has proven impossible to extend the technique to neutral-neutral reactions.

The ***merged beam experiment*** of Gerlich and co-workers[27] can be considered as belonging to both groups of techniques: skimmed supersonic expansion is used here to produce a dense beam of neutrals with cooled internal states. It is merged with an ion beam which makes extensive use of cryogenically cooled traps for ion production and relaxation. By matching the ion beam velocity to the hypersonic terminal velocity of the neutral beam, extremely low $KE_{cm}$ can be achieved. The use of radiofrequency ion guides makes possible the production of intense slow ion beams. Again the reader is referred to chapter 3 by Gerlich.

#### 2.2.2.2. *Electron Attachment*

Techniques which allow electron attachment to be studied at very low electron energy have been described in several reviews (see for example[13]). They can be separated into two categories: single collision, and multiple collision techniques. In the former class, the electrons are produced at very low energy and collide with the molecules. The latter class includes LTE techniques which produce electrons that are injected into a buffer gas and are thermalised by collisions with this gas. In this section only the techniques allowing the measurement of attachment rate coefficients of electrons with energies below 100 meV (corresponding to 770 K) are considered.

In experiments performed under single collision conditions, the use of Rydberg atoms and threshold photoionisation techniques of rare gases are the two most representative techniques for producing electrons. The former experiment does not use low energy electrons but instead uses the weakly bonded electron of a high Rydberg level of an atom. High Rydberg states of atoms, such as xenon or rubidium, are generated by laser irradiation and react directly with molecules via electron transfer.[13] Due to the weak bonding of the electron, the reaction rate of the Rydberg atom with the molecules is identical to the attachment rate of electrons with the molecules, assuming that the electrons have the same velocity distribution as the reacting Rydberg atoms. The number density of the Rydberg atoms is followed in time by a selective ionising field. Such techniques allow measurement within an energy range from 0.05 meV to 40 meV. On the other hand, photoionisation techniques produce low energy free electrons and measure the direct attachment rate.[13] Electrons are generated either by direct near threshold photoionisation of a gas using VUV light or by exciting argon atoms by electron impact, then using a two photon process to generate argon ions and threshold electrons. In both cases the energy of the resulting electrons is the difference between the total energy provided to ionise the atom and its ionisation energy. After reactions with the low energy electrons, the negative ions formed are detected by mass spectrometry. Better energy resolution is obtained using the multi-step process.

The flowing-afterglow/Langmuir-probe (FALP) technique is characteristic of experiments performed at LTE. This technique was developed by D. Smith, Adams and co-workers to measure ion-molecule, ion-ion, electron-ion and electron-molecule collision phenomena down to 200 K.[28] In the FALP approach an afterglow plasma is generated in a fast-flowing helium carrier gas by means of an upstream microwave discharge. Argon atoms are added to the plasma to quench any metastable helium atoms, and a small quantity of the target gas is added downstream. At the point of this addition a large axial gradient of the electron density is generated due to continual depletion by attachment to the molecule. Neglecting ambipolar diffusion and knowing the flow velocity, measurement of the attachment rate is achieved by measuring the electron density along the axis of the flow, using a Langmuir probe, for different molecular densities.

Other techniques are available to study electron attachment. However, most of them do not permit one to vary the temperature of the molecular species from room temperature to low temperature in true LTE. For the latter purpose, the CRESU apparatus may be the most suitable technique. Details are presented in Section 2.3.2.

#### 2.2.2.3. *Neutral-Neutral Reactions*

Cryogenic cooling is certainly the most intuitive technique to be used for the study of radical-neutral reactions at sub-ambient temperatures. However condensation is a major problem and therefore the method cannot be applied to gas phase reactions at temperatures which are relevant to interstellar chemistry. At 77 K, which is the minimum temperature achieved to date using this technique (for reactive systems, at least), experiments have been conducted on only a limited number of reactions involving gases such as CO and $H_2$ which have high enough vapour pressures at such temperatures. The cryogenic method is well designed for studies in the range 180–300 K and has been widely used for improving the understanding of atmospheric chemistry. The technique is often well suited for the study of slow reactions (typ. $k \sim 10^{-15}$–$10^{-13}$ cm$^3$ molecule$^{-1}$ s$^{-1}$) because reaction times of several milliseconds can commonly be achieved and relatively high concentrations of reagents introduced into the cell. Pulsed photolysis or microwave discharges are used to produce radicals which are then detected by the use of either optical methods such as Laser Induced Fluorescence (LIF) or mass spectrometry. It is worth mentioning here that cryogenic cooling techniques have been used to obtain rate coefficients for reactions between free radicals such as OH + O and OH + N for which it is necessary to produce one of them in large excess and in a variable and well-defined concentration. Although these methods are limited in temperature, they provide a complement to other techniques that will be discussed below. Further details can be found in a recent review by I.W.M. Smith.[1]

As previously mentioned, the free jet technique is not well suited to the study of neutral-neutral reaction kinetics. Molecular beams, which are an extension of this technique, have been used for a long time however to obtain information on the mechanisms driving such processes and are very useful in studies of collisional dynamics. More particularly, the technique provides valuable information with respect to the nature of the products of the reaction and its mechanism.[29] As the frequency of collisions is very small in a molecular beam, data are obtained from experiments in which two collimated reagent beams intersect at a fixed angle in a scattering region where single collisions occur between the components of the beams. The species of each beam are characterised by a well-defined velocity and usually also internal quantum states. By varying the velocity of the molecular beams and their intersection angle, differential cross sections (product angular and velocity distributions) can be measured at different collision energies.

Most crossed molecular beam machines however work with beams at right angles and therefore the usual collisional energies are rather high with respect to those encountered in the molecular rich regions of the ISM. However some devices are designed so that the beam intersection angle can be varied enabling the relative collision energy to be reduced. The lowest collisional energy presently available was obtained by Naulin and Costes[30] who were able to perform experiments using an apparatus for which the intersection angle can be varied from 90° down to 22°, the latter corresponding to a collision energy of 0.35 kJ/mol (i.e. $\sim$30 K). In this experiment, the relative total cross sections for reaction (also referred to as the excitation function in the literature) can be integrated assuming a Maxwell distribution of relative velocities at temperature $T$ to yield an estimate of the thermal rate coefficient, $k(T)$. This, however, is a relative value and must be normalised using data obtained from other techniques. It is interesting to compare these cross sections obtained at low collision energies with rate coefficients obtained at low temperatures by the CRESU technique. If there is a negligible rotational state dependence of the reaction cross section/rate coefficient, and, for multiple product channels, if the branching ratio for the monitored product in the case of the crossed beam experiments is collision energy independent, then the temperature dependence $n$ of the thermal rate coefficient $k$ (where $k \propto T^n$) and the collision energy dependence $\alpha$ of the reaction cross section $\sigma$ (where $\sigma \propto E_T^\alpha$) should be related by the simple formula $n = \alpha + 1/2$. This has been found to be the case, within experimental error, for a number of reactions, including for example the reactions of $C(^3P)$ atoms with $O_2$ and NO.[31] When similar agreement was not found for the reaction of $C(^3P)$ atoms with $C_2H_2$,[32] the discrepancy was attributed to a temperature-dependent branching to a spin-forbidden product channel ($C_3 + H_2$ in this case).

## 2.3. Adaptation of Uniform Supersonic Flows to the Study of Low Temperature Chemical Kinetics: The CRESU Technique

### 2.3.1. *General Aspects and Limitations of the Technique*

It was realised by Rowe in the early 1980s that the uniform supersonic flows obtained by the correct design of a Laval nozzle and used for decades in rarefied wind tunnels for aerodynamic studies[15] could provide an ideal flow reactor for the study of chemical reactivity at low and very low temperature. This was the cornerstone around which the CRESU technique has been developed. At the exit of the Laval nozzle, as there is no further expansion downstream of the nozzle exit, the flow parameters (i.e. temperature, density, pressure and velocity) do not exhibit any axial and radial variations at least in the centre of the jet (typically 10 to 20 mm in diameter) where the flow is isentropic for several tens of centimetres. The diffusion velocity is always negligible with respect to the bulk velocity therefore avoiding the major problem of condensation associated with the use of cryogenically cooled cells. As a consequence, in such expansions, heavily supersaturated conditions prevail and condensable species such as water, ammonia or even polycyclic aromatic hydrocarbons (PAHs hereafter), can be maintained in the gas phase at very low temperatures.

In contrast to free jet expansions and molecular beams, where the concept of temperature is not really valid, the relatively high gas density ($10^{16}$–$10^{17}$ molecules cm$^{-3}$) in the uniform supersonic flow ensures that frequent collisions take place during the expansion and subsequent flow, maintaining thermal equilibrium.

For a given Laval nozzle, uniformity of the flow parameters in the isentropic core of the supersonic flow can be achieved under suitable conditions that depend on the inner shape of the Laval nozzle. Due to the viscous nature of gases, this isentropic kernel is embedded within a boundary layer which develops during expansion in the divergent part of the Laval nozzle. In this layer large radial gradients exist because at the wall of the Laval nozzle the flow velocity must go down to zero whereas the temperature must reach that of the wall. Suitable conditions therefore mean that the viscous boundary layers downstream of the Laval nozzle exit do not merge in the centre of the jet, destroying the flow uniformity, slowing down the flow and increasing the temperature, due to transport phenomena. Since the jet downstream of the nozzle exit can be thought as a flow in a duct without walls, the problem of "suitable conditions" can be discussed in terms of the "flow entrance length" $L$,[33] the length at which boundary layers merge. The thickness of the boundary layer along the entrance of a planar wall can be estimated approximately as:

$$\delta(x) = 5x\left(\frac{\rho u x}{\eta}\right)^{-(1/2)} \tag{2.4}$$

where $x$ is the length along the plane wall, $\eta$ the viscosity of the gas at the outer edge of the boundary layer and $\rho$ the mass concentration. By definition, in a duct of diameter $D$ boundary layers will merge if $\delta = \mathrm{D}/2$. The flow entrance length $L$ can then be represented by the following expression:

$$\frac{L}{D} = \frac{1}{100}\frac{\rho u D}{\eta} \tag{2.5}$$

where $\rho u x/\eta$ and $\rho u D/\eta$ are Reynolds numbers. A more rigorous analysis for the duct boundary layer gives:

$$\frac{L}{D} = \frac{1}{20}\frac{\rho u D}{\eta}$$

which can be immediately converted into:

$$L = 0.06\frac{Q_m}{\eta} \tag{2.6}$$

where $Q_m$ is the mass flow rate: $\rho u \pi D^2/4$. It follows from this very simple discussion that the order of magnitude of the length over which the flow can sustain uniformity is only a function of the nature of the gas through the viscosity and of the total mass flow rate. This applies, of course, to continuous as well as pulsed flows. The pressure at which the experiment needs to be conducted then determines the size of the apparatus and of the pumps. The typical required value for $Q_m$ corresponds to a few moles per minute and consequently, at the relatively low pressure needed to avoid too high clustering, a continuous apparatus requires large pumping capacities. This requirement, as long as it was not fully understood, probably inhibited earlier use of Laval nozzles in chemistry.

Beyond the significant mass flow rates that must be introduced to generate a uniform supersonic flow downstream of a Laval nozzle, it is also important to stress that the inner shape of the divergent part of the Laval nozzle and the temperature of the reservoir completely constrain the flow conditions; i.e. nature of the buffer gas, gas flow rate, supersonic temperature and pressure. In other words, for a given Laval nozzle, the temperature in the supersonic flow is not a tuneable parameter. Hence, a series of different Laval nozzles are required to match the range of temperature that needs to be explored. The typical temperatures that can be achieved in the present working CRESU apparatuses are usually in the range 15–300 K. This temperature is directly linked to the reservoir temperature by the relation:

$$\frac{T_0}{T} = 1 + \frac{\gamma - 1}{2} M^2 \tag{2.7}$$

which is a direct consequence of Eq. (2.2) $M$ is the Mach number which is a characteristic of each Laval nozzle and $\gamma$ is the ratio of specific heats of the buffer gas at constant pressure and volume. Whatever the reservoir temperature, the temperature ratio is constant and temperatures lower than 15 K can be obtained by pre-cooling the reservoir, using liquid nitrogen. In these conditions, supersonic flows with a temperature as low as 7 K can be achieved.[34] In this case however, the choice of the reagent is limited to those having a vapour pressure high enough to avoid condensation on the walls of the reservoir where the flow is almost stagnant. On the other hand, heating the reservoir can be of interest when it is desirable to study reagents with very low vapour pressures such as PAHs. In recent studies, which will be described in Section 2.5.3, the reservoir temperature was raised to 470 K using a circulating flow of heated oil. The investigation of the reaction kinetics of anthracene in the gas phase then became feasible down to 58 K.[35]

For the determination of rate coefficients, a limitation of the CRESU technique lies in the supersonic nature of the flow which restricts the available hydrodynamic time to a few hundreds of microseconds. In principle, this could be compensated by increasing the concentration of the reagent because what is usually measured is a first-order rate coefficient, which is then plotted as a function of the reagent concentration to obtain the second-order rate coefficient. In the CRESU technique however, introduction of a reagent is usually restricted to a few percent of the buffer gas flow rate. For high mole fractions, the expanding flow must be considered as a gas mixture and then the viscosity has to be calculated in a different way and is usually found to be different from the case of a single buffer gas. This may generate a modification of the supersonic flow conditions and even its own destruction. As a consequence of the limitations in hydrodynamic time and in reagent concentrations, only rate coefficients $>5 \times 10^{-13}\,\mathrm{cm}^3$ molecule$^{-1}\,\mathrm{s}^{-1}$ can usually be measured.

As mentioned briefly above, the size of the apparatus is another significant constraint. Since Laval nozzles are usually designed to obtain a low pressure in the supersonic flow (typically of the order of 0.1 mbar or even less for ion-molecule studies), large pumping capacities are necessary. Typically, studying binary reactions involving neutral partners, at temperatures as low as 15 K, requires pumping capacities of about $30000\,\mathrm{m}^3\,\mathrm{hr}^{-1}$. For ion-molecule reactions, even larger pumping capacities are necessary to avoid too high clustering of ions that sometimes prevent the study of binary ion-molecule reactions. Experiments carried out in Meudon usually needed pumping capacities close to $100000\,\mathrm{m}^3\,\mathrm{hr}^{-1}$. The size of pumps directly determines the minimum size of the chamber. The Meudon apparatus was 2 m in diameter and 10 m long while the five Rennes chambers are typically 0.5 m in diameter and 3 m long.

If pressures of about 1 mbar are acceptable in the supersonic flow for the study of a given process, then a significant reduction in pumping capacity can be accepted as is the case for the mini-CRESU that has been constructed at the Université de Bordeaux 1 by M. Costes and co-workers.[18] This apparatus however cannot reproduce temperatures lower than 50 K in the supersonic flow because of its limited pumping capacities. Another way to reduce the size of a CRESU apparatus is to develop pulsed supersonic flows. A special section will be dedicated to this evolution of the CRESU technique in Section 2.3.4.

### 2.3.2. *Reaction Kinetics Involving Charged Species*

The first CRESU apparatus[15] dedicated to the study of ion-molecule reactions took advantage of a very large rarefied wind tunnel facility with exceptional pumping capacity ($144000\,\mathrm{m}^3\,\mathrm{h}^{-1}$) over a rather low pressure range, i.e. $10^{-4}$–$10^{-1}$ mbar. In the very first design,[15] ions were continuously generated in the flow by adding a suitable parent molecule in the reservoir of the nozzle and ionising it in the supersonic flow making use of a high energy (10 kV, 10–100 $\mu$A) electron beam crossing the supersonic jet at right angles. These ions reacted with a neutral reagent also added in the reservoir at various flow rates. A moveable mass spectrometer probe allowed monitoring of both reagent and product ions along the flow axis, downstream of the electron beam. This was achieved by skimming the flow and introducing the ions into a quadrupole mass filter by means of electrostatic lenses. After deflection by suitable electrostatic fields the ions were detected by a channeltron set at right angles to the quadrupole axis. The exponential variations of the reagent ion concentration with distance to the ion source and with neutral reagent concentration directly yielded the reaction rate coefficient, using the simplest standard flow reactor analysis.[24]

In an attempt to measure exact branching ratios of reactions a selective ion injection source was added to this apparatus in 1988.[36] It made use of a quadrupole filter to select reagent ions that were injected in the nozzle boundary layer and subsequently drifted into the core of the flow by suitable electric fields. The great advantage of this source was that the neutral parent of the reagent ion was not present in the flow thus avoiding an additional destruction route of the ion.

In 1993 part of this apparatus was moved to the University of Rennes 1 where it was rebuilt and adapted to the use of slightly higher pressures due to the lower pumping capacities that were available there. A sketch of the present apparatus is shown in Fig. 2.1. In addition, a pulsed ion source[37] was further developed, allowing the injection of ions into the flow by drifting them in electric fields and avoiding the presence of parent neutrals, as well as a variety of excited species initially present in the flow with the electron beam technique. By pulsing the drifting field, ions penetrated the core of the flow where they could cool down to the flow temperature prior to reaction, thereby avoiding the field heating problem common in drift techniques.[22,37]

Subsequently it was realised that this apparatus could be used for studies of electron attachment. The electron beam creates a plasma with a moderate level of ionisation ($10^8$ to $10^9$ electrons $cm^{-3}$). The electron density can be monitored easily by a Langmuir probe and the produced anions by the mass spectrometer. A great advantage of the CRESU apparatus is that the internal states of the neutral attaching gas are usually very efficiently cooled down, due to the large number of collisions with the buffer gas. Nitrogen was known to be very efficient at relaxing electrons towards the flow temperature[38] and was therefore used as the buffer. However, careful studies[39] have shown that, in this experiment, the electrons relax to a temperature higher than that in the flow.

### 2.3.3. *Radical-Neutral Reactions*

During the late 1980s, I.W.M. Smith realised that the CRESU technique could be adapted to study rapid radical-neutral reactions, notably the

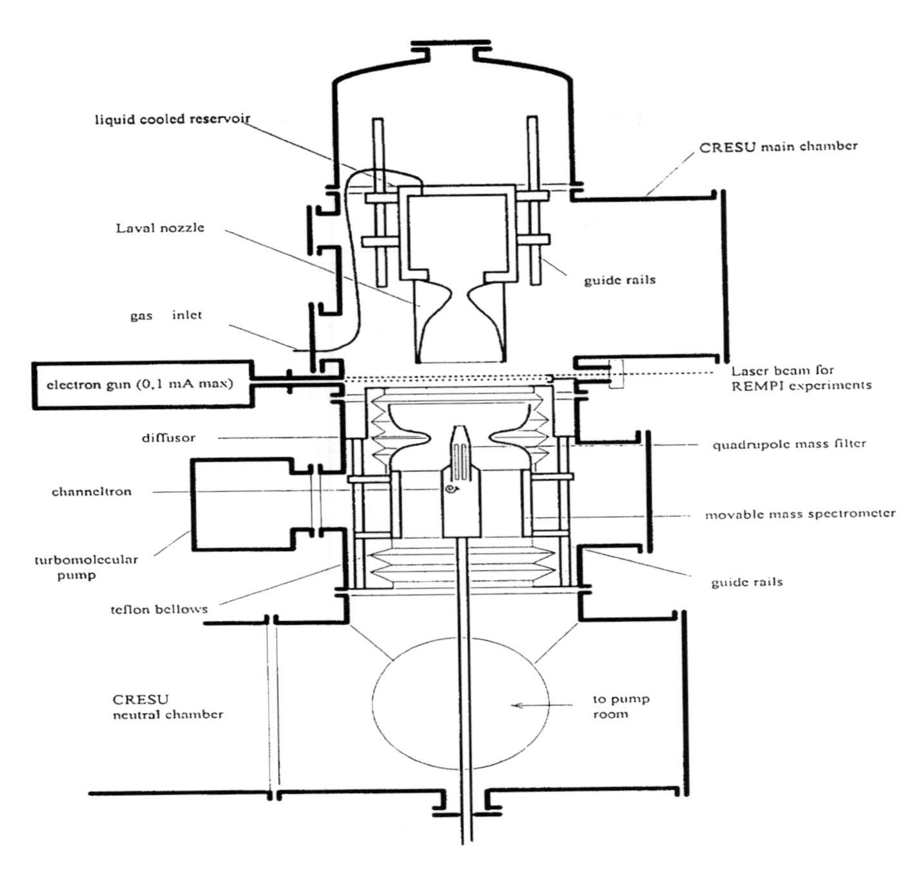


Fig. 2.1. Sketch of the CRESU apparatus devoted to the study of ion-molecule reactions.

reaction of CN with $O_2$ whose rate had just been measured and shown to be fast down to 99 K.[40] In a joint project between Rennes and Birmingham, the pulsed laser photolysis — laser induced fluorescence (PLP-LIF) technique was combined with the new CRESU apparatus that had just been constructed in Rennes, and the first measurements on the $CN + O_2$ reaction down to 26 K were published in 1992,[16] and later down to 13 K with precooling of the reservoir by liquid nitrogen.[41]

A sketch of the apparatus is presented in Fig. 2.2. The method to obtain second-order rate coefficients is very similar to that in cryogenic experiments. It is essentially based on the establishment of pseudo-first-order conditions in the supersonic flow. For a reaction $R + M \rightarrow$ products, of a radical R with a molecule M introduced in large excess, we have the bimolecular

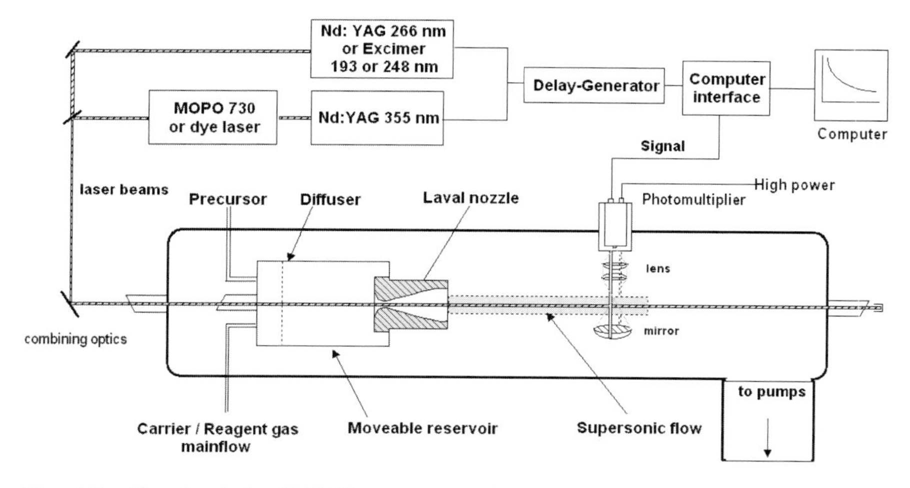


Fig. 2.2. Sketch of the CRESU apparatus adapted for the study of neutral-neutral reactions and inelastic collisions.

rate equation

$$-\frac{d[\mathrm{R}]}{dt} = k_{1\mathrm{st}}[\mathrm{R}] \tag{2.8}$$

where $k_{1\mathrm{st}}$ is the pseudo-first-order rate coefficient (equal to $k[\mathrm{M}]$). This pseudo-first-order rate equation may easily be solved to yield

$$\frac{[\mathrm{R}]_t}{[\mathrm{R}]_{t=0}} = \exp(-k_{1\mathrm{st}}t) \tag{2.9}$$

It can be seen that the determination of $k_{1\mathrm{st}}$ does not require a knowledge of the absolute concentration of R: a time-resolved relative measurement will suffice.

Although for specific studies, chemiluminescence marker techniques can be employed (see Section 2.5.4) the PLP-LIF technique is the method used most often in CRESU experiments. This method makes use of two pulsed lasers which are combined and co-propagate along the axis of the supersonic flow. They enter the CRESU apparatus through a Brewster angle window, pass through another such window mounted on the back of the reservoir, and co-propagate out through the throat of the Laval nozzle and along the axis of the flow, before leaving the vacuum chamber via a third Brewster angle window. Note here that it is also possible to counter-propagate the lasers or to cross them at right angles in the detection plane. The first laser produces an homogeneous concentration of the desired radical by photolysis of a suitable precursor which is introduced in the reservoir in addition to the buffer gas and the reagent of interest. Then, after a variable and controlled time delay, some of the radicals are excited to an upper electronic state by a tuneable laser acting as a probe. Its fluorescence is collected at a known distance downstream of the Laval nozzle exit (usually 10 to 50 cm) by means of an optically fast telescope-mirror combination mounted inside the main vacuum chamber at right angles to the supersonic flow, focused through a slit to reduce scattered light and directed onto the photocathode of a (V)UV-sensitive photomultiplier tube after passing through an interference filter.

For a given concentration of the reagent, the rate of loss of the radical is observed by scanning the delay time between the pulses from the photolysis and probe lasers using a delay generator. As the time delay between lasers increases the fluorescence signal decreases because radicals have been consumed in the reaction during their transit along the flow. A typical trace of the LIF decay can be seen in Fig. 2.3 (upper panel) as well as the exponential fit which allows one to determine a first-order rate coefficient. Of course, in order to extract the required bimolecular rate coefficient $k$, we do need to measure $k_{1st}$ as a function of the absolute molecular reagent concentration [M]. Plotting this rate as a function of [M], which can be varied by means of flow controllers, leads to a straight line whose slope gives the required second-order rate coefficient. A typical second-order plot is shown in Fig. 2.3 (lower panel) as an illustration. It is worth stressing that this kind of study only probes the destruction of the reagent radical, hence no information can be extracted about the nature of the products and the branching ratios when several exit channels are open.

### 2.3.4. *The Pulsed Version of the CRESU Technique*

In order to reduce the high cost for the development and use of a continuous flow CRESU apparatus, another alternative can be to pulse the supersonic flow. Using pulsed injection valves the total gas flow rate, and therefore the pumping capacity, can be considerably reduced. In this case the large flow rate through the Laval nozzle is achieved for only a short time, typically a few milliseconds.

Basically the pulsed Laval apparatus consists of one or two pulsed valves supplying gas to a reservoir on which is mounted a Laval nozzle. The pulsed supersonic expansion is generated in the main chamber exhausted by a mechanical pump to pressures of 0.1–1 mbar. The uniformity of the Laval expansion is achieved by the appropriate background pressure, consisting of the pulsed gases and a "slip gas," which collimates the expansion along the axis of the gas flow. After a short transition time due to the valve cycle, the uniform supersonic flow is established in the chamber. As

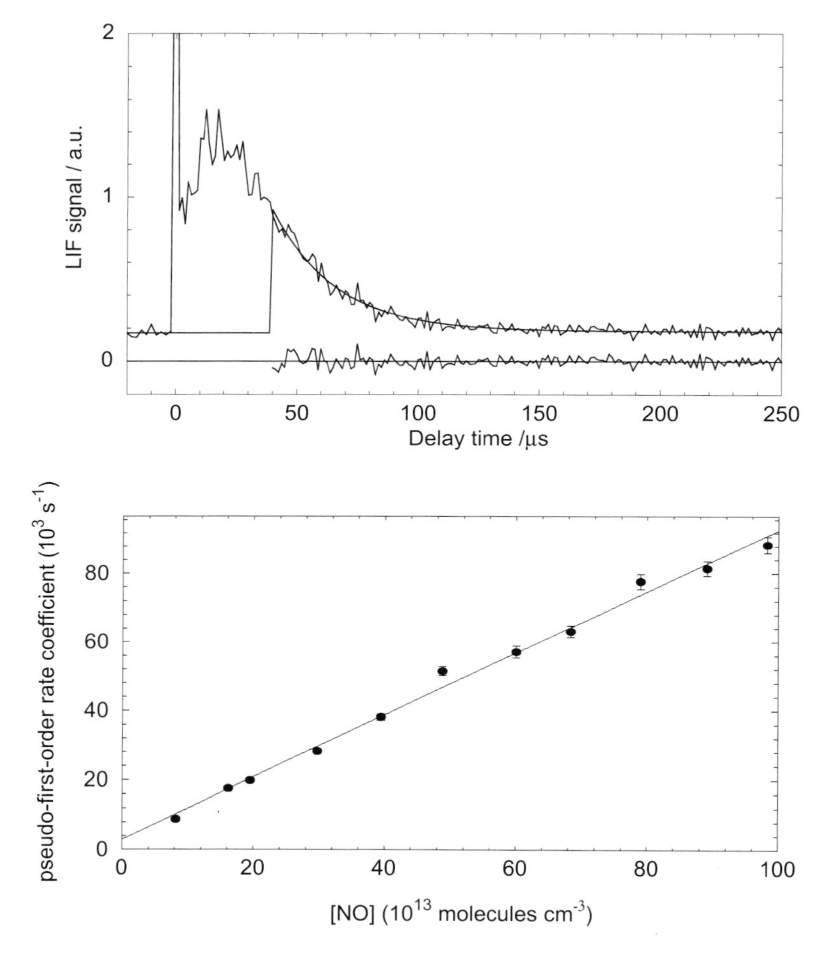


Fig. 2.3. Upper panel: First-order decay of the LIF signal from $C_2(a^3\Pi_u)$ in the presence of $1.6 \times 10^{14}$ molecules $cm^{-3}$ of NO at 145 K in $N_2$, fitted to a single-exponential decay, with residual shown below. The abscissa corresponds to the delay time between the photolysis and probe laser pulses. Lower panel: First-order decay constants for $C_2(a^3\Pi_u)$ at 145 K in $N_2$ plotted against the concentration of NO.

shown in Fig. 2.4, the overall time during which a homogeneous flow is formed is at least one order of magnitude longer than the hydrodynamic time (typically $300\,\mu s$) of the supersonic expansion, making this technique suitable for kinetics studies. To date, the lowest temperature achieved with a pulsed Laval nozzle, (53 K with pre-cooling[42,43]), remains higher than those obtained with a continuous apparatus. This is due in part to

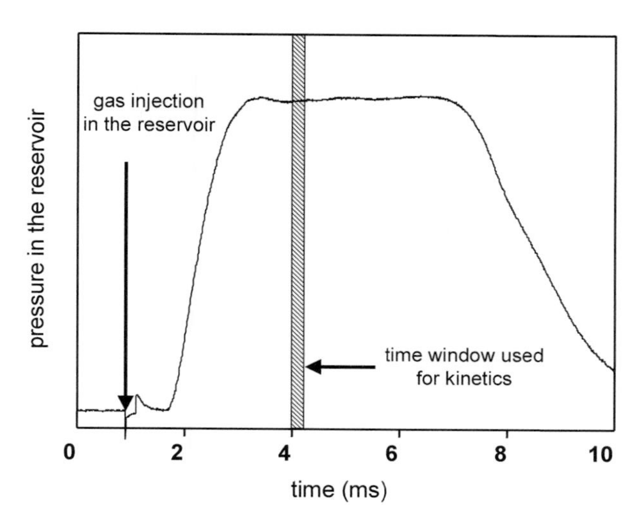


Fig. 2.4. Time dependence of the reservoir pressure in the case of a pulsed Laval nozzle apparatus.

the small pumping capacity of the existing pulsed experiments (typically $500\,m^3\,hr^{-1}$). But the existing pulsed Laval nozzle systems are all based on a design[19] which employs a reservoir at the inlet side of the Laval nozzle which has been made as small as possible (typically only $1\,cm^3$) in order to minimise the time needed for the reservoir pressure to stabilise after the pulsed valve opens. This design results in a gas at the inlet of the Laval nozzle that is far from being at rest, leading to limitations on both the low temperatures and uniformity of flow that can ultimately be obtained downstream.

The pulsed Laval nozzle technique was first developed by M.A. Smith and co-workers.[19] Using this apparatus they reported investigations of the OH + HBr reaction, as well as the reactions of the NH radical with NO and hydrocarbons down to 53 K.[42,43] In the late 1990s a similar apparatus was developed by the group of Leone.[20] It has been mostly used to study reactions of the $C_2H$ radical with a wide range of hydrocarbons for Titan chemistry. More recently, in Troe's group,[44] a pulsed Laval nozzle has been used to study the reactivity of the OH radical with important atmospheric compounds. In all these experiments laser techniques are employed, in the same way as in experiments using the continuous flow CRESU apparatuses.

Although many kinetics studies can be performed in both the continuous and the pulsed versions of the CRESU apparatus, some specific reactions, for which one of the reagents is not available in large quantity, require the use of a pulsed Laval nozzle. This is the case for unstable species that need to be synthesised and hence that cannot be produced in large amounts and also for isotopically enriched molecules which are usually expensive. The latter are often used for measuring the kinetic isotope effect (KIE) which is a variation in the rate coefficient of a reaction when an atom in one of the reagents is replaced by one of its isotopes. Measuring the KIE is a good test of the reaction mechanism when the main component of the reaction coordinate is associated with motion of a light atom such as hydrogen.

## 2.4. What Have We Learned from CRESU Experiments About Ion-Molecule and Electron Molecule Reactions?

In this section we discuss what we have learned from CRESU work for reactions of charged species at low temperature. The case of cations reacting with molecules is examined first, starting with binary reactions which are fast at any temperature; i.e., with a rate close to the collision limit. Slow reactions are then considered and some recent results on association reactions are also pointed out. Finally, electron attachment investigations are described with special attention paid to the electron temperature characterisation inside the supersonic flow.

### 2.4.1. *Fast Binary Reactions of Ions with Molecules*

The rate coefficient of a binary reaction can be viewed as the product of a reaction probability and a capture rate coefficient $k_c$. The value of $k_c$ depends only on the long-range part of the interaction potential between the reagents which can be expressed as the sum of various electrostatic terms: ion-induced dipole, ion-dipole, ion-quadrupole, dipole-dipole, etc. On the basis of this potential a collision is considered as leading to capture when there is enough energy along the line of centres to overcome the so-called centrifugal barrier.[3] Clearly such a model is less questionable for ion-molecule reactions than for neutral-neutral reactions due to the magnitude of the long range electrostatic term in the former case. In fact, for a large majority of exothermic ion-molecule reactions, it is found that the reaction probability is close to unity, yielding $k = k_c$. It is only in a few cases that some other mechanisms, controlled by short range intermolecular forces, are involved and the reaction proceeds at a rate slower than capture. Also when more than one potential energy surface correlates with the separated reagents, some of these surfaces may exhibit barriers to reaction, and the actual rate coefficient at low temperature is then a fraction of the capture one[3]. We first examine cases of ion-molecule capture for both non-polar and polar molecules.

### 2.4.2. *Fast Reactions of Ions with Non-Polar Molecules*

In this case, the intermolecular potential depends only on the separation $R_{\mathrm{AB}}$ of the two reagents and takes the very simple form:

$$V(R_{\mathrm{AB}}) = -\alpha\frac{q^2}{2R_{\mathrm{AB}}^4} \quad \text{in CGS units.} \tag{2.10}$$

where $\alpha$ is the neutral polarisability and $q$ the charge of the ion. Simple calculations[3] then lead to a cross section:

$$\sigma(\mathrm{KE_{cm}}) = \pi\sqrt{\frac{2\alpha q^2}{\mathrm{KE_{cm}}}} \tag{2.11}$$

which integrated over a thermal distribution of collision energies leads to the well known Langevin formula:

$$\begin{aligned} k(T) &= 2\pi q\sqrt{\frac{\alpha}{\mu}} && \text{in CGS units} \quad \text{or} \\ k(T) &= 2\pi q\sqrt{\frac{\alpha'}{(4\pi\varepsilon_0)^2\mu}} && \text{in SI units} \end{aligned} \tag{2.12}$$

where $\mu$ is the reduced mass and $\alpha'$ the polarisability expressed in SI units.

This rate coefficient is independent of temperature. Table 2.1 summarises a large fraction of the data obtained with the CRESU for fast reactions involving non-polar molecules, together with the Langevin values. A glance at this table clearly shows that the value of the rate coefficient is mostly independent of temperature and indeed very close to the Langevin estimate. Note that the effect of large quadrupole moments on the rate coefficient has been investigated experimentally, using molecules such as $c$-$C_6H_{12}$ and $C_6F_6$, and found to be negligible.[24]

Table 2.1. Rate coefficients (in units of $10^{-9}$ cm$^3$ molecule$^{-1}$ s$^{-1}$) for ion-non-polar molecule reactions measured with the CRESU technique. Rate coefficients are compared to the Langevin values.

| Reaction | $T = 8$ K | $T = 20$ K | $T = 30$ K | Langevin rate |
|---|---|---|---|---|
| $He^+ + N_2$ | 1.2 | 1.3 | 1.3 | 1.7 |
| $He^+ + O_2$ | 1.0 | 0.85 | | 1.6 |
| $He^+ + CO$ | 1.5 | 1.4 | | 1.8 |
| $N^+ + O_2$ | 0.55 | | | 0.95 |
| $N^+ + CH_4$ | 0.82 | | | 1.4 |
| $N^+ + CO$ | 1.1 | | | 1.1 |
| $H_3^+ + CO$ | | | 1.7 | 2.0 |
| $H_3^+ + N_2$ | | | 1.3 | 1.9 |
| $H_3^+ + CH_4$ | | | 1.9 | 2.4 |

### 2.4.3. *Fast Reactions of Ions with Polar Molecules*

When the neutral reagent molecule has a permanent dipole $\mu_D$ the long range part of the interaction potential becomes:

$$V(R_{\mathrm{AB}}) = -\alpha \frac{q^2}{2R_{\mathrm{AB}}^4} - \frac{q\mu_D \cos\theta}{2R_{\mathrm{AB}}^2} \tag{2.13}$$

where $\theta$ is the angle that the dipole makes with the line joining the centres of the two species. Following the pioneering work of Su and Bowers[45] leading to the average dipole orientation (ADO) theory, many calculations have been performed in order to improve the capture model.[46] Some of the most refined treatments have been made nearly simultaneously by Clary (ACCSA model[47]) and Troe (SACM model[48]). Both calculate rate coefficients for individual rotational levels $j$ of the neutral molecule and predict that they increase more strongly for low $j$ when the temperature is lowered. Consequently most of the increase in $k(T)$ at low $T$ results from the change in rotational population.[49] At temperatures higher than 10 K, most of the models yield very similar results, in good agreement with the early classical trajectory calculation of Su and Chesnavitch,[8] which has led to a very simple analytical formulation of the rate coefficient. Most often there is also an excellent agreement with the experimental findings as for the $N^+ + NH_3$[50] and the $N^+ + H_2O$ reactions.[49] Table 2.2 summarises some of the CRESU data obtained for a choice of polar molecules, showing clearly the strong increase of the rate coefficients at low temperature.

Table 2.2. Rate coefficients (in units of $10^{-9}$ cm$^3$ molecule$^{-1}$ s$^{-1}$) for ion-polar molecule reactions measured with the CRESU technique.

| Reaction | $T = 27$ K | $T = 30$ K | $T = 68$ K |
|---|---|---|---|
| $He^+ + HCl$ | 11.0 | | 4.6 |
| $He^+ + SO_2$ | 8.2 | | 6.5 |
| $He^+ + H_2S$ | 5.5 | | 4.6 |
| $He^+ + NH_3$ | 4.5 | | 3.0 |
| $He^+ + H_2O$ | 4.3 | | 1.8 |
| $C^+ + HCl$ | 3.8 | | 1.9 |
| $C^+ + SO_2$ | 5.7 | | 4.1 |
| $C^+ + H_2S$ | 4.8 | | 3.0 |
| $C^+ + NH_3$ | 4.6 | | 3.2 |
| $C^+ + H_2O$ | 12.0 | | 5.2 |
| $H_3^+ + SO_2$ | | 11.0 | |
| $H_3^+ + H_2S$ | | 6.5 | |
| $H_3^+ + NH_3$ | | 9.1 | |
| $N^+ + NH_3$ | 5.2 | | 3.2 |
| $N^+ + H_2O$ | 9.9 | | 6.0 |

### 2.4.4. *Slow Bimolecular Reactions*

While the behaviour of the majority of ion-molecule reactions can be adequately represented using capture theories, there are numerous reactions for which a more sophisticated treatment is required, namely those reactions which are "slow" at room temperature, that is whose reaction probability is much lower than unity. CRESU measurements have shown that, in several cases, the rate coefficients increase at lower temperatures, sometimes approaching $k_c$ when extrapolated towards 0 K. This has been shown for the reaction of $O_2^+ + CH_4$, $N_2^+ + O_2$, and of $Ar^+$ with $N_2$ and $O_2$.[51] Such behaviour can be generally understood as indicative of complex formation, whose lifetime with respect to re-dissociation increases when the temperature is lowered. However, it has to be kept in mind that no general "rule of thumb" can be given; some reactions, such as $He^+ + H_2$, of potential importance for astrochemistry have been found to remain very slow when the temperature is lowered. In some cases there is an increase of the rate coefficient but the ratio of the experimental value to the capture one remains constant and therefore there is no increase of the reaction probability at low temperatures. This can be understood as a consequence of the multi-surface nature of open-shell systems, as in the reaction of $C^+$ with HCl.[51]

### 2.4.5. *Association Reactions*

Radiative and termolecular association of ions with neutral molecules play important roles in the chemistry of ISCs and planetary atmospheres, respectively. Such reactions and the reverse thermal dissociation can be represented within the framework of unimolecular rate theory [52, and references therein]. While radiative association can be studied using trap techniques at very low pressures (see chapter 3 by D. Gerlich) the CRESU technique allows only the study of the termolecular process. By contrast with two-body processes, the adduct formed by association of two reagents may have no exothermic dissociation pathway. The reaction can therefore only proceed by collisional stabilisation of the adduct. This mechanism is both temperature and pressure-dependent. Evolution with pressure of the second-order rate coefficient for three-body processes is rather complicated and has stimulated a lot of theoretical work. In a few words, in the low pressure limit, evolution is linear with pressure whereas the rate coefficient is constant in the high pressure limit. The pressure range in between these two limits, which depend on the system of interest, is known as the fall-off region. In principle, a complete study of a termolecular association reaction requires measurements of the rate coefficient over a wide pressure and temperature range $k(T, p)$ in order to construct the fall-off curves and to extract the low pressure $k_0$ and high pressure $k_{\rm inf}$ rate coefficients, as

Table 2.3. Rate coefficients (in units of $10^{-27}$ cm$^6$ molecule$^{-2}$ s$^{-1}$) for ion–molecule association reactions measured with the CRESU technique.

| Reaction | 15 K | ~20 K | ~50 K | ~70 K | ~150 K |
|---|---|---|---|---|---|
| $N_2^+ + N_2 + N_2$ | | | 1.5 | 1 | |
| $O_2^+ + O_2 + O_2$ | | >1 | >0.3 | 0.2 | |
| $CH_3^+ + H_2 + He$ | | 3.4 | | 1.9 | |
| $CH_3^+ + CO + He$ | | 85 | | 34 | |
| $CH_3^+ + N_2 + He$ | | 12 | | 1.5 | |
| $CH_3^+ + H_2O + He$ | | >70 | | >40 | |
| $C^+ + H_2 + He$ | | <0.1 | | | |
| $Ar^+ + Ar + Ar$ | | ~0.005 | ~0.001 | | |
| $H_3O^+ + H_2O + He$ | | 27 | 8.7 | | 2.1 |
| $H_3O^+ + H_2O + N_2$ | | | 17 | 10 | |
| $NH_4^+ + NH_3 + He$ | 93 | 51 | 17 | | 2.3 |
| $NH_4^+ + NH_3 + Ar$ | | | 29 | | |
| $NH_4^+ + NH_3 + N_2$ | | | 124 | 50 | 8.7 |

well as the broadening factor which characterises the deviation of the actual fall-off curve from the Lindemann–Hinshelwood curve.[53] Investigating the full $T$, $p$ range in CRESU experiments would be extremely time and money consuming, as a change of pressure implies a change of nozzle as mentioned previously. Nevertheless invaluable results have been obtained for such reactions as regards the temperature range. In some cases like the $Ar^+$ association with its parent neutral[4] the CRESU and free jet reactor data agree very well and have allowed measurements over an extraordinary temperature range (0.5–300 K). When linked to other experimental data the CRESU results have sometimes allowed a complete theoretical understanding of the association process.[52]

Table 2.3 summarises the ion association reactions that have been studied in CRESU experiments. Due to the low pressure in the CRESU apparatus, especially in its first Meudon version, many of these results are very close to the low pressure limit. However, as the temperature is lowered, association becomes so efficient that the reaction often enters the fall-off zone. Indeed in this case for some reactions, such as $CH_3^+ + H_2O$, the reaction proceeds nearly at the capture rate; i.e., at the high pressure limit.

### 2.4.6. *Electron-Molecule Reactions: The Thermalisation Problem*

When studying an electron-molecule reaction one must consider two different temperatures: the temperature of the gas ($T_g$) and the temperature of the electrons ($T_e$), assuming they have a Maxwell-Boltzmann velocity distribution. For many polyatomic molecules, for example $SF_6$, the capture process is dominated at low energies by an s-wave threshold law.[13] This law predicts inverted square root electron energy dependence for the cross section for capture of an electron by a neutral molecule. This behaviour of the cross section leads to a rate coefficient independent of $T_e$. At the lower $T_g$ studied by the CRESU technique, only the ground vibrational state of the target molecule is significantly populated, while for the 300 K experimental results several vibrational states can be populated. If one examines the potential energy curves describing systems like $CF_3Br$ and $CF_3 + Br^-$, it is seen that the intersection of these two curves occurs away from the ground vibrational state and therefore one can expect that the attachment rate coefficient will depend very sensitively upon the level of internal excitation of the target. Such an effect of $T_g$ has been demonstrated for the cases of attachment to $CF_3Br$ and $CCl_2F_2$ for which increasing the gas temperature from 48 to 170 K led to an increase in the attachment rate of over two orders of magnitude (see Fig. 2.5).[38] Other interesting results concern the

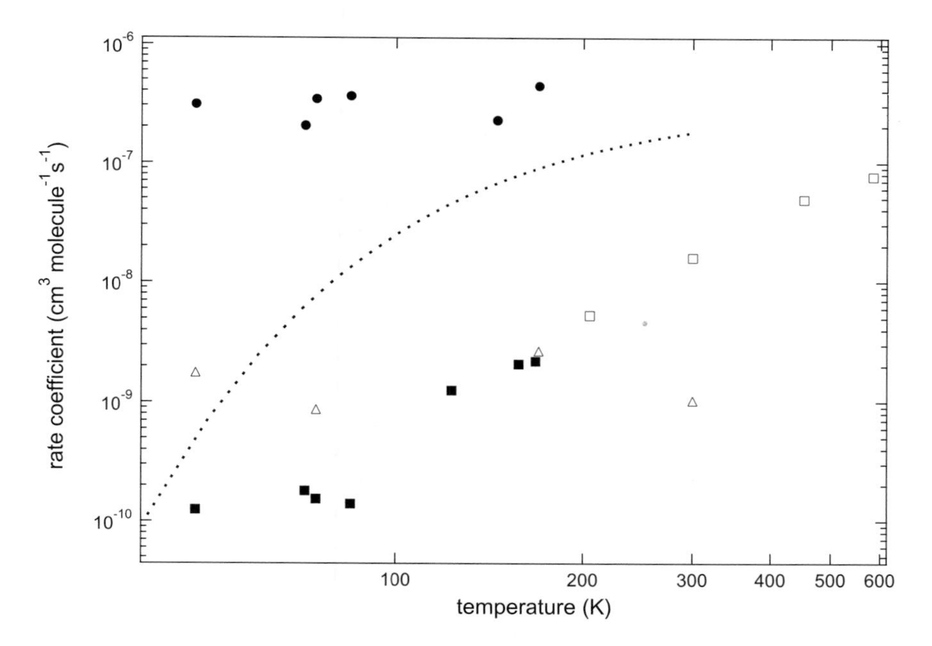


Fig. 2.5. Rate coefficient for electron attachment measured as a function of the temperature. The experimental attachment rate on DI (•) measured using the CRESU technique is compared to the predicted rate (dashed line). The attachment rate on anthracene (△) does not exhibit any temperature dependence contrary to the attachment rate on $CF_3Br$ measured using the CRESU technique (■) and a FALP apparatus (□).

dissociative electron attachment on HCl and HBr. Although both reactions are endothermic, an increase of the attachment rates has been observed at temperatures below 170 K. The reason for such behaviour is that HBr and HCl are cooled in the supersonic flow to a point where neutral cluster formation begins to occur. These clusters exhibit rapid electron attachment with rate coefficients greater than $10^{-8}\,\mathrm{cm}^3\,\mathrm{molecule}^{-1}\,\mathrm{s}^{-1}$.[54]

Considering that $T_e = T_g$ in the CRESU apparatus, surprising results were obtained for the rate coefficient of electron attachment to anthracene ($C_{14}H_{10}$)[55] and $CH_3I$.[56] These are displayed in Fig 2.5. Although the rate of coefficient for the electron attachment to anthracene is several orders of magnitude below the capture limit, no apparent temperature dependence was observed from room temperature down to 48 K. Also, the results on dissociative electron attachment on $CH_3I$ were found to disagree with results obtained using a laser photoionisation attachment technique combined with free jet cooling of $CH_3I$, as well as with experiments using Rydberg atoms. For this reaction, the rate coefficient is expected to increase towards lower $T_e$ as well as being strongly dependent on $T_g$. Such behaviour for the attachment rate on $CH_3I$ was not observed with the CRESU apparatus.

Whether the temperatures of the electrons and the neutral molecules are the same within the CRESU jet was always a matter of concern, as direct measurements of electron temperature are not reliable in this temperature range. Equilibrium was, however, expected because experiments were performed either in nitrogen or helium as the buffer gas: both gases relax electrons efficiently, nitrogen by rotational transfer and helium by translational transfer. In addition, the excellent agreement with previous FALP data in the temperature range common to both experiments, sometimes showing similar strong temperature dependence, was another argument in favour of true thermal conditions in the CRESU experiments. Unfortunately most of these first results concerned reactions which are rather insensitive to electron temperature but strongly dependent on the internal states of the molecule, and therefore cannot be taken as a probe for electron temperature. The fact that electron relaxation is very fast towards the cold rotation/translation bath is not a definitive proof of equilibrium in the CRESU experiment: if an efficient heating process competes with the cooling process, electrons will be maintained in a steady state at some intermediate temperature. In order to probe the electron temperature it is necessary to study an attachment reaction which is predicted to be strongly dependent on the electron energy, even at low temperatures, that is, a slightly endothermic reaction. The dissociative electron attachment of deuterium iodide, DI, is therefore the best candidate. Only the vibrational ground state is populated in CRESU experiments. As the attachment reaction is endothermic for the low-lying rotational states, and becomes exothermic for levels greater than J = 9, strong positive temperature dependence is therefore expected, provided the electrons are thermalised. By contrast, electron attachment to HI is exothermic. In the 23–300 K range, however, the measured rate coefficients for DI do not exhibit the expected temperature dependence, as shown in Fig. 2.5.[39] This can be taken as a proof that in the CRESU flow, when an electron beam is used as a plasma source, the electron temperature is not the same as the neutral one. The only possible fast process for electron heating consists in superelastic collisions with excited vibrational states of $N_2$, since electric field heating has been excluded. Vibrational excitation of the nitrogen buffer gas in the electron source could heat the electrons to some intermediate temperature. However, a small value for the attachment rate coefficient to HBr shows that the electron energy is not greater than 0.4 eV in the attachment zone as the reaction is endothermic by 0.4 eV. Moreover, the calculated electron relaxation times are small compared to the hydrodynamic time, and the electron temperature reaches a steady state in the CRESU experiment. The experimental results for electron attachment on DI, however, cannot be explained if one does not assume that the electron temperature is higher than the neutral one. The work performed with HBr and DI show that electron temperature ranges between the neutral temperature and 500 K.[39] The problem of electron thermalisation cannot be solved by choosing helium as buffer gas because metastable helium atoms are produced by the electron beam which ionises molecules in the supersonic flow. These metastable atoms can be destroyed by adding some argon to the flow, but this quenching reaction forms metastable argon atoms, which undergo superelastic collisions with electrons too. Several results obtained with a CRESU apparatus highlight the strong effect of the cooling of the internal states of the neutral molecules or of the presence of neutral dimers and higher oligomers within the flow and are still valid.[38,54] However the above technique is not an efficient way to study the influence of $T_e$ on the attachment rate.

There is still a great interest in developing the CRESU method for studies of electron attachment. This technique allows one to cool the neutral molecules down to very low temperatures in conditions of true thermodynamic equilibrium. New developments are in progress to produce low energy electrons and avoid the buffer gas excitation using laser photoionisation of a precursor which has an ionisation potential close to the laser photon energy.

## 2.5. What Have We Learned From Continuous and Pulsed CRESU Experiments About Neutral — Neutral Reactions

### 2.5.1. *Overview of Neutral-Neutral Reaction Studies*

Since 1992 and the first success of combining laser techniques with the CRESU apparatus, a good number of neutral-neutral processes have been studied.[3,4] These include bimolecular reactions, three-body reactions, and inelastic collisions, such as spin orbit relaxation of atoms (Al, Si and C), rotational relaxation of molecules such as NO and CO, and even for some specific cases vibrational relaxation (CH, NO and toluene). In this section we will only concentrate on reactive processes.

Table 2.4 lists the two-body neutral-neutral reactions that have been experimentally studied using the continuous or pulsed version of the CRESU method. The minimum temperatures at which reactions have been studied are given. Among the 93 reported reactions, most (73) were studied at the Universities of Rennes 1 (France), Birmingham (UK) and a few at Bordeaux (France) using the continuous version of the CRESU apparatus. A significant number (20), however, have been investigated with the pulsed version of the device. Almost all processes that are presented in Table 2.4 are fast at room temperature ($k > 10^{-11}\,\mathrm{cm^3\ molecule^{-1}\,s^{-1}}$). Only a few of them could not be studied at very low temperature because of the presence of an activation barrier which led to the decrease of the rate coefficient below the CRESU sensitivity when the temperature was lowered. For these particular cases ($C_2H + CH_3CN$, $C_2(^1\Sigma_g^+)$+$O_2$, $C_2(^3\Pi_u) + C_2H_6$ and $C_4H + CH_4$) data are only available at about 200 K and above and the rate coefficients at room temperature are typically close to $10^{-12}\,\mathrm{cm^3}$ molecule$^{-1}$ s$^{-1}$. The majority of the reactions studied revealed a rate coefficient that remains large even at very low temperature indicating that the reactions must proceed over a potential energy surface with no barrier. The dependence with temperature, however, was quite variable and four typical examples are shown in Fig. 2.6. In general, the measured rate coefficients increase at sub-ambient temperature, either monotonically and following a $T^{-n}$ power law (Fig. 2.6a) or reaching a plateau or a maximum followed by a slight decrease. This may reveal the presence of a very small energy barrier as, for instance, for the reaction between boron atoms and $C_2H_2$ (Fig. 2.6b). In several cases, the rate coefficient was found to be essentially temperature independent (Fig. 2.6d) and in few examples (e.g. $CN + C_2H_6$: Fig. 2.6(c) and $O(^3P)$ + $C_3H_6$: Fig. 2.12) the rate coefficient reached a minimum at about 200 K increasing to both higher and lower temperatures. Rationalising these very different behaviours is challenging and until

Table 2.4. Bimolecular neutral-neutral reactions studied using the continuous and pulsed versions of the CRESU apparatus. The numbers indicate the minimum temperature at which the experiment has been carried out.

| Reagent\Radical | CN | OH | CH(v = 0) | NH | $C_2H$ | $^1C_2(^1\Sigma_g^+)$ | $^3C_2(^3\Pi_u)$ | $C_4H$ | $C(^3P_J)$ | $Al(^2P_J)$ | $Si(^3P_J)$ | $B(^2P_J)$ | $O(^3P_J)$ |
|---|---|---|---|---|---|---|---|---|---|---|---|---|---|
| $O_2$ | 13 K | | 13 K | | 15 K | 145 K | | | 15 K | 23 K | 15 K | 24 K | |
| NO | | | 13 K | 53 K | | 24 K | 24 K | | 15 K | | 15 K | | |
| $O(^3P)$ | | 39 K | | | | | | | | | | | |
| $D_2$ | | | 13 K | | | | | | | | | | |
| $H_2O_2$ | | 96 K | | | | | | | | | | | |
| HBr | | 23 K | | | | | | | | | | | |
| $NH_3$ | 25 K | | 23 K | | 104 K | | | | | | | | |
| $CH_4$ | | | 23 K | 53 K | | 24 K | | 200 K | | | | | |
| $C_2H_2$ | 25 K | | 23 K | [a]53 K | 15 K | 24 K | 24 K | 39 K | 15 K | | 15 K | 23 K | |
| $C_2H_4$ | 25 K | | 23 K | [a]53 K | 15 K | 24 K | 24 K | 39 K | 15 K | | 15 K | 23 K | 39 K |
| $C_2H_6$ | 25 K | | 23 K | 53 K | 96 K | 24 K | 200 K | 39 K | | | | | |
| $CH_3C{\equiv}CH$ | 15 K | | 77 K | | 63 K | | | 39 K | 15 K | | | | |
| $H_2C{=}C{=}CH_2$ | 15 K | | 77 K | | 63 K | | | [b]39 K | 15 K | | | | |
| $C_3H_6$ | | | 77 K | [a]53 K | 15 K | | | [b]39 K | 15 K | | | | 23 K |
| $C_3H_8$ | | | | | 96 K | 24 K | 36 K | 39 K | | | | | |
| $C_4H_2$ | | | | [a]53 K | | | | | | | | | |
| $C_4H_6$ butyne | | | | | | | | [b]39 K | | | | | |
| 1,3 butadiene | | | | | 104 K | | | [b]39 K | | | | | |

*(Continued)*

Table 2.4. (Continued)

| Reagent\Radical | CN | OH | CH(v = 0) | NH | $C_2H$ | $^1C_2(^1\Sigma_g^+)$ | $^3C_2(^3\Pi_u)$ | $C_4H$ | $C(^3P_J)$ | $Al(^2P_J)$ | $Si(^3P_J)$ | $B(^2P_J)$ | $O(^3P_J)$ |
|---|---|---|---|---|---|---|---|---|---|---|---|---|---|
| $C_4H_8$ *cis*-butene | | | | | | | | | | | | | 27 K |
| but-1-ene | | | 23 K | | 103 K | | | [b]39 K | | | | | 23 K |
| *iso*-butene | | | | | 104 K | | | | | | | | 27 K |
| *trans*-butene | | | | | | | | | | | | | 27 K |
| $C_4H_{10}$ *n*-butane | | | | | 96 K | | | [b]39 K | | | | | |
| *iso*-butane | | | | | 104 K | | | | | | | | |
| benzene | | | | | 105 K | | | | | | | | |
| anthracene | | | 58 K | | | | | | | | | | |
| $CH_3CHO$ | | 58 K | | | | | | | | | | | |
| $CH_3CN$ | | | | | 165 K | | | | | | | | |
| $C_2H_5CN$ | | | | | 104 K | | | | | | | | |
| $C_3H_7CN$ | | | | | 104 K | | | | | | | | |

[a] For these reactions, the binary nature is not firmly established or is still under discussion.
[b] Not yet published.

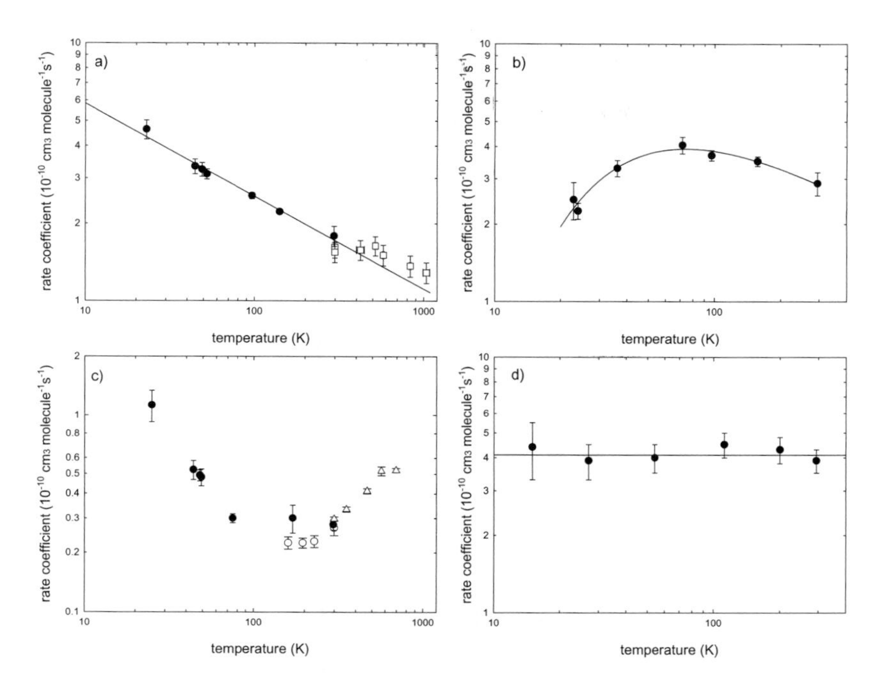


Fig. 2.6. Rate coefficients for the reaction of (a) aluminium $Al(^2P_J)$ atoms with $O_2$, (b) boron atoms $B(^2P_J)$ with acetylene, (c) the cyanogen radical CN with $C_2H_6$ and (d) with allene $CH_2{=}C{=}CH_2$ as a function of temperature, displayed on a log-log scale. The filled circles represent the results obtained in the CRESU experiment. Results from Garland *et al.*[89] at higher temperatures are also shown in Fig. 2.6(a): (□) as well as those from Herbert *et al.*[90]: (△) and data from cooled cell experiments[77]: (○) in Fig 2.6(c). CRESU results are fitted using the following formula: $k(T) = 1.72 \times 10^{-10}$ $(T/298\,\mathrm{K})^{-0.36}\,\mathrm{cm^3}$ molecule$^{-1}\,\mathrm{s^{-1}}$ in Fig. 2.6(a); $k(T) = 3.21 \times 10^{-10}$ $(T/300\,\mathrm{K})^{-0.5}$ $\exp(-37/T)\,\mathrm{cm^3}$ molecule$^{-1}\,\mathrm{s^{-1}}$ in Fig. 2.6(b) and $k(T) = 4.1 \times 10^{-10}\,\mathrm{cm^3}$ molecule$^{-1}$ $\mathrm{s^{-1}}$ in Fig. 2.6(d).

recently it was not possible to predict the most probable evolution of a rate coefficient when the temperature changes. The contribution by Klippenstein and Georgievskii in this book presents new theoretical developments which go a long way towards understanding these phenomena.

Because of their interest in interstellar chemistry and for planetary atmospheres, many carbon-based species are present in the table either as radicals or stable reagents. The most intensively studied radicals are $C_2H$, CH and $C_4H$ whereas acetylene and ethylene are the most investigated reagents. For this reason it is interesting to gather several data from different publications and show how the reactivity of stable molecules is influenced by the nature of the radical. This is presented in Fig. 2.7 for

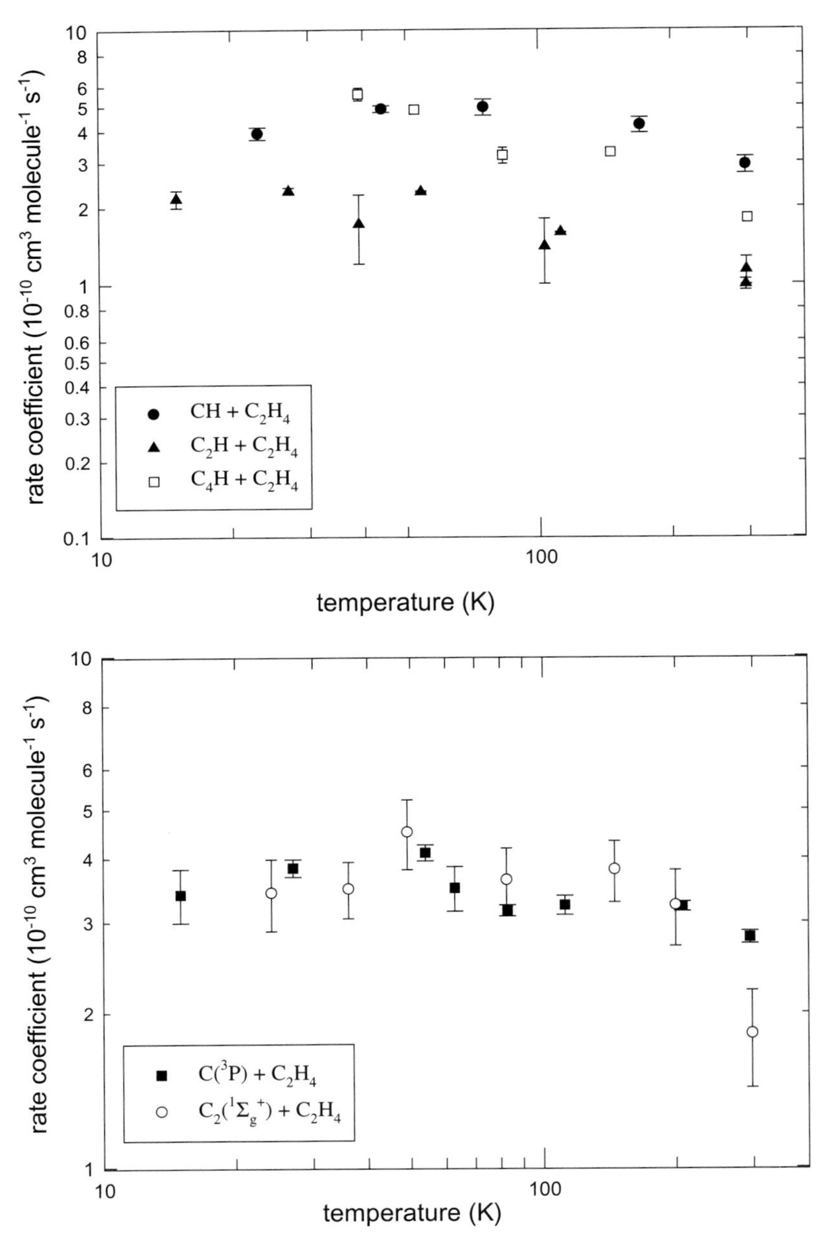


Fig. 2.7. Rate coefficients $k$ for the reaction of $C_2H_4$ with CH (●)[91]; $C_2H$ (▲)[92]; $C_4H$ (□)[59]; $C(^3P)$ (○)[62] and $C_2(X^1\Sigma_g^+)$ (■)[93] measured as a function of the temperature.

$C_2H_4$ for a variety of carbon bearing radicals (C, CH, $C_2$, $C_2H$ and $C_4H$) demonstrating that the rate coefficients are rapid and depend very little, if at all, on temperature. The same trends are observed for $C_2H_2$. A value of $k = 2\text{–}3 \times 10^{-10}\,\text{cm}^3\text{molecule}^{-1}\,\text{s}^{-1}$ can be considered as very representative of these classes of reactions. This may help in making more sensible guesses when neither experiments nor theoretical data are available for reactions involving those reagents and other carbon-based radicals.

As a case in point, for quite a long time no data were available for reactions involving $C_{2n}H$ radicals ($n > 1$). In their pioneering photochemical model of the atmosphere of Titan in 1984, Yung *et al.*[57] proposed that "*the higher polyacetylene radicals are probably less reactive than $C_2H$, and their rate coefficients have been adjusted according to $k(C_{2n}H) = 3^{1-n}k(C_2H)$.*" This empirical guess was then taken up by new generations of photochemical models without further reconsideration. This assumption would predict that the rate coefficient for a reaction involving $C_4H$ radicals will be three times less than that for reaction of $C_2H$ with the same reagent. In a more recent model, Wilson and Atreya[58] modified this relation and decided to adopt the rate coefficients recently obtained for $C_2H$ for reactions involving $C_{2n}H$ radicals. Although this new option seemed to be more realistic in view of the growing body of low temperature experimental data, very recent measurements for the $C_4H$ radical[59] reveal that the assumed values were typically a factor of two below the experimentally determined ones; in other words, experiments were producing rate coefficients typically six times greater than the initial recommendation by Yung *et al.*

Another tempting shortcut that some modellers may wish to consider is to suppose that the chemistry of atoms of the same Group in the periodic table of the elements will present the same reactivity with a given reagent. The case of silicon is interesting because it is in the same Group as carbon. Recent results obtained for $\text{Si}(^3\text{P})$[60,61] and for $\text{C}(^3\text{P})$[31,62] clearly show that the temperature dependence of the rate coefficients and sometimes their order of magnitude are very different. On the one hand, the rates of reactions of carbon atoms with $C_2H_2$ and $C_2H_4$ were found to be almost independent of temperature (see Fig. 2.7 for $C_2H_4$) whereas the reaction with $O_2$ monotonically increased when the temperature was lowered (Fig. 2.8). On the other hand, the analogous reactions with silicon atoms presented a well defined maximum of the rate coefficient at about 50 K.

For the clarity of Table 2.4, studies involving deuterated species have been omitted with the exception of $D_2$. It is worth however stressing some recent results that were obtained by pulsed CRESU devices which are more suitable than the continuous version for these studies, (see Section 2.3.4). Two experimental studies have been performed concerning the KIE at low temperature: the reaction of $C_2H$ with $NH_3$ and $ND_3$ from 104 K to 294 K[63]

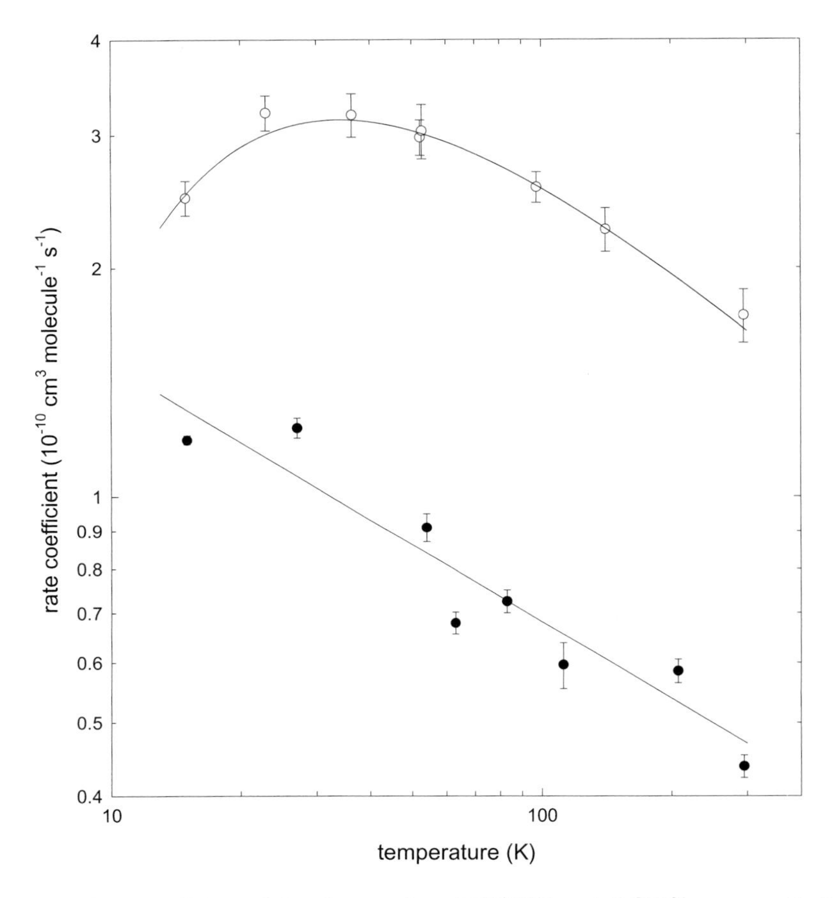


Fig. 2.8. Rate coefficients $k$ for the reaction of $Si(^3P)$[61] and $C(^3P)$[31] atoms with $O_2$ plotted on a log-log scale against temperature. The circle symbols show the results obtained with the CRESU technique. The solid lines show the result of a non-linear least-squares fit to the data giving $k = 1.75 \times 10^{-10}\ (T/298\,\text{K})^{-0.50} \exp^{-17/T}$ cm$^3$ molecule$^{-1}$ s$^{-1}$ for $Si(^3P)$ and $k = 4.7 \times 10^{-11}\ (T/298\,\text{K})^{-0.34}$ cm$^3$ molecule$^{-1}$ s$^{-1}$ for $C(^3P)$.

and the reaction of OH/OD with HBr/DBr from 53 K to 135 K.[43] The KIE is said to be "primary" when the isotopic replacement is involved in a chemical bond that is broken or formed and "secondary" when the substitution is not directly involved in the formed or broken bond. The KIE is usually measured as the ratio of the rate coefficient for the unsubstituted molecules ($k_H$) over the rate coefficient for the reaction involving the isotopically substituted molecules ($k_D$). When this ratio is lower than unity, an inverse KIE is said to have been observed. In the case of OH with HBr, the observed inverse secondary KIE is reasonably well reproduced by theory and is understood to occur owing to the difference in the DOH and HOH bending frequencies in the transition state. However the absolute value of the primary KIE and its temperature independence observed for the $C_2H + NH_3$ reaction as well as $OH + HBr$ reaction are not reproduced by models based on transition state theory. Further theoretical studies are therefore needed to explain the observed temperature independence of the primary KIE.

Among recent studies with the pulsed CRESU version, that of OH radicals with acetaldehyde $CH_3CHO$ deserves a special attention. Vohringer–Martínez *et al.* measured the rate coefficient for the above reaction from 58 K to 300 K with and without the addition of water to the flow.[64] In both cases the reaction becomes faster as the temperature is lowered. The most interesting result is the observation, at low temperature only, of a temperature dependent enhancement of the reaction rate coefficient when adding a small amount of water. Theoretical calculations show that the reaction of acetaldehyde with OH proceeds via formation of a pre-reaction complex which cannot be stabilised by collision with the nitrogen buffer gas employed in their study. The exothermic exit channel exhibits a barrier that remains below the reagent energy. The association complex can decompose either back to the initial reagents or to the final products: the $CH_3CO$ radical and the water molecule. When considering a hydrogen-bounded $CH_3CHO–H_2O$ complex reacting with the OH radical the calculated barrier of the reaction exit channel is more than an order of magnitude lower than the aforementioned barrier. This very low energy barrier combined with the low probability of going back to the initial reagents of the water-containing complex increases the reaction rate coefficient toward the temperature independent association rate coefficient value. At room temperature no enhancement of the reaction rate coefficient is observed as the density of complexes is too small compared to the total density of acetaldehyde molecules. When decreasing the temperature, the $CH_3CHO$-$H_2O$ complex density increases to reach a maximum value at a temperature well below 100 K. The increase of the rate coefficient enhancement when going toward lower temperatures is therefore due to the increasing contribution of complex reaction to the removal of the OH radicals.

Other interesting results obtained with a pulsed Laval nozzle apparatus concern product detection down to 90 K. In an innovative experimental development, Leone's group has used 118 nm single photon ionisation and time of flight mass spectrometry to probe the reaction products after sampling of the pulsed supersonic flow. Diacetylene ($C_4H_2$) and the ethynyloxy radical (HCCO) have been detected after reaction of $C_2H$ with respectively $C_2H_2$ and $O_2$ down to 90 K.[20,65] However, the small 118 nm photon flux obtained by frequency tripling laser radiation in a rare gas does not allow an accurate measurement of product branching ratios.

Table 2.5. Three-body neutral-neutral reactions studied using the continuous and pulsed versions of the CRESU apparatus. The numbers indicate the minimum temperature at which the experiment has been carried out.

| | Probed species | | |
|---|---|---|---|
| Reagent | OH | CH(v = 0) | $C_6H_6$ |
| $H_2$ | | 53 K | |
| $N_2$ | | 53 K | |
| NO | 23 K | | |
| CO | | 53 K | |
| $C_2H_4$ | 96 K | | |
| $C_3H_6$ | 58 K | | |
| *cis*-butene | 23 K | | |
| *trans*-butene | 23 K | | |
| 1-butene | 23 K | | |
| $C_5H_8$ isoprene | 58 K | | |
| $C_6H_6$ | | 25 K | 15 K |
| anthracene | 58 K | | |

Beyond the study of bimolecular processes, a significant although limited number of association reactions have been investigated by the CRESU technique. These are summarised in Table 2.5. Ideally, it would be necessary to study the pressure evolution of the second-order rate coefficient for a series of temperatures. This is not presently possible with the CRESU apparatus as the number of available Laval nozzles is not sufficient to provide an adequate temperature-pressure grid. At 53 K, however, a series of five nozzles with densities of the argon buffer gas ranging from $5.15 \times 10^{16}$ to $8.2 \times 10^{17}$ atoms $cm^{-3}$ have been constructed and used, especially for studies involving the CH radical and reagents such as $H_2$, CO and $N_2$.[66] In these examples evidence of fall-off behaviour has been observed for $H_2$ and CO whereas with $N_2$ the rate coefficients were in the low pressure limit. For larger adducts, the kinetics will proceed in the high pressure limit even at the relatively low pressures available in the CRESU experiments. An example of three-body association will be presented in Section 2.5.5.

### 2.5.2. *The Radical-Radical Reaction O + OH: An Experimental (and Theoretical) Challenge*

The first neutral-neutral reaction to be studied by the CRESU technique, the reaction of CN radicals with $O_2$,[41] is, of course, a radical-radical reaction, as are a number of other reactions subsequently studied in Rennes and in Birmingham using the same technique (CN + NO, CH + NO and

$O_2$, $C_2H + O_2$, $C + O_2$, etc.).[3] However, in all of these studies, one of the radical partners is actually a stable species, either $O_2$ or NO. The establishment of pseudo-first-order conditions posed no particular difficulty in these cases beyond the creation and time resolved detection of the unstable radical species.

If we wish to extend studies to the measurement of reaction rate coefficients between *two* unstable species we are faced with a rather unpalatable choice. In general, radical species are difficult both to produce *and* to quantify. If we are not able to establish pseudo-first-order conditions, the solution of the second-order rate equation will require us to record the time-resolved behaviour of the *absolute* concentrations of *both* reagents. Furthermore, the characteristic reaction time may become much longer, which could preclude CRESU studies. Clearly we would wish to employ pseudo-first-order conditions. However, this requires us to be able to place one of the two unstable reagents in large excess, and to be able to measure its absolute concentration. In experiments using cryogenic cooling, this has been achieved using the discharge flow technique coupled with gas titration to determine the radical concentration. The transfer of such experiments into the CRESU apparatus has not yet been achieved, due to a number of technical difficulties including the much higher gas flows which would necessitate more powerful discharges, the problem of establishing an even distribution of radical species across the reaction zone downstream of the microwave discharge when diffusion in the supersonic flow is negligible, and the difficulty of applying atom titration techniques in a supersonic flow. Theoretical calculations on such systems are also notoriously difficult. The open-shell nature of the two reagents results in multiple potential energy surfaces, and the treatment of reactions occurring on such surfaces is still a subject for much current debate.

Despite these difficulties, radical-radical reactions are of significant interest in low temperature chemistry, owing in part, of course, to their rapidity at low temperatures because of the expected lack of barrier on at least the lowest electronic potential energy surface for reaction. The reaction (R1) of ground state oxygen atoms with ground state OH radicals to yield molecular oxygen and atomic hydrogen serves as a prototypical example:

$$O(^3P_J) + OH(X^2\Pi_\Omega) \rightarrow O_2(X^3\Sigma_g^-) + H(^2S); \quad \Delta_r H_\theta = -67.2\,\text{kJ mol} \qquad \text{(R1)}$$

This reaction has been the target of numerous experimental and theoretical studies over a period of more than 20 years. In the late 1970s and early 1980s, using the discharge flow technique to generate $O(^3P)$ atoms and pulsed laser photolysis and laser-induced fluorescence to generate and observe OH radicals, I.W.M. Smith's group measured the rate coefficient of this reaction down to 250 K, and subsequently in the mid 1990s down to 158 K. A number of other less extensive measurements have also been made, as well as numerous theoretical calculations, and references to most of these studies may be found in Ref. 67. Based on this work, astrochemical modellers had been using, by extrapolation from the lowest temperature values a rate coefficient for this reaction at 10 K in excess of $10^{-10}\,cm^3$ molecule$^{-1}\,s^{-1}$. However, results from satellite observations of dense ISCs — the non detection of $O_2$ by the SWAS (sub-millimetre wave astronomy satellite) and later the tentative detection of very low $O_2$ concentrations by the Odin satellite, for example in the rho Ophiuchi cloud[68] — have resulted in significant interest in reaction R1 as it is thought to be the principal source of $O_2$ in these environments. Viti *et al.*[69] have suggested that an activation energy $E_{act}/R = 80\,K$ would be necessary to explain the low abundance of $O_2$ in ISCs, resulting in a rate coefficient at 10 K around four orders of magnitude below the previously assumed value. Partly in response to this debate it was decided to undertake the difficult task of measuring this rate coefficient at lower temperatures.

A novel VUV co-photolysis technique was chosen to generate the necessary concentrations of ground state atomic oxygen and the low concentrations of OH radicals. Using Laval nozzles designed to work with $N_2$ buffer gas (see below) at temperatures down to 39 K, Carty *et al.*[67] introduced a mixture of known concentrations of $O_2$ and trace quantities of water vapour into the supersonic flow. An $F_2$ excimer laser operating at 157 nm was then used to produce oxygen atoms by photolysis of molecular oxygen, and OH radicals via photolysis of water. At each $O_2$ concentration used, the fluence of the VUV laser was carefully measured *in situ* and this, coupled with a knowledge of the $O_2$ photodissociation cross section at 157 nm, enabled the O atom concentration to be calculated. The photodissociation of $O_2$ at 157 nm produces both $O(^3P)$ and $O(^1D)$ atoms, and so the use of $N_2$ as a buffer gas was essential in order to relax rapidly the excited state $O(^1D)$ atoms to $O(^3P)$. This co-photolysis thus created the necessary pseudo-first-order conditions and the subsequent decay of OH radical concentration was followed by LIF using a second pulsed tuneable dye laser operating at 282 nm. An example of the decay of OH radicals in the presence of oxygen atoms at 47 K is shown in Fig. 2.9 along with the associated second-order plot. The results obtained required correction for expansion of the gas caused by absorption of the photolysis beam, as well as relaxation of vibrationally excited OH by $O_2$ and $N_2$ during the reaction. Once these (relatively minor) corrections had been applied, the rate coefficient for the reaction between $OH(X^2\Pi_\Omega)$ and $O(^3P_J)$ shows no significant variation between 142 and 39 K with a value of $(3.5 \pm 1.0) \times 10^{-11}\,cm^3$ molecule$^{-1}\,s^{-1}$ through this range of temperature. This result compares well with the calculations of Troe and co-workers, as well as the more recent calculations

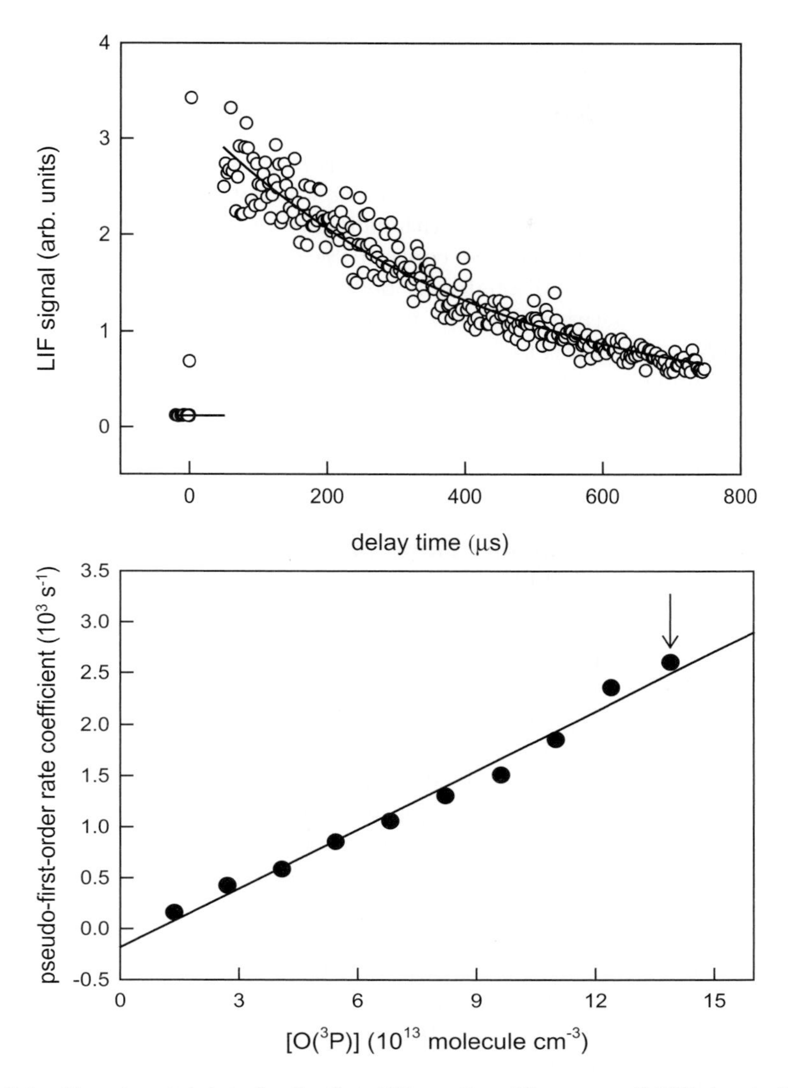


Fig. 2.9. Experimental data for the O + OH reaction: (Upper panel) LIF signals from OH at different delay times between the pulses from the photolysis and probe lasers at 47 K and $[O(^3P)] = 13.9 \times 10^{13}$ molecules $cm^{-3}$. (Lower panel) Plot of the pseudo-first-order rate coefficients for decay of the OH concentration at 47 K plotted against the concentration of $O(^3P)$ atoms, uncorrected for the heating/expansion effect mentioned in the text. The statistical errors are smaller than the size of the points. The arrow indicates the value of $k_{1st}$ derived from the fit shown in the upper panel.

of Honvault and co-workers,[70] and moderately well with the previous measurements of Smith and Stewart at temperatures down to 158 K, and calculations by Guo and co-workers[71] based on an earlier DMBE potential (see Ref. 67 see for further details).

These new measurements on the O + OH reaction have important astrochemical implications. The rate coefficient was not observed to fall as the temperature is lowered to 39 K. Although this temperature remains higher than those found in the coldest cores in ISCs, it appears unlikely that the rate coefficient will fall dramatically between 39 K and those temperatures of 10–20 K. Any activation energy must be appreciably less than the value of 80 K ($E_{act}/R$) that Viti *et al.* suggested would be necessary to explain the low abundance of $O_2$ in ISCs.[69] It would seem that the explanation for the low interstellar abundance of $O_2$ must lie elsewhere than in the slow rate of the reaction between OH radicals and $O(^3P_J)$ atoms at very low temperatures.

### 2.5.3. *Another Challenge: Large Molecules, the Case of PAHs*

PAHs are organic molecules that consist of fused aromatic rings. In their neutral or ionic forms PAHs are believed to be ubiquitous in the Universe.[72] They are thought to be present in cold extraterrestrial media like ISCs and planetary atmospheres as well as in the atmosphere of Titan. If the presence of such molecules is now accepted in various cold media, their form is still unidentified. Large neutral molecules, cations, anions and clusters are possible PAH forms. Their formation is likely to occur in hot environments, such as circumstellar envelopes, or in gas phase media subjected to ionising radiations. Then they can be transported to cold media. In non-ionised media, it is assumed that PAH formation is accomplished through the attachment of the ethynyl radical ($C_2H$) to benzene or acetylene to phenyl. Considering this mechanism and the fact that benzene has been widely detected in the ISM and planetary atmospheres it is expected that relatively small PAHs will also be present. After their formation, these molecules can further react with radicals present, such as CH, OH, $C_2H$ and CN. The addition of the hydroxyl radical on the aromatic ring represents the first step of a complex mechanism leading to the formation of oxygenated compounds. Reactions of $C_2H$ or CH with aromatic species lead to the growth of the carbon skeleton. Reaction of CN radicals with polycyclic aromatic molecules may lead to the formation of polycyclic aromatic nitrogen heterocycles. Although it has yet to be fully established, it is believed[73] that these reactions initiate the complex chemical scheme at the origin of carbon nanoparticle production.

Despite the importance of these reactions there is a lack of experimental studies due to the fact that all PAHs are solids at room temperature with relatively low vapour pressures. Experimental studies involving PAHs require the capability to inject a known density of molecules at low temperature and to avoid condensation. Goulay *et al.*[35] have developed a dedicated CRESU apparatus for cooling PAH molecules in the gas phase to temperatures as low as 58 K in order to study their reaction. In this special CRESU apparatus the reservoir is heated to 470 K by circulating heated oil from a temperature-regulated bath through the double jacketed walls of the reservoir. The aluminium Laval nozzle is embedded in a conical hollow aluminium support mounted on the reservoir and heated by thermal conduction. The PAH molecules are vaporised upstream of the reservoir in an oven, whose temperature is maintained constant in the 500–550 K range. Then the PAH vapour is flushed into the nozzle reservoir by a controlled flow of carrier gas. To avoid condensation downstream of the oven, the gas line between the oven and the reservoir is heated to a higher temperature than the oven. Kinetics measurements have been performed on reactions of anthracene ($C_{14}H_{10}$) with OH and CH radicals. Both radicals were generated in the flow by pulsed laser photolysis of a gaseous precursor and the

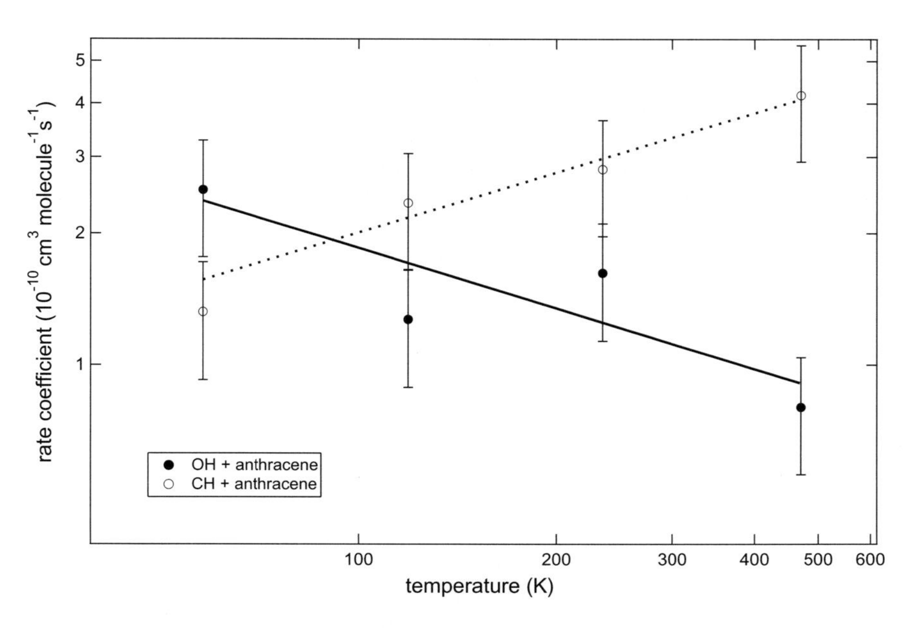


Fig. 2.10. Rate coefficients for reaction of the OH (•) and CH radicals (○) with anthracene. The experimental values are fitted by a power law over the experimental temperature range, full line for the OH reaction and dashed line for the CH reaction.

radical concentration followed by LIF. Results for the 58–470 K temperature range are displayed in Fig. 2.10. In these experiments, the uncertainty of the rate coefficient (30%) is greater than for studies of radical reactions with volatile molecules, because the PAH density in the flow is more difficult to evaluate.

For the reaction of OH with anthracene the rate coefficient increases monotonically as the temperature is lowered and its dependence on temperature can be fitted to the expression $k(T) = 1.12 \times 10^{-10}\ (T/298\,\mathrm{K})^{-0.46}$ between 58 and 470 K in good agreement with indirect and direct measurements at room temperature and higher. It is generally believed that two reaction pathways compete in the reaction of OH with aromatic rings: OH addition to the aromatic ring to form an OH-aromatic adduct, and H-atom abstraction to form a water molecule and a radical. The initially energy-rich OH-aromatic adduct can either decompose back to reagents, be collisionally stabilised or undergo H-elimination, leading to the formation of an alcohol. All channels do not have the same importance and this will depend on temperature. For example, it is known that for benzene, the H-abstraction reaction is slow at room temperature, and becomes an important channel only at temperatures greater than 500–600 K where the adduct decomposes very rapidly back to reagents. The same has been observed for the reaction OH + naphthalene: the abstraction channel becomes important for $T > 410$ K and has a rate which is one order of magnitude smaller than the rate of formation of the adduct, which also mostly decomposes back to the initial reagent. Theoretical calculations on OH + naphthalene indicate that the formation of the adduct is barrierless,[74] and that further reaction of the adduct to form the alcohol has a barrier ranging from 105 to 174 kJ/mol, depending on which isomer is formed. No data are available concerning the reaction with anthracene, but if one supposes that it behaves like benzene and naphthalene, the reaction of anthracene with OH at low temperature proceeds via the formation of the adduct that can be stabilised by the surrounding gas. This is supported by two facts. First, there is not a sharp decrease of the reaction rate coefficient when the temperature increases, as expected if the abstraction channel became dominant. Second, the half-life of the OH-benzene adduct towards decomposition has been estimated to be in the range 5–10 ms.[75] If at least a similar life time is assumed for the anthracene-OH adduct then the adduct should not decompose but will be collisionally stabilised. The magnitude of the rate coefficients, and the negative dependence of the rate coefficients on temperature, provide evidence for the absence of any maximum of electronic potential energy along the minimum energy path leading from separated reagents to the adduct. Furthermore it is expected that the main product at low temperatures will be that displayed, as the most reactive carbon atoms are those in the central ring bound to H-atoms.

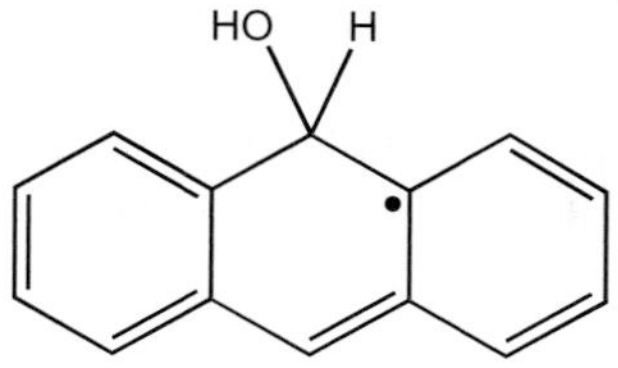

HO H

The rate coefficient for the reaction of CH with anthracene increases monotonically as the temperature is increased from 58 K to 470 K. In this temperature range, the dependence of the rate coefficient on the temperature can be fitted to the expression $k(T) = 3.32 \times 10^{-10}\ (T/298\,\text{K})^{0.46}\ \text{cm}^3$ molecule$^{-1}$ s$^{-1}$. For this reaction, the kinetic behaviour at low temperature is different to that observed for the reaction of CH with linear unsaturated hydrocarbons, the rate coefficients usually increase below room temperature (see for example Fig. 2.7) but then appear to reach a limiting value determined by the long range electrostatic forces between the reagents. The CH + anthracene rate coefficient monotonically decreases when the temperature is lowered. This behaviour may be due to the presence of a small entrance barrier. But the rate coefficient of the reaction remains very high even at low temperature with a value close to the gas kinetic rate coefficient (i.e. the capture rate). On the basis of this behaviour and in the absence of pressure dependence and product studies it is difficult to draw firm conclusions regarding the true mechanism of the reaction. However the reactivity of anthracene with CH at low temperature can be discussed by comparison to other studies.

It is commonly accepted that reactions of CH with unsaturated hydrocarbons proceed without any entrance barrier via the formation of an initial intermediate which decomposes via hydrogen elimination. These reactions are exothermic and show several exit pathways due to the existence of isomers of final products. However there is debate on the exact nature of the reaction mechanism. As for the reactions of CH with acetylene and ethylene, the most likely mechanism seems to be the addition of the CH radical on the carbon multiple bonds to form a 3-carbon-atom cycle. It is believed that this cyclic intermediate isomerises to give linear intermediates which decompose to give the final products. The insertion of the CH radical in the C-H bond of the molecule is not thought to be a favourable entrance channel. For the reactions of CH with aromatic compounds a similar mechanism is expected. With respect to anthracene the observed behaviour of the rate coefficient is usually indicative of a reaction which does not involve a collisionally-stabilised association process as proposed for the reaction of OH with anthracene. It is therefore consistent to assume that the CH + anthracene reaction proceeds by addition of the radical in the $\pi$-system of the molecule followed by H-elimination. However, it is not possible to calculate the exothermicity of the reaction due to the lack of data for the $C_{15}H_{10}$ compound. The main product for the CH reaction with anthracene is expected to be the stable norcaratriene-like compound cyclopropa[b]anthracene:

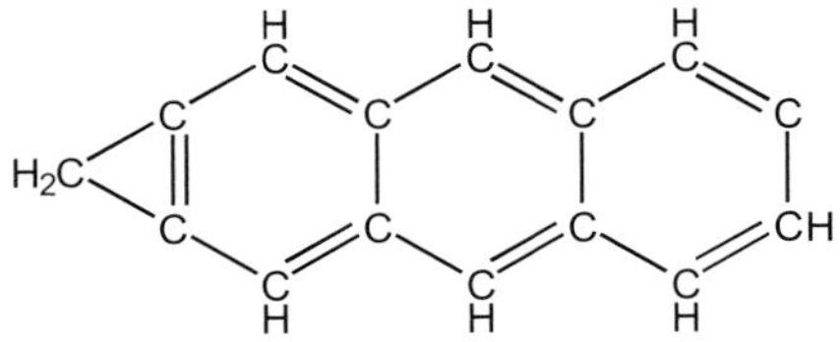

An insertion mechanism would lead to a benzyl like radical that would need to be stabilised. In order to confirm the mechanism of this reaction and the formation of cyclopropa[b]anthracene, *ab initio* calculations are needed.

### 2.5.4. *Rationalising Low Temperature Reactivity for Radical — Molecule Reactions: A Case Study of the Reactions of* $O(^3P)$ *with Alkenes*

As can be appreciated from this review, a substantial body of data (~100 reactions) has now been accumulated on rapid neutral-neutral reactions at very low temperatures (see Tables 2.4 and 2.5). These data have had a significant impact on models of low temperature chemical environments such as dense ISCs and planetary atmospheres. However, the number of potentially important neutral-neutral reactions in such environments is not only much greater that the number studied to date, but it includes many reactions whose rate coefficients are very hard to measure (for example, most reactions between two unstable species). While for many ion-molecule reactions, reliable methods exist for predicting rate coefficients at low temperatures (see Section 2.4), the same cannot be said for neutral-neutral reactions. In most cases, computationally expensive *ab initio* potential energy surface calculations combined with close-coupled quantum dynamics calculations may provide theoretical estimates of rate coefficients but, as at very low temperatures, rate coefficients are extremely sensitive to the presence or absence of very small barriers along the minimum energy path (MEP) leading from reagents to products, the accuracy of even the best calculations is often far from satisfactory. It is clear, therefore, that empirical or semi-empirical methods for estimating whether a reaction is likely to be rapid at *ca.* 20 K would be of great value to modellers of interstellar chemistry.

Motivated by such observations, and a survey of the existing measurements on very low temperature reactivity, I.W.M. Smith *et al.*[76] have proposed two methods for estimating low temperature reactivity. The first, purely empirical, method states that any radical-molecule reaction is likely to be rapid down to 10–20 K if its room temperature rate coefficient, $k(298\,\mathrm{K})$, is greater than *ca.* $5 \times 10^{-12}\,\mathrm{cm^3\,molecule^{-1}\,s^{-1}}$, and if it has an activation energy $E_{act}$ at 298 K and above which is zero or negative. Conversely, reactions with $k(298\,\mathrm{K}) < 5 \times 10^{-12}\,\mathrm{cm^3\,molecule^{-1}\,s^{-1}}$ and $E_{act} > 0$ are likely to become too slow at 10–20 K to be of importance in ISC chemistry. This 'rule of thumb' appears to work well for the majority of low temperature reactions studied to date. However, there are some notable exceptions. The strange behaviour of the CN + $C_2H_6$ reaction[77] is a case in point. This reaction has a relatively rapid room temperature rate coefficient (*ca.* $3\times10^{-11}\,\mathrm{cm^3\,molecule^{-1}\,s^{-1}}$), greater than *ca.* $5 \times 10^{-12}\,\mathrm{cm^3\,molecule^{-1}\,s^{-1}}$, but displays a *positive* activation energy of *ca.* $2\,\mathrm{kJ\,mol^{-1}}$ at and above room temperature. However, low temperature measurements in the CRESU apparatus showed that its rate coefficient passes through a minimum, rising to its fastest measured value of $1.1 \times 10^{-10}\,\mathrm{cm^3\,molecule^{-1}\,s^{-1}}$ at 25 K (see Fig. 2.6c).

Various explanations have been proposed to account for this behaviour, but it is now generally agreed that it can be understood in terms of a two transition state model[78] as first proposed[79] by Klippenstein and co-workers for the reaction OH+$C_2H_4$. Briefly, the MEP for radical-molecule reactions can be represented schematically as in Fig. 2.11: at long range, attractive

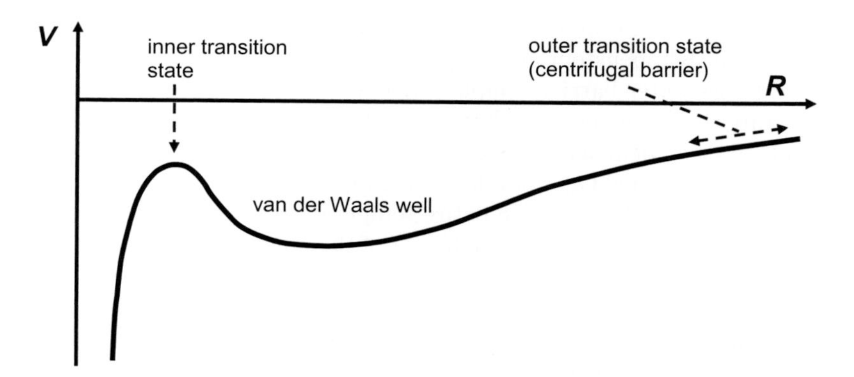


Fig. 2.11. Schematic diagram of the minimum energy path for the reaction of a radical and a molecule in the two transition state model of Georgievskii and Klippenstein.[78]

(van der Waals and other electrostatic) forces dominate, and the potential falls with decreasing separation between the reagents. Centrifugal barriers arise on this long-range part of the potential as a result of the need to conserve angular momentum. These barriers correspond to an outer, loose transition state. Such barriers also exist of course in the case of radical-radical reactions, giving rise to the now well-established negative temperature dependence of such reactions. At shorter range, however, an avoided curve crossing between reagent ionic and covalent states gives rise to a definite maximum in the intermolecular potential along the MEP, resulting in an inner barrier, which corresponds to a tight transition state. This inner transition state may be above the energy of the separated reagents, at which point it corresponds to an absolute or 'true' barrier along the MEP resulting in classic activated behaviour and a negligible rate coefficient at 20 K. But in many cases (as illustrated in Fig. 2.11), this inner transition state is submerged below the energy of the separated reagents, at which point the outer, loose transition state controls the reaction, resulting in a similar negative temperature dependence to that seen for radical-radical reactions, with a rate coefficient in excess of $10^{-10}\,\text{cm}^3\,\text{molecule}^{-1}\,\text{s}^{-1}$ at *ca.* 20 K. However, if the inner barrier is only just submerged below the energy of the separated reagents, as seems to be the case for $CN + C_2H_6$, an interesting situation can arise. At low temperatures, and consequently low average total angular momentum for collision, the outer barrier controls the reaction, resulting in a negative temperature dependence. As the temperature rises, so does the average total angular momentum for collision. At the inner barrier (corresponding to a 'tight' transition state), the internal states at this smaller inter-reagent separation are more widely spaced than at the outer, 'loose', transition state. This results in the effective appearance of a real barrier to reaction and a transition towards classic Arrhenius-type positive temperature dependence of the rate coefficient. In the approach pioneered by Klippenstein and co-workers[78,79] a form of capture theory is applied to the outer barrier, coupled with a version of microcanonical transition state theory at the inner barrier in order to calculate the temperature dependent rate coefficient. Coupled with high level *ab initio* calculations this approach enabled Georgievskii and Klippenstein to model the rate coefficient for $CN + C_2H_6$ over the entire temperature range 25–1000 K.[78]

It should now be clear that the height of the inner barrier is likely to play a key role in providing reliable estimates of low temperature rate coefficients for radical-molecule reactions. Donahue and co-workers have demonstrated that not only does the inner barrier in radical-molecule reactions arise from an avoided curve crossing between reagent ionic and covalent states, but that both the energy and other transition state properties are strongly correlated to reagent ionic energy.[80] On this basis, coupled to an analysis of a large number of radical-unsaturated molecule reactions studied to date at very low temperatures, I.W.M. Smith *et al.*, proposed a second, semi-empirical method for deciding whether rate coefficients would be substantial at the temperature of ISCs,[76] whereby, when the difference between the ionisation energy of the molecular reagent (I.E.) and the electron affinity of the radical (E.A.), (I.E. − E.A.), is significantly greater than 8.75 eV, the reaction is likely to possess a 'true' barrier and will therefore become negligibly slow at 20 K. On the other hand, reactions with (I.E. − E.A.) $< \sim 8.75$ eV are likely to be characterised by, at most, inner barriers that are submerged below the asymptotic reagent energy, and therefore be rapid at 20 K and of potential importance in ISC chemistry. Smith *et al.*, further pointed out that the reactions between ground state oxygen atoms, $O(^3P)$, and simple unsaturated hydrocarbons possess values of (I.E. − E.A.) that bridge the 'critical' value of 8.75 eV (see Table 2.6) and would therefore provide an important test of their hypothesis.

Motivated by this proposal, Sabbah *et al.*[81] performed low temperature measurements of the rate coefficients for these reactions. $O(^3P)$ atoms were generated in the cold CRESU flow via 355 nm laser photolysis of $NO_2$, and their concentration followed using chemiluminescence from excited $NO_2$, formed in the association of $O(^3P)$ with NO, as a 'marker' for the oxygen atom concentrations, which decayed exponentially as a result of reaction with the added alkene. Results were obtained in most cases down to 23 K or 27 K, except for the $O(^3P)$ + ethene reaction, for which only an upper limit to the rate coefficient could be measured at 39 K.

Sabbah *et al.*[81] also reported calculations of the same sort as described above for $CN + C_2H_6$, employing second-order multi-reference perturbation theory (CASPT2) to determine the inner barrier heights (the results are shown in Table 2.6). For O + propene, the CASPT2 predicted barrier was

Table 2.6. Comparison of rate coefficients, $k(298\,K)$, and activation energies, $E_{act}$, from Cvetanovic[82] with (I.E. − E.A.) for the reactions of $O(^3P)$ atoms with alkenes (the electron affinity of the $O(^3P)$ atoms is 1.46 eV). Also given are $E_{\text{inner}}$, CASPT2 energies (before adjustment) on the minimum energy path (all relative to the energy of the separated reagents in their ground state) for electronic + zero-point energy at the inner transition state. If there were multiple addition sites, only the lowest energy is given here: for full information, the reader is referred to the article by Sabbah *et al.*[81]

| | ethene | propene | 1-butene | *cis*-butene | *iso*-butene | *trans*-butene |
|---|---|---|---|---|---|---|
| $k(298\,K)/10^{-11}$ $cm^3\,molecule^{-1}\,s^{-1}$ | 0.073 | 0.40 | 0.42 | 1.8 | 1.7 | 2.2 |
| $E_{act}$ / kJ mol$^{-1}$ | 6.65 | 2.3 | 2.9 | −1.2 | −0.1 | −0.1 |
| (I.E. − E.A.)/eV | 9.05 | 8.27 | 8.09 | 7.65 | 7.76 | 7.64 |
| $E_{inner}$ / kJ mol$^{-1}$ | 5.4 | 0.9 | −0.5 | −2.2 | −2.5 | −3.1 |

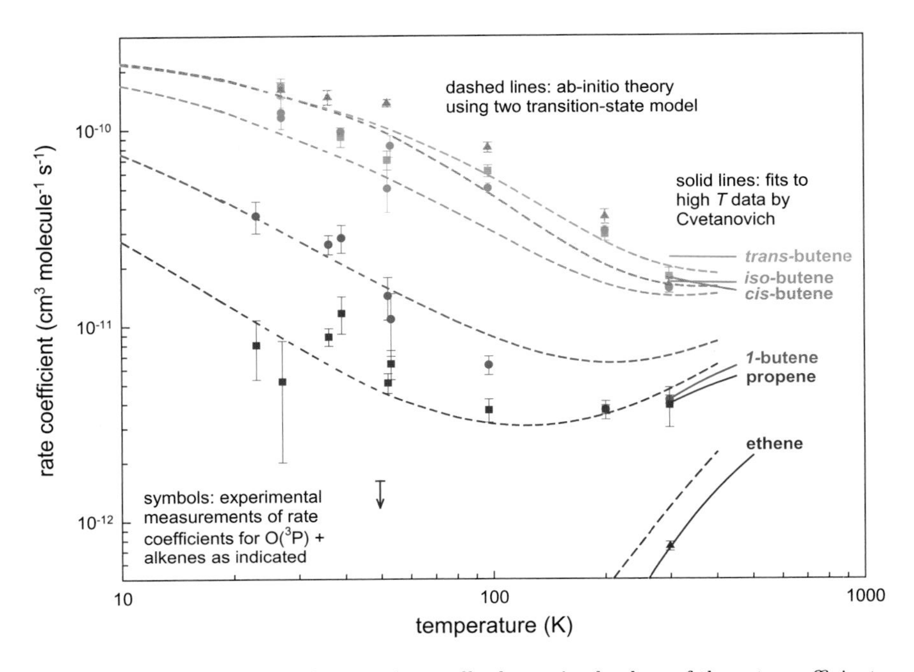


Fig. 2.12. The points show the experimentally determined values of the rate coefficients for the reactions of $O(^3P)$ atoms with alkenes at different temperatures, whilst the dashed lines show the results of the theoretical calculations described in the text.[81] The solid lines to the right of the diagram represent the Arrhenius expressions recommended by Cvetanovic[82] to fit kinetic data between 300 and 700 K.

adjusted downwards by $0.6\,\mathrm{kJ\,mol^{-1}}$ to obtain optimum agreement with experiment. For consistency, this adjustment was applied to the other reactions as well. For the butene cases, this adjustment had an insignificant effect on the predictions. The experimental low temperature data, high temperature literature data from Cvetanovich[82] and the theoretical calculations are all displayed in Fig. 2.12. The experimental results clearly fall into the three categories alluded to above: $O(^3P)$ + *cis-*, *trans-* and *iso*-butene possess fully submerged barriers, and display rapid rate coefficients with a negative temperature dependence at all temperatures. $O(^3P)$ + ethene possesses a 'true' non-submerged barrier, and its rate coefficient falls below the minimum value measurable in the CRESU of *ca.* $10^{-12}\,\mathrm{cm^3\,molecule^{-1}\,s^{-1}}$ at 39 K. $O(^3P)$ + 1-butene and $O(^3P)$ + propene appear to fall into the category of a barely submerged inner barrier, and display the characteristic minimum first observed at low temperatures for the $CN + C_2H_6$ reaction. In all cases the calculations achieve a remarkably good agreement with experiment. In particular, the semi-empirical predictions of I.W.M. Smith *et al.*[76] are strongly supported by these data. All of the reactions possessing (I.E. – E.A.) $< 8.75$ eV are found to be rapid at low temperatures, and the two cases showing a turnaround from positive to negative temperature dependence as the temperature falls are identified by their values of (I.E. – E.A.) only slightly below this threshold (8.09 and 8.27 eV for $O(^3P)$ + 1-butene and $O(^3P)$ + propene respectively).

To what extent can these predictions be applied more widely? It is interesting to compare the value of (I.E. – E.A.) for the reaction $O(^3P)$ + cis-butene, 7.65 eV, with that for the reaction $CN + C_2H_6$, 7.66 eV. The former clearly has a fully submerged barrier, while the latter's kinetic behaviour places it into the same category of a barely submerged barrier displayed by the $O(^3P)$ + 1-butene and $O(^3P)$ + propene reactions. Clearly, the rule is less than perfect outside of an homologous series of reagents such as $O(^3P)$ + alkenes, but it still should prove very valuable for reactions where (I.E. – E.A.) is either much greater than 8.75 eV (e.g. $CN+H_2$ (I.E. – E.A.) $=$ 11.56 eV, negligibly slow at low temperatures), or much less than this value (e.g. $CN+C_2H_2$ (I.E. – E.A.) $=$ 6.65 eV, rapid at low temperatures[77]). Furthermore, Sabbah *et al.*, propose a refinement to the rules of I.W.M. Smith *et al.*, compare theoretical and experimental values of the room temperature rate coefficient in order to estimate the barrier height at the inner transition state, and when this barrier is submerged one can reasonably expect the reaction to be rapid at the low temperatures of ISCs. However, the most important conclusion that can be drawn is that both the empirical and semi-empirical rules proposed by Smith *et al.*, and based respectively on kinetic behaviour at room temperature and above, and on the values of (I.E. – E.A.) appear to perform well and should provide a reasonably firm basis for predicting the low temperature rate coefficients for (as yet) unmeasured reactions.

### 2.5.5. *The Onset of Condensation: Dimer Formation in the CRESU Experiment*

The rate and mechanism by which gas phase individual molecules coalesce to form liquid or solid macroscopic particles are both very important and poorly understood. For a supersaturated gas, classical nucleation theory proposes that homogeneous condensation occurs via the formation of spherical droplets and defines a critical radius at which free energy begins to decrease with size.[83] The free energy at this critical size can be viewed as an energy barrier to the nucleation process. At temperatures well below that of liquid nitrogen most gases are in a condition of heavy supersaturation. It is clear that in this case classical nucleation theory no longer applies, as the size of the critical nucleus moves towards molecular size, and hence can no longer be treated as a macroscopic object. The barrier to nucleation arises from the fact that the negative entropy change $\Delta S$ for the addition of a monomer to a cluster can lead to positive values of the free energy change $\Delta G = \Delta H - T\Delta S$, up to a given size of the molecular cluster, despite the exothermic nature of the reaction. The formation of the dimer in these conditions will be inhibited by a fast re-dissociation. However, below a given temperature the $T\Delta S$ term will be smaller than the absolute enthalpy change at any size for many gases and therefore there is no longer a thermodynamic barrier to dimer formation and nucleation. Condensation is then entirely controlled by the kinetic rate at which dimers and higher oligomers form.

To our knowledge, despite considerable interest in the problem of nucleation and particle formation, there are no direct measurements of rate coefficient for dimer formation reported in the literature, except that reported by Hamon *et al.*[84] concerning benzene dimer formation using the CRESU technique. Benzene was chosen in this experiment because the monomer is easily detected by LIF in its intense $6_0^1 1_0^1$ band in the $^1B_{2u} \leftarrow ^1A_{1g}$ electronic transition. The aim of this study was to determine second-order rate coefficients for the dimerisation of benzene monomers at the temperature and helium densities provided by a variety of Laval nozzles. Benzene monomer concentrations were monitored by LIF versus benzene flow rate at various distances from the nozzle exit. At low concentrations, the increase of the LIF signal is perfectly linear with benzene flow rate, but increasing the flow rate leads to a saturation of the signal and even to a decrease at the highest flow rate. The saturation effect is stronger when moving downstream of the nozzle exit. These observations were attributed to the onset of condensation, which was confirmed by various observations.[84] Therefore, measurements were performed under conditions such that the loss of benzene monomer did not exceed 20–30%, so that processes involving the formation of higher oligomers could be neglected. Under these conditions, the benzene concentration $[C_6H_6]$ follows the very simple equation:

$$-\frac{d[C_6H_6]}{dx} = 2k_{2nd}\frac{[C_6H_6]^2}{v} \tag{2.15}$$

where $x$ is distance along the flow, $v$ the velocity and $k_{2nd}$ the second-order rate coefficient for dimer formation. A plot of the reciprocal of benzene monomer concentration versus distance yields directly $k_{2nd}$.

This study was conducted only with helium as a buffer gas in order to avoid possible problems resulting from a chaperone mechanism, which could be possible in argon or nitrogen. The behaviour of the third-order rate coefficient, which exhibits a strong negative temperature dependence

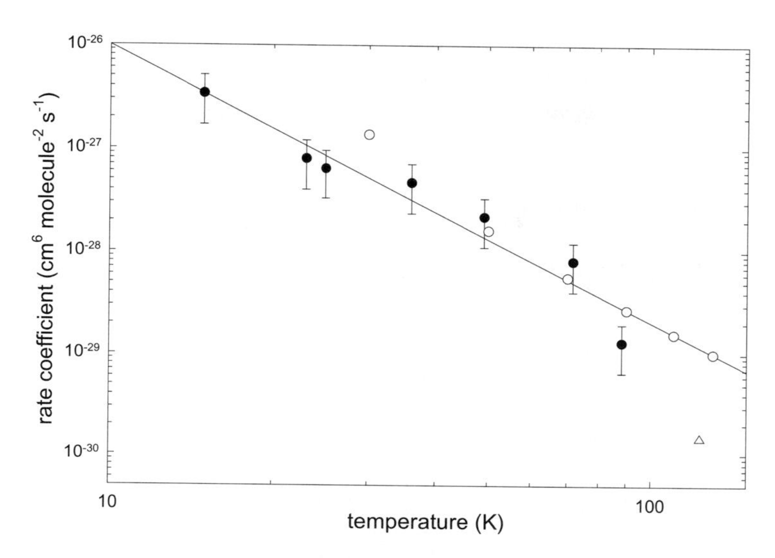


Fig. 2.13. The solid circles show the variation with temperature of the experimental effective third-order rate coefficients (i.e. $k_{2nd}/[He]$) of the benzene dimerization. The open triangle for $T = 123.0$ K represents an upper limit to the rate coefficient at this temperature. The open circles show the calculated third-order rate coefficients.

of $T^{-2.8}$, is presented in Fig. 2.13. Also shown on the figure are the results of a calculation of the limiting low pressure rate coefficient, estimated following a procedure proposed by Troe.[53] The agreement with experiment is remarkably good but given the number of approximations surely fortuitous and must be taken essentially as a check that the experimental results look reasonable. It has also to be kept in mind that at the lowest temperatures available in this experiment the binary rate coefficients approach the capture values and therefore must be well into the fall-off regime.

It is worth noting that at 123 K, the highest temperature at which measurements were attempted, it was impossible to observe any saturation of the monomer signal, even at the highest benzene flow rate. Therefore only an upper limit of the rate coefficient was deduced. This inability to measure any loss of the monomer was attributed to the strong onset of the reverse dissociation reaction due to a large change in the free energy. From these assumptions it is possible to estimate a value of the binding energy of the dimer of approximately 12 kJ mol$^{-1}$ compared with theoretical values ranging from 7.1 to 13.9 kJ mol$^{-1}$.[84]

Most recently in Rennes, the use of time of flight mass spectrometry techniques has allowed us to extend this kind of measurement to PAHs. Pyrene exhibits a behaviour very close to that of benzene, except that the temperature at which no dimerisation occurs is higher. Such results are potentially of great importance for the understanding of soot formation in combustion.

## 2.6. Future Developments

### 2.6.1. *Charged Species*

The very recent discovery of polyyne radical anions $C_nH^-$ ($n = 4$, 6 and 8) in the ISM has excited the astrochemical community and it is clear that there is now a real desire to include electron attachment reactions in the models based on reliable laboratory data. The field is completely virgin at the temperatures of dense ISCs and present developments of the CRESU technique should allow us to tackle this interesting topic. One of the main goals will be to study the electron attachment of PAHs using a cold source of electrons (see Section 2.4.4). By coupling laser and mass spectrometry techniques it should be then possible to study attachment to radicals such as those observed in the ISM, including smaller ones such as CN. Mastering the production of cold electrons in the CRESU supersonic flow could also open up the possibility of measuring thermal rate coefficients for the dissociative recombination of polyatomic ions at temperatures below 77 K, which will be of considerable interest for interstellar chemistry.

With respect to ion-molecule reactions, the development of pulsed supersonic flows in Rennes (see Section 2.6.5) will also offer the possibility of achieving very low pressures in the supersonic flow and therefore will open the way to a revival of the ion-molecule CRESU apparatus by the use of a selective injection of ions which will be directly derived from expertise gained previously.[36,37] Studies at temperatures close to 1 K will be a decisive test in comparing present collision theories and are obviously of major importance.

### 2.6.2. *Simple Radical-Neutral Reactions Including Radical-Radical Processes*

Most of the presently available studies concern small systems — i.e. atomic radical or simple molecular reagents (see Table 2.4) — but there are still several aspects that remain unexplored. As a case in point, the chemistry of sulfur is completely unknown at very low temperatures, although molecules such as SO or $SO_2$ are routinely used to trace shocks in the ISM. The reactivity of sulfur atoms is certainly of interest in the production of more complex sulfur-bearing molecules. The chemistry of metastable atoms is also important in our atmosphere and in other planetary atmospheres. Only a few data relating to the reactivity of $O(^1D)$ are available at temperatures close to 200 K. Extension to the lower temperatures of Titan should be valuable as well as determining rate coefficients for other atoms such as $N(^2D)$, $C(^1D)$ or $S(^1D)$. A much more difficult task is the experimental study of radical-radical reactions. Future work may focus on reactions such as N + OH, O + CN, O + CH, or simpler ones such as N + NO, which are all of particular interest in interstellar chemistry.

### 2.6.3. *Increasing the Complexity*

Another important aspect revealed by modern observational astronomy is the complexity of the existing interstellar molecules. It is important to understand how this complexity may be generated and to what level it can be brought. Although PAHs have not yet been directly observed in the ISM, it is now generally accepted that they are widely present. Their reactivity with other molecules is then of particular interest because insertion mechanisms could contribute to the building up of new species containing an increasing number of aromatic cycles. The study of dimerisation and eventually nucleation of PAHs can also give new insights in the generation of nanoparticles and by extension interstellar dust. Finally, reactions of simple stable $C_2$, $C_3$ and $C_4$-based hydrocarbons with radicals should also be completed although a substantial body of data does already exist.

The pioneering work of Goulay *et al.*[35] on a heated CRESU apparatus has paved the way for the study of very large molecules in the gas phase at temperature below that of liquid nitrogen. The recent implementation of a time of flight mass spectrometer on the heated CRESU chamber has allowed large molecules such as PAHs to be detected. The dimerisation of pyrene is currently under study. There is a complete lack of data for reactions of large radical species with large molecules in the gas phase, at low temperatures and above. This new apparatus will allow us to measure rate coefficients for such reactions.

### 2.6.4. *Product Detection*

Measuring rate coefficients is only a partial answer to the understanding of a chemical process. Obviously, identification of the products and the quantification of the branching ratios are vital as well. This aspect is much more difficult to study using the available experimental techniques dedicated worldwide to the investigation of gas phase kinetics. Although products can be identified by several methods such as LIF or mass spectrometry the determination of their concentration is not usually possible, except in some 'favourable' cases. For example, reactions for which only one exit channel is open can be used as references to determine the product branching ratio of another process. This idea has been used in the groups of Seakins[85] and Loison[86] to obtain the branching ratios of channels producing hydrogen atoms. Calibration reactions such as $CN+H_2 \rightarrow HCN+H$, $CH + CH_4 \rightarrow C_2H_4 + H$ or $C + C_2H_4 \rightarrow C_3H_3 + H$ can be used where H atoms are detected by VUV LIF. These studies however are only available at room temperature and it would be worthwhile to extend them to interstellar temperatures.

The development of infrared lasers opens new possibilities for the determination of branching ratios. Multipass cells can be coupled to infrared light to detect molecular radicals by absorption techniques and quantify their concentration. Using this method, Hershberger and his co-workers obtained a significant amount of results at room temperature and above.[87] One of the major problems here lies in the vibrational population of the product molecules which may not be in equilibrium, thus perturbing the infrared absorption measurements. This has been solved by using $SF_6$ as a relaxant, a technique which is inapplicable to CRESU supersonic flows because of potential polymerisation of large concentrations of $SF_6$ during the expansion.

A more promising technique lies in the photoionisation of products at threshold using VUV light and collection of the ions by a time of flight mass spectrometer. VUV can be generated by non-linear mixing laser techniques. However, this method is rather limited because of the restricted tunability of the generated light. Furthermore, the photon flux is generally rather small. A more appropriate tool for this kind of study is light produced from a synchrotron. Recent studies have been undertaken at room temperature coupling a slow flow cell reactor to the ALS synchrotron in Berkeley.[88] By scanning the VUV light it is possible to detect products according to their ionisation potential which allows one to identify isomers. Provided that ionisation cross sections are known, the relative concentration of the products and therefore the branching ratios can be extracted from the experiment. Using a reduced version of the CRESU apparatus, it could be possible to couple it to a synchrotron (e.g. SOLEIL in France) and then perform isomer sensitive product detection at low temperature.

### 2.6.5. *Pulsed Laval Nozzles: Toward Absolute Zero?*

Although most of the applications of low temperature chemistry to natural environments lie in the 10–300 K range, it would be of considerable interest from a theoretical point of view to extend the temperature range toward the 0 K limit. Indeed, theoretical predictions indicate that peculiar behaviours of the rate coefficients below 10 K should occur. For example, capture theories predict a maximum of the rate coefficient close to that temperature and then a significant decrease for colder conditions. However, the ability to study real chemical reactions in the gas phase at temperatures well below 10 K remains a considerable challenge. The lowest temperature achieved in this field, 0.3 K, was for reactions of atomic ions with atoms in the free jet flow reactor of Smith[12] and have not been extended further. For reasons discussed earlier in this chapter, the technique cannot be employed for the study of neutral-neutral chemistry.

There are no fundamental reasons why the CRESU technique could not be extended towards temperatures closer to absolute zero. However chemistry requires collisions and it is clear that the condensation of the helium buffer will then fix the ultimate limit of the temperature achieved by this technique. Values of the vapour pressure of helium show that it is not unreasonable that this limit will be a fraction of Kelvin. To obtain such low temperatures a very high Mach number at the exit of the Laval nozzle will be necessary which will require, in turn, a very large pressure ratio between the nozzle reservoir and the experimental chamber. If one wants to avoid problems of cluster formation during the expansion, a moderate reservoir pressure is required. This implies a very low pressure in the experimental chamber, i.e. around $10^{-4}$ mbar. As mentioned above, since the mass flow rate in the Laval nozzle has to be kept sufficiently large to obtain good flow conditions, it is clear that the size of a continuous apparatus and especially the pumping capacity will be unreasonably large. A way to overcome this problem would be to cool down the reservoir to liquid helium temperatures. Since the viscosity is greatly reduced in this case, the requirement for mass flow rate will be much less (at least an order of magnitude lower). However, cooling down the reservoir with liquid helium has a great disadvantage: most gases will instantaneously freeze on its walls. Although there are some ways to overcome this problem, it is presently thought that the use of a pulsed CRESU remains the best way to overcome the pumping capacity problem. It is clear that the size of the apparatus will remain very large, as flow rate and pressure fix it in any mode, continuous or pulsed, but is not intractable (the diameter of the nozzle will be around one metre). Finally a difficult problem of fluid mechanics will have to be solved for the design of the Laval nozzle: the core of the flow will remain in the continuous regime since the mean free path stays small despite the low pressure resulting from the extremely low temperature, but the boundary layer which is hotter will be in an intermediate case close to the molecular free regime for which the Navier Stokes equations that describe the CRESU supersonic expansions will no longer be valid. A special technique of pulsing the flow is under development in the laboratory of the University of Rennes 1 in order to achieve such a goal. It is anticipated that it will allow the temperature range of the CRESU to be extended down close to 1 K. It will also be advantageous at any temperature, as the reduction in the mean flow rate will enable expensive gases, such as deuterated isotopologues, to be used.

## 2.7. Conclusions

The experimental advances detailed in this chapter have enabled a significant body of experimental data on rate coefficients at very low temperatures to be built up over the last 15 years. However, as has been pointed out, a considerable amount of work remains. The field of gas kinetics at low temperature is very active and still possesses many unexplored areas. The development of new technologies will, no doubt, enable experimentalists to push the limits even further, and exciting results are to be expected in the near future.

## References

1. Smith IWM. (2003) Laboratory studies of atmospheric reactions at low temperatures. *Chem. Rev.* **103**: 4549–4564.
2. Smith IWM. (2002) The Liversidge Lecture 2001–2002. Chemistry amongst the stars: Reaction kinetics at a new frontier. *Chem. Soc. Rev.* **31**: 137–146.
3. Smith IWM. (2006) Reactions at very low temperatures: Gas kinetics at a new frontier. *Angew. Chem. Int. Ed.* **45**: 2842–2861.
4. Smith IWM, Rowe BR. (2000) Reaction kinetics at very low temperatures: Laboratory studies and interstellar chemistry. *Acc. Chem. Res.* **33**: 261–268.
5. Smith IWM, Herbst E, Chang Q. (2004) Rapid neutral-neutral reactions at low temperatures: A new network and first results for TMC-1. *Mon. Not. R. Astron. Soc.* **350**: 323–330.
6. Althorpe SC, Clary DC. (2003) Quantum scattering calculations on chemical reactions. *Annu. Rev. Phys. Chem.* **54**: 493–529; Harding LB, Maergoiz AI, Troe J, Ushakov VG. (2000) Statistical rate theory for the $HO+O \leftrightarrow HO_2 \leftrightarrow H + O_2$ reaction system: SACM/CT calculations between 0 and 5000 K. *J. Chem. Phys.* **113**: 11019–11034.
7. Troe J. (1996) Statistical adiabatic channel model for ion-molecule capture processes. 2. Analytical treatment of ion-dipole capture. *J. Chem. Phys.* **105**: 6249–6262.
8. Su T, Chesnavich WJ. (1982) Parametrization of the ion-polar molecule collision rate-constant by trajectory calculations. *J. Chem. Phys.* **76**: 5183–5185.
9. Troe J, Lorquet JC, Manz J, Marcus RA, Herman M. (1997) Recent advances in statistical adiabatic channel calculations of state-specific dissociation dynamics. *Chemical Reactions and Their Control on the Femtosecond Time Scale XXth Solvay Conference on Chemistry,* Vol. 101, pp. 819–851.

10. Maergoiz AI, Nikitin EE, Troe J. (1996) Adiabatic channel study of the capture of nitrogen and oxygen molecules by an ion: Effect of nuclear symmetry and spin-spin interaction. *Z. Phys. D: At. Mol. Clusters.* **36**: 339–347.
11. Herbst E, Klemperer W. (1973) The formation and depletion of molecules in dense interstellar clouds. *Astrophys. J.* **185**: 505–533.
12. Smith MA, Ng CY, Baer T, Powis I. (1994) Ion molecule reaction dynamics at very low temperatures. *Unimolecular and Bimolecular Reaction Dynamics,* pp. 183–251. John Wiley & Sons Ltd, New York.
13. Chutjian A, Garscadden A, Wadehra JM. (1996) Electron attachment to molecules at low electron energies. *Phys. Rep.* **264**: 393–470.
14. Randeniya LK, Smith MA. (1990) A study of molecular supersonic flow using the generalized Boltzmann equation. *J. Chem. Phys.* **93**: 661–673.
15. Dupeyrat G, Marquette JB, Rowe BR. (1985) Design and testing of axisymmetric nozzles for ion molecule reaction studies between 20 K and 160 K. *Phys. Fluids.* **28**: 1273–1279.
16. Sims IR, Queffelec JL, Defrance A, Rebrion-Rowe C, Travers D, Rowe BR, Smith IWM. (1992) Ultralow temperature kinetics of neutral-neutral reactions: The reaction $CN + O_2$ down to 26 K. *J. Chem. Phys.* **97**: 8798–8800.
17. James PL, Sims IR, Smith IWM. (1997) Total and state-to-state rate coefficients for rotational energy transfer in collisions between $NO(X^2\Pi)$ and He at temperatures down to 15 K. *Chem. Phys. Lett.* **272**: 412–418.
18. Daugey N, Caubet P, Retail B, Costes M, Bergeat A, Dorthe G. (2005) Kinetic measurements on methylidyne radical reactions with several hydrocarbons at low temperatures. *Phys. Chem. Chem. Phys.* **7**: 2921–2927.
19. Atkinson DB, Smith MA. (1995) Design and characterization of pulsed uniform supersonic expansions for chemical applications. *Rev. Sci. Instrum.* **66**: 4434–4446.
20. Lee S, Hoobler RJ, Leone SR. (2000) A pulsed Laval nozzle apparatus with laser ionization mass spectrometry for direct measurements of rate coefficients at low temperatures with condensable gases. *Rev. Sci. Instrum.* **71**: 1816–1823.
21. Spangenberg T, Kohler S, Hansmann B, Wachsmuth U, Abel B, Smith MA. (2004) Low-temperature reactions of OH radicals with propene and isoprene in pulsed Laval nozzle expansions. *J. Phys. Chem. A.* **108**: 7527–7534.
22. Bohringer H, Arnold F. (1983) Studies of ion molecule reactions, ion mobilities, and their temperature-dependence to very low-temperatures using a liquid-helium-cooled ion drift tube. *Int. J. Mass Spectrom Ion Processes* **49**: 61–83.
23. Barlow SE, Dunn GH, Schauer M. (1984) Radiative association of $CH_3^+$ and $H_2$ at 13 K. *Phys. Rev. Lett.* **52**: 902–905.
24. Rowe BR, Marquette JB. (1987) CRESU Study of ion-molecule reactions. *Int. J. Mass Spectrom Ion Processes* **80**: 239–254.
25. Hawley M, Mazely TL, Randeniya LK, Smith RS, Zeng XK, Smith MA. (1990) A free jet flow reactor for ion molecule reaction studies at very low energies. *Int. J. Mass Spectrom Ion Processes* **97**: 55–86.

26. Smith MA, Hawley M. (1992) Ion chemistry at extremely low temperatures: a free Jet expansion approach. *Adv. Gas Phase Ion Chem.* **1**: 167–202.
27. Gerlich D. (1993) Experimental investigations of ion-molecule reactions relevant to interstellar chemistry. *J. Chem. Soc, Faraday Trans.* **89**: 2199–2208.
28. Smith D, Adams NG, Alge E. (1984) Attachment coefficients for the reactions of electrons with $CCl_4$, $CCl_3F$, $CCl_2F_2$, $CHCl_3$, $Cl_2$ and $SF_6$ determined between 200 K and 600 K using the FALP technique. *J. Phys. B: At. Mol. Opt. Phys.* **17**: 461–472.
29. Balucani N, Capozza G, Leonori F, Segoloni E, Casavecchia P. (2006) Crossed molecular beam reactive scattering: from simple triatomic to multichannel polyatomic reactions. *Int. Rev. Phys. Chem.* **25**: 109–163.
30. Naulin C, Costes M. (1999) Crossed beam study of the $Al(^2P_{1/2,3/2})$ + $O_2(X^3\Sigma_g^-) \rightarrow AlO(X^2\Sigma^+) + O(^3P_J)$. *Chem. Phys. Lett.* **310**: 231–239.
31. Geppert WD, Reignier D, Stoecklin T, Naulin C, Costes M, Chastaing D, Le Picard SD, Sims IR, Smith IWM. (2000) Comparison of the cross sections and thermal rate constants for the reactions of $C(^3P_J)$ atoms with $O_2$ and NO. *Phys. Chem. Chem. Phys.* **2**: 2873–2881.
32. Clary DC, Buonomo E, Sims IR, Smith IWM, Geppert WD, Naulin C, Costes M, Cartechini L, Casavecchia P. (2002) $C + C_2H_2$: A key reaction in interstellar chemistry. *J. Phys. Chem. A.* **106**: 5541–5552.
33. Bejan A. (1993) *Heat Transfer.* John Wiley & Sons, New York.
34. James PL, Sims IR, Smith IWM. (1997) Rate coefficients for the vibrational self-relaxation of $NO(X^2\Pi, v = 3)$ at temperatures down to 7 K. *Chem. Phys. Lett.* **276**: 423–429.
35. Goulay F, Rebrion-Rowe C, Biennier L, Le Picard SD, Canosa A, Rowe BR. (2006) The reaction of anthracene with CH radicals: An experimental study of the kinetics between 58 K and 470 K. *J. Phys. Chem. A.* **110**: 3132–3137; Goulay F, Rebrion-Rowe C, Le Garrec JL, Le Picard SD, Canosa A, Rowe BR. (2005) The reaction of anthracene with OH radicals: an experimental study of the kinetics between 58 K and 470 K. *J. Chem. Phys.* **122**: 104308.
36. Rowe BR, Marquette JB, Rebrion C. (1989) Mass selected ion-molecule reactions at very low temperatures: The CRESUS apparatus. *J. Chem. Soc. Faraday Trans. II* **85**: 1631–1641.
37. Speck T, Mostefaoui T, Travers D, Rowe BR. (2001) Pulsed injection of ions into the CRESU experiment. *Int. J. Mass Spectrom* **208**: 73–80.
38. Le Garrec JL, Sidko O, Queffelec JL, Hamon S, Mitchell JBA, Rowe BR. (1997) Experimental studies of cold electron attachment to $SF_6$, $CF_3Br$ and $CCl_2F_2$. *J. Chem. Phys.* **107**: 54–63.
39. Goulay F, Rebrion-Rowe C, Carles S, Le Garrec JL, Rowe BR. (2004) The electron attachment rate coefficient of HI and DI in uniform supersonic flow. *J. Chem. Phys.* **121**: 1303–1308.
40. Sims IR, Smith IWM. (1988) Rate constants for the radical-radical reaction between CN and $O_2$ at temperatures down to 99 K. *Chem. Phys. Lett.* **151**: 481–484.

41. Sims IR, Queffelec JL, Defrance A, Rebrion-Rowe C, Travers D, Bocherel P, Rowe BR, Smith IWM. (1994) Ultra-low temperature kinetics of neutral-neutral reactions: The technique, and results for the reactions $CN + O_2$ down to 13 K and $CN + NH_3$ down to 25 K. *J. Chem. Phys.* **100**: 4229–4241.
42. Mullen C, Smith MA. (2005) Low temperature $NH(X^3\Sigma^-)$ radical reactions with NO, saturated, and unsaturated hydrocarbons studied in a pulsed supersonic Laval nozzle flow reactor between 53 and 188 K. *J. Phys. Chem. A.* **109**: 1391–1399.
43. Mullen C, Smith MA. (2005) Temperature dependence and kinetic isotope effects for the OH + HBr reaction and H/D isotopic variants at low temperatures (53–135 K) measured using a pulsed supersonic Laval nozzle flow reactor. *J. Phys. Chem. A* **109**: 3893–3902.
44. Hansmann B, Abel B. (2007) Kinetics in cold Laval nozzle expansions: From atmospheric chemistry to oxidation of biomolecules in the gas phase. *Chem. phys. chem.* **8**: 343–356.
45. Su T, Bowers MT. (1973) Theory of ion-polar molecule collisions. Comparison with experimental charge transfer reactions of rare gas ions to geometric isomers of difluorobenzene and dichloroethylene. *J. Chem. Phys.* **58**: 3027–3037.
46. Georgievskii Y, Klippenstein SJ. (2005) Long-range transition state theory. *J. Chem. Phys.* **122**: 194103.
47. Clary DC. (1990) Fast chemical reactions — Theory challenges experiment. *Annu. Rev. Phys. Chem.* **41**: 61–90.
48. Troe J. (1985) Statistical Adiabatic Channel Model of ion neutral dipole capture rate constants. *Chem. Phys. Lett.* **122**: 425–430.
49. Clary DC. (1987) Rate constants for the reactions of ions with dipolar polyatomic-molecules. *J. Chem. Soc, Faraday Trans. II* **83**: 139–148.
50. Rowe BR, Canosa A, Le Page V. (1995) FALP and CRESU studies of ionic reactions. *Int. J. Mass Spectrom Ion Processes* **149/150**: 573–596.
51. Rowe BR, Rebrion C. (1991) Recent Laboratory works toward Astrochemistry. *Trends. Chem. Phys.* **1**: 367–389.
52. Troe J. (2005) Temperature and pressure dependence of ion-molecule association and dissociation reactions: The $N_2^+ + N_2 + M \leftrightarrow N_4^+ + M$ reaction. *Phys. Chem. Chem. Phys.* **7**: 1560–1567.
53. Troe J. (1979) Predictive possibilities of unimolecular rate theory. *J. Phys. Chem.* **83**: 114–126.
54. Speck T, Le Garrec JL, Le Picard S, Canosa A, Mitchell JBA, Rowe BR. (2001) Electron attachment in HBr and HCl. *J. Chem. Phys.* **114**: 8303–8309.
55. Moustefaoui T, Rebrion-Rowe C, Le Garrec JL, Rowe BR, Mitchell JBA. (1998) Low temperature electron attachment to polycyclic aromatic hydrocarbons. *Faraday Discuss* **109**: 71–82.
56. Speck T, Mostefaoui T, Rebrion-Rowe C, Mitchell JBA, Rowe BR. (2000) Low temperature electron attachment to $CH_3I$. *J. Phys. B.* **33**: 3575–3582.
57. Yung YL, Allen M, Pinto JP. (1984) Photochemistry of the atmosphere of Titan — Comparison between model and observations. *Astrophys. J. Suppl. Ser.* **55**: 465–506.

58. Wilson EH, Atreya SK. (2003) Chemical sources of haze formation in Titan's atmosphere. *Planet Space Sci.* **51**: 1017–1033.
59. Berteloite C, Le Picard SD, Birza P, Gazeau MC, Canosa A, Benilan Y, Sims IR. (2008) Low temperature (39 K–298 K) kinetic study of the reactions of $C_4H$ radical with various hydrocarbons observed in Titan's atmosphere: $CH_4$, $C_2H_2$, $C_2H_4$, $C_2H_6$, $C_3H_4$ and $C_3H_8$ *Icarus* **194**: 746–757.
60. Canosa A, Le Picard SD, Gougeon S, Rebrion-Rowe C, Travers D, Rowe BR. (2001) Rate coefficients for the reactions of $Si(^3P_J)$ with $C_2H_2$ and $C_2H_4$: Experimental results down to 15 K. *J. Chem. Phys.* **115**: 6495–6503.
61. Le Picard SD, Canosa A, Pineau des Forêts G, Rebrion-Rowe C, Rowe BR. (2001) The Si + $O_2$ reaction: a fast source of SiO at very low temperature; CRESU measurements and interstellar consequences. *Astron. Astrophys.* **372**: 1064–1070.
62. Chastaing D, Le Picard SD, Sims IR, Smith IWM. (2001) Rate Coefficients for the Reactions of $C(^3P_J)$ atoms with $C_2H_2$, $C_2H_4$, $CH_3C{\equiv}CH$ and $CH_2{=}C{=}CH_2$ at temperatures down 15 K. *Astron. Astrophys.* **365**: 241–247.
63. Nizamov B, Leone SR. (2004) Rate coefficients and kinetic isotope effect for the $C_2H$ reactions with $NH_3$ and $ND_3$ in the 104–294 K temperature range. *J. Phys. Chem. A.* **108**: 3766–3771.
64. Vohringer-Martínez E, Hansmann B, Hernandez H, Francisco JS, Troe J, Abel B. (2007) Water catalysis of a radical-molecule gas-phase reaction. *Science* **315**: 497–501.
65. Lee S, Leone SR. (2000) Rate coefficients for the reaction of $C_2H$ with $O_2$ at 90 K and 120 K using a pulsed Laval nozzle apparatus. *Chem. Phys. Lett.* **329**: 443–449.
66. Brownsword RA, Canosa A, Rowe BR, Sims IR, Smith IWM, Stewart DWA, Symonds AC, Travers D. (1997) Kinetics over a wide range of temperature (13–744 K): rate constants for the reactions $CH(v = 0)$ with $H_2$ and $D_2$ and for the removal of $CH(v = 1)$ by $H_2$ and $D_2$ *J. Chem. Phys.* **106**: 7662–7677; Le Picard SD, Canosa A, Rowe BR, Brownsword RA, Smith IWM. (1998) Determination of the limiting low pressure rate constants of the association reactions of CH with $N_2$ and CO: A CRESU measurement at 53 K. *J. Chem. Soc, Faraday Trans.* **94**: 2889–2893.
67. Carty D, Goddard A, Köhler S, Sims IR, Smith IWM. (2006) Kinetics of the Radical - Radical Reaction, $O(^3P_J)+OH(X^2\Pi_\Omega) \rightarrow O_2+H$, at Temperatures down to 39 K. *J. Phys. Chem. A.* **110**: 3101–3109.
68. Larsson B, Liseau R, Pagani L, Bergman P, Bernath P, Biver N, Black JH, Booth RS, Buat V, Crovisier J, *et al.* (2007) Molecular oxygen in the rho Ophiuchi cloud. *Astron. Astrophys.* **466**: 999–U158.
69. Viti S, Roueff E, Hartquist TW, des Forêts GP, Williams D. (2001) Interstellar oxygen chemistry. *Astron. Astrophys.* **370**: 557–569.
70. Xu CX, Xie DQ, Honvault P, Lin SY, Guo H. (2007) Rate constant for $OH(^2\Pi)$ + $O(^3P)$ $\rightarrow$ $H(^2S)$ + $O_2(^3\Sigma_g^-)$ reaction on an improved ab initio potential energy surface and implications for the interstellar oxygen problem. *J. Chem. Phys.* **127**: 024304.

71. Lin SY, Rackham EJ, Guo H. (2006) Quantum mechanical rate constants for $H + O_2 \leftrightarrow O + OH$ and $H + O_2 \rightarrow HO_2$ reactions. *J. Phys. Chem. A.* **110**: 1534–1540.
72. Dalgarno A. (2006) Introductory lecture: The growth of molecular complexity in the Universe. *Faraday Discuss* **133**: 9–25.
73. Schuetz CA, Frenklach M. (2002) Nucleation of soot: Molecular dynamics simulations of pyrene dimerization. *Proc. Combust. Inst.* **29**: 2307–2314.
74. Ricca A, Bauschlicher CW. (2000) The reactions of polycyclic aromatic hydrocarbons with OH. *Chem. Phys. Lett.* **328**: 396–402.
75. Atkinson R. (1986) Kinetics and mechanisms of the gas-phase reactions of the hydroxyl radical with organic-compounds under atmospheric conditions. *Chem. Rev.* **86**: 69–201.
76. Smith IWM, Sage AM, Donahue NM, Herbst E, Quan D. (2006) The temperature-dependence of rapid low temperature reactions: experiment, understanding and prediction. *Faraday Discuss* **133**: 137–156.
77. Sims IR, Queffelec JL, Travers D, Rowe BR, Herbert LB, Karthäuser J, Smith IWM. (1993) Rate constants for the reactions of CN with hydrocarbons at low and ultra-low temperature. *Chem. Phys. Lett.* **211**: 461–468.
78. Georgievskii Y, Klippenstein SJ. (2007) Strange kinetics of the $C_2H_6$ + CN reaction explained. *J. Phys. Chem. A* **111**: 3802–3811.
79. Greenwald EE, North SW, Georgievskii Y, Klippenstein SJ. (2005) A two transition state model for radical-molecule reactions: A case study of the addition of OH to $C_2H_4$. *J. Phys. Chem. A.* **109**: 6031–6044.
80. Donahue NM. (2003) Reaction barriers: Origin and evolution. *Chem. Rev.* **103**: 4593–4604.
81. Sabbah H, Biennier L, Sims IR, Georgievskii Y, Klippenstein SJ, Smith IWM. (2007) Understanding reactivity at very low temperatures: The reactions of oxygen atoms with alkenes. *Science* **317**: 102–105.
82. Cvetanovic RJ. (1987) Evaluated chemical kinetic data for the reactions of atomic oxygen $O(^3P)$ with unsaturated-hydrocarbons. *J. Phys. Chem. Ref. Data* **16**: 261–326.
83. Abraham FF. (1974) *Homogeneous nucleation theory: The pretransition theory of vapor condensation, supplement I* Academic Press, New York.
84. Hamon S, Le Picard SD, Canosa A, Rowe BR, Smith IWM. (2000) Low temperature measurements of the rate of association to benzene dimers in helium. *J. Chem. Phys.* **112**: 4506–4516.
85. Choi N, Blitz MA, McKee K, Pilling MJ, Seakins PW. (2004) H atom branching ratios from the reactions of CN radicals with $C_2H_2$ and $C_2H_4$. *Chem. Phys. Lett.* **384**: 68–72.
86. Loison JC, Bergeat A. (2004) Reaction of carbon atoms, C ($2p^2$, $^3P$) with $C_3H_4$ (allene and methylacetylene), $C_3H_6$ (propylene) and $C_4H_8$ (trans-butene): Overall rate constants and atomic hydrogen branching ratios. *Phys. Chem. Chem. Phys.* **6**: 5396–5401.
87. Taatjes CA, Hershberger JF. (2001) Recent progress in infrared absorption techniques for elementary gas-phase reaction kinetics. *Annu. Rev. Phys. Chem.* **52**: 41–70.

88. Goulay F, Osborn DL, Taatjes CA, Zou P, Meloni G, Leone SR. (2007) Direct detection of polyynes formation from the reaction of ethynyl radical ($C_2H$) with propyne ($CH_3$–C≡CH) and allene ($CH_2$=C=$CH_2$). *Phys. Chem. Chem. Phys.* **9**: 4291–4300.
89. Garland NL, Nelson HH. (1992) Temperature-dependence of the kinetics of the reaction $Al + O_2 \rightarrow AlO + O$. *Chem. Phys. Lett.* **191**: 269–272.
90. Herbert L, Smith IWM, Rowland D, Spencer-Smith RD. (1992) Rate constants for the elementary reactions between CN radicals and $CH_4$, $C_2H_6$, $C_2H_4$, $C_3H_6$, and $C_2H_2$ in the range $-295 < T/K < 700$ K. *Int. J. Chem. Kinet.* **24**: 791–802.
91. Canosa A, Sims IR, Travers D, Smith IWM, Rowe BR. (1997) Reactions of the methylidine radical with $CH_4$, $C_2H_2$, $C_2H_4$, $C_2H_6$ and but-1-ene studied between 23 and 295 K with a CRESU apparatus. *Astron. Astrophys.* **323**: 644–651.
92. Chastaing D, James PL, Sims IR, Smith IWM. (1998) Neutral-neutral reactions at the temperatures of interstellar clouds - Rate coefficients for reactions of $C_2H$ radicals with $O_2$, $C_2H_2$, $C_2H_4$ and $C_3H_6$ down to 15 K. *Faraday Discuss* **109**: 165–181; Vakhtin AB, Heard DE, Smith IWM, Leone SR. (2001) Kinetics of $C_2H$ radical reactions with ethene, propene and 1-butene measured in a pulsed Laval nozzle apparatus at T=103 and 296 K. *Chem. Phys. Lett.* **348**: 21–26.
93. Canosa A, Páramo A, Le Picard SD, Sims IR. (2007) An experimental study of the reaction kinetics of $C_2(X^1\Sigma_g^+)$ with hydrocarbons ($CH_4$, $C_2H_2$, $C_2H_4$, $C_2H_6$ and $C_3H_8$) over the temperature range 24–300 K. Implications for the atmospheres of Titan and the Giant Planets. *Icarus* **187**: 558–568.

# Chapter 3

# The Study of Cold Collisions Using Ion Guides and Traps

D. Gerlich

*Faculty of Natural Science*
*Technical University, 09107 Chemnitz, Germany*
*gerlich@physik.tu-chemnitz.de*

## Contents

## 3.1. Introduction

### 3.1.1. *Cold Interstellar Clouds*

It has been emphasized by Herbst and Millar in the first chapter of this book that studies of cold collisions in the gas phase and on grains are important for understanding the physics and chemistry of cold interstellar clouds. However, the experimental or theoretical determination of pure *thermal* rate coefficients may not be sufficient since the production and the destruction of matter is a complex result of many elementary processes. In general, the matter in interstellar or circumstellar space is not in thermal equilibrium. Klemperer[1] likes to illustrate this with the conversion of hydrogen and carbon monoxide into methane and water. Simple thermodynamics predict an equilibrium ratio for $CO + 3H_2 \leftrightarrow CH_4 + H_2O$ which differs from interstellar observation by more than 500 orders of magnitude. Another example is the population of the rotational states of CH, $CH^+$, and CN, the absorption spectra of which were already observed many decades ago. Usually rotational motion can couple efficiently via collisions to the translational temperature of the surrounding gas (5 K or higher); however, polar species can become colder since they interact efficiently with the cosmic background radiation of 2.73 K, an electromagnetic microwave field which is associated with the origin of the universe. Other examples include the non-thermal abundances of isomers (e.g. $HCO^+$ and $HOC^+$, see Smith *et al.*[2]), heating of particles via cosmic rays, stimulated emission from the inversion transition of ammonia, or the role of o-$H_2$ in creating non-thermal populations in cold environments.[3] One of the conclusions from this is that for astrochemistry, often cold state specific cross sections are needed instead of thermally averaged rate coefficients.

### 3.1.2. *Ultracold Collisions and Quantum Matter*

For performing precision measurements of their properties, ultracold atoms and molecules are one of the central themes of current fundamental research in chemical and optical physics. Laser cooling of atoms allows micro-Kelvin temperatures to be reached. Innovative techniques allow both the quantum-statistical behavior of condensates of dilute, weakly interacting atoms to be studied and coherent matter–waves to be produced, creating a new field of material science. Various approaches are being developed to produce ultracold molecular systems; however, molecules are much harder to manipulate with optical or other schemes. Today there are many activities in one subfield where the *ultracold molecules* are just homonuclear dimers, formed via photoassociation (see e.g. the chapter by Weiner). In addition, methods are under development and discussed in this book which allow ultracold molecules with a few atoms to be created; however, we are far from freezing all degrees of freedom of large molecules. The work described by the Rennes group in Chapter 2 and in this contribution is far from temperatures where coherent interaction of matter–waves plays a role. However, there is huge interest in the study of collisions under interstellar conditions and how to cool large biological molecules to a few K or below. Another interesting experimental method to cool molecular ions is based on sympathetic cooling by mixing them into a cloud of laser cooled atomic ions. Ions prepared in this way have been used successfully for spectroscopy and high-precision measurements in fundamental physics as discussed in more detail in Chapter 6. Below the question is raised whether these *so-called* ultracold ions really open up new routes for the study of state-selective chemical reactions at low collision energies.

### 3.1.3. *Definition of "Temperature", Coherent Motion*

A key problem in many applications is to define what one means by "cold" or "ultracold". What is the definition of temperature and what is a suitable thermometer? Presently the definition of the unit 1 Kelvin uses the triple point of water; however, in a few years, the Boltzmann constant $k_B$ will be fixed and the energy $k_BT$ will be the basis for defining the temperature scale. Most probably one will use the average translational energy associated with the disordered motion of Ar atoms in an acoustical gas thermometer. In most applications discussed in this book, the mean kinetic energy is used for defining the temperature; however, the velocity distribution of a beam of slowed particles or of an actively cooled ensemble is not necessarily a Maxwell-Boltzmann distribution. For trapped particles the boundary conditions lead to quantization of the translational degrees of freedom. The population of these states provides information on the mean translational energy of the system. In the case of a single trapped ion one can extract the information from a time-averaged measurement instead of using an ensemble average.

The definition of temperature in traps where ions are confined with time-varying external forces is problematic. A special case is a radio frequency quadrupole ion trap, in which ions are laser cooled into an ordered Coulomb crystal. This structure is characteristic of a temperature in the sub-mK range. Nonetheless the kinetic energy of the trapped ions is up to seven orders of magnitude higher, as discussed in detail by Schiffer *et al.*[4] The explanation of this seeming contradiction is the separation of the motion into the fast coherent motion driven by the oscillating electric field and a superimposed random part. This definition of temperature is in good agreement with the observed phase transition and also with the Doppler width if the laser is oriented orthogonally to the secular motion; however, a collision between the so-called ultracold ion and a neutral target is very *hot*, the mean translational energy may correspond to thousands of K. In addition, as will be discussed below, sympathetic cooling of molecular ions does not work for the internal degrees of freedom.

### 3.1.4. *How to Obtain Low Collision Energies or to Prepare a Cold Ensemble*

Progress in experimental science is closely related to experimental innovations. Starting in the 1980s, several groups have developed sophisticated instruments for studying ion-molecule and neutral-neutral reactions over a wide range of temperatures. Presently techniques are under development both for cooling ions below one K and for heating them with lasers to thousands of K.[5] An introduction to the various methods used today in low temperature chemistry is given in Chapter 2. There are various strategies to prepare cold or slow ensembles of atoms or molecules. The traditional and rather general techniques are based on collisional cooling with cold buffer gas while modern techniques make use of laser cooling or slowing neutrals with suitable electromagnetic fields (see the chapter by van de Meerakker *et al.*). The present chapter focuses exclusively on instruments which make use of inhomogeneous fast oscillating electric fields. Special ion guides and an innovative merged beam arrangement have opened up the meV collision energy range and are, in principle, capable of reaching sub-meV. The most successful and versatile solutions for reaching low ion temperatures are based on multi-electrode ion traps combined with cryogenic cooling. In order to overcome the limitations imposed by condensation of neutral species, traps also are combined with cold effusive or supersonic beams.

### 3.1.5. *Reactions at Low Temperatures*

Several aspects of the dependence of chemical reactions on temperature are discussed in other chapters of this book. For low relative velocities and a long-range attractive interaction potential, it is common to use capture models (see Chapter 4 by Klippenstein and Georgievskii). The dynamics of complex formation in collisions of rotationally excited diatomic molecules with ions at very low collision energies and translational temperatures have been discussed recently.[6] Special effects occur at very low collision energy where only a few partial waves contribute. Below a certain limit, s-wave scattering prevails. In general, lowering the total energies leads to an increase in the lifetime of the intermediate complex. This often allows for slow processes such as tunneling through barriers or emission of photons. As emphasized in Chapter 1, in the context of the astrochemically important process of deuterium fractionation, minor differences in zero-point energies caused by isotopic rearrangement can play a key role in low temperature chemistry. There are also several ion-molecule reactions which are nearly thermoneutral or which are influenced by small barriers. Excitation of reactants to low-lying fine structure or rotational states can change the outcome of a reaction significantly. In such cases it is advantageous if one can measure the rate coefficients for state specific reactions instead of working with a Boltzmann population. Combining ion traps with photoionization or optical pumping of stored ions with lasers opens up many related possibilities.

### 3.1.6. *Preview*

This chapter is organized as follows. After a short reminder of the basics of the ion guiding and trapping technique and a survey of the relevant kinematics, several instruments are described in detail together with characteristic measuring procedures. The versatility of the machines and modules is illustrated with a selection of experimental results, including meV collisions with cold polar molecules, cluster formation, radiative association, isotopic enrichment of deuterated molecules. Very recent activities aim at understanding reactions of cold ions with slow H atoms. In the outlook it will be mentioned that trapping charged objects is rather general. In principle there are no restrictions concerning mass, temperature or storage time. Present activities include cooling ions below one K or heating nanoparticles to very high temperatures.

## 3.2. Fast Oscillating Forces

### 3.2.1. *Inhomogeneous Electric Fields*

All experimental setups described in this chapter are based on the use of specific inhomogeneous, electrical fields $\boldsymbol{E}_0(\boldsymbol{r}, t)$ which vary in space and time. Usually one uses

$$\boldsymbol{E}_0(\boldsymbol{r},t) = \boldsymbol{E}_0(\boldsymbol{r})\cos(\Omega t), \tag{3.1}$$

i.e., a harmonic oscillation with the frequency $\Omega$. The method to trap or guide ions, the basic theory, and a variety of applications in ion physics and chemistry has been thoroughly documented previously.[7] In what follows more recent publications are reviewed. To better understand the method and its limitations, one needs some basic theory describing the motion of a particle under the influence of a fast oscillating force. The most important precondition is that, if the frequency is high enough, the main influence of the oscillatory force can be described by the so called effective potential

$$V^* = q^2\boldsymbol{E}_0(\boldsymbol{r})^2/4m\Omega^2, \tag{3.2}$$

where $m$ and $q$ are the particles' mass and charge, respectively. In order to get a quantitative estimate of the range of validity of the high frequency or adiabatic approximation, one introduces an adiabaticity parameter defined by

$$\eta = 2q|\boldsymbol{\nabla}\boldsymbol{E}_0(\boldsymbol{r})|/m\Omega^2. \tag{3.3}$$

This useful dimensionless parameter can be introduced in various ways; for example, by postulating that the change of the field over the oscillation must be much smaller than the field itself. Experience based on experimental tests and on numerical calculations has shown that one should use operating conditions such that $\eta$ remains everywhere smaller than 0.3. This condition can always be achieved by using a high enough frequency.

The basic principle is quite general and can be applied in many fields. It is important to realize that $V^*$ and $\eta$ scale inversely with $\mathrm{m}\Omega^2$. The effective potential $V^*$can be used for confining electrons ($\Omega$ must be in the GHz range), ions (radio frequency range) and clusters (audio frequency). For trapping charged microscopic particles one can use AC voltages. Typical applications include strong focusing lenses in accelerator physics, the well-known Paul trap invented in the 1950's or the Guided Ion Beam technique developed in the 1970's. For a detailed review of the historical development and of specific applications see Ref. 7.

### 3.2.2. *Multipole Fields*

In any real experimental device, the electric field $\boldsymbol{E}_0(\boldsymbol{r})$ is determined by the boundary conditions imposed by suitable electrodes. For guiding and trapping ions, a variety of electrode arrangements have been applied. Many theoretical and practical hints are given in Ref. 7. One of the best characterized examples is a linear multipole. The potential of a $2n$ pole is given by

$$\Phi(r, \phi) = \Phi_0 \hat{r}^{\mathrm{n}} \cos(n\phi) \tag{3.4}$$

where $\Phi_0 = U_0 - V_0 \cos \Omega t$ is the applied dc and rf potential, $2n$ corresponds to the number of poles, and $\hat{r} = r/r_0$ is a reduced radius. In most applications one operates a multipole "rf-only", i.e., $U_0 = 0$. Calculating the electric field $|E_0|$ from $\Phi_0$ and using Eq. (3.2) one obtains the effective potential

$$V^*(r) = \frac{n^2}{4} \frac{q^2 V_0^2}{m\Omega^2\, r_0^2} \hat{r}^{2n-2}. \tag{3.5}$$

From Eq. (3.3) one gets the adiabaticity parameter

$$\eta(r) = 2n(n-1) \frac{q\, V_0}{m\Omega^2 r_0^2} \hat{r}^{n-2}. \tag{3.6}$$

For the case $U_0 \neq 0$ one has to add in Eq. (3.3) the dc term $q\ U_0 \hat{r}^n$ $\cos(n\phi)$. The formula become even simpler if one introduces the characteristic energy $\varepsilon$,

$$\varepsilon = 1/2n^2 m\Omega^2\ r_0^2. \tag{3.7}$$

For numerical calculations it is useful to work with the units cm, MHz, V, the atomic mass unit u, the charge unit e, and the energy unit eV. Using these units, $\varepsilon$ can be simply calculated from

$$\varepsilon = 1.036/2n^2 m\Omega^2\ r_0^2 \ (\text{units: eV}, u, \text{cm, MHz}). \tag{3.8}$$

### 3.2.3. *Selecting Parameters*

For selecting the proper ion guide and the optimum operating parameters for a special purpose, different criteria have to be taken into account. Parameters of the multipole include the number of poles, $2n$, the scale length $r_0$, the amplitude $V_0$, the frequency $\Omega = 2\pi f$, and eventually the dc component $U_0$. The ions are characterized by the charge $q$ and the mass $m$. Adiabatic conservation of energy ensures that safe transmission does not depend on the individual initial conditions, but only on the maximum transverse energy $E_m$. It is easy to show that a quadrupole ion guide is the optimum choice if one just intends to confine ions with a given $E_m$ to a specific maximum radius $r_m$ with a minimal amplitude $V_0$. From Eqs. (3.5) and (3.6) and $\eta < 0.3$, analytical formulas have been derived that permit one to calculate "safe operating conditions".[7] Using, for example, $n = 4$

and $E_m = 1\,\text{eV}$, it is found that an amplitude $V_0 = 48.8\,\text{V}$ is sufficient to guide a singly charged particle. This result is independent of $m$, i.e., it holds for electrons, protons, or charged nanoparticles. For simultaneously guiding ions with $E_m = 0.1\,\text{eV}$ and masses ranging from $\text{m}_1 = 1\,\text{u}$ to $\text{m}_2 = 100\,\text{u}$, the values needed are $V_0 = 105\,\text{V}$ and $f = 17.8\,\text{MHz}$ for an octopole ($n = 4$, $r_0 = 0.3\,\text{cm}$). It also can be shown that for a wider mass range, e.g., a factor of $10^4$, a decapole is preferred. More detailed examples can be found elsewhere.[7]

### 3.2.4. *Some Features*

For $n = 2$ (a quadrupole), inspection of Eqs. (3.5) and (3.6) reveals that the effective potential $V^*$ is harmonic and that $\eta$ does not depend on $\boldsymbol{r}$. This leads to special properties; e.g., focusing and phase space conserving transmission. Moreover, the equation of motion of an ion, moving in a quadrupole, can be solved without approximation by transforming it into the well-known Mathieu differential equation. This allows for a mathematically precise classification of stable and unstable operating conditions. A complete theoretical description of rf quadrupoles is given by Dawson.[8] Concerning the use of a quadrupole as a mass filter or as an ion guide, it must be noted that this primarily depends on the stability parameter $\eta$ (called $q$ in the Mathieu equation). For mass-selective transmission, one operates usually with $\eta = 0.706$; whereas for guiding ions, conservation of the kinetic energy is desirable, i.e., one uses $\eta < 0.3$. This distinction is independent on whether one operates in the rf-only mode or with a superimposed dc field.

Some features of the motion of ions in an octopole are illustrated in Fig. 3.1. All sample trajectories start in the center with different initial conditions. First it can be seen that the ions move, rather unperturbed, on straight trajectories in the inner part. The three trajectories in the upper half illustrate the importance of selecting a small stability parameter. For $\eta_m = 0.1$ there are many oscillations during the reflection, i.e., the adiabatic approximation holds, while for $\eta_m = 0.7$ the ion absorbs energy from the rf field. The three examples on the lower right side illustrate that ions with a transverse energy of $E_m = 0.1\,\text{eV}$ are safely stored while, using the same amplitude and frequency, $10\,\text{eV}$ ions are also confined. However, the reflection law does not hold; i.e., impulse is transferred to the ion. The two examples in the lower left region compare two ions which are moving towards a rod or towards the gap between two rods. The effective potential (see Eq. (3.5)) is rotationally symmetric; however the oscillatory micromotion is different, in one case it is longitudinal in the other one transverse.

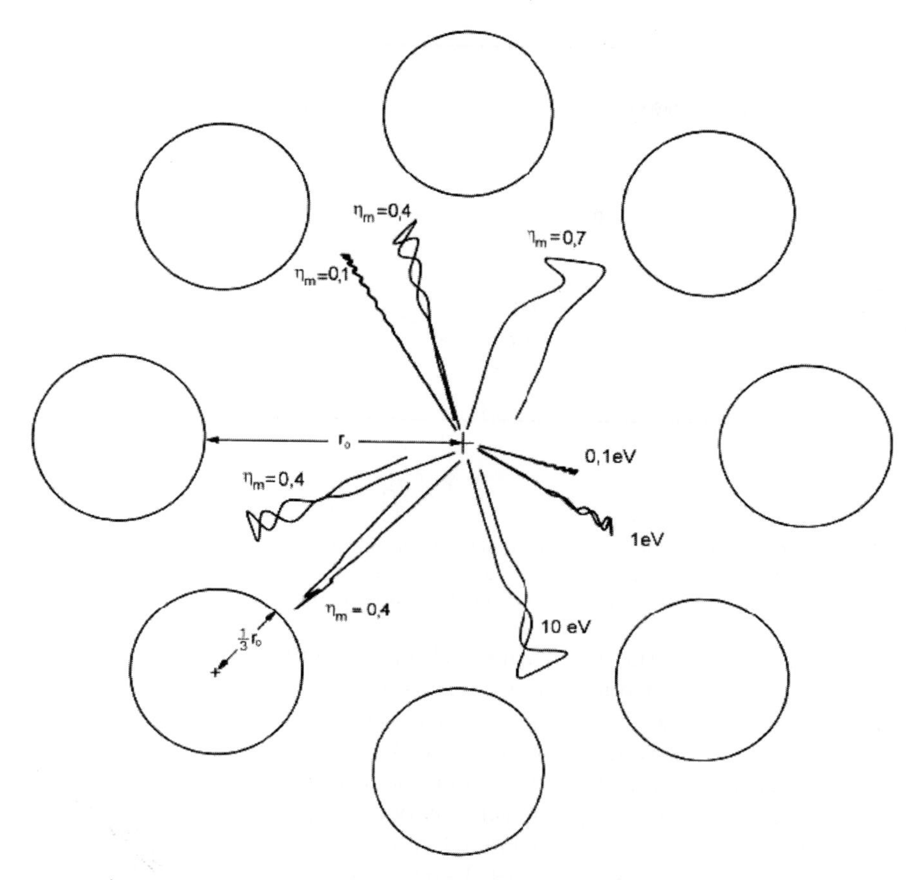


Fig. 3.1. Selected ion trajectories calculated for an octopole illustrating the influence of the rf field. The ions start in the center, but with different stability parameters $\eta_m$, initial energies $E_m$, and two directions (between and towards the rods). Details are explained in the text.

### 3.2.5. *Cooling Ions*

For cooling ions and for studying low energy collisions it is necessary to understand in more detail the impairment of the ion kinetic energy by the time varying forces acting on the stored ions. It is rather obvious that electrode arrangements with wide field-free regions are better suited for this purpose. The effective potentials plotted in Fig. 3.2 illustrate that a 22-pole has significantly steeper walls than a quadrupole; it is even superior to a ring electrode trap with similar electrode dimensions. More information is gained if one calculates the kinetic energy of a stored ion as a function of

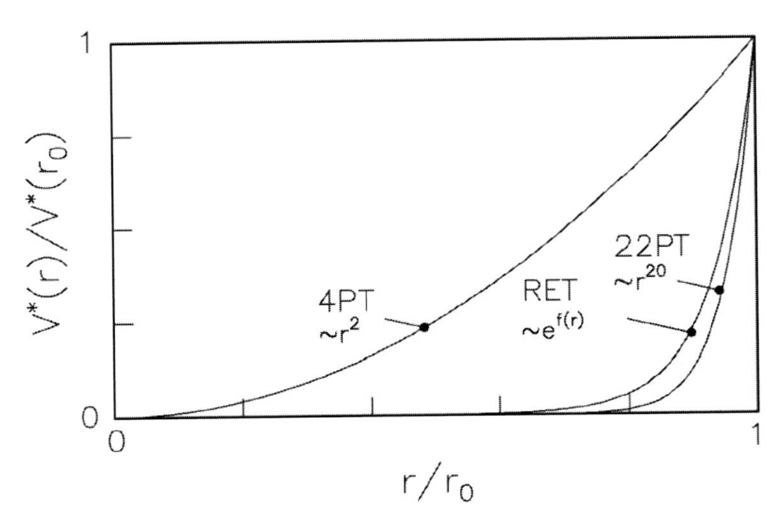


Fig. 3.2. Comparison of the r-dependence of the normalized effective potentials of a linear quadrupole, a 22-pole and an arrangement of ring electrodes. In the limit of an extremely large number of electrodes one obtains a box potential.

time. Fig. 3.3 shows the modulation of $E$ in units of the initial energy $E_m$ if it approaches an electrode. The two results, one has been calculated numerically exact the other one in the adiabatic approximation, reveal that the energy reaches any value between 0 and 3 $E_m$. Comparison of the initial and final energy reveals that the system behaves on average like a conservative one if the conditions for the adiabatic approximation are fulfilled. In order to estimate the time averaged influence, energy distributions have been calculated over many rf cycles. Two typical results are given in Fig. 3.4. The distributions are very narrow peaking at the nominal energy $E_m$. This result confirms that the influence of the rf field on the kinetic energy of the stored ions is smaller the higher the number of electrodes.

An additional criterion comes into the play if one intends to cool ions with buffer gas or if one wants to study collision processes between neutrals and ions confined in rf fields. In this case one of the most important features is that, in general, both $V^*(\boldsymbol{r})$ and $\eta(\boldsymbol{r})$ are functions of the coordinate $\boldsymbol{r}$ and that both go to 0 for $|\boldsymbol{r}| \to 0$. This means that, with decreasing kinetic energy $E_m$, the adiabatic approximation becomes better. In the case of a multipole ($2n$ pole) one can derive the relation

$$\eta \sim E_m^{(n-2)/(2n-2)}. \tag{3.9}$$

During collisional cooling of ions, this reduction of $\eta$ reduces simultaneously more and more the impairment via the rf field. As mentioned above and

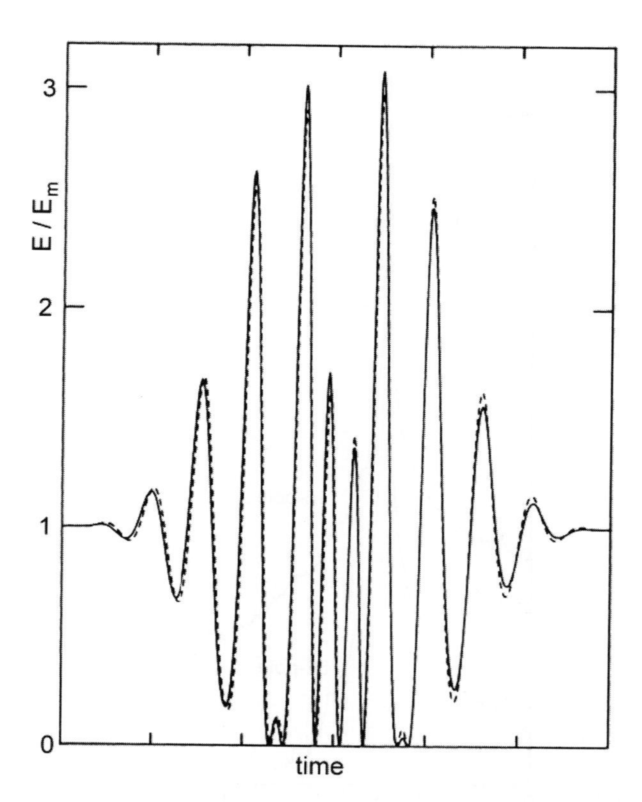


Fig. 3.3. Modulation of the kinetic energy of an ion which starts with the energy $E_m$ in the field free region and which is heading towards an electrode (see Fig. 3.1). The parameter $\eta_m$ has been chosen 0.3. For a short moment, the kinetic energy can rise up to three times the initial energy, $3E_m$. The solid line is a numerical solution of the exact equation of motion, the dashed line has been calculated within the adiabatic approximation. For more details see Ref. 7.

as can be seen from Eq. (3.9), there is only one example, the quadrupole. For $2n = 4$, the stability parameter $\eta$ is constant. Therefore a quadrupole trap is poorly suited for cooling ions to very slow motion. One has to use at least a hexapole. In other words: collisional relaxation is only efficient in higher order electrode arrangements.

Many numerical simulations have been performed for understanding in more detail the process of collisional relaxation in rf traps. A summary can be found in Ref. 7. In principle, an ion can always gain energy if it undergoes a collision during its interaction with the rf field. This effect depends strongly on the mass ratio. For example it is much easier to cool heavy ions in a light buffer gas than the other way around. Very problematic

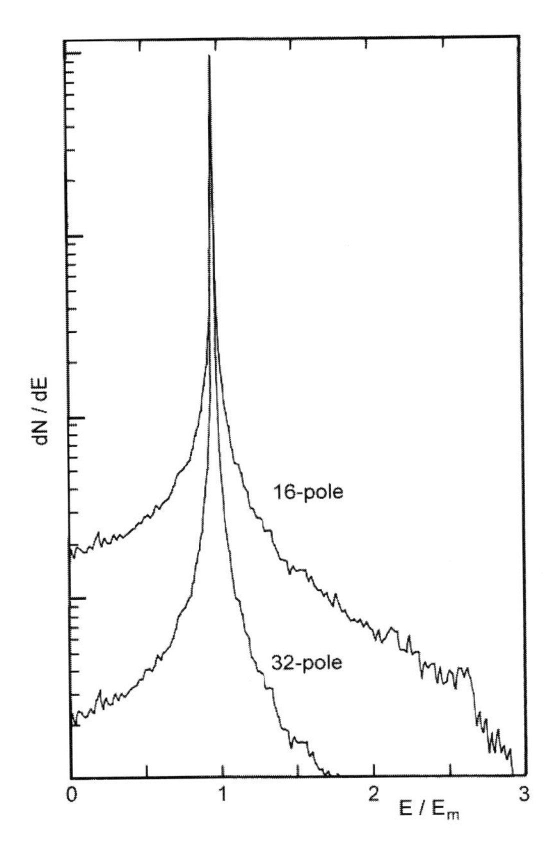


Fig. 3.4. Time averaged kinetic energy distribution of an ion stored in a 16-pole and in a 32-pole trap. The results have been calculated using numerically exact trajectory calculations. For a better comparison the probability distributions are peak normalized.

are local potential distortions on electrodes ("patch effects") which can pull the ions into regions of strong rf fields. An innovative approach, which avoids collisions during the reflection of the ions from the rf walls and which is presently under development, is to use a cold skimmed beam of neutrals for collisional cooling. More details are discussed in Chapter 6.

## 3.3. Kinematics

The chapter by Strecker and Chandler is called *Kinematic cooling of molecules*. Their method relies on single, well-defined collisions in which just the correct amount of momentum is transferred for one of the particles to

Table 3.1.

| | |
|---|---|
| $m_1 m_2$ | mass (ion, neutral) |
| $M = m_1 + m_2 = m_1' + m_2'$ | total mass |
| $\mu = m_1 \cdot m_2/M$ | reduced mass |
| $v_1 v_2$ | velocity in the LAB frame |
| $g = v_1 - v_2$ | relative velocity |
| $v_C(m_1 v_1 + m_2 v_2)/M$ | velocity of the CM of the system |
| $u_1 = v_1 - v_C$ | velocity of the CM frame |
| $\Lambda \leq (v_1, v_2)$ | intersection angle $v_1 v_2$ |
| $E_1 E_2$ | energy in the LAB frame |
| $E_T = \mu/2g^2$ | translational energy (collision energy) |
| $\Delta E_T = E_T' - E_T$ | translational exoergicity |
| $E_{\text{int}}$ | internal energy (vib, rot, electr. ) |
| $-\Delta E_0$ | reaction exo-ergicity |
| $E_{\text{int}} + E_T - \Delta E_0 = E_{\text{int}}' + E_T'$ | total available energy |

remain at rest after the interaction. In more detail, they distinguish between *slowing* and *cooling*, to which one might add *stopping*. Our approaches to reach meV collision energies or sub-K collision temperatures are different. These latter methods rely on many collisions, either for preparing two cold beams or for cooling one trapped ensemble of ions, if necessary with thousands of collisions. In order to assess the capabilities and limitations of beam-beam and beam-cell arrangements in reaching meV or even sub-meV collision energies, a detailed analysis of kinematic averaging is required. In the following we treat the kinematics of a beam-beam and a beam-cloud arrangement. The second situation can be either an ion beam — neutral target cell arrangement (e.g. a standard guided ion beam apparatus) or a set-up where a neutral beam interacts with a cold ensemble of stored ions. The most important abbreviations used in this section are collected in Table 3.1.

### 3.3.1. *Beam-Beam Arrangements*

In a standard treatment of kinematics (see for example Ref. 9) one starts with two well-prepared beams of colliding partners which interact with each other in some spatial region. If the species in the two beams are characterized by individual velocity distributions, $f_1(\boldsymbol{v}_l)$ and $f_2(\boldsymbol{v}_2)$, one measures, instead of the elementary cross section $\sigma(g)$, an effective cross section defined by

$$\sigma_{\text{eff}}(\langle g \rangle) = \int_0^\infty \frac{g}{\langle g \rangle} \sigma(g) f(g) dg \tag{3.10}$$

Here, $g = |\boldsymbol{g}| = |\boldsymbol{v}_1 - \boldsymbol{v}_2|$ is the relative velocity and the distribution function $f(g)$ is defined by

$$f(g)dg = \int dv_1 \int dv_2 f_1(\mathrm{v}_1) f_2(\mathrm{v}_2) \tag{3.11}$$

The integration must be restricted to that subspace $(\boldsymbol{v}_1, \boldsymbol{v}_2)$ where $|\boldsymbol{v}_1 - \boldsymbol{v}_2| \in [g, g + dg]$. The mean relative velocity, $\langle g \rangle$, is defined via the integration

$$\langle g \rangle = \int_0^\infty g f(g) dg \tag{3.12}$$

Note that $\langle \boldsymbol{g} \rangle$ and $\langle \boldsymbol{g}' \rangle = \langle \boldsymbol{v}_1 \rangle - \langle \boldsymbol{v}_2 \rangle$ are not necessarily identical. The number of products, $dN_\mathrm{p}$, produced per unit time in a small scattering volume $d\tau$ at location $\boldsymbol{r}$ is given by

$$dN_P = \langle g \rangle \sigma_{\mathrm{eff}}(\langle g \rangle) n_1(\mathrm{r}) n_2(\mathrm{r}) d\tau \tag{3.13}$$

To precisely determine absolute integral cross sections from measured quantities, the spatial overlap of the two reactant beams and their local densities must be known. Integration of $dN_\mathrm{p}$ in accordance with the merged beam geometry leads, for weak attenuation of the primary beam, to

$$\sigma_{\mathrm{eff}} = \frac{N_P}{N_1} \frac{1}{\langle n_2 \rangle L} \frac{v_1}{\langle g \rangle} \tag{3.14}$$

where the ion density $n_1$ has been replaced by the ion flux $N_1 \sim n_1 \langle v_l \rangle$, $\langle n_2 \rangle$ is the mean density of the neutral target, and $L$ the length of the overlapping region. In many experiments the product $\langle n_2 \rangle L$ can be determined with a known reaction, e.g., for hydrogen target one can use the $Ar^+ + H_2$ reaction. It is useful to compare Eq. (3.14) with the result for a beam — isotropic target arrangement where $\langle v_2 \rangle = 0$ and, therefore $\langle v_1 \rangle = \langle g \rangle$. This leads to

$$\sigma_{\mathrm{eff}} = \frac{N_P}{N_1} \frac{1}{n_2 L} \tag{3.15}$$

where $n_2$ is the directly measurable density of the target. The length of the overlapping region is in this case identical with the effective length of the scattering cell or of the ion trap.

### 3.3.2. *Collision Energy*

Usually, measured effective cross-sections are reported as a function of the mean collision energy, $\langle E_T \rangle$. For two well-collimated monoenergetic beams crossing at an angle $\Lambda$, $E_T$ is given by

$$E_T = \frac{1}{2}\mu g^2 = \frac{1}{2}\mu(v_1^2 + v_1^2 - 2 v_1 v_2 \cos\Lambda) \tag{3.16}$$

where $\mu = m_1 m_2/(m_1 + m_2)$ is the reduced mass and $m_1$ and $m_2$ are the masses of the ion and the neutral, respectively. For a crossed beam arrangement ($\Lambda = 90°$) the mean collision energy is related to the mean ion and neutral velocities by

$$\langle E_T \rangle = \frac{1}{2}\mu(\langle v_1 \rangle^2 + \langle v_2 \rangle^2) \tag{3.17}$$

whereas for a merged geometry ($\Lambda = 0^0$) one obtains

$$\langle E_T \rangle = \frac{1}{2}\mu(\langle v_1 \rangle - \langle v_2 \rangle)^2 \tag{3.18}$$

Comparison of these two results reveals that the lowest accessible energy is limited by the mean energy of the neutrals for beams crossing at 90° whereas a nominal energy $\langle E_T \rangle = 0$ can be obtained with the collinear arrangement. The lowest collision energy which can be obtained is determined by the energy spread of the two beams.

### 3.3.3. *Energy Resolution*

In general, the distribution of the relative velocity $f(g)$ must be determined by numerical integration. In many cases it is sufficient to approximate $f_1(\boldsymbol{v}_1)$ and $f_2(\boldsymbol{v}_2)$ with Gaussians, using experimentally determined parameters. For a coaxial beam-beam arrangement the six-dimensional integral over $(\boldsymbol{v}_1, \boldsymbol{v}_2)$ is reduced to a three-dimensional one over $(v_1, v_2, \Lambda)$. To make a quick analytical estimate of the attainable energy resolution, error propagation can be used. For the merged beam geometry one obtains:

$$\Delta E_T = \left[\left(\frac{\mu g}{m_1 v_1}\Delta E_1\right)^2 + (\mu g \Delta v_2)^2 + \left(\frac{\mu}{2} v_1 v_2 \Delta\Lambda^2 \cos\Lambda\right)^2\right]^{1/2} \tag{3.19}$$

where $\Delta E_1$ is the energy half-width of the ion beam, $\Delta v_2$ the velocity half-width of the supersonic beam, and $\Delta\Lambda$ the mean relative angular divergence between the two beams. For a crossed beam geometry one gets a slightly different result,

$$\Delta E_T = \left[\left(\frac{\mu}{m_1}\Delta E_1\right)^2 + (\mu v_2 \Delta v_2)^2 + (\mu v_1 v_2 \Delta\Lambda \sin\Lambda)^2\right]^{1/2} \tag{3.20}$$

Comparison of these two formulas reveals that, under identical beam conditions, the resolving power of the merged beam experiment is superior to that using crossed beams. This is partly due to the kinematic compression and partly due to the angular divergence, affecting $\Delta E_T$ in first-order in the crossed, but only in second-order in the merged beam arrangement.

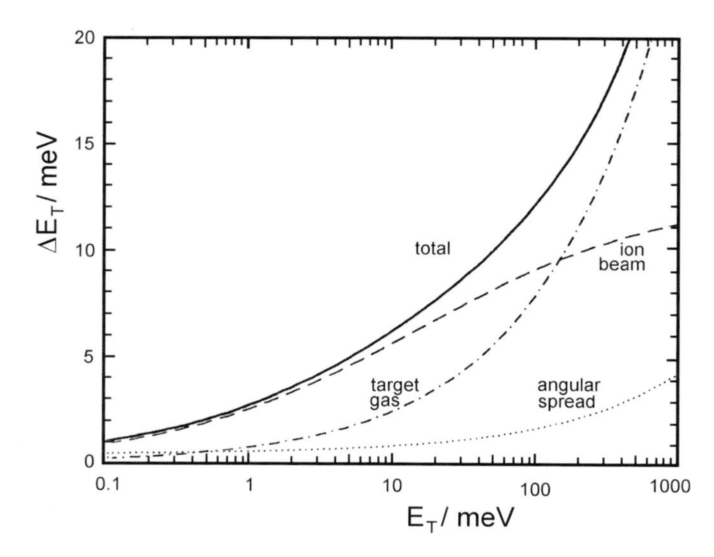


Fig. 3.5. Energy resolution in a merged beam arrangement for the collision system $N^+ + H_2$. The energy half width of the ion beam is $\Delta E_1 = 100\,\mathrm{meV}$, the speed ratio of the $H_2$ beam SR = 10 and the angular spread is $\Delta\Lambda = 10°$. With these conditions collision energies as low as 1 meV have been achieved.

Two examples are used to illustrate the capabilities of a merged beam experiment in which a supersonic beam and a guided ion beam are superimposed. Fig. 3.5 shows the total energy resolution which has been obtained in a merged beam arrangement for the collision system $N^+ + H_2$. At low collision energies the ion beam half-width ($\Delta E_1 = 100\,\mathrm{meV}$) contributes most while at higher energies the neutral target beam (speed ratio SR = 10) determines the energy resolution. The angular spread of the intersecting beams ($\Delta\Lambda = 10°$) is negligible. With these conditions, collision energies as low as 1 meV have been achieved. Another example is given in Fig. 3.6 for $Ar^+ + H_2$ and the conditions ($\Delta E_1 = 100\,\mathrm{meV}$, SR = 10, $\Delta\Lambda = 10°$). The distributions of the collision energy $\mathrm{d}N/\mathrm{d}E_T$ have been obtained by numerical integration over the measured distributions. It can be seen that at the lowest energy, the mean energy is slightly larger than the nominal energy.

### 3.3.4. *Slow Beam — Thermal Target*

To describe the special case of a monoenergetic ion beam passing through a scattering cell containing target gas at the temperature $T_2$, the resulting integral can be reduced analytically, leading to the well-known generalized

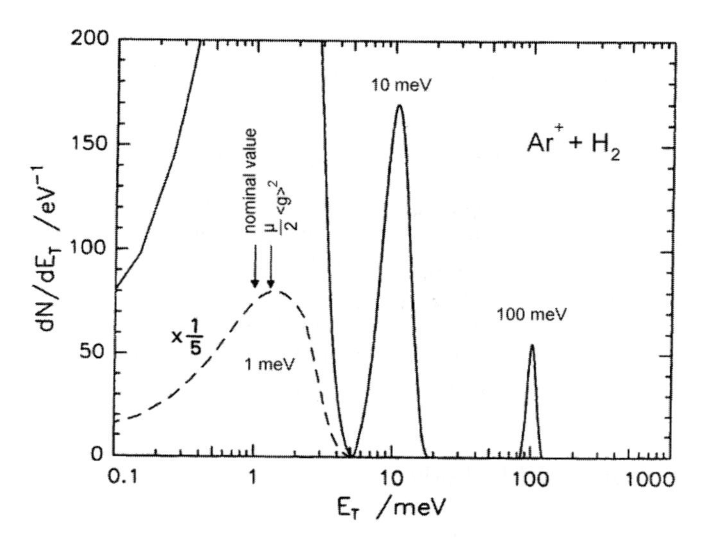


Fig. 3.6. Distributions of the collision energy $dN/dE_T$ obtained by numerical integration over the measured distributions for the nominal collision energies 1, 10, and 100 meV (merged beam arrangement, collision system $Ar^+ + H_2$, $\Delta E_1 = 100$ meV, SR = 10, $\Delta\Lambda = 10°$). The functions are area normalized. At 1 meV the mean collision energy, derived from the mean relative velocity $\langle g \rangle$ is slightly larger than the nominal collision energy.

Maxwell-Boltzmann distribution[7]

$$\begin{aligned} f(g) &= f^*(g; v_1, T_2) \\ &= (m_2/2\pi k T_2)^{1/2} \frac{g}{v_1} \left[ \exp\left( -\frac{m_2}{2kT_2}(g - v_1) \right)^2 \right. \\ &\quad \left. - \exp\left( -\frac{m_2}{2kT_2}(g + v_1)^2 \right) \right]. \end{aligned} \tag{3.21}$$

At low ion velocities, this function approaches a normal Maxwellian $f_M(g; \mu_2, T_C)$ with a reduced temperature $T_C = \{m_1/(m_1 + m_2)\}T_2$. In the majority of ion beam experiments this limit has only been reached for heavy ion masses, i.e., for $m_1 \gg m_2$. In such a situation the lowest attainable collision temperature is equal to the target temperature, $T_C \sim T_2$.

A special solution for obtaining very slow ions is the so-called trapped ion beam technique, a method which has been described in Ref. 10. Using an ion guide, e.g. an octopole, together with meV potential barriers, one can store nearly monoenergetic ions at meV laboratory energies. So far this method has been used only in a few cases, e.g. for $He^+$ stored in $O_2$. For this system, experiments have been performed with $He^+$ ions trapped at laboratory energies as low as 9 meV. In the limit of very slow beams, one can reach a collision temperature of 33 K with 300 K oxygen ($4/36 \times 300$ K).

### 3.3.5. *Cold Effusive Beam — Trapped Ions*

From a kinematic point of view, a similar situation is the combination of a cold effusive or supersonic neutral beam interacting with a stored ion cloud. If the trapped ions are thermalized to a temperature $T_2$, the collision energy distribution is also given by Eq. (3.21), if one assumes a monoenergetic beam. Otherwise one has to integrate in addition over the relevant distribution of $v_1$. This strategy is presently used for cooling trapped ions to temperatures below 1 K. More details are given in Chapter 6.

## 3.4. Experimental Methods

As already mentioned in the Introduction, several sophisticated technologies have been developed for cooling neutral or ionized gases. The evolution of the various methods and the techniques used today in low temperature ion and radical chemistry have been summarized and compared in Chapter 2. Many results have been obtained with ion drift-tubes, supersonic flows or free jet expansions.[11] Other contributions to this book describe techniques which make use of laser cooling, slowing neutrals with suitable electromagnetic fields, or with just one single collision. In the following sections we describe first our strategy to prepare and utilize a slow beam of molecular ions. It is based on the guided ion beam technique. For reaching low collision energies these ion beams are superimposed to a cold neutral beam. Our approach to reach low temperatures is based on multi-electrode traps. As will be seen from the examples, the most important representative is the 22-pole trap.

### 3.4.1. *Guided Ion Beam Merged with a Neutral Beam*

In principle, it is easy to decelerate ions and, therefore, it seems to be straightforward to obtain a very slow ion beam. However, electric or magnetic stray fields, surface patch effects and space charge problems have held up progress in extending standard techniques towards meV collision energies. Another basic problem, which is often overlooked, is the Liouville theorem. Special rf ion guides and an innovative merged beam arrangement have contributed to overcome such difficulties. The first breakthrough to get near-thermal ion beams without buffer gas was due to the guided ion beam technique.[12] Today there are many applications of this technique including the determination of differential cross sections.[13] Despite all these technical advances, it is not straightforward to create an intense slow ion beam. Phase space conservation requires a well-defined ion ensemble to be prepared. There are many ion sources based on electron bombardment, corona-discharge followed by a supersonic expansion or state selective ionization via single or multiple photoionization. Nonetheless the most universal way for cooling and phase space compression is the use of rf traps.

There are several procedures to determine the kinetic energy of ions moving in an ion guide. An accuracy of sub-meV is possible. The elegant oscillating guided ion beam method uses on one side a reflecting wall and towards the detector a semi-transparent one. Fig. 3.7 shows as a typical example a trapped $He^+$ beam. From the distance (in the example shown 32 mm) and the oscillation period the kinetic energy can be determined. The lowest laboratory energy obtained in this way is 2 meV. For more details see Ref. 10. The standard method to calibrate a pulsed guided ion beam is based on time of flight analysis. Usually one measures the mean kinetic energy $\langle E_1 \rangle$ as a function of the voltage U applied to the ion guide, see Fig. 3.8. The deviations from the linear dependence at low energies are due to the energy spread of the ion beam. For a detailed description of the calibration procedure see Ref. 7. Most of the guided-ion-beam (GIB) experiments have been performed by surrounding an octopole with a scattering cell. These arrangements allow one to determine integral cross sections with high accuracy and sensitivity. However, the influence of the random motion of the thermal target gas, usually 300 K, leads to a broadening of the

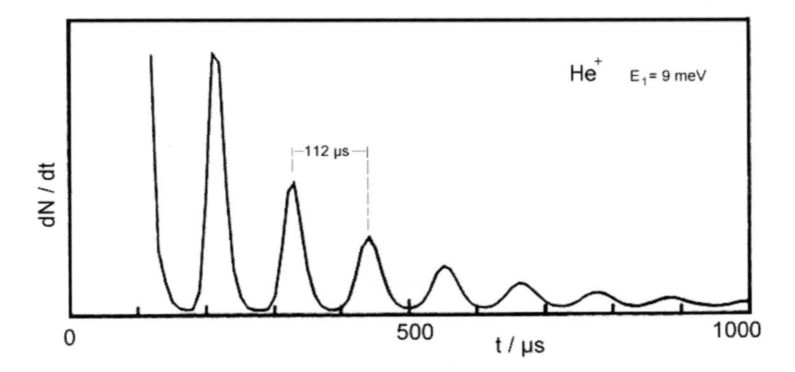


Fig. 3.7. Measured time distribution of a $He^+$ ion beam which is trapped between two electrostatic barriers in an octopole (distance 32 mm). One of them is semitransparent leading to the shown distribution. The voltages on external ring electrodes have been set such that only ions with a kinetic energy close to 9 meV are oscillating. For a detailed description of the technique see Ref. 7.

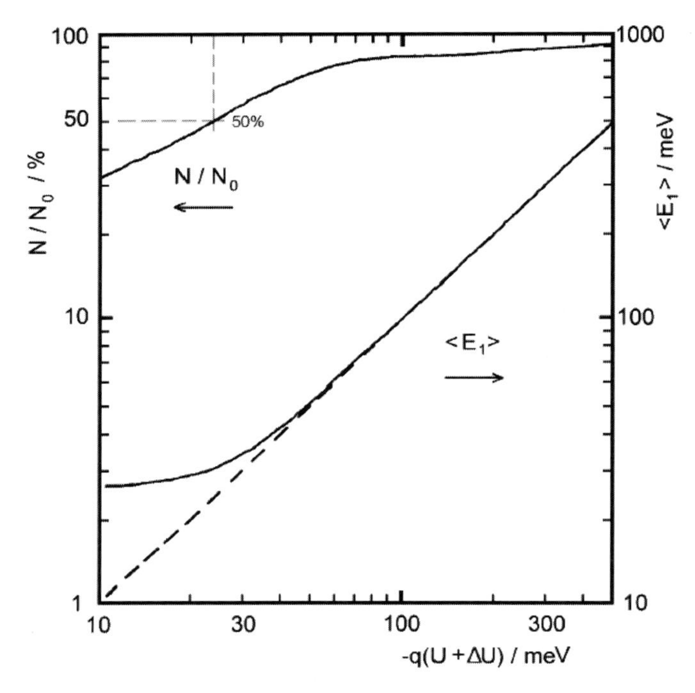


Fig. 3.8. Calibration of the laboratory energy of an ion beam using the retarding field (upper curve) and the time of flight method. Over a wide range the mean kinetic $\langle E_1 \rangle$ is proportional to the voltage $U$ applied to the ion guide. $\Delta U$ is an usually small additive correction accounting for space or surface charges. At low energies, the energy spread of the ion beam plays a role (Liouville theorem). For a detailed description of the technique see Ref. 7.

collision energy (see Eq. (3.21)) and limits the lowest energy. Combinations of a well-characterized ion beam with a supersonic beam allow collisions at very low relative energies to be studied. Collision energies as low as 1 meV have been achieved, with the molecular beam and ion guiding techniques available today sub-meV energies are achievable.

Merging two beams for reaching low relative velocities with high energy resolution is a powerful method. However, despite many advances in technologies, such experiments remain a challenge due to their complexity. Pioneering work started in the 1960s. A recent review explains the fundamental principles of the method and summarizes the worldwide efforts in atomic and molecular physics.[14] In all the instruments described by Phaneuf *et al.*, two fast beams, with parallel velocity vectors having nearly the same magnitude, are superimposed to obtain very low relative energies. It is usually emphasized that kinematic compression allows meV collision energies to be reached, even with beam energy spreads of several eV, if one operates at keV laboratory energies; however, the critical dependence of the energy resolution on the angular spread is often overlooked (see Eq. (3.19)). Owing to the high laboratory energies, it is necessary to create the fast neutral target beam via charge exchange neutralization of an ion beam. All this leads to a number of difficulties and problems, ranging from low reactant densities, a small and uncertain interaction volume, short interaction times, to unknown internal excitation of the reactants. All these drawbacks have impeded widespread use of this method.

Instead of two fast beams, combining a supersonic neutral beam with a slow guided ion beam solves many problems. The two situations are compared in Fig. 3.9 for $H_3^+$ colliding with $D_2$ at a relative energy of 10 meV. In order to reach the required velocity difference of 1.1 km/s, the pair 1.8 km/s and 2.9 km/s is used on the left side while 547.7 and 548.8 km/s in the case of fast beams. The 5 orders of magnitude higher target density and the 100 times longer interaction time makes the slow merged beam technique very efficient. Despite the low laboratory energies involved, the kinematic compression resulting from the merged-beam geometry leads to surprising results. For advantageous systems collision energies below 1 meV can be obtained. The first realization of a merged guided ion and neutral beam apparatus has been described in some detail elsewhere,[7] together with typical results measured for $H_3^+ + D_2$ and $D_3^+ + H_2$. Later improvements include a variable-temperature ion trap for separately cooling the primary ions before they enter the reaction region and an intense continuous supersonic beam, which can be chopped mechanically (see Fig. 3.3 of Ref. 15).

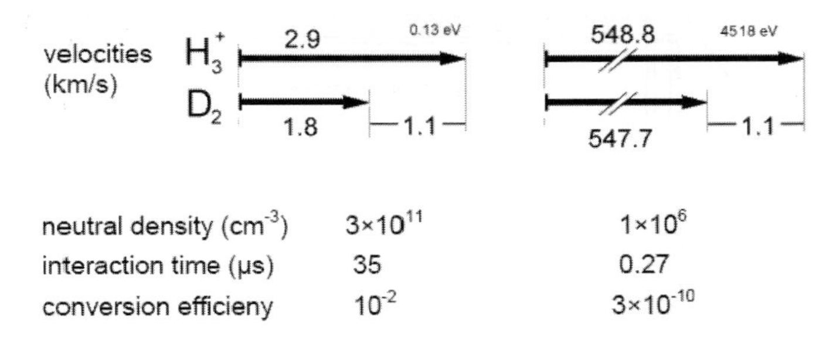


Fig. 3.9. Schematic comparison of two merged beam arrangements for studying collisions between $H_3^+$ and $D_2$ at a selected relative velocity of 1.1 km/s, corresponding to a collision energy of 10 meV. On the left side the laboratory energy of the ion is 0.13 eV, on the right side 4518 eV. The low target density and the short interaction time makes the fast merged beams method very inefficient while the use of a supersonic beam superimposed to a slow guided ion beam can allow to convert a few % of the primary ions into products.

For rotational relaxation of hydrogen, a para-hydrogen generator has been used and the nozzle temperature has been cooled down to 10 K. Several differential pumping stages are used to maintain a sufficiently low background pressure of the target gas in the interaction chamber. Unfortunately, it is not straightforward to develop in an university laboratory routines and strategies for operating and maintaining such a complex instrument in a professional way.

A rather simple version of a slow merged-beam apparatus is depicted in Fig. 3.10. Here the slow ion beam is guided in a wire quadrupole. With such a multi-electrode structure the electrodes of which are at large distances, extremely slow and narrow ion beams can be prepared.[16] For creation

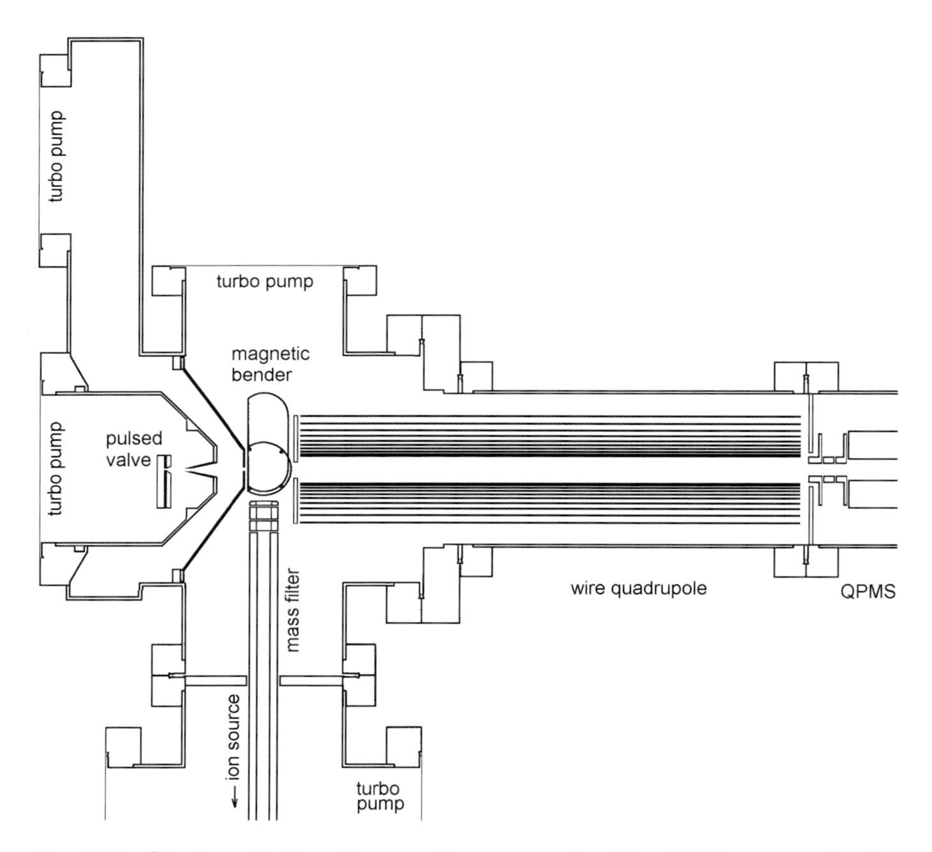


Fig. 3.10. Simple realization of a merged beam apparatus. For obtaining a very slow ion beam a wire quadrupole is used as ion guide. The pulsed neutral beam is produced using a piezo driven valve. A rather complicated arrangement where both ions and neutrals can be cooled to a few K has been presented in Ref. 44.

of the supersonic beam a pulsed valve is used. Two differential pumping stages maintain a sufficiently low background pressure of the target gas in the interaction chamber. The well-collimated neutral beam has a density of up to $10^{11}\,\mathrm{cm}^{-3}$ and an angular divergence of much less than $10°$. For preparing ions the standard combination of a storage ion source and focusing quadrupole is used. The ions are pulsed synchronously with the neutral beam. The mass- and energy-selected ions are deflected by $90°$ in a weak magnetic field of less than 1 kG and then injected at the desired energy into a weak rf guiding field. There, they propagate in the same direction as the neutral beam and react at low relative energies. Calibration of the ion axial energy, $E_1$, and determination of the energy distribution is performed using the method described above. To measure the mean velocity $\langle v_2 \rangle$ and the spread of the neutral beam (FWHM, $\Delta v_2$), an additional differentially pumped ionization detector is mounted subsequent to the ion detector at a distance of about 1 m from the nozzle. For this application, a quadrupole ion guide is preferred over a higher order multipole, since the harmonic effective potential confines the ion beam so close to the axis that full overlap with the neutral beam can be achieved. The impairment of the collision energy is negligible since only the transverse velocity is modulated. In such an arrangement one can operate routinely with ion laboratory energies down to 50 meV. For many mass ratios a nominal relative velocity of zero can be obtained. In some cases, it is advantageous to seed the neutrals in order to operate at higher laboratory velocities.

As a typical set of raw data recorded with a merged beam arrangement, Fig. 3.11 shows an $H_2O^+$ signal produced via electron transfer from $H_2O$ to $Ar^+$. The uncorrected, directly measured product intensity increases very steeply at an $Ar^+$ laboratory energy of 0.62 eV. At this energy the ion beam and the $H_2O$ beam move with the same velocity. In the present example the steep increase of the product rate versus zero relative velocity, $|\boldsymbol{g}| = 0$, depends very critically on the operating conditions of the supersonic beam. The reason is that the high pressure expansion not only leads to a well-defined collision energy, characterized in our kinematic considerations with the speed ratio, but also to rotational cooling of the target molecules. In order to analyze such results quantitatively, the resulting rotational distribution of the $H_2O$ in the beam has to be known. First attempts have been made.[17] Since cooling of molecules to translational temperatures around 1 K has become routine in supersonic jets, collisions in the sub-meV range are possible.

Much effort has been put into the development of special merged beam arrangements which allow one to study the interaction between electrons and cold ions at meV collision energies. One example is the ion storage ring TSR where fast ions are cooled with an intense cold electron

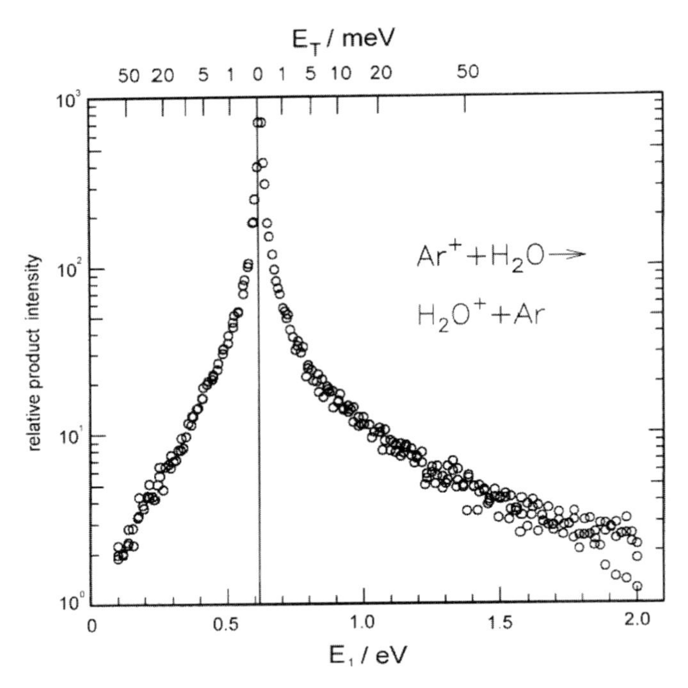


Fig. 3.11. Electron transfer from $H_2O$ to $Ar^+$. The uncorrected, directly measured product intensity increases very steeply at an $Ar^+$ laboratory energy of 0.62 eV. As can be seen from the upper scale, this corresponds to a nominal collision energies of 0 eV, e.g. the two beams move with the same velocity. The steepness depends critically on the expansion conditions since it influences the rotational population of the $H_2O$ molecules in the beam.[17]

beam and where they can interact with a separate electron target. In the recombination of electrons with hydrogen molecular ions, rovibrational resonances have been measured down to a few meV.[18] Electrons with translational temperatures of a few K have been created using a cryogenic photocathode. To cool the ions, e.g. $H_3^+$, internally a low temperature 22-pole trap has been used as ion source. First steps towards state selective measurements of dissociative recombination rate coefficients have been made, more work is in progress. The set-up also allows for an *in-situ* diagnostic of the population of low-lying rovibrational states using chemical probing in the radiofrequency multipole ion trap.[19]

### 3.4.2. *RF Ion Traps, Buffer Gas Cooling*

Various ion traps have been used since the 1960's for studying ion-molecule reactions. With suitable electric or magnetic fields, ions can be confined for long times and, with the use of buffer gas, thermal conditions can be attained. The most common techniques used in chemistry are the ion cyclotron resonance (ICR) machines and Paul traps. Much of our own work has been stimulated by the liquid helium-cooled Penning ion-trap experiment of Dunn and co-workers.[20] Despite the fact that magnetic confinement is a dynamic trap, i.e., kinetic energy is needed for trapping, collision temperatures below 10 K have been reached for collisions between hydrocarbon ions and hydrogen; i.e., for favorable mass ratios.[21]

As emphasized above, fast oscillating forces on the boundary of a wide field free region are the best solution for confining and cooling charged particles. In contrast to storage rings, fast beams trapped between electrostatic mirrors or Penning traps, the most important feature of such traps is the possibility of relaxing ions by collisions with buffer gas. Very important in this context is Eq. (3.9). Since the principle is very general, one can tailor the device to specific boundary conditions imposed by neutral or laser beams and by optical detectors. An example for a specific geometrical solution is the U-shaped storage ion source.[7] Very successful and versatile solutions utilize multi-electrode arrangements, such as stacks of ring electrodes or linear multi-pole traps, combined with cryogenic cooling. For many applications, the work horse has become the 22-pole ion trap (22PT). First results from this device have been reported in a review on radiative association.[22] A thorough description has been presented at the Nobel Symposium 91 on *Trapped Charged Particles and Related Fundamental Physics*.[23] As shown in Fig. 3.12 the trapping electrodes are mounted onto a closed cycle helium refrigerator in order to obtain a cold environment. When the trap is surrounded by two radiation shields, temperatures as low as 3.6 K have been reached with modern cold heads.

Today some ten 22-pole traps are in operation worldwide. They are used for reaction dynamics,[24,25] laser induced reactions,[26] or as an ion source for preparing cold ions for the electrostatic storage ring in Aarhus. As already mentioned above, a 22PT is used to cool $H_3^+$ ions at the ion storage ring TSR in Heidelberg for measuring state specific rate coefficients for dissociative recombination.[19] In a new 22PT machine, absolute photodetachment cross sections of negative ions have been determined.[27] Two 22PT based instruments became operational recently aimed at measuring spectra of large cold ions of relevance to astronomical observations[28] or of very cold protonated biomolecules.[29] One typical arrangement of an ion-trapping machine is shown schematically in Fig. 3.13. In such a "standard arrangement" the primary ions are created externally by electron bombardment and prethermalized in a high-pressure storage ion source. For selecting the mass of the primary ions a quadrupole filter is used. It is best to operate this filter in the high frequency, low mass filter mode to avoid rf heating of the kinetic

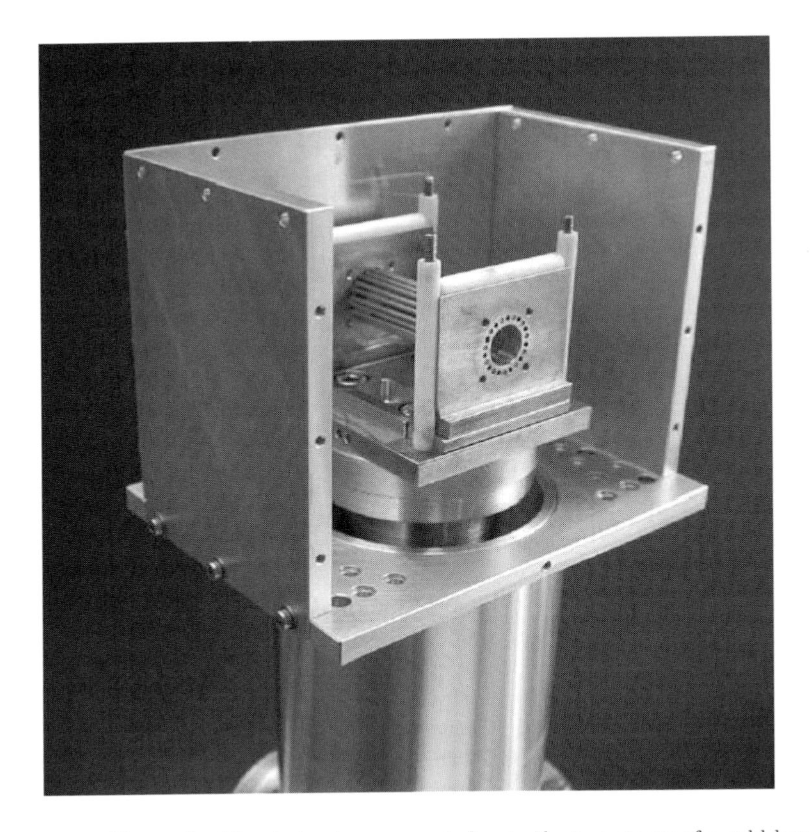

Fig. 3.12. Photo of a 22-pole ion trap, mounted onto the two stages of a cold head.

energy of the ions. There is also progress with *in situ* synthesis of reactants within the trap. Most trapping machines are optically transparent in the axial direction, at least from one side, for facilitating applications of laser methods.

Upon entering the trap, the ions are thermalized by injecting them together with a very intense, short He or $H_2$ gas pulse directly into the inner box (see Fig. 3.12). It is also possible (but not efficient) to operate at very low number densities and to wait long enough for the trapped ions to be internally cooled by radiation. As can be seen from Fig. 3.14 a fast piezo-valve makes it possible to raise the pressure in less than 1 ms. Peak number densities up to $10^{15}\,\mathrm{cm}^{-3}$ have been reached. With higher pumping speed or with mechanical shutters closing the trap, one can go much higher. The dense buffer gas leads to thousands of collisions in a few ms and, therefore, to a very efficient relaxation of all degrees of freedom of the stored ions.

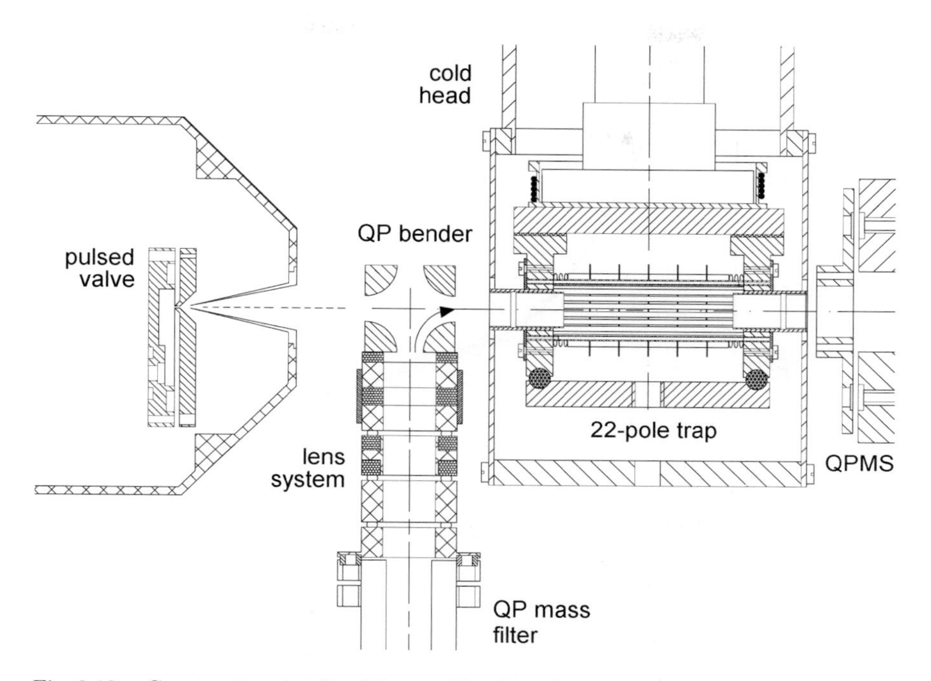


Fig. 3.13. Construction details of the combination of a temperature variable 22-pole ion trap and a pulsed supersonic beam. Primary ions are produced in a storage ion source and mass selected in a quadrupole mass filter (from the bottom, not shown). After deflection in an electrostatic quadrupole bender the ions are injected into the trap (see Fig. 3.12). There the ions are confined in radial direction by the rf field and in axial direction with dc voltages applied to the entrance and exit electrodes (less than 100 mV). Buffer gas is used, usually He, for thermalizing the ions. From the left a skimmed molecular beam traverses the trap without colliding with the cold surfaces. For detection, the stored ions are extracted to the right, mass selected and counted via a Daly detector. A laser beam can be injected via the detector tract for state selective excitation of the trapped ions.

The FWHM of the gas pulse is 5 ms and, within 20 ms, the number density drops with a fast time constant of 3.5 ms by 2 orders of magnitude. The subsequent decay has a larger time constant of 23 ms; however, in most cases the rates of ternary processes involving the buffer gas atoms are negligible after 100 ms.

In typical experiments, reactions are studied by periodically filling the trap with ions, buffer gas and reactants. Depending on the process to be studied, repetition periods between ms and many minutes are chosen. After a certain storage time, primary and product ions are extracted from the trap, mass analyzed, and detected with a scintillation detector. The delay of the extraction pulse can be varied within the repetition period, i.e., from $\mu$s to min or longer. In the absence of target gas, the mean decay

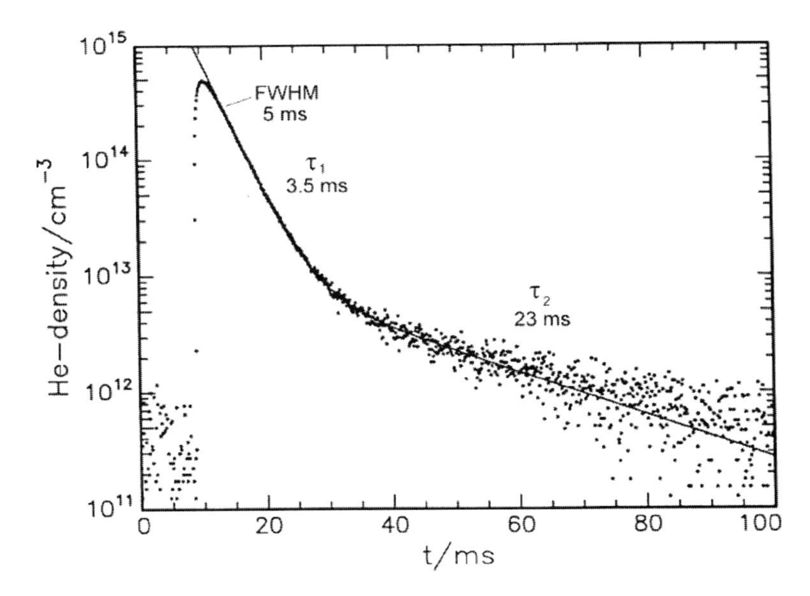


Fig. 3.14. Time dependence of the intense gas pulse injected into the ion trap using a piezo valve. During their transfer into the trap, ions are cooled with buffer gas, in most cases He. The width of the pulse (FWHM = 5 ms) and the two decay times (3.5 ms and 23 ms) are determined by the geometry of the small housing surrounding the trap, the volume of the vacuum chamber and the pumping speeds of the pumps.

time of the number of specific ions is determined by reactions with background gas, the pressure of which can be below $10^{-12}$ mbar, especially at low temperatures. For the study of specific collision processes, target gas is added continuously or in a pulsed mode at number densities varying from below $10^9\,\mathrm{cm}^{-3}$ to above $10^{14}\,\mathrm{cm}^{-3}$. By choosing a suitable combination of number density and interaction time, the rates of rather fast as well as very slow bimolecular reactions can be measured. The wide range of operating parameters makes the trapping device unique. For example, very slow radiative association processes have been separated from ternary association. For comparing forward-backward reactions, e.g. in the case of isotopic enrichment, a special feature of the trapping technique is that one can choose a well-defined gas mixture and evaluate the stationary equilibrium reached after a sufficiently long storage time. At very low collision rates and long storage times, the coupling of the internal degrees of freedom to the low temperature black-body radiation field has also been observed in some cases.

A state of the art trapping machine, the Atomic Beam 22-Pole Trap (AB-22PT) is shown schematically in Fig. 3.15. This instrument has been

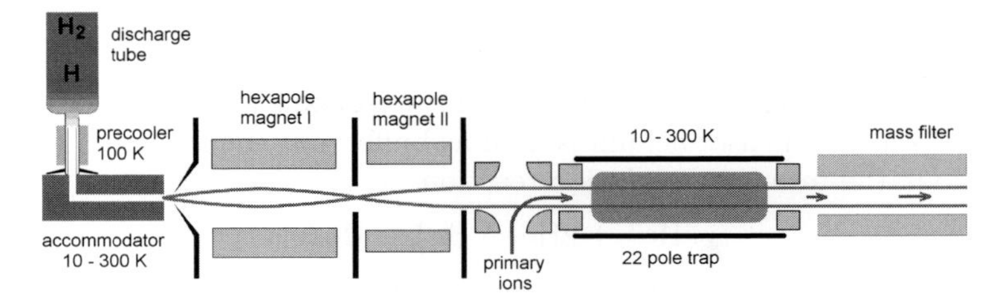


Fig. 3.15. Schematic diagram of the Atomic Beam 22-Pole Trap Apparatus (AB-22PT). This instrument has been developed for exposing cold trapped ions to an effusive beam of H atoms. The velocity distribution of the hydrogen beam depends on the temperature of the accommodator and the transmission features of the two hexapole magnets. A short description of this setup has been given recently.[24] A detailed documentation of this sophisticated instrument is in preparation.[30]

developed for exposing cold trapped ions to a slow effusive beam of H atoms. A first description of this set-up can be found in a conference proceedings.[24] The crucial part of this machine is the differentially pumped effusive target beam. In the machine shown in Fig. 3.15, hydrogen atoms are produced via an rf discharge. After their precooling to 100 K, the atoms are thermalized in an accommodator to a temperature which can be as low as 10 K. As tested by time of flight measurements, the final velocity distribution of the hydrogen atoms is strongly influenced by the transmission features of the two hexapole magnets. A publication describing this sophisticated instrument in detail is in preparation.[30]

### 3.4.3. *How to Determine Temperatures*

As discussed above, there is always a competition between buffer gas cooling and rf heating if one uses oscillatory fields for trapping the ions. Many collisions or long storage times lead to a stationary equilibrium but not necessarily to the temperature of the ambient buffer gas or the black-body radiation field penetrating the ion cloud. Therefore one has to provide tools for measuring the actual temperature of the trapped ions. It is well known from experiments in Paul-traps that ion-neutral collisions lead to a significant transfer of energy from the radio-frequency field to the motion of the ions. One of the proofs has been the appearance of products from reactions with rather high endothermicities. As discussed above the heating process is significantly reduced in traps with steep rf potential, e.g. defined by a stack of ring electrodes or high order multi-poles. In addition, the wide field free regions support cooling collisions to the temperature of the buffer gas. Numerical simulations and experimental tests quantitatively support these considerations.

Various experimental tests have been performed to determine the translational energy distribution of the trapped ions or the population of internal states. One of them is based on temperature dependent rate coefficients. Since in almost all cases helium is used as buffer gas, the formation of $He_2^+$ dimers via the ternary association reaction

$$He^+ + 2He \rightarrow He_2^+ + He \tag{3.22}$$

is a good test reaction for characterizing a low temperature trap. This reaction has been studied over a wide temperature range in a selected ion drift tube (SIDT) apparatus by Böhringer *et al.*[31] Their measurements, plotted in Fig. 3.16 as open squares, have been extended towards lower temperatures operating the 22-pole trap at [He] number densities in the range of $10^{14}\,\text{cm}^{-3}$. The results have been fitted with the function

$$k_3 = 1.4 \times 10^{-31}(300\text{K}/T)^{0.6}\,\text{cm}^6\text{s}^{-1}. \tag{3.23}$$

Presently this temperature dependence is used as a calibration standard above 10 K. However, for extending it towards lower temperatures, more precise experiments are needed in order to separate the influence of ion temperature, density of the neutrals and the temperature dependence of the ternary association reaction.

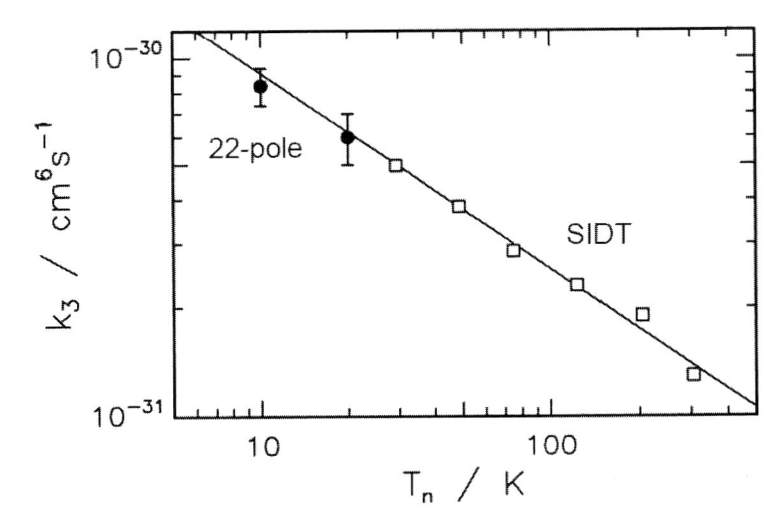


Fig. 3.16. Temperature dependence of the ternary rate coefficients for the reaction $He^+ + 2He \rightarrow He_2^+ + He$. The filled circles are data obtained with a 22-pole trap, the open squares have been measured using a selected ion drift tube.[31] The fit $k_3 = 1.4 \times 10^{-31}(300\,\text{K}/\text{T})^{0.6}\,\text{cm}^6/\text{s}$ is used as a "thermometer". Note, however, that more precise experiments are needed in order to separate the influence of ion temperature, density of the neutrals and the temperature dependence of $k_3$.

The *in situ* growth of weakly bound clusters, e.g. hydrogen clusters $H_n^+$, also depends on the temperature.[32] A related sensitive quantity is the final cluster size distribution as discussed in Ref. 23. Problematic is that the thermodynamics of weakly bound clusters are only known from high pressure measurements and more studies are needed at low temperatures and low pressures. Other possible processes include the formation and decay of protonated or deprotonated water clusters[33] or the enrichment of ions with isotopes, e.g. H-D exchange. First tests have been made with isotope exchange using $^3He/^4He$ or $^{12}C/^{13}C$. In general, there is not yet any ideal chemical reaction with a well-characterized low temperature behavior. Even the process of deuteration, where the energetics are best known from the differences of zero point energies, causes problems.[3] An interesting collision system with a strong temperature dependence is $N^+ + H_2$. However, as discussed below, the formation of $NH^+$ is not yet suitable for ion thermometry. At the moment this reaction is used as a sensitive probe for minor traces of ortho-hydrogen.

In order to determine ion temperatures more directly, laser based techniques have been used and new schemes of "action spectroscopy" are under development. The Doppler profile of an observed transition provides direct information on the ion velocity distribution along the laser beam. In addition a quantitative analysis of the population of rotational or other suitable internal states can provide information about how cold the ions are internally. The first successful application based on the laser induced reaction

$$N_2^+ + h\nu + Ar \rightarrow Ar^+ + N_2 \tag{3.24}$$

has been discussed in detail by Schlemmer *et al.*[34] In order to determine the translational temperature of the ion cloud the line shapes of several rovibrational transitions were recorded and fitted with Gaussians. The temperature of the fitted Maxwell-Boltzmann velocity distributions has been in good overall agreement with the nominal temperature of the trap. Also the rotational population of the $N_2^+$ ions was in accord with the wall temperatures. So far, these measurements have not been extended to temperatures below 50 K because of condensation of the Ar gas on the walls; however, work is in progress to use target beams for chemical probing.

The first Doppler profile at the present temperature limit of the trapping technology, 10 K, has been recently reported by Glosik *et al.*[35] In this experiment an overtone transition in $H_2D^+$ ions, which have been trapped in hydrogen, has been excited at $6536.317(8)\,cm^{-1}$. This additional energy allows the $H_2D^+$ ions to reconvert in collisions with $H_2$ into $H_3^+$ ions, an endothermic process. The kinetic temperature, derived from the Doppler profile was $(9 \pm 4.5)$ K. More details and examples are given in chapter 6 which describes the production, control and study of ultra-cold molecular ions.

### 3.4.4. *Sub-Kelvin Cooling of Trapped Ions*

It has often been stated that an experimental limitation of low temperature ion traps is condensation of the neutral species on the walls. Since usually only very low number densities are required one can operate the trap — at least in principle — at such a low temperature that the resulting vapor pressure is just sufficient. The example shown in Fig. 3.17. illustrates that one can operate a trap at 38 K and simultaneously maintain a nitrogen number density of $10^9\,\mathrm{cm}^{-3}$ by leaking a stationary gas flow of $3 \times 10^{-8}$ mbar $\mathrm{l\,s}^{-1}$ into the trap. In this test, the dissociative charge transfer $\mathrm{He}^+ + \mathrm{N}_2 \rightarrow \mathrm{N}^+ + \mathrm{He} + \mathrm{N}$ has been used to follow *in situ* the time dependence of the nitrogen density. The example shows that it takes about 1 hr to reach stationary conditions. At the beginning of the experiment pumping of the cold clean surfaces maintains a stationary number density below $10^7\,\mathrm{cm}^{-3}$. From the constant flow one can conclude that cryopumping has a pumping speed of a few l/s.

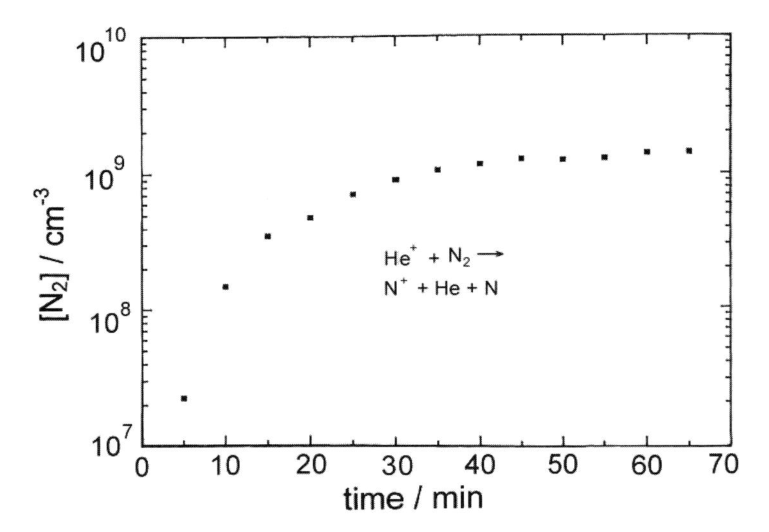


Fig. 3.17. The dissociative charge transfer $\mathrm{He}^+ + \mathrm{N}_2 \rightarrow \mathrm{N}^+ + \mathrm{He} + \mathrm{N}$ has been used to follow the time dependence of the nitrogen density in the trap at a temperature of 38 K. The gas flow has been hold constant gas at $3 \times 10^{-8}$ mbar $\mathrm{l\,s}^{-1}$. At the beginning of the experiment, the pumping speed of the cold clean surfaces maintains a rather low stationary density. With increasing time, saturation reduces the sticking efficiency.

A more versatile solution to extend the ion trapping method towards lower temperatures and to use also condensable gases for reactions is to combine linear traps with skimmed beams of molecules, atoms or radicals. Geometrical arrangements are similar to those shown in Figs. 3.13 and 3.15. Presently special beam sources are under development for cooling trapped ions to temperatures below 1 K. The method is based on the simple idea of using only the slow Boltzmann tail of a cold effusive beam (e.g. He or p-$H_2$ at a few K). This can be achieved by pulsing the beam and chopping the entrance of the trap. Simple calculations show that there are still enough cooling collisions if one cuts off 90% of the thermal distribution. The remaining distribution is *very cold*. For specific purposes and depending on the mass ratio, supersonic beams may also be used to collisionally cool the ions. As already mentioned above, a unique advantage of the skimmed beam is that cooling collisions take place only in the field free region. With efficient cryopumping there are practically no collisions during the interaction of the ions with the rf wall of the trap. Another strategy is based on a cold effusive H-atom beam (see for example Fig. 3.15) in combination with a weak magnetic field which guides slow H atoms. The ultimate solution is to superimpose a magnetic trap for ultracold H atoms with the multipole ion trap.

There are a variety of *in situ* experiments (spectroscopy) planned which are of fundamental importance for characterizing such a trap for creating brilliant beams of ultracold molecular ions, including biomolecules (e.g. from an electrospray ion source). It is also possible to synthesize directly in the trap special ions such as clusters or molecules with van der Waals bound ligands. More details, e.g. how to record a pure rotational spectrum of ultracold $CH_5^+$ ions, are discussed in Chapter 6.

### 3.4.5. *Collisions in a Sympathetically Cooled Ion Crystal*

Another strategy to reduce the translational energy of ions in traps is based on laser cooling. For more than 20 years atomic ions, which are confined in Penning traps or radio frequency quadrupole ion traps, have been cooled with lasers to such low temperatures that they form regular structures.[36] This process, called Coulomb crystallization, has attracted a lot of experimental and theoretical interest. As already mentioned in the Introduction, the condensation to such structures is characteristic for *temperatures* in the sub-mK or even $\mu$K range. With the discovery of sympathetic cooling of other ions mixed into the cloud of atomic ions the application of Coulomb crystals has been significantly extended.[37] Some more aspects of this very interesting subject are discussed in Chapter 6. Here we just want to mention briefly that this fascinating technique is able to observe ion-neutral reactions with single-particle resolution as described recently by Roth *et al.*;[38] however it is misleading to call such processes ultracold collisions.

There has been significant progress in recent years in using sympathetically cooled ions in linear quadrupole traps.[39,40] An interesting publication discusses the production of ultracold diatomic and triatomic molecular ions of spectroscopic and astrophysical interest.[41] With clever experimental detection schemes in combination with molecular dynamics simulations one can study chemical reactions without seeing the reactants or the products. The information is derived from images of the fluorescent atomic ion ensemble and their perturbation by the dark ions. The localization of these ions in specific regions together with mass selective excitation of the secular motion of the trapped ions allows one to follow chemical reactions. For small Coulomb crystals changes in the ion cloud can be detected at a single-ion level. For example, formation of localized $MgH^+$ and $MgD^+$ molecules in collisions of $Mg^+$ ions with hydrogen and deuterium has been reported.[39] A recent publication reports rate coefficients for reactions of $H_3^+$ ions with various neutrals.[38]

Despite the fact that in all these examples so-called *ultra-cold* ions are involved in the reactions, the chemistry occurs at rather high collision energies and most probably with hot reactants. The problem is that one has to distinguish between the *collision temperature*, i.e., the energy distributions in the center of mass frame which is determined by the rf driven motion of the ions and the temperature of the neutrals and the *optical temperature*, derived from the first order Doppler effect. It seems to be contradictory but in certain processes, e.g. absorption of a photon, the ions confined in a Coulomb crystal can be characterized with mK or even $\mu$K temperatures, while in other processes, e.g. chemical reactions, they contribute energy which is equivalent to several 1000 K. The explanation is the huge kinetic energy associated with the oscillating motion. This velocity component is not seen in the optical transitions as long as it is orthogonal to the laser beam and if there is no coupling of the radial and axial motion of the ions. An example illustrating nicely the overall situation is the recent high-resolution detection of rovibrational infrared transitions in $HD^+$.[42] A Doppler width corresponding to 0.2 K has been achieved with ions which have been sympathetically cooled by $Be^+$ to 20 mK.

A quite different problem impeding the study of *cold chemical reactions* in ion crystals is the internal energy of the stored ions. While Coulomb interactions are very efficient at sympathetically cooling translation, the internal degrees of freedom of molecular ions are not coupled to the laser cooled atomic ions. The creation of ultracold ions in rf traps is further discussed in Chapter 6.

### 3.5. Selected Results

The various methods described above, and also in Chapter 2, have been used to study a large number of ion-molecule reactions over a wide range of densities, temperatures or energies. The first application of a temperature variable multi-electrode trap, consisting of a stack of ring electrodes, has been the study of the association reactions of $C^+$ and $H^+$ with $H_2$ at 230 and 320 K.[43] Results from an early version of the merged beam apparatus have been mentioned in Ref. 7. Evidence for the unique sensitivity of ion traps is provided by the systematic studies performed on radiative association, a reaction mechanism of astrochemical importance.[22] Examples such as $NHe^+$ formation in $N^+ + 2He$ collisions have illustrated that one can measure extremely small rate coefficients with a few hundred ions.[44] An interesting application of traps is to study the dynamics of the growth of clusters. Starting from $H_3^+$, hydrogen cluster ions up to $H_{23}^+$ have been formed in collisions with para- and normal-hydrogen at 10 K. Ion trapping allows one to perform a systematic study of the dynamics of the association, cooling and fragmentation processes under well defined thermal conditions.[32] Especially important is that the choice of an adequate density enables one to stretch the time scale for the growth processes such that individual steps can be traced in great detail. Establishing stationary equilibria in a trap, thermodynamic data can be determined. A typical result has been published for protonated and deprotonated water cluster ions.[33]

A first low temperature study of reactions between stored ions and condensable gases, the formation of protonated methanol via radiative association, has been reported.[45] The trapping method also allows for a lot of additional tests. For example the method of chemical probing has been used to determine the rate coefficient for isomerization of $HOC^+$ in collisions with $H_2$. The resulting rate coefficient obtained at 25 K is surprisingly large, $3.8 \times 10^{-10}\,\mathrm{cm^3\,s^{-1}}$.[2] Some very recent applications for addressing key problems in laboratory astrochemistry have been summarized by Gerlich and Smith.[25] In the following, a few selected topics are discussed in more detail.

#### 3.5.1. *The Role of Fine-Structure, Rotational, and Zero Point Energy*

Chemistry at low temperatures can be significantly influenced by the excitation of low lying electronic states, e.g. fine-structure or even hyperfine-structure levels, by rotational and soft vibrational motions, and by gain or loss of a small amount of energy caused by isotope substitution. The interpretation of low energy collisions often suffers from the fact that one does not know how these different forms of energy help to promote the reaction in the endothermic direction. It is also often unclear whether the formation of a new product really needs energy or whether a small activation barrier has to be overcome. Quantum chemical calculations are not yet sufficiently accurate to predict the energies which are required for understanding collisions at a few K. A typical example is the hydrogen abstraction reaction

$$N^+ + H_2 \rightarrow NH^+ + H. \tag{3.25}$$

Since this simple triatomic reaction is of fundamental importance and since formation of interstellar ammonia occurs via this process and subsequent hydrogen abstraction reactions, it has been studied very often. Nonetheless, there are still many open questions. It is hard to believe that one does not yet knows the energy balance, although one could get it with spectroscopic accuracy just by measuring the dissociation energy of $NH^+$. The best value which is presently available for the endothermicity of reaction (3.25) is $\Delta H = 11\,\mathrm{meV}$ for $H_2$ and $\Delta H = 29\,\mathrm{meV}$ for $D_2$. Note that the differences of the two values are not in accord with the differences of the asymptotic zero point energies. This indicates that there are intermediate barriers slowing down the reaction. The $\Delta H$-values have been derived from a variety of low temperature measurements most of which are more than 10 years old. According to the best of our knowledge there are no recent measurements or new calculations.

As explained in Ref. 44, the temperature dependence of rate coefficients can be measured rather quickly during the cooling of the trap, which typically takes 1 hr. Fitting experimental results measured for n-$H_2$ and p-$H_2$ has resulted in temperature dependent rate coefficients,

$$k(\text{n-H}_2) = 1.1 \times 10^{-11} \exp(-26\,\mathrm{K/T})\mathrm{cm^3s^{-1}}, \tag{3.26}$$

$$k(\text{p-H}_2) = 1.4 \times 10^{-9} \exp(-230\mathrm{K/T})\mathrm{cm^3s^{-1}}. \tag{3.27}$$

The lowest rate coefficient, which has been measured with p-$H_2$ at 15 K ($k$(15 K, 99% p-$H_2$) $= 1.4 \times 10^{-13}\,\mathrm{cm^3s^{-1}}$), has been explained with an o-$H_2$ impurity of $\sim$1%. For pure para-hydrogen calculations predict $1 \times 10^{-14}\,\mathrm{cm^3s^{-1}}$.

The interpretation of such thermal rate coefficients is complicated by the fact that the endothermicities $\Delta H$ or the barriers involved are comparable with the rotational energy of $H_2$ (14.4 meV for j = 1) and with the fine structure energy of $N^+$($^3P_0$ 0 meV, $^3P_1$ 6.1 meV, $^3P_2$ 16.2 meV). As long as there are no measured state specific cross sections, assumptions have to be made about the role of these different energy forms in driving the reaction. Therefore, the best evaluation one can perform is based on state specific cross sections determined with a statistical theory.[46] The situation

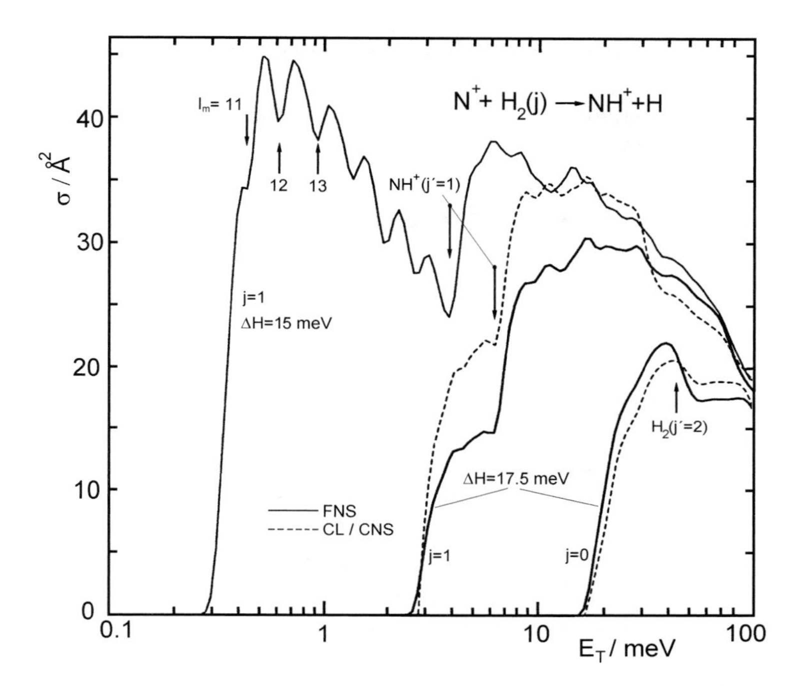


Fig. 3.18. Phase space calculation of state specific cross sections for the reaction $N(^3P_0)^+ + H_2(j) \rightarrow NH^+ + H$. The endothermicity which is not yet known with the required accuracy, has been assumed to be $\Delta H = 17.5$ meV and 15 meV. For more details see the text.

is illustrated in Fig. 3.18 where calculated state-specific cross sections are plotted as a function of the collision energy. It is obvious that, ideally, one would have to perform state specific experiments at meV collision energies.

In order to study separately the influence of translational, rotational and fine-structure energy on the threshold onset of a few meV, guided ion beam experiments have been performed for $N^+ + D_2 \rightarrow ND^+ + D$.[47] High-energy resolution has been achieved by combining a slow guided ion beam with a supersonic deuterium beam, in one case by crossing the two beams, in the other case by merging them. Fig. 3.19 shows a typical result from the merged beam arrangement. In these measurements every effort has been made to approximate "ultracold" conditions; i.e., to prepare both reactants exclusively in their ground state. For the neutral target, para-hydrogen has been expanded from a liquid-nitrogen cooled nozzle, the $N^+(^3P_J)$ ions have been relaxed in a trap using a 20 K He buffer. As can be seen from the figure, the reduction of rotational energy has a significant influence while

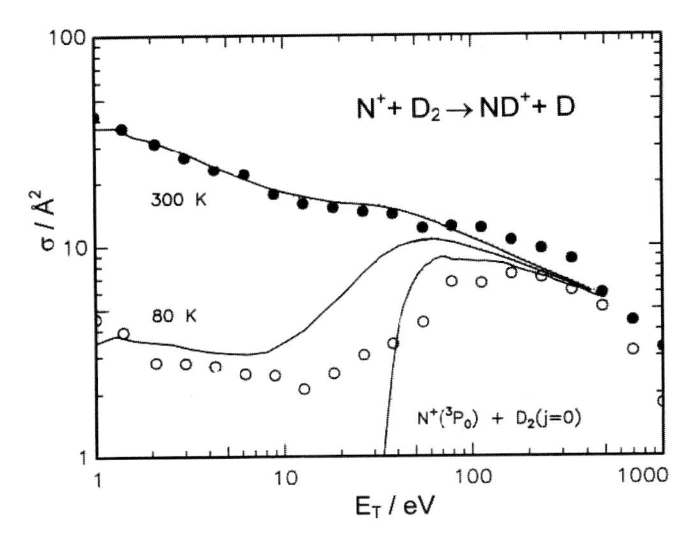


Fig. 3.19. Merged beam results for the reaction $N^+ + D_2 \rightarrow ND^+ + D$. The experimental data which have been measured for various target temperatures, are compared with theoretical cross sections, based on phase space calculation.[17]

no changes in the reactivity could be detected as a function of the "ion temperature".[47]

This is a hint that the $N^+$ fine structure energy may not be available for promoting the reaction. A careful trapping experiment corroborates this. In this experiment the $N^+$ ions have been created by electron bombardment in a 350 K ion source. Under such conditions it is safe to assume that this leads to a statistical population of the three fine- structure states $^3P_0$, $^3P_1$, and $^3P_2$, i.e. to the ratio 1 : 3 : 5. If the excited ions react faster, one would get two breaks in the slope of the reaction decay curve. In the experiment the number of $N^+$ ions decreases with a single time constant over several orders of magnitude. Also the competition between collisional relaxation and reaction has been excluded.[47] For more critical tests, experiments with state selectively prepared $N^+(^3P)$ reactants are needed or one has to find an independent method to prepare and test a low temperature fine structure ensemble (see Chapter 6).

Isotope exchange has often been used in experiments in order to get detailed information on the microscopic dynamics. Since most of such experiments have been performed at high energies, it was assumed that the minor differences in zero point energies could be ignored. At low temperatures, however, these differences become important, especially if one replaces an H atom by a D atom in a molecule. This can lead to isotope enrichment, a special field which has attracted a lot of interest in recent years. The relevance for astrochemistry has been emphasized in Chapter 1. In addition to D, important isotopes which are routinely detected in interstellar molecules are $^{13}C$, $^{15}N$. In order to correlate quantitatively isotope ratios in molecules with the isotope ratios of the elements one has to know in detail the kinetics and dynamics of a variety of processes including chemical reactions, inelastic collisions and radiative transitions.

Low temperature ion traps are ideal tools for studying such processes and have been used rather often in recent years.[26] Selected ion trap studies on deuterium fractionation have been reviewed by Gerlich and Schlemmer.[48] Recent results for the fundamental $H_mD_n^+$ collision systems have been discussed by Gerlich *et al.*[3] and are briefly mentioned below. Many laboratory studies have concentrated on the deuteration of hydrocarbon ions in collisions with HD. For example, sequential deuteration of or $CH_3^+$ or $C_2H_2^+$ in collisions with HD has been found to be rather fast, although there are deviations from simple statistical expectations.[49] In collisions of $CH_4^+$ with HD, the $CH_5^+$ product ion (68% at 15 K) prevails over $CH_4D^+$ (32%).[50] The scrambling reaction

$$CH_5^+ + HD \rightarrow CH_4D^{++} + H_2, \tag{3.28}$$

is extremely slow if it occurs at all. For the rate coefficient an upper limit of $5 \times 10^{-18}$ cm$^3$ s$^{-1}$ has been determined,[49] a nice illustration of the sensitivity of the 22PT based instruments.

Early ion trap studies of isotopic variants of the $C_3H^+ + H_2$ reaction have indicated that this system is very complicated.[51] More detailed studies of $C_3H_n^+$ colliding in a trap with HD have been performed recently.[52] Many surprises have been found. For example the rate coefficient for the reaction

$$C_3^+ + HD \rightarrow C_3D^+ + H, \tag{3.29}$$

$k = 9.3 \times 10^{-10}$ cm$^3$ s$^{-1}$, is six times larger than assumed in astrochemical models. Surprisingly, direct production of $C_3HD^+$ via radiative association,

$$C_3^+ + HD \rightarrow C_3HD^+ + h\nu, \tag{3.30}$$

has been observed to be fast ($k_r = 6.0 \times 10^{-11}$ cm$^3$s$^{-1}$). Another unexpected result is the branching in

$$C_3H^+ + HD \rightarrow C_3HD^+ + H \quad \text{and} \tag{3.31}$$

$$\rightarrow C_3H_2^+ + D. \tag{3.32}$$

Forming the deuterated ion (3.31) is over hundred times faster than channel (3.32). A final example is the exothermic H-D exchange reaction

$$C_3H_3^+ + HD \rightarrow C_3H_2D^+ + H_2. \tag{3.33}$$

A careful analysis of the data reveals that the 15 K rate coefficient is smaller than $4 \times 10^{-16}\,\mathrm{cm^3\,s^{-1}}$. This is in accord with theoretical predictions. All these $C_3H_n^+ + HD$ ion trap studies have been thoroughly discussed by Savic *et al.*[53]

### 3.5.2. *Reactions with Significant Temperature Dependences*

It is still generally assumed that the rate coefficients for exothermic ion-molecule reactions are almost independent of temperature; however, experimental studies at low temperatures often reveal more and more exceptions. One typical example is the reaction

$$CH_4^+ + H_2 \rightarrow CH_5^+ + H \tag{3.34}$$

which has been discussed by Asvany *et al.*,[50] another one is

$$C_3^+ + H_2 \rightarrow C_3H^+ + H. \tag{3.35}$$

Although exothermic, reaction (3.35) is rather slow at room temperature but becomes faster with decreasing temperature.[53] In addition to the increasing lifetime of the collision complex, this behavior may be caused by the floppy structure of $C_3^+$.

It is well known, at least qualitatively, that the rate coefficients for ions colliding with polar molecules increase with decreasing collision energies. Fig. 3.20 shows a result measured in a merged beam arrangement for the charge transfer reaction

$$Kr^+ + H_2O \rightarrow H_2O^+ + Kr. \tag{3.36}$$

The $H_2O$ beam has been seeded in He, the nozzle temperature was 300 K. Comparison of the two energy scales (lower scale: laboratory energy of the ion, $E_1$, upper scale: collision energy $E_T$) reveals that the $Kr^+$ ions have been accelerated to 1.3 eV in order to match the velocity of the seeded $H_2O$ molecules. The Gaussian-like curves show two distributions of the collision energy. They have been determined numerically for the nominal collision energies 0 and 50 meV. The dashed curve is an empirical fit indicating that the rate coefficient increases proportional to $E_T^{-0.5}$. Accounting for the energy resolution of the experimental set-up leads to the solid line. This simulated curve is in good agreement with the measured data. The minor deviations at laboratory energies below 0.5 eV are probably due to experimental difficulties, such as changes in the ion beam shape.

Another reaction where the rate coefficient exhibits an interesting temperature dependence is the charge transfer process

$$Ar^+ + N_2 \rightarrow N_2^+ + Ar. \tag{3.37}$$

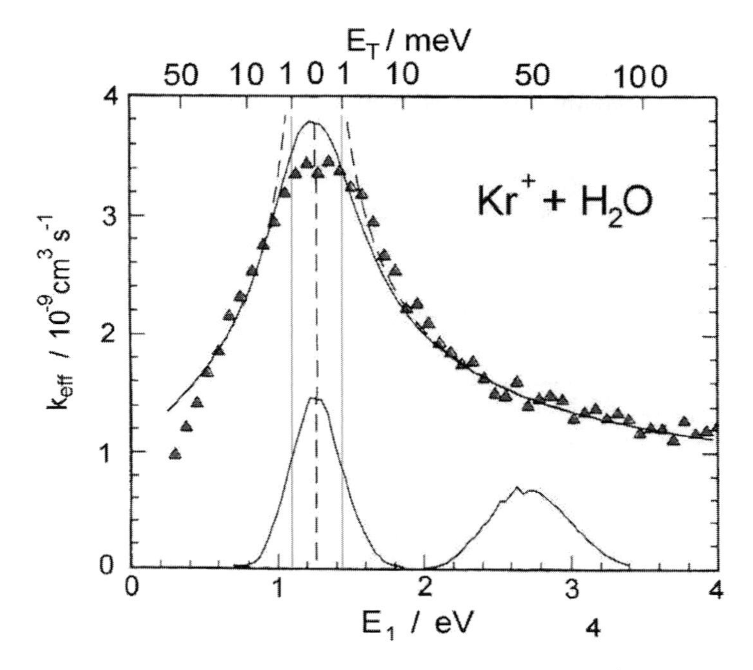


Fig. 3.20. Merged beam results for the charge transfer reaction $Kr^+ + H_2O \rightarrow H_2O^+ + Kr$.[17] The rate coefficient is plotted as a function of the laboratory energy of the ion, $E_1$ (lower scale). Accounting for the velocity of the water beam which has been seeded in He, one obtains the collision energy $E_T$ (upper scale). The two Gaussian-like curves represent the numerically determined energy distributions at the nominal energies 0 and 50 meV. The dashed curve is an empirical fit.

It has been studied very often and in great detail, both experimentally and theoretically; however, the low temperature behavior is not yet understood. Although it is 0.179 eV exothermic, it proceeds very slowly at collision energies below 100 meV as can be seen from Fig. 3.21. The merged beam results (open circles) have been measured with a supersonic beam. Above 100 meV the merged beam data are in good agreement with GIB and temperature variable DRIFT tube data (see Refs. 17 and 54). The low energy points are from a free jet[11] and a CRESU experiment (see Chapter 2). The increase of the rate coefficient at low temperatures may be taken as an indication that the longer lifetime of the collision complex allows for non-adiabatic transitions. A similar reaction mechanism, where the increasing lifetime of the collision complex allows for tunneling, has been discussed for the temperature dependence of the reaction

$$NH_3^+ + H_2 \rightarrow NH_4^+ + H. \tag{3.38}$$

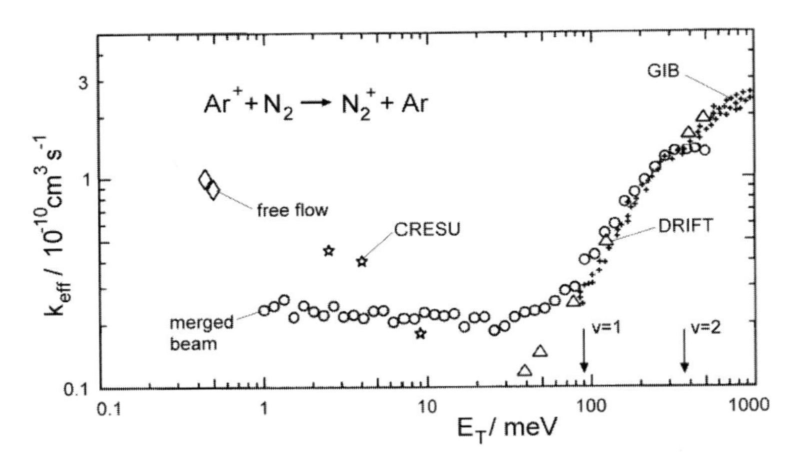


Fig. 3.21. Energy dependence of the effective rate coefficients for the exothermic charge transfer reaction $Ar^+ + N_2 \rightarrow N_2^+ + Ar$. Above 100 meV the merged beam data (open circles) are in good agreement with results which have been obtained with the Guided Ion Beam technique (GIB: crosses) and a temperature variable DRIFT tube (open triangles, see Ref. 54). The increase of the rate coefficient with increasing energy coincides with the energy needed for forming $N_2^+$ in the first vibrational state, v = 1. The points at low energies are from a free jet[11] and a CRESU experiment (see Chapter 2).

A short summary of this reaction and some ion trap results can be found in Ref. 44.

### 3.5.3. *Formation and Destruction of Hydrocarbons*

Elementary steps in forming and destroying small hydrocarbons in the ISM include collisions of $C^+$ ions with H and $H_2$, proton transfer from $H_3^+$ to C or collisions of $CH_n^+$ with H atoms. In the last few years, many relevant studies have been performed using ion traps. Of fundamental importance is the radiative association reaction

$$C^+ + H_2 \rightarrow CH_2^+ + h\nu. \tag{3.39}$$

This reaction has been studied several times in rf ion traps with increasing accuracy. The results are summarized in Fig. 3.22 (see also Refs. 15 and 55). For p-$H_2$ the rate coefficient for radiative association is $1.7 \times 10^{-15}\,\mathrm{cm^3\,s^{-1}}$ at 10 K, while the value for n-$H_2$ is 2.5 times smaller. As discussed in detail in Ref. 22, much more has been learned about such processes, e.g. the competition between complex life time and radiative decay, by comparing ternary and radiative association and by isotopic substitution.

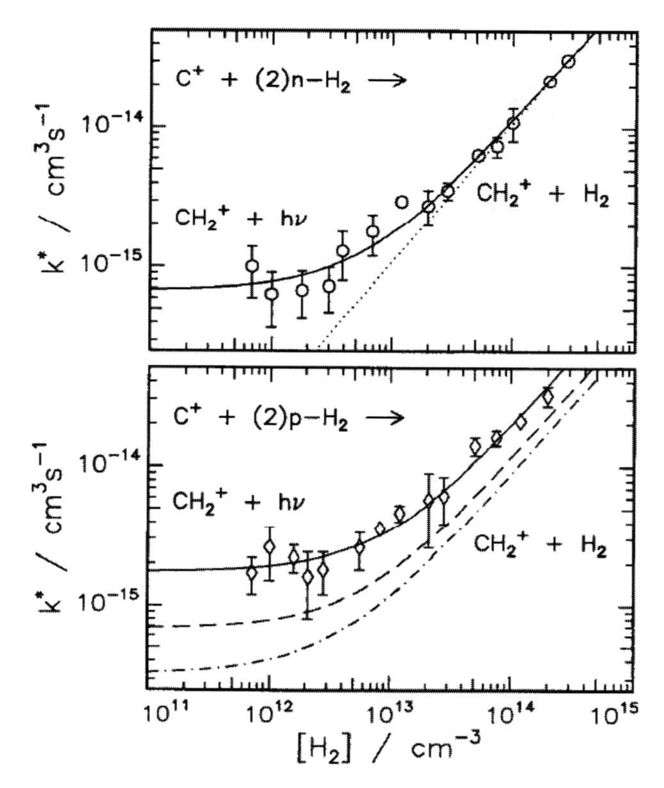


Fig. 3.22. Ternary and radiative association of $C^+$ with n-$H_2$ (upper part) and p-$H_2$ (lower part) at a nominal temperature of 10 K, measured over a wide range of hydrogen number densities. The solid lines represent fits of $k^* = [H_2]k_3 + k_r$ to the experimental data. The resulting parameters are given in the text. In the lower figure the dashed line is the n-$H_2$ result from above while the dash-dotted line predicts the dependence for pure o-$H_2$.

An interesting hydrocarbon system, where different low temperature experiments have resulted in quite different results, is the reaction

$$C_2H_2^+ + H_2 \rightarrow C_2H_3^+ + H. \tag{3.40}$$

This reaction is already slow at room temperature and the rate coefficient falls below $10^{-13}\,cm^3\,s^{-1}$ at 100 K as determined with the 22-pole ion trap. Evaluation of several measurements has lead to an endothermicity of $(50 \pm 20)$ meV in good accordance with earlier conclusions. The question whether this endothermicity is in reality a barrier was raised by measurements performed with a low temperature flow reactor (for details see Ref. 55). In this instrument a rate coefficient of $10^{-12}\,cm^3\,s^{-1}$ was measured

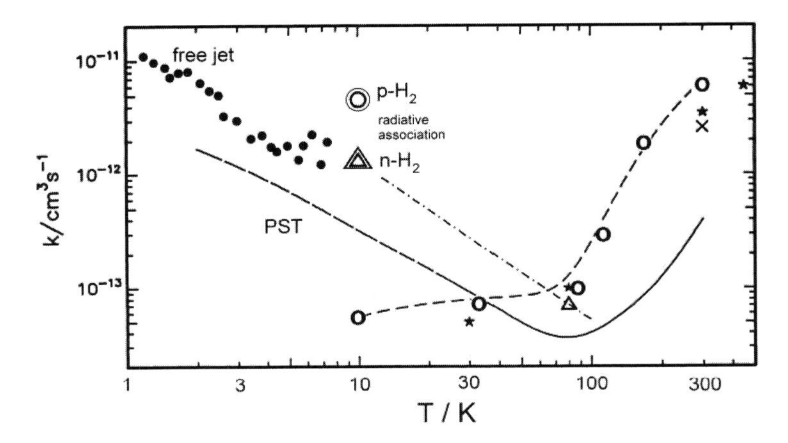


Fig. 3.23. Temperature dependence of the rate coefficient for the reaction $C_2H_2^+ + H_2 \rightarrow C_2H_3^+ + H$. As discussed by Gerlich,[55] there is some disagreement between the free jet and the ion trap experiments. While, at low temperatures, the high pressure experiments seems to indicate an increasing rate coefficient, the 10 K ion trap results proof that the rate coefficient for the abstraction reaction is much smaller than $10^{-13}\,cm^3s^{-1}$. A possible explanation is the fast radiative association process ($k_r$(p-$H_2$) $= 5\times10^{-12}\,cm^3s^{-1}$), which has been observed in the ion trap experiment (open circle and triangle). The phase space calculations (solid line) have used adjusted parameters for getting the low temperature behavior.

at 10 K and an increasing tendency towards $10^{-11}\,cm^3\;s^{-1}$ was observed when the temperature was lowered to 1 K. This increase was interpreted with a tunneling mechanism. None of the ion trap studies could corroborate these conclusions. The main fraction of mass 27 products could be attributed to $C_2HD^+$ which are formed via isotope enrichment in collisions with the natural abundant HD. One of the possible explanations for the signal found in the free jet experiment is based on the large radiative association rate coefficient detected in the trap. This process together with ternary association occurring in the high pressure flow can lead to highly excited $C_2H_4^+$ products which are fragmented during the extraction and detection. Studying the reverse reaction $C_2H_3^+ + H$ will shed some more light onto the energy balance of this system.

Ion traps also have been combined with beams of radicals. For studying collisions between small neutral carbon molecules $C_n$ (n = 1–3) and stored ions a special experimental setup has been developed.[56] In this instrument ions are confined in a ring electrode trap (RET) at temperatures between 80 K and 600 K. There they interact with an effusive carbon beam, which is produced via high-temperature vaporization of a carbon rod. For the reaction

$$D_3^+ + C \rightarrow CD^+ + H_2 \qquad (3.41)$$

the importance of which is mentioned in chapter 1, first results have been obtained. The reaction rate coefficients measured for forming $C_nD^+$ are almost a factor two smaller than values presently used in astrochemical models. Another important class of reactions concerns the growth of pure carbon chains via radiative association,

$$C_m^+ + C_n \rightarrow C_{m+n}^+ + h\nu. \qquad (3.42)$$

First results have indicated that the rate coefficients are slower than generally assumed in astrochemical models; however, all the experiments have to be extended towards lower total energies.[56] While it is easy to change the temperature of the ions, the neutral carbon target needs to be cooled, especially in the case of the dominating $C_3$ molecule. This is only possible by replacing the high temperature vaporization carbon source with a high pressure laser ablation source.

One of the most important radicals is the hydrogen atom. Hydrogen is the most abundant baryonic species in the universe and it plays therefore, in both its atomic and molecular form, a central role in many regions of the ISM. It was therefore a challenge to combine a low temperature ion trap with a beam of H-atoms the velocity distribution of which can be varied. A first version of the experimental setup is shown schematically in Fig. 3.15.

Collisions of hydrocarbon ions with H often lead to dehydrogenation. One example is the reaction

$$CH_5^+ + H \rightarrow CH_4^+ + H_2 \qquad (3.43)$$

The rate coefficients for this reaction, which have been presented at a conference,[24] are compared in Fig. 3.24 with those for the reverse reaction which also has been mentioned above.[50] A detailed discussion of all results together with a critical analysis of the kinematics is in preparation.[57] A question which is related to reaction (3.43) is whether catalytic cycles such as $XY^+ + H \rightarrow XYH^+$ followed by $XYH^+ + H \rightarrow XY^+ + H_2$ can contribute to the formation of molecular hydrogen under interstellar conditions.

### 3.5.4. *Pure Hydrogen Chemistry*

One motivation for studying pure hydrogen chemistry, i.e., reactions between hydrogen ions, atoms, molecules and deuterated variants, originates from cold pre-protostellar cores. As described for example by Walmsley *et al.*,[58] such processes become of central importance if, at very

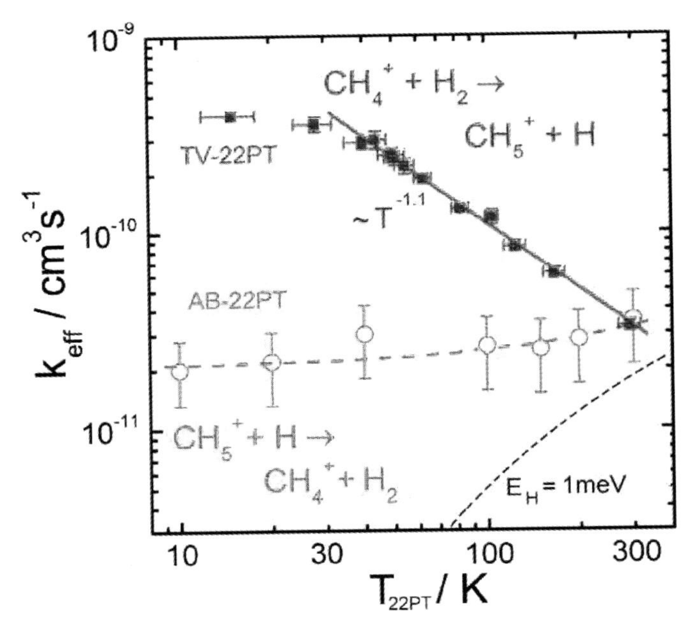


Fig. 3.24. Temperature dependent rate coefficients for the $CH_6^+$ collision system. The reaction $CH_4^+ + H_2 \rightarrow CH_5^+ + H$ has been reported by Asvany *et al.*[50] First results for the $CH_5^+ + H$ collision system have been presented on a conference.[24] The results depend critically on the energy distribution of the H-atom beam. A detailed analysis is in preparation.[30,57] The dashed line predicts the rate coefficient as a function of the ion temperature, $T_{22PT}$, for an hydrogen beam with 1 meV kinetic energy.

low temperatures, all heavy elements become integrated into ice layers on dust grains, i.e., vanish from the gas phase. In a recent summary of our present knowledge of various $H_mD_n^+$ collision systems, it has been emphasized that systems with $m + n \leq 3$ are reasonably well understood, while much more work needs to be done on systems involving more than three atoms.[3] For example the reaction

$$H_3^+ + HD \rightarrow DH_2^+ + H_2 \tag{3.44}$$

is still intriguing.[35] An experimental difficulty is that, at low temperatures, very small traces of o-$H_2$ have a significant but rather unknown influence on the measured $H_3^+/DH_2^+$ ratio.[59]

Another motivation for studying reactions involving several H and D atoms at low temperatures is connected with the fact that they are fermions and bosons, respectively. The question is whether, at low enough energies, one can detect experimentally consequences from what one could call Bose- or Fermi-chemistry. "Simple" consequences of the exchange symmetry in

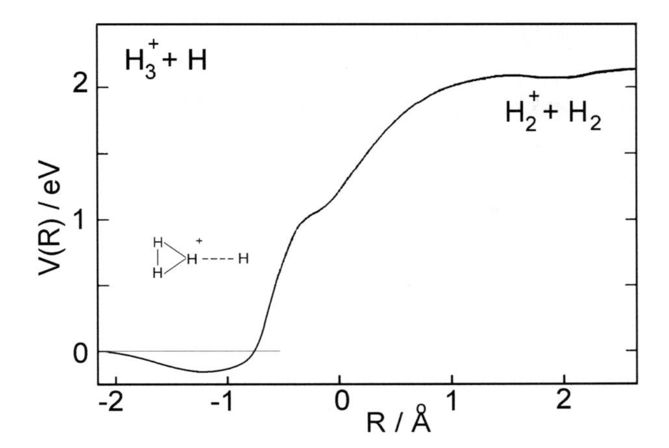


Fig. 3.25. Schematic illustration of the $H_4^+$ potential energy surface. Well characterized are $H_2^+ + H_2$ collisions (right side); however, in order to understand scrambling in low energy $H_3^+ + H$ collisions (left side) e.g. ortho-para transitions or H-D exchange in partly deuterated combinations, quantum calculations must be performed on a precise potential energy surface (see Ref. 60).

molecules containing identical atoms and the rather stringent restrictions of the conservation of the total nuclear spin in a scrambling collision have been discussed by Gerlich *et al.*[3] So far only statistical models have been used for making predictions although it must be expected that dynamical constraints may also play a role. Special effects can be expected for the reaction

$$H_3^+ + H \rightarrow H_3^+ + H. \tag{3.45}$$

The relevant $H_4^+$ potential energy surface is shown schematically in Fig. 3.25. The $H_2^+ + H_2$ part (right side) is probably characterized well enough for understanding formation of $H_3^+ + H$.[60] However, in order to model scrambling in sub-K $H_3^+ + H$ collisions correctly, a very precise potential energy surface is required on the left side. Astrochemically important variants of reaction (3.45) are H-D exchanges, e.g. in $H_3^+ + D$, or ortho-para conversion of $H_3^+$ via H atom scrambling. In all these cases a very good potential energy surface is needed in order to distinguish between effects caused by the van der Waals type interaction, the zero point energies and the dynamics of the mixed 4-center fermion/boson system. Corresponding experiments involving cold ions and slow H or D atoms are in preparation.

### 3.5.5. *State Specific Reactions*

There have been so far only a few approaches for studying ion-molecule reactions at collision energies of a few meV with state selected reactants. One of the early studies at low temperatures was the association reaction

$$CO^+(v = 0, j) + 2CO \rightarrow (CO)_2^+ + CO. \qquad (3.46)$$

The experiments have been performed in a free CO expansion using multiphoton ionization for preparing $CO^+(v = 0, j)$ ions.[61] Propensity rules favor the formation of specific rotational states. Several aspects of the kinematics in single beam and merged beam arrangements have been discussed in Ref. 62. In addition to integral cross sections which have been measured for the reaction

$$H_2^+ + H_2 \rightarrow H_3^+ + H \qquad (3.47)$$

at collision energies between 5 meV and 5 eV with a merged beam arrangement, state specific information has been obtained at meV energies by preparing $H_2^+$ ions in different rotational states via (3 + 1) REMPI (resonance enhanced multi photon ionization).[62] REMPI also has been used to create $NO^+$ ions directly inside an rf trap; however, no reactions have been studied.

In recent years, there have been several successful attempts to use lasers for re-exciting ions after they have been cooled down. Since the ion cloud consists of a limited number, various strategies are possible for deriving information on state specific collision dynamics. The methods range from a dedicated perturbation of a stationary low temperature equilibrium via burning a hole into the state population to two colour pump — probe experiments. Most of such activities are used for spectroscopy or for understanding the low temperature populations of trapped hydrogen ions. They are discussed in Chapter 6.

One example for a low temperature state specific reaction is the hydrogen abstraction reaction

$$C_2H_2^+ + H_2 \rightarrow C_2H_3^+ + H, \qquad (3.48)$$

which has been measured in a trap by infrared excitation of the acetylene ion.[63] From the measured intensities and their dependence on parameters such as storage time, laser fluence and target gas density, information on state specific rate coefficients has been obtained. Vibrational excitation of $C_2H_2^+$ increases the rate of reaction (3.48) by more than three orders of magnitude, while rotation hinders the reaction. The fine-structure state of the parent ion does not affect its reactivity.

Important contributions towards a detailed state specific understanding of H-D exchange reactions in collisions of $H_3^+$ with H or $H_2$ in all possible deuterated variants have been made in the last years in several laboratories. The activities, which all use low temperature 22-pole traps in combination with the method of laser induced reactions, have been summarized in Ref. 3. The feasibility of analyzing cold $H_3^+$ via laser excitation, followed by proton transfer to Ar, has been first demonstrated in a compact trapping machine which has been mainly developed as an ion source.[64,65] Meanwhile high power CW diode lasers are applied to excite $H_3^+$ ions and isotopic variants in specific states via overtone transitions.[35] Also the first spectra with a free electron laser have been reported.[26]

In order to extract from such laser induced processes quantitative information such as state populations or reliable state specific rate coefficients, it is mandatory to separate relaxation, reactions and laser induced processes. This is possible by combining the flexible trapping method with short gas pulses, chopped or modulated CW lasers, and pulsed effusive or supersonic beams.

### 3.6. Conclusions and Outlook

Trapping techniques have been and still are the basis of many new experiments in physics and chemistry. All these experiments make use of inherent advantages such as extremely long interaction times, the possibility to accumulate weak beams, phase space compression, laser cooling or interaction with buffer gas. This contribution has focused on the use of rf fields to explore collisions at low temperatures or with low relative velocities. The examples have shown that it is now possible to study collisions at energies of 1 meV or at temperatures of 10 K. As already mentioned, there are activities to cool ions in traps to temperatures below 1 K using the slow tail of a cold effusive beam for buffer gas cooling. There are also efforts to heat ions with a laser in order to access temperatures above 2000 K.[5]

There are many technical improvements and innovative applications possible. Especially important is the development of methods for non-destructive detection of the stored objects, e.g. based on image current or optical methods. More work needs to be done in order to optimize the combination of an ion trap with a time of flight mass spectrometer.[66] Low temperatures collisions also could be used successfully in several fields of analytical chemistry. As proposed recently for applications in elemental analysis,[67] the dependence of ternary association on the collision temperature and the number of internal degrees of freedom can be used efficiently for distinguishing between atomic ions and molecular ions with the same mass which otherwise create a background. The addition of helium to trapped seed ions is possible if one uses a chopped beam of very slow He atoms or He clusters. Among many interesting questions concerning sub-K cooling, such experiments can provide results similar to those of the He droplet experiments described by Slenczka and Toennies in Chapter 7. The extension of the rf trapping technique toward anions is obvious and causes no problems with rf trapping since the effective potential is proportional to the square of the charge, $q^2$. Nonetheless such activities have been started only very recently.[27]

For astrochemistry more laboratory data are needed in order to understand and model the formation and destruction of matter in a variety of environments, ranging from very cold dense interstellar clouds via planetary atmospheres to the high temperatures of circumstellar environments. At the University of Arizona, a new instrument, the central part of which is a temperature variable multi-electrode trap, is close to completion. In combination with radical sources, complex gas phase reactions can be studied over a wide range of temperature. Based on the fact that the effective potential can be tailored for many purposes, special traps can be designed for observing and characterizing one *single* nanoparticle over long times. This opens up the possibility of studying cold or also hot interstellar grain equivalents.

Cooling all degrees of freedom of molecular ions down to a few K or even in the sub-K range has many obvious applications in spectroscopy. One example is to study rotational transitions in floppy molecular ions such as $CH_5^+$. Applications of rf traps in the analysis of molecular structures together with planned extensions towards infrared or microwave absorption spectroscopy on ultracold ions are discussed in Chapter 6.

## References

1. Klemperer W. (2006) Interstellar chemistry. *PNAS* **103**: 12232–12234.
2. Smith MA, Schlemmer S, von Richthofen J, Gerlich D. (2002) $HOC^+ + H_2$ isomerization Rate at 25 K: Implications for the observed $[HCO^+]/[HOC^+]$. *Ratios in the Interstellar Medium ApJL* **578**: 87–90.
3. Gerlich D, Windisch F, Hlavenka P, Plašil R, Glosik J. (2006) Dynamical constraints and nuclear spin caused restrictions in $H_m D_n^+$ collision systems and deuterated variants. *Phil. Trans. R. Soc. Lond. A* **364**: 3007–3034.
4. Schiffer JP, Drewsen M, Hangst JS, Hornekær L. (2000) Temperature, ordering, and equilibrium with time-dependent confining forces. *PNAS* **97**: 10697–10700.
5. Decker S, Savi I, Gerlich D. (2007) Astrochemistry in ion traps: From cold hydrogen to hot carbon Molecules. In Lemaire JL, Combes F. (eds). *Space & Laboratory, Paris.*

6. Nikitin EE, Troe J. (2005) Dynamics of ion–molecule complex formation at very low energies and temperatures. *Phys. Chem. Chem. Phys.* **7**: 1540–1551.
7. Gerlich D. (1992) Inhomogeneous electrical radio frequency fields: A versatile tool for the study of processes with slow ions. *Adv. Chem. Phys.* **LXXXII**: 1–176.
8. Dawson PH. (1976) *Quadrupole Mass Spectrometry*, Elsevier ScientificPublishing, Amsterdam.
9. Gerlich D. (1989) Kinematic averaging effects in thermal and low energy ion-molecule collisions: Influence on product ion kinetic energy distributions. *J. Chem. Phys.* **90**: 127–139.
10. Gerlich D. (2003) *Molecular Ions and Nanoparticles in rf and AC Traps Hyperfine Interactions* **146–147:** 293–306.
11. Smith MA. (1993) Ion molecule reactions dynamics at very low temperatures ion. *Chem. Phys.* **2:** 183–251.
12. Teloy E, Gerlich D. (1974) Integral cross sections for ion-molecule reactions: The guided beam technique. *Chem. Phys.* **4**: 417–427.
13. Mark S, Gerlich D. (1996) Differential cross sections, measured with guided-ion-beams. Applications to $N^+ + N_2$ and $C_2H_2^+ + C_2D_4$ Collisions. *Chem. Phys.* **209**: 235–258.
14. Phaneuf RA, Havener CC, Dunn GH, Müller A. (1999) Merged-beams experiments in atomic and molecular physics. *Rep. Prog. Phys.* **63**: 1143–1180.
15. Gerlich D. (1993) Guided ion beams, rf ion traps, and merged beams: State specific ion-molecule reactions at meV energies XVIII. In Andersen T, Fastrup B, Folkmann F, Knudsen H. (eds.), *International Conference on Physics of Electronic and Atomic Collisions*, AIP, New York, pp. 607–622.
16. Mark S, Glenewinkel-Meyer T, Gerlich D. (1996) REMPI in a focusing rf-quadrupole: A new source for mass and state selected ions. *Int. Rev. Phys. Chem.* **15**: 283–298.
17. Wick O. (1995) *Untersuchung von Ionen-Molekülreaktionen bei extrem niedrigen Energien in einer Merged Beam Apparatur Doktorarbeit Freiburg.*
18. Wolf A, Buhr H, Grieser M, von Hahn R, Lestinsky M, Lindroth E, Orlov DA, Schippers S, Schneider IF. (2006) Progress in stored ion beam experiments on atomic and molecular processes. *Hyperfine Interact.* **172**: 111–124.
19. Kreckel H, Mikosch J, Wester R, Glosik J, Plasil R, Motsch M, Gerlich D, Schwalm D, Zajfman D, Wolf A. (2005) Towards state selective measurements of the $H_3^+$ dissociative recombination rate coefficient *J. Phys.* **4**: 126–133.
20. Luine JAG, Dunn H. (1985) Ion-molecule reaction probabilities near 10. *Ap. J.* **299**: L67–0.
21. Barlow SE, Luine JA, Dunn GH. (1986) Measurement of ion/molecule reactions between 10 and 20 K. *Int. J. Mass Spectrom. Ion Proc.* **74**: 97–128.
22. Gerlich D, Horning S. (1982) Experimental investigations of radiative association processes as related to interstellar chemistry. *Chem. Rev.* **92**: 1509–1539.
23. Gerlich D. (1995) Ion-Neutral Collisions in a 22-pole trap at very low energies. *Physica Scripta* **T59**: 256–263.

24. Luca A, Borodi G, Gerlich D. (2005) Interactions of ions with Hydrogen atoms Progress report in XXIV ICPEAC 2005, Rosario, Argentina, July 20–26, 2005, In Colavecchia FD, Fainstein PD, Fiol J, Lima MAP, Miraglia JE, Montenegro EC, Rivarola RD. (eds) **XXIV**, pp. 20–26.
25. Gerlich D, Smith M. (2006) Laboratory astrochemistry: Studying molecules under inter- and circumstellar conditions. *Physica Scripta* **73**: C25–C31.
26. Schlemmer S, Asvany O, Hugo E, Gerlich D. (2005) Deuterium fractionation and ion-molecule reaction at low temperatures. In Lis DC, Blake GA, Herbst E. (eds). *Proceedings IAU Symp.* No 231, pp. 125–134.
27. Trippel S, Mikosch J, Berhane R, Otto R, Weidemüller M, Wester R. (2006) Photodetachment of cold $OH^-$ in a multipole ion trap. *Phys. Rev. Lett.* **97**: 193003-1–193003-4.
28. Dzhonson A, Gerlich D, Bieske EJ, Maier JP. (2006) Apparatus for the study of electronic spectra of collisionally cooled cations: Para-dichlorobenzene. *Journal of Molecular Structure* **795**: 93–97.
29. Boyarkin OV, Mercier SR, Kamariotis A, Rizzo TR. (2006) Electronic spectroscopy of cold, protonated tryptophan and tyrosine. *J. Am. Chem. Soc* **128**: 2816–2817.
30. Borodi G, Luca A, Mogo C, Smith M, Gerlich D. (2007) On the combination of a low energy hydrogen atom beam with a cold multipole ion trap to be submitted to *Rev. Sci. Instr.*
31. H. Böhringer, Glebe W, Arnold F. (1983) $He^+ + 2$ He ternary association. *J. Phys. B: At. Mol. Phys* **16**: 2619–2626.
32. Paul W, Lücke, B, Schlemmer S, Gerlich D. (1995) On the dynamics of the reaction of positive hydrogen cluster ions ($H_5^+$to $H_{23}^+$) with para and normal hydrogen at 10 K. *Int. J. Mass Spectrom. Ion Proc.* **150**: 373–387.
33. Wang Y-S, Tsai C-H, Lee YT, Chang H-C, Jiang JC, Asvany O, Schlemmer S, Gerlich D. (2003) Investigations of protonated and deprotonated water clusters using a low-temperature 22-pole ion trap. *J. Phys. Chem. A* **107**: 4217–4225.
34. Schlemmer S, Kuhn T, Lescop E, Gerlich D. (1999) Laser excited $N_2^+$ in a 22-pole trap, experimental studies of rotational relaxation processes. *Int. J. Mass Spectrom. Ion Proc.* **185**: 589–602.
35. Glosik J, Hlavenka P, Plašil R, Windisch F, Gerlich D. (2006) Action spectroscopy of $H_3^+$ using overtone excitation. *Phil. Trans. R. Soc. Lond. A* **364**: 2931–2942.
36. Wineland DJ, Bergquist JC, Itano WM, Bollinger JJ, Manney CH. (1987) Atomic-ion coulomb clusters in an ion trap. *Phys. Rev. Lett.* **59**: 2935–2938.
37. Bowe P, Hornekaer L, Brodersen C, Drewsen M, Hangst JS. (1999) Sympathetics crystallization of trapped ions. *Phys. Rev. Lett.* **82**: 2071–2074.
38. Roth B, Blythe P, Wenz H, Daerr H, Schiller S. (2006) Ion-neutral chemical reactions between ultracold localized ions and neutral molecules with single-particle resolution. *Phys. Rev. A* **73**: 42712-1–42712-9.
39. Molhave K, Drewsen M. (2000) Formation of translationally cold $MgH^+$ and $MgD^+$molecules in an ion trap. *Phys. Rev. A* **62**: 11401–11405.

40. Hornekaer L, Kjaergaard N, Thommesen AM, Drewsen M. (2001) Structural properties of two-component coulomb crystals in linear paul traps. *Phys. Rev. Lett.* **86**: 1994–1997.
41. Roth B, Blythe P, Daerr H, Patacchini L, Schiller S. (2006) Production of ultracold diatomic and triatomic molecular ions of spectroscopic and astrophysical interest. *J. Phys. B* **39**: 1241–1258.
42. Roth B, Koelemeij JCJ, Daerr H, Schiller S. (2006) Rovibrational spectroscopy of trapped molecular hydrogen ions at millikelvin temperatures. *Phys. Rev. A* **74**: 040501-1–040501-4.
43. Gerlich D, Kaefer G. (1987) Measurements of extremely small rate coefficients in an rf trap: The association reaction of $C^+$ and $H^+$ with $H_2$ at 230 and 320 K. In Adams NG, Smith D. (eds.), *5th Int. Swarm Seminar, Proceedings*, Birmingham University, Vol. 133.
44. Gerlich D. (1993) Experimental investigations of ion molecule reactions relevant to interstellar chemistry. *J. Chem. Soc., Faraday Trans.* **89**: 2199–2208.
45. Luca A, Voulot D, Gerlich D. (2002) Low temperature reactions between stored ions and condensable gases: Formation of protonated methanol via radiative association *WDS'02*. In Safrankova (ed.), *Proceedings of Contributed Papers*, Part II, Matfyzpress, pp. 294–300.
46. Gerlich D. (1989) Reactive scattering of $N^+ + H_2$ and deuterated analogs: Statistical calculation of cross sections and rate coefficient. *J. Chem. Phys.* **90**: 3574–3581.
47. Tosi P, Dmitriev O, Bassi D, Wick O, Gerlich D. (1994) Experimental observation of the energy threshold in the ion-molecule reaction $N^+ + D_2 \rightarrow ND^+ + D$ *J. Chem. Phys.* **6**: 4300–4307.
48. Gerlich D, Schlemmer S. (2002) Deuterium fractionation in gas phase reactions measured in the laboratory. *Plan. Sp. Sci.* **50**: 1287–1297.
49. Asvany O, Schlemmer S, Gerlich D. (2004) Deuteration of $CH_n^+$ $(n = 3 - 5)$ in collisions with HD measured in a low temperature ion trap. *Ap. J.* **617**: 685–692.
50. Asvany O, Savic I, Schlemmer S, Gerlich D. (2004) Variable temperature ion trap studies of $CH_4^+ + H_2$, HD and $D_2$: Negative temperature dependence and significant isotope effect. *Chem. Phys.* **298**: 97–105.
51. Sorgenfrei A. (1994) Ion-Molekülreaktionen kleiner Kohlenwasserstoffe in einem gekühlten Ionenspeicher Doktorarbeit Freiburg.
52. Savić I. (2004) Formation of Small Hydrocarbon Ions Under Inter- and Circumstellar Conditions: Experiments in Ion Traps. *Doktorarbeit TU Chemnitz.*
53. Savić I, Gerlich D. (2005) Temperature variable ion trap studies of $C_3H_n^+$ with $H_2$and HD. *Phys. Chem. Chem. Phys.* **7**: 1026–1035.
54. Viggiano AA, Robert Morris A. (1993) Rate constants for the reaction of $Ar^+(^2P_{3/2})$ with $N_2$ *as a function of* $N_2$ vibrational temperature and energy level. *J. Chem. Phys.* **99**: 3526–3530.
55. Gerlich D. (1994) Recent Progress in Experimental Studies of Ion-Molecule Reactions Relevant to Interstellar Chemistry. In Nenner I. (ed.), "*Molecules and Grains in Space*", AIP Press, New York, pp. 489–500.

56. Savić I, Čermak I, Gerlich D. (2005) Reactions of $C_n$ (n = 1 − 3) with ions stored in a temperature-variable radio frequency trap. *Int. J. Mass* **240**: 139–147.
57. Borodi G, Luca A, Mogo C, Gerlich D. (2007) Collisions of cold trapped $CH_5^+$ with slow hydrogen atoms in preparation.
58. Walmsley CM, Flower DR, Pineau des Forets G. (2004) Complete depletion in prestellar cores. *Astron. Astroph.* **418**: 1035–1043.
59. Gerlich D, Herbst E, Roueff E. (2002) $H_3^+ + HD \rightarrow H_2D^+ + H_2$: Low-temperature laboratory measurements and interstellar implications. *Plan. Sp. Sci.* **50**: 1275–1285.
60. Eaker CW, Schatz GC. (1986) A surface hopping quasiclassical trajectory study of the $H_2^+ + H_2$ and $(H_2 + D_2)^+$ systems. *Chem. Phys. Lett.* **127**: 343–346. For a more recent calculation see: Moyano GE, Pearson D, Collins MA. (2004) Interpolated potential energy surfaces and dynamics for atom exchange between *H* and $H_3^+$, and D and $H_3^+$. *J. Chem. Phys.* **121**: 12396–12401.
61. Gerlich D, Rox T. (1989) Association reactions with state selected ions at meV collision energies: $CO^+ + 2CO \rightarrow (CO)_2^+ + CO$ Z. *Phys. D.* **13**: 259–268.
62. Glenewinkel-Meyer Th, Gerlich D. (1997) Single and Merged Beam Studies of the Reaction $H_2^+ + H_2 \rightarrow H_3^+ + H$. *Israel Journal of Chemistry* **37**: 343–352.
63. Schlemmer S, Lescop E, von Richthofen J, Gerlich D, Smith M. (2002) Laser induced reactions in a 22-Pole ion trap: $C_2H_2^+ + h\nu_3 + H_2 \rightarrow C_2H_3^+ + H$, *Chem. Phys* **117**: 2068–2075.
64. Mikosch J, Kreckel H, Wester R, Plasil R, Glosik J, Gerlich D, Schwalm D, Wolf A. (2004) Action spectroscopy and temperature diagnostics of $H_3^+$ by chemical probing. *J. Chem. Phys.* **121**: 11030–11037.
65. Wolf A, Krekel H, Lammich L, Strasser D, Mikosch J, Glosik J, Plasil R, Altevogt S, Andrianarijaona V, Buhr H, Hoffmann J, Lestinsky M, Nevo I, Novotny S, Orlov DA, Pedersen HB, Terekhov AS, Toker J, Wester R, Gerlich D, Schwalm D, Zajfman D. (2006) Effects of molecular rotation in low-energy electron collisions of $H_3^+$. *Phil. Trans. R. Soc. Lond. A* **364**: 2981–2997.
66. Luca A, Schlemmer S, Čermak I, Gerlich D. (2001) On the combination of a linear field free trap with a time-of-flight mass spectrometer. *Rev. Sci Instr.* **72**: 2900–2908.
67. Gerlich D. (2004) Applications of rf fields and collision dynamics in atomic mass spectrometry. *J. Anal. At. Spectrom.* **19**: 581–590.

# Chapter 4

# Theory of Low Temperature Gas-Phase Reactions

Stephen J. Klippenstein

*Chemical Sciences and Engineering Division*
*Argonne National Laboratory*
*Argonne, IL 60439, USA*

Yuri Georgievskii

*Combustion Research Facility, Sandia National Laboratories*
*Livermore, CA 94551-0969, USA*

## Contents

### 4.1. Introduction

In this chapter, we describe the current status of theoretical kinetics for chemical reactions at low temperature, i.e., from ~1 to ~200 K. The desire to understand the chemistry of interstellar space and of low temperature planetary atmospheres provides the general motivation for studying chemical kinetics at such temperatures. For example, the chemistry of Titan's atmosphere is currently a topic of considerable interest. This motivation led to the development of novel experimental techniques, such as the CRESU (cinetique de reaction en ecoulement supersonique uniforme) method,[1] which allows for the measurement of rate coefficients at temperatures as low as 10 K (see Chapter 2 by Canosa *et al.*). Such measurements provide important tests for theory and have sparked a renewed interest in theoretical analyses for this temperature range.[2]

Experimental studies of low temperature kinetics are generally limited to reactions that are quite rapid. Furthermore, only such rapid reactions are generally of importance in low temperature chemical environments. Rapid low temperature reactions generally have no barrier (or only one that lies below reactants) along some reaction pathway that leads to the production of exothermic products. For such reactions, it would seem appropriate to focus on the kinetics of the capture process while considering only the long-range contributions to the interaction potential. Indeed, for ionic reactions, this long-range capture rate often provides an adequate description of the low temperature experimental observations. However, for neutral reactions, perhaps somewhat surprisingly, the experimentally observed rate coefficients are seldom in agreement with such long-range capture rate predictions, often being considerably lower than expected.[3,4]

In some cases, the reduction in the rate coefficient may be related to reflection by a low entropy transition state in the exit channel.[5] Another general explanation, which is confirmed by detailed *ab initio* transition state theory calculations,[6–9] is that the decrease in entropy that occurs with decreasing separation yields an "inner" transition state. This inner transition state can significantly reduce the entrance flux below that predicted by the long-range capture process, even at temperatures as low as a few K. The properties of this inner transition state are determined by a delicate balance between chemical bonding interactions and short-range repulsions, both of which increase in magnitude with decreasing separation. Importantly, the location of this inner transition state is generally distinct (i.e., well separated) from that for the long-range capture process. As a result, the effects of this inner transition state can be significant even when the long-range (or "outer") transition state lies at a separation where bonding and repulsion interactions are irrelevant.

Theoretical studies of the long-range capture rate have a long history.[4,10,11] A combination of classical and quantum statistical theory analyses, as well as classical and quantum dynamical simulations, have yielded a clear qualitative and quantitative understanding of this process. Only the contribution of nearly degenerate electronic states remains somewhat poorly understood. Otherwise, it is relatively straightforward to predict the long-range capture rate to within $\sim$10%. The present review begins with a pedagogical summary of the contributions from the different theoretical methods to our understanding of long-range capture rates.

Theoretical efforts at quantitatively predicting the effects of short-range interactions on low temperature rate coefficients have been more limited. This neglect is due in part to the perceived lack of importance of these effects and in part to the difficulty of their study, which must rely heavily on *ab initio* electronic structure calculations. Fortunately, the remarkable advances in *ab initio* quantum chemistry, which are due to methodological developments as well as to the ever increasing computational capabilities, now allow for the quantitative study of the inner transition state region. Indeed, the coupling of these advances in quantum chemistry with continuing developments in transition state and related kinetic theories have transformed theoretical chemical kinetics from a simply interpretive and modeling tool to a truly predictive one. The next section briefly discusses the application of *ab initio* simulations to the prediction of rate coefficients for rapid low temperature reactions.

Subsequently, the inner transition state, and its importance for different classes of reactions, is reviewed. The radical-molecule, radical-radical, and ion-neutral cases are each discussed in some detail. A two transition state model, which accounts for the combined effect of the inner and outer transition states acting in series, is summarized. Both adiabatic and statistical models for the electronic motions are also briefly discussed.

*Ab initio* based transition state theory studies are used to illustrate the transition from long-range dominated capture to short-range dominated capture for sample reaction classes. The presence, or lack, of two minima in the reactive flux is demonstrated with illustrative plots for each class of reaction. With modest adjustments in saddlepoint energies, the predicted rate coefficients are seen to be in quantitative agreement with CRESU experimental results for a number of reactions. This success is a testament to the high accuracy of current *ab initio* methods as well as the underlying kinetic assumptions. These illustrative applications also highlight the need for an improved understanding of non-adiabaticities in (i) the electronic states at long range and (ii) the dynamics between the two transition states.

The review closes with a brief summary of the current status of theoretical treatments for both long-range and short-range transition states.

## 4.2. Capture Rates for Long-Range Potentials

Classical statistical theories provide the simplest procedures for predicting the capture rate. They continue to be widely employed because they are of sufficient accuracy for applied purposes and require significantly less computational resources than dynamical theories. They also provide a useful reference framework for more accurate dynamical theories. Statistical theories are particularly useful in predicting the energy and angular momentum dependence of the reactive flux for use in the calculation of rate constants for multichannel reactions.

Classical trajectory simulations provide accurate predictions of the capture rate as long as (i) an accurate interaction potential is employed, (ii) the temperature is not too low, and (iii) the reaction occurs on only a single electronic state or the electronic dynamics is adiabatic. Such simulations have provided important validation tests for classical statistical theories. Parameterized fits provide useful representations of the trajectory results for a number of specific interaction potentials.

Quantum mechanical approaches are required for low temperatures (e.g., $<10\,\mathrm{K}$), where the quantization of rotational and electronic energy levels becomes important. Quantization of translational motion is only of importance at very low temperatures ($\sim 10^{-3}\,\mathrm{K}$). Fortunately, the vibrational motions of the reactants factor out of the problem, and so the quantization of vibrations is irrelevant. Quantum dynamical studies of some limiting cases provide the most definitive tests for low temperature. In many instances, adiabatic channel approaches, which can be viewed as quantized statistical approaches, provide a sufficiently accurate treatment of the quantum effects. Fortunately, since there are relatively few accessible energy levels at low temperatures, such quantum approaches become readily applicable at precisely the temperature where quantum effects become important.

Taken together, the combination of classical and quantum statistical and dynamical approaches yields a quantitative description of the capture process. The following subsections review the contributions of each of these methods to our understanding of the capture rate.

### 4.2.1. *Statistical Theories for Isotropic Potentials*

The phase space theory (PST)[12,13] assumption of an isotropic interaction between two reacting fragments greatly simplifies the theoretical analysis of their capture process. The capture problem then reduces to the central force problem on an effective potential. This effective potential, $V_{\text{eff}}(R)$, is given by the sum of the isotropic long-range interaction potential, $V_{\text{LR}}(R)$, and the orbital angular momentum, $l$, dependent centrifugal energies:

$$V_{\text{eff}}(R) = V_{\text{LR}}(R) + \frac{\hbar^2 l(l+1)}{2\mu R^2}, \tag{4.1}$$

where $R$ is the distance between the centers-of-mass of the two reacting moieties, $\mu$ is the reduced mass, $m_1 m_2/(m_1+m_2)$, and $\hbar$ is Planck's constant, $h$, divided by $2\pi$. Writing the isotropic interaction potential as $-C_n/R^n$, yields $l$-dependent centrifugal barriers, $E_l^\dagger$, given by

$$E_l^\dagger = \left(\frac{n}{2} - 1\right) C_n \left[\frac{\hbar^2 l(l+1)}{\mu n C_n}\right]^{n/n-2}. \tag{4.2}$$

A classical treatment of the central-force problem implies that the probability of reaction for each of the reactant states is simply a step function in the translational energy relative to the centrifugal barrier height. The capture rate for a given total energy, $E$, and total angular momentum, $J$, is then proportional to the number of states, $N(E,J)$, whose translational energy exceeds $E_l^\dagger$. For the most general case of two non-linear fragments this "phase space theory" number of available states is given by

$$N(E,J) = (2J+1) \sum_{j_1,k_1,j_2,k_2,l,j,v_i} \Theta(E - E_{j_1k_1} - E_{j_2k_2} - E_{v_i} - E_l^\dagger) \times \Delta(J,j,l)\Delta(j,j_1,j_2) \tag{4.3}$$

where $E_{j_ik_i}$ is the rotational energy of fragment $i$ in rotational state $j_ik_i$, $E_{v_i}$ is the vibrational energy of the fragments in vibrational state $v_i$, $\Theta$ is the step function, and $\Delta$ is the triangle inequality.

The thermal capture rate may be written as

$$k(T) = \frac{1}{hQ}\int N(E,J)\exp(-\beta E)dEdJ, \tag{4.4}$$

where $Q$ is the canonical partition function for the reactants including relative translation, and $\beta = 1/\text{k}_B T$, with $T$ as the temperature, and $k_B$ as Boltzmann's constant. If the vibrations of the fragments are assumed to be decoupled from the reactive motions, then the sums over vibrational states in Eq. (4.3) cancel with the corresponding contribution to the canonical partition function of the reactants in Eq. (4.4) and so can be removed from further consideration.

The expression in Eq. (4.3) assumes quantized rotational states. Alternatively, the sum over quantum states may be replaced with a classical phase space integral:

$$N(E,J) = \frac{\hbar}{h^{\nu}} \int d\Omega_{12} d\Omega_1 d\Omega_2 dP_{\Omega_{12}} dP_{\Omega_1} dP_{\Omega_2} \times \Theta[E - V_{LR}(R^{\dagger}) - T_{12}(R^{\dagger}) - T_1 - T_2]\delta(\hat{J} - J\hbar), \quad (4.5)$$

where the contribution from the vibrational states of the fragments has been removed. In Eq. (4.5) $\nu$ is the number of rotational degrees of freedom, $\Omega$, $P_{\Omega}$, and $T$, denote the orientational coordinates, conjugate momenta, and kinetic energy, respectively, for the orbital motion (subscript $_{12}$) and the reacting fragments (subscripts $_1$ and $_2$), and $R^{\dagger}$ denotes the value of $R$ correlating with the maximum in the effective potential, Eq. (4.1). This classical expression, Eq. (4.5), provides the conceptual basis for variational transition state theory treatments, whereas the quantum expression, Eq. (4.3), provides the conceptual basis for adiabatic channel treatments.

The PST assumptions of isotropic interactions and corresponding rovibrationally adiabatic dynamics are particularly appropriate for the reaction of ionic species with nonpolar molecules, for which the ion induced dipole interaction is the dominant long-range interaction. For a spherical molecule, the ion induced dipole interaction (in CGS units) takes the simple form:

$$V_{\mathrm{LR}}^{iid}(R) = -\frac{\alpha q^2}{2R^4}, \quad (4.6)$$

where $\alpha$ is the polarizability of the molecule, and $q$ is the charge on the ion. For this potential, the PST calculated thermal capture rate is equal to the classical Langevin expression[14]

$$k(T) = 2\pi q \sqrt{\alpha/\mu}. \quad (4.7)$$

The generally non-spherical nature of a molecule correlates with asymmetries in the polarizability, and thus in the interaction potential. However, the effect of this asymmetry tends to be quite minor[15,16] and the simple replacement of the polarizability, $\alpha$, with the mean polarizability $(\alpha_{||} + 2\alpha_{\perp})/3$ is quite effective. The polarizability can generally be estimated quite accurately as a sum of bond polarizabilities.[17] Alternatively, accurate values can be obtained from straightforward electronic structure calculations (e.g., density functional theory or second order Moller Plesset perturbation theory, MP2, calculations with augmented double-zeta basis sets or larger).

The ion induced-dipole potential and the corresponding Langevin rate expression have proven to be of great value in the interpretation of numerous studies of ion-molecule reaction kinetics (see chapters 2 and 3 by Canosa *et al.*, and by Gerlich). For neutral reactions, a $1/R^6$ potential, which can be taken to represent dispersion and/or dipole induced-dipole interactions, has provided a similarly important reference potential. The Gorin model,[18] which is based on PST like assumptions for this potential, provides a simple expression for the capture rate:[4,19]

$$k = \Gamma(2/3)2^{11/6}\pi^{1/2}\frac{C_6^{1/3}(k_BT)^{1/6}}{\mu^{1/2}} \tag{4.8}$$

where $C_6$ is the coefficient in the potential. The 'modified' Gorin model, which has been used, especially by Benson and Golden,[20,21] to treat barrierless radical-radical reactions, incorporates an empirical steric reduction factor to account for the nonreactivity of certain orientations of the fragments.

### 4.2.2. *Statistical Theories for Anisotropic Potentials*

Unfortunately, the assumption of an isotropic interaction seldom holds for either ionic or neutral reactions. For example, for an ion-molecule reaction, the presence of a dipole, $d$, in the neutral fragment implies that the dominant long-range term is the ion-dipole term,

$$V_{LR}^{id}(R) = -\frac{qd\cos\theta}{R^2}, \tag{4.9}$$

which is clearly anisotropic. The dominance of this term at long-range implies that it is the determining factor for the kinetics in the low temperature limit. Furthermore, even in the absence of a dipole, the anisotropic ion-quadrupole interaction, which is essentially always present, has a $1/R^3$ behavior and thus becomes the dominant interaction at low temperature.

Similarly, for neutral reactions, the dipole-dipole interaction with a $1/R^3$ form, or the dipole-quadrupole interaction with a $1/R^4$ form, or even the quadrupole-quadrupole interaction with a $1/R^5$ form, are each of longer range than the dispersion and dipole induced-dipole terms that contribute to the $1/R^6$ potential. Thus, the isotropic assumptions of PST are essentially never valid when considering low temperature kinetics. At higher temperatures chemical bonding interactions and short range repulsions, which are even more highly anisotropic, become important. As a result, at least for neutral reactions, there is essentially no temperature range where PST is generally applicable. For ionic reactions, the relevance of PST and the corresponding Langevin rate prediction at higher temperatures (e.g. near room temperature) depends on the magnitude of the polarizability relative to the dipole and/or quadrupole moments. For example, if $d^2/(2\alpha k_B T) < 1$ the deviations from the Langevin predictions can be quite modest.[22]

These anisotropies lead to a coupling of the fragment rotational and orbital angular momenta. The quantities $l$, $j_1$, $j_2, k_1$, and $k_2$ are no longer strictly conserved. This coupling is manifested in the addition of an orientational dependence to $V_{LR}(R)$ in Eq. (4.5) or equivalently to the replacement of $E_{j1k1} + E_{j2k2} + E_l$ in Eq. (4.3) with the energy $E_i(R)$ for the fully coupled motion. There have been numerous attempts to generalize PST to account for the effects of this coupling, with these treatments based on either Eq. (4.3) or Eq. (4.5). Here we summarize the work based on Eq. (4.5), with a discussion of adiabatic capture treatments left for Section 4.2.4.

Much of the work at treating the effects of anisotropies in the potential has focused on ion-molecule reactions, and particularly on the incorporation of corrections for the ion-dipole and/or ion-quadrupole interactions. Various "effective potential" approaches have employed an isotropic potential that is obtained from some sort of averaging of the real anisotropic potential. The average dipole orientation approach of Bowers and coworkers[23] is based on a dynamical average of the orientational angle. In the extended Langevin approach[24] the capture rate is instead calculated for fixed angle and then averaged over angle. Markovic and Nordholm discussed the relation of the free energy based effective potential approach[25] to canonical variational transition state theory and incorporated microcanonical based restrictions on the energies.[26] Most recently, Rayez and coworkers have considered a simple orientationally averaged effective potential.[27]

Variational transition state theory (VTST) provides a rigorous upper bound to the capture rate and also allows for the simple incorporation of energy and angular momentum conservation. Chesnavich and coworkers developed a sophisticated VTST treatment for the ion-dipole + ion-induced dipole case.[22] In a series of articles Niedzielski and coworkers generalized and extended this work to the detailed consideration of a number of specific ion-molecule cases.[28,29] Smith and Troe applied a VTST treatment, based on the assumption of $J \approx l$, to the potential given as the sum of the ion-dipole, ion-quadrupole, and anisotropic ion-induced dipole potentials.[30] Hase and coworkers considered the application of traditional reaction path Hamiltonian approaches, which assume harmonic vibrators for all the vibrational modes including the low frequency interfragment bending motions.[31]

A very general approach was considered by Wardlaw and Marcus,[32] who proposed the evaluation of the number of states in Eq. (4.5) via Monte Carlo integration, but with the inclusion of the full orientation dependent $V_{LR}$. With the advances in computational abilities at the time, this allowed for the straightforward prediction of the capture rate for arbitrary potentials and for arbitrary rotors (atomic, linear, spherical, symmetric, or asymmetric). This "flexible" TST approach was readily applicable to arbitrary long-range potentials for both ion-molecule and radical-radical reactions. Although very general, the flexible TST approach did not readily yield analytic expressions for the rate coefficients, which is one of the great values of many of the earlier VTST studies.

In recent work, we have developed a "long-range" form of transition state theory that is readily applied to arbitrary potential forms, but also immediately yields analytic results for many simple potential forms.[4] These analytic results have previously been derived in a variety of ways. The long-range TST approach clarifies the results that have been obtained in these prior studies. For example, it demonstrates that the inadequacies of some of the early transition state theory studies were largely due to the neglect of the conservation of energy and/or angular momentum.

This "long-range" transition state theory is based on the simple assumption that the orbital moment of inertia is much greater than each of the fragment moments of inertia. This assumption is valid when focusing on the long-range part of the potential since the orbital moment of inertia is proportional to $R^2$. Generally, the quantitative validity of this assumption becomes debatable for separations of about 5 Å or less. Coincidentally, long-range expansions of the potential also become inappropriate at this same separation.

With this assumption of a large orbital moment of inertia, and the vibrational decoupling assumption, the capture rate can be written as

$$k(T) = \left(\frac{8\pi}{\mu(k_B T)^\nu}\right)^{1/2} \frac{1}{\Gamma(\nu/2 - 1/2)} \int_0^\infty dE e^{-E/k_B T} \int_0^\infty dX N_x^\dagger(E, X), \tag{4.10}$$

where

$$X = \frac{J^2\hbar^2}{2\mu}, \tag{4.11}$$

$$N_x(E, X|R) = \langle [E - V(O, R) - X/R^2]^{(\nu-3)/2} \rangle, \tag{4.12}$$

and

$$N_x^\dagger(E, X) = \min_R N_x(E, X|R). \tag{4.13}$$

In these expressions, $\nu$ is the dimensionality of the system given by $\nu_1 + \nu_2 + 3$, where $\nu_i$ is the number of rotational degrees of freedom for reactant $i$. Also, the average is taken with respect to the orientational space

$O = (\Omega_{12}, \Omega_1, \Omega_2,)$ of the rotors, and its argument is set to zero when the expression is negative.

The above result for the capture rate is based on the proper conservation of energy and total angular momentum. Simpler, more transparent, expressions were also derived for the less accurate microcanonical and canonical cases. In particular, for the microcanonical case the rate is given by

$$k(T) = \left(\frac{8\pi}{\mu\,(k_B T)^\nu}\right)^{1/2} \frac{1}{\Gamma(\nu/2+1/2)} \int_0^\infty dE e^{-E/k_B T} N_E^\dagger(E), \quad (4.14)$$

where

$$N_E(E|R) = R^2 \langle [E - V(O,R) - X/R^2]^{(\nu-1)/2} \rangle, \quad (4.15)$$

and

$$N_E^\dagger(E) = \min_R N_E(E|R). \quad (4.16)$$

For the canonical case, the rate is simply

$$k(T) = \left(\frac{8\pi k_B T}{\mu}\right)^{1/2} N_T^\dagger, \quad (4.17)$$

where

$$N_T(R) = R^2 \langle e^{-V(O,R)/k_B T} \rangle, \quad (4.18)$$

and

$$N_T^\dagger = \min_R N_T(R). \quad (4.19)$$

Typically, neglecting either the conservation of energy or angular momentum yields an overestimate in the capture rate coefficient of about 5–10%. Notably, however, the conservation of $E$ and $J$ can be much more significant when considering multiple transition states.

In the low temperature limit only the longest ranged term in the expansion is relevant. In this case, analytic expressions for the rate coefficient are readily derived. In particular, for arbitrary single term long-range potentials,

$$V_{LR} = \frac{V_0 f(O)}{R^n} \quad (4.20)$$

the long-range transition state theory expression reduces to

$$k(T) = C_1 \mu^{-1/2} V_0^{2/n} (k_B T)^{1/2-2/n}, \quad (4.21)$$

where $C_1$ is a constant that is obtained from a well specified orientational average of the long-range potential (*cf.* Eq. 4.12), and $V_0$ is a dimensional constant characterizing the strength of the potential. The function $f(O)$ describes the orientational dependence of the potential. In general, the constant $C_1$ depends on the dimensionality of the system, the rotational constants of the fragments, and whether or not energy and angular momentum conservation are accounted for.

The analytic results obtained for a number of standard cases are summarized in Table 4.1.

Unfortunately, these results for the single term long-range expansions of the interaction potential are of very limited practical applicability, particularly for neutral reactions. At temperatures below ~10 K the classical treatment of the rotational motions becomes inaccurate. Meanwhile, for neutral reactions at temperatures above ~10 K, the long-range transition states tend to be at separations where multiple terms in the long-range expansion are of similar magnitude.[4] For ionic reactions, the lowest order term in the expansion is often dominant until about 100 K, providing a window of applicability from ~10 to ~100 K. When multiple terms in the potential are of similar magnitude, the long-range capture rate can readily be evaluated from Eqs. (4.10)–(4.12), but now employing the sum of all the relevant contributions to the potential. Nevertheless, the interplay between the different contributions complicates the dependence on temperature and interaction potential.

The maximum attractive contributions from the ion-dipole, ion induced-dipole, and ion-quadrupole terms to the $H_3^+$ + HCl interaction potential

Table 4.1. Components of the analytic expression for the long-range transition state theory capture rate for various asymptotic expansions of the potential.

| Potential | Order $n$ | $V_0$[a] | $C_1$[b] |
|---|---|---|---|
| Ion-Dipole | 2 | $q_1 d_2$ | 2.5–2.7 |
| Ion-Quadrupole | 3 | $q_1 Q_2$ | 5.3–5.4 |
| Ion-Induced-Dipole | 4 | $\alpha\ 2ql^2/2$ | 8.9 |
| Dipole-Dipole | 3 | $d_1 d_2$ | 4.8–5.0 |
| Dipole-Quadrupole | 4 | $d_1 Q_2$ | 4.4–4.5 |
| Dipole-Induced-Dipole | 6 | $d_1^2 \alpha_2$ | 8.7 |
| Dispersion | 6 | $C_6$ | 8.55 |

[a] The terms $q_i, d_i, Q_i, \alpha_i$, and $C_6$ represent the molecular charge, dipole moment, quadrupole moment, polarizability for fragment $i$ and the dispersion coefficient, where the latter is often taken to be $3\alpha_1\alpha_2 E_1 E_2/(2(E_1 + E_2))$, where $E_i$ is the ionization energy of fragment $i$.
[b] The coefficent $C_1$ is generally within the specified range of values, with the actual value depending on the parameters for the specific system.

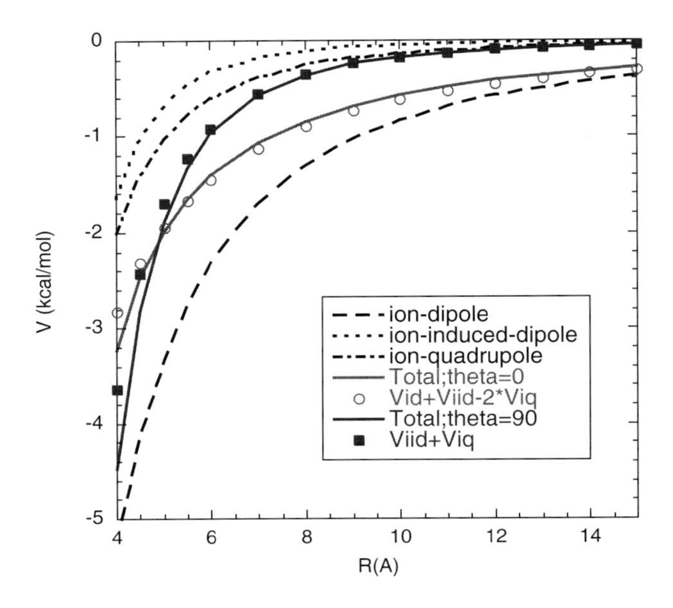


Fig. 4.1. Plot of the contributions to the long-range potential for $H_3^+$ + HCl.

are illustrated in Fig. 4.1. The ion-dipole term clearly dominates, but the other two terms approach a similar magnitude as the separation distances $R$ decreases toward 4 Å. Notably, a number of the earlier statistical theory studies of ion-molecule reactions recognized the importance of multiple terms in the long-range potential. Although these studies were generally unable to directly obtain analytic expressions, they did often provide empirical fits to their numerical simulations.

The total potential obtained from MP2/aug-cc-pvtz calculations is illustrated with symbols for dipolar angles of 0 and 90°. The corresponding long-range sums, represented with solid lines, accurately reproduce the *ab initio* potential for separations of ~5 Å and larger. The increasing deviations from the long-range representation at separations below ~5 Å are expected to be fairly representative for both neutral and ionic reactions.

Explicit prescriptions for the radial dependence of the long-range transition state location were provided in Ref. 4 for a number of representative cases. For ionic reactions, the transition state tends to approach the 5 Å region at about 300–500 K. For neutral reactions, the failures of the long-range estimates are more dramatic and often arise at much lower temperatures due to the much weaker attractiveness of the potential. A more detailed discussion of the failures for neutral reactions, which can extend down to a few K, will be provided in Section 4.4.

### 4.2.3. *Trajectory Simulations*

Trajectory simulations of the long-range capture rate were pioneered by Su and coworkers in the 1980's.[22,33] The reacting fragments were considered as rigid bodies with the analysis considering only the rotational and translational dynamics. These authors focused on ion-molecule reactions, considering the effect of various ion polar interactions on the ion-induced dipole rate coefficients. These trajectory simulations have provided important data for the testing of other more approximate theories. Su and coworkers have also made efforts to parameterize their results[34] in order to facilitate their use by other workers. More recently, Troe and coworkers have expanded on this work, avoiding some numerical problems, considering a larger range of parameters, and considering neutral in addition to ionic reactions.[35–39] They have also provided a parameterization for a large number of cases.[19]

One conclusion from these studies is that trajectory simulations are now quite straightforward and can be routinely performed for any analytic potential energy surface. However, there is still interest in statistical theories for a variety of reasons. In particular, when studying multi-channel (or multi-transition state) reactions it is important to have the capture flux at a wide range of $E$ and $J$. The evaluation of this capture flux from trajectory simulations is still somewhat time-consuming. More importantly, accurate evaluations of the capture rate often require *ab initio* quantum chemical evaluations of the potential energy surface. These *ab initio* evaluations can be quite intensive computationally. The direct coupling of such *ab initio* evaluations with statistical theories requires many fewer potential evaluations than does the corresponding coupling with trajectory simulations. For these reasons it is still important to have some understanding of the quantitative accuracy of statistical predictions for the capture rate. Our own trajectory simulations indicate that the long-range transition state theory predictions for the capture rate are generally accurate to 10% or better.[4]

### 4.2.4. *Adiabatic Channel Calculations and Quantum Dynamics*

Adiabatic channel (or equivalently adiabatic capture) calculations provide a generalization of the quantized representation of phase space theory, Eq. (4.3), to account for the effect of anisotropies in the potential.[40] Such calculations focus on the evaluation of the quantized energy levels, $E_i^J(R)$, for the coupled fragment and orbital rotational motions at a given center-of-mass separation, $R$, and a given total angular momentum $J$. The sums over the independent fragment and orbital energy levels are replaced with a sum over the energy levels, $i$, of the coupled system. For a given energy

$E$, the reactive flux is given by the number of channels whose adiabatic maxima, $E_i^{J,\max}$ lie below that energy,

$$N(E, J) = (2J + 1) \sum_i \Theta(E - E_i^{J,\max}), \quad (4.22)$$

Notably, adiabatic channel models and variational transition state theory yield identical rate estimates as long as they employ the same assumptions for the energy levels.[41] However, the implementations of the two theories generally employ different assumptions (e.g., classical versus quantum) and so yield modestly different rate estimates. Furthermore, the adiabatic models sometimes implement additional constraints such as an assumed conservation of the orbital angular momentum.

The explicit consideration of the quantized nature of the rotational motions makes adiabatic channel treatments particularly appropriate for examining the kinetics at low temperature. Furthermore, adiabatic channel treatments are most effectively implemented when there are not too many states, i.e., at low temperatures. Classical treatments of the rotational states, as in long-range TST, are accurate when $k_BT/B \gg 1$, where $B$ is the largest rotational constant of the reactants. For typical molecular dimensions, classical partition functions become inaccurate somewhere in the 0.1 to 50 K range.

The accurate evaluation of the quantized rotational energy levels can be a computationally intensive task. For this reason, numerous authors have introduced various approximate schemes. Clary, who pioneered the implementation of adiabatic channel (AC) models for the long-range capture rate, introduced three primary approaches, the ACIOSA, ACPCSA, and ACCSA methods.[10,42–45] The ACCSA method, which is the most accurate, employs the centrifugal sudden approximation (CSA). This approximation ignores the coupling between different values of the $z$-projection of the body — fixed total angular momentum. Comparison with close coupled scattering calculations demonstrated the accuracy of the adiabatic capture and centrifugal sudden approximations for the $He^+ + HCl$ reaction.[46] The ACIOSA, which is the least accurate approach, is based on the infinite order sudden approximation (IOSA). This approximation includes the CSA and also ignores the coupling between different angular momentum states of the fragments. With the ACIOSA method, the coupling of the angular momenta reduces to an orientational average, which greatly simplifies the calculations. However, the resulting predictions are often in significant error. The ACPCSA, where P denotes partial, employs the CSA for the fragment with the larger rotational constant, and the IOSA for the fragment with the smaller rotational constant.

Initial applications of the ACCSA method to a number of ionic reactions yielded generally good agreement with experimental observations, but there were modest discrepancies for a few cases.[47–49] Subsequent application to a series of first and second row atomic cation reactions found generally good agreement for the first row cations.[50] For the second row cations discrepancies of a factor of two or three were typical. These and other discrepancies could often be explained at least qualitatively by the presence of high exit channel barriers. Explicit *ab initio* calculations of the potential for a few other reactions indicated modest deviations due to the contribution from additional terms beyond the ion-dipole and ion-induced dipole terms.[51] Unfortunately, these *ab initio* calculations were performed at the SCF level, which does not treat dispersion effects.

Open shell species lead to considerable complications in the analysis related to the presence of multiple electronic states and the effect of rovibronic couplings due to spin-orbit interactions. The separation into ortho and para rotational manifolds have similar effects. Graff and Wagner provided a pioneering study of such effects for the prototypical $O(^3P)+OH(^2\Pi)$ reaction.[52] Clary and coworkers provided a series of analyses for the reactions of both neutral and ionic $^2\Pi$ radicals.[49,51,53] These studies focused on the low temperature limit where rovibronic coupling effects become increasingly important. In this limit, only the ground spin-orbit state contributes, and the analysis focuses on the accurate evaluation of the rovibronically coupled energy levels via perturbation theory. This work has been extended by Stoecklin and coworkers to include the reaction of open shell atoms ($^2P$ or $^3P$).[54] These rotationally adiabatic perturbation theory studies yield useful analytic representations of the zero temperature limit classical adiabatic capture rate for a variety of long-range potentials.

At higher temperatures there can be some contribution from low lying excited electronic states (e.g., a second spin-orbit state). In this case, the multiple surface effects are generally treated via the consideration of the number of surfaces on which the potential is attractive. The single surface rate coefficient is then multiplied by the ratio $f_e$ given by the electronic partition function for the reactive surfaces divided by that for the reactants.[55] This result corresponds to an adiabatic approximation for the electronic dynamics.

Important further work by Troe and Nikitin and coworkers[19,30,35–39,56,57] considered the calculation of the capture rate from the perspective of the statistical adiabatic channel model (SACM). Ramillon and McCarroll[58] demonstrated that the adiabatic capture method of Clary and the SACM method of Troe are identical in concept. However, there are still some minor differences in the approaches used to evaluate the rotational energies. Direct comparison for a number of ion-dipole capture rates found good agreement down to about 50 K, but there were increasing discrepancies at lower temperatures. These discrepancies are apparently a sign of numerical problems. Related methods include the perturbed rotational state approximation of Sakimoto[59] and the adiabatic invariance method of Bates and coworkers.[60]

These SACM studies of Nikitin and Troe and coworkers derived general expressions for a wide variety of specific cases and also explored various limitations. Troe's initial focus was on ion-molecule reactions.[30,35,36,56] Other work focused on linear-linear neutral reactions, initially considering only a dipole-dipole potential.[37,57] Subsequently, the effects of valence interactions were also considered from a general perspective,[38,39] and via a detailed study of the OH + OH recombination.[61] These studies each contain useful comparisons with trajectory simulations, with phase space theory calculations, and with zero temperature limiting expressions.

The classical SACM and ACCSA methods fail at very low temperature and sometimes also at intermediate temperatures. In the low temperature limit, both the adiabatic assumption and the classical treatment of the relative translational motion fail. Vogt and Wannier[62] have shown that the zero energy limit of the quantum cross section for an ion-induced dipole potential is twice that predicted by the classical Langevin model. The classical SACM and ACCSA methods both predict the Langevin result. An empirical study by Hessler[63] of the CN + $O_2$ reaction interpolated from classical high energy cross sections to the Bethe and Wigner threshold behaviors. These empirical calculations suggest that quantum translational effects become important at about 7 K. However, subsequent detailed analyses by Nikitin and Troe and coworkers indicate that the classical SACM model is actually accurate down to $\sim 10^{-3}$ K.[64,65] For even lower temperatures they have developed an axially non-adiabatic channel approximation and have also performed explicit quantum coupled channel calculations.

The higher temperature failures arise solely from non-adiabatic transitions, with the global Massey parameter, $\xi$, being a key indicator.[19] This parameter is defined as the ratio of the average collision time to the average rotation time. When $\xi \gg 1$, the dynamics is globally adiabatic. Nikitin and Troe and coworkers have estimated the Massey parameter and non-adiabatic effects for a range of potentials and other parameters. This analysis shows that the dynamics, sometimes, but not always, become non-adiabatic at some temperature that is almost always high relative to the rotational temperatures of the reactants. Then, at some even higher temperature, the dynamics returns to being globally adiabatic. For this intermediate temperature regime, where the dynamics is non-adiabatic, Troe and coworkers have advocated the use of trajectory simulations. In some cases, the dynamics remains adiabatic throughout the regime where classical rotational descriptions are appropriate and the classical SACM models are applicable. Alternatively, the use of a post-adiabatic representation yields a description in terms of uncoupled dynamic states.[66]

Notably, their discussions about non-adiabaticities have been for a reaction coordinate assumed to be the separation between the centers-of-mass of the two reactants. The use of a more general reaction coordinate, as in some VTST methods, may account for some of the non-adiabaticities. Indeed, reaction coordinate optimizations in the variable reaction coordinate (VRC)-TST approach[67] have often found a reduction by a factor of two or more relative to that obtained for the separation between the centers-of-mass. However, these reductions are generally for short range separations where the long-range potential expansions are inappropriate.

The adiabatic channel model directly yields estimates for the rate coefficient $k_j(T)$ for the capture rate as a function of the rotational quantum state $j$ of the reactants.[42] For a given $j$, such rate coefficients generally increase with temperature. Meanwhile, at a given temperature, the $k_j(T)$ generally decrease with increasing $j$ and the population of high $j$ states increases with increasing temperatures. Due to these two competing factors, the rate coefficient can either rise or decline with increasing temperature depending on their relative importance. Interestingly, a recent experiment of Olkhov and Smith provides detailed information on the $j$ dependence of the capture rate for the reaction of CN with acetylene.[68]

### 4.2.5. *Illustrative Comparisons*

Some long-range TST predictions for the $H_3^+$ + HCl capture rate are illustrated in Fig. 4.2. The capture rate obtained by considering solely the ion-dipole potential is in good agreement with that for the total potential, consisting of the sum of the ion-dipole, ion induced-dipole and ion-quadrupole potentials, for temperatures up to about 100 K. At higher temperatures, the other terms in the potential yield an increasing contribution to the long-range capture rate. By 1000 K, the capture rate for the total potential is about twice as large as that for the ion-dipole potential. However, in reality, by 1000 K the long-range capture rate theory is no longer applicable, because the transition state is at too short a separation. Many more potential terms need to be included, and a long-range expansion is no longer effective.

Also shown in Fig. 4.2 are the predictions from ACCSA[48] and SACM calculations,[69] and the experimental results from Adams and coworkers.[48] Above 20 K, the long-range TST and ACCSA results are within 10% of the SACM results, which in turn are in excellent agreement with the experimental results. At lower temperatures the long-range TST results overpredict the SACM results, which are expected to be accurate. This

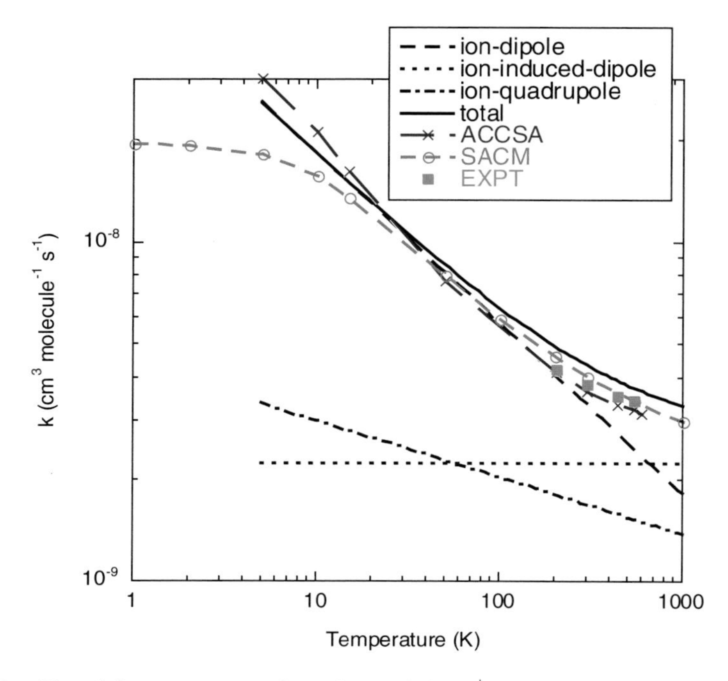


Fig. 4.2. Plot of the temperature dependence of the $H_3^+ + HCl$ capture rate for different terms in the potential and for different theoretical methods and for experimental data from Ref. 48.

deviation is likely due to inadequacies in the classical treatment of the HCl and $H_3^+$ rotational motions, which have rotational constants corresponding to 15 and 63 K, respectively. The low temperature deviations of the ACCSA results from the SACM results arise from numerical inaccuracies in their earliest implementation.

In Fig. 4.3, the long-range TST and SACM[69] results for the capture rate in $N^+ + H_2O$ are compared with experimental[70,71] results that extend down to 27 K. Both theoretical predictions are again in good agreement with experiment, with a maximum discrepancy of only 20%. The deviation between the two theoretical predictions at low temperature is again indicative of important quantum effects. The rotational constants for $H_2O$ are equivalent to temperatures of 14, 21, and 39 K.

The quantum effects are complicated by the nearly degenerate $^3P_0$, $^3P_1$, and $^3P_2$, spin-orbit states of $N^+$. It is not clear how many electronic states are reactive. The red curve assumes that there are two reactive triplet states, and that the electronic motion is adiabatic. This assumption seems reasonable, given the two unpaired electrons in $N^+$. If, instead, only one

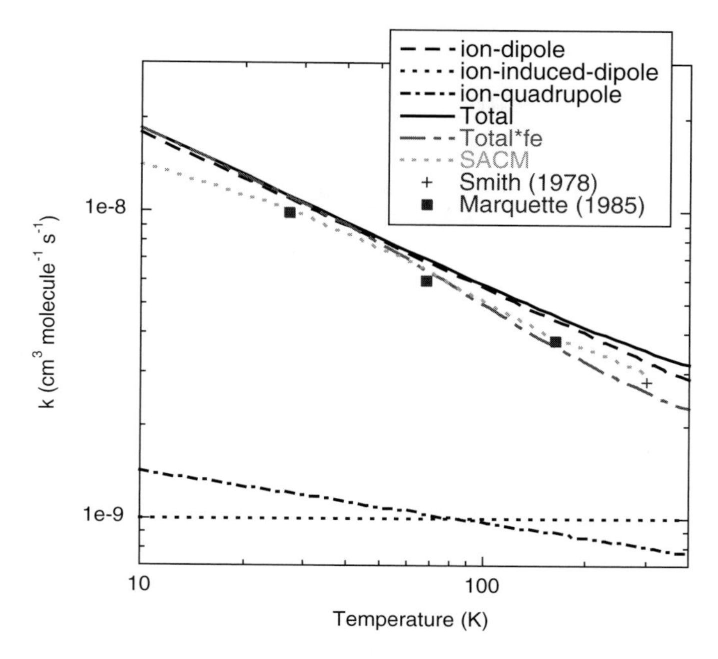


Fig. 4.3. Plot of the temperature dependence of the $N^+ + H_2O$ capture rate for different terms in the potential and for different theoretical methods and for experimental data from Refs. 70 and 71.

triplet state was reactive, the predicted rate would be another factor of two lower at 300 K, well below the experimental result. Such ambiguities in the long-range predictions are unfortunately quite commonplace. The proper treatment of the electronic degeneracy factor $f_e$, requires detailed *ab initio* simulations of the minimum energy paths, and also hinges on the adiabatic or statistical nature of the electronic dynamics.

CRESU experiments indicate that a number of radicals react with unsaturated hydrocarbons at close to the capture rate, especially at low temperature.[72] In Fig. 4.4, long-range TST predictions for the CH + $C_2H_2$ capture rate are illustrated together with the experimental data from Refs. 73–75. The prediction for the full potential, given by the sum of the dipole-quadrupole, dispersion, and dipole induced-dipole potentials, already deviates from the dominant dispersion term at 10 K. By 200 K, the dipole-quadrupole term has become the dominant factor. The different contributions to the rate coefficient over the 0 to 20 K region for the CN + $O_2$ and CH(A) + CO reactions were studied with ACCSA calculations by Tan *et al.*[76]

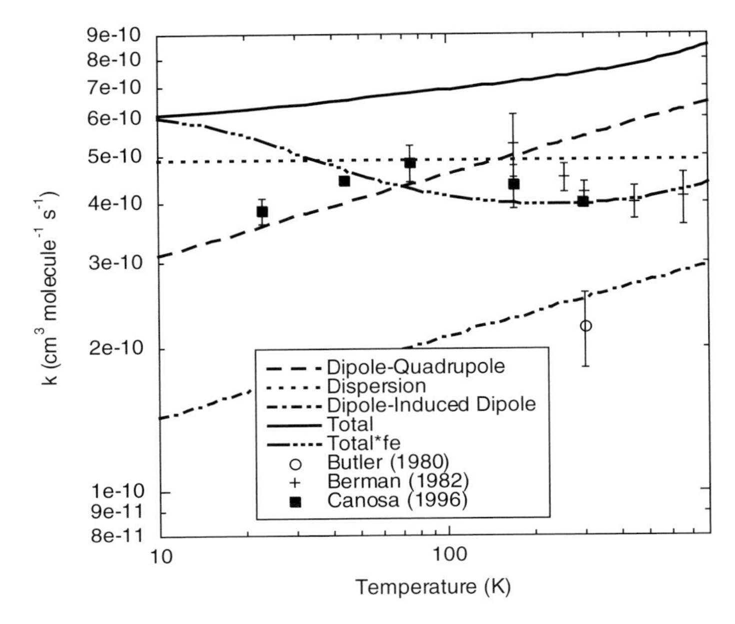


Fig. 4.4. Plot of the temperature dependence of the capture rate for different terms in the potential in comparison with experimental data[73–75] for $CH + C_2H_2$.

The CH radical has only a $28\,\text{cm}^{-1}$ spin-orbit coupling leading to the $^2\Pi_{1/2}$ and $^2\Pi_{3/2}$ states. Sample CASPT2/aug-cc-pvtz//CASPT2//aug-cc-pvdz calculations indicate that only one of the two doublet states has a barrierless reaction path. Thus, the long-range TST predictions should be multiplied by the electronic degeneracy factor $f_e$ given by $2/(2 + 2\exp(-40/T(K)))$. The dash-dot-dot-dot line corrects the long-range TST prediction for the capture rate by this factor, and is in remarkably good agreement with experiment. The deviations for the lowest temperature, 23 K, are likely indicative of quantum effects given that the rotational constant of CH corresponds to 21 K. Notably, the quantum adiabatic capture rate for the dipole-quadrupole potential, which is the dominant term at low-temperature, is predicted to have a $T^{1/4}$ dependence as $T \rightarrow 0$.[53]

Similar agreement between theory and experiment has been found for ACCSA calculations of the capture rate for a number of atom diatom reactions. In particular, the predictions for the $B + O_2$,[77] $Al + O_2$,[78] $Si + O_2$ and NO[79] reactions accurately reproduce the temperature behavior and the numerical values are generally within about 30% of the experimental data. The incorporation of spin-orbit effects on the rovibronic energy levels is a key component of these studies. For the C + NO reaction, trajectory simulations[80] are in good agreement with experiment,[81] whereas ACCSA predictions[82] appear to be too low by about a factor of 3. Some of the improvement in the trajectory result arises from the replacement of the form for $f_e$ employed in the ACCSA treatment with an adiabatic form.

Other classes of neutral reactions that appear to proceed at close to the capture rate over the full low temperature range (10 to 200 K) include the reactions of atomic $B(^3P)$,[83,84] $C(^3P)$,[85,86] and $Si(^3P)$[87] atoms with $C_2H_2$ and $C_2H_4$ and the reactions of CH,[73,88,89] and CN[90,91] with unsaturated hydrocarbons ranging from acetylene and ethylene to anthracene. The reactions of $C_2H$ with unsaturated hydrocarbons[92–95] appear to proceed at close to but not quite the capture rate.[4]

### 4.3. *Ab Initio* Simulations

The quantitative exploration of the effect of short-range interactions on the low temperature kinetics requires detailed *ab initio* simulations. We have recently provided an extensive review of *ab initio* methods for studying reactive potential energy surfaces.[96] Here, for completeness, we provide a brief review of some of the key aspects relevant to the treatment of rapid low temperature reactions.

#### 4.3.1. *Radical-Molecule Systems*

##### 4.3.1.1. *Inner Transition State*

For radical-molecule reactions the number of states at the inner transition state is largely determined by the energy and rovibrational properties of the saddle point. Simple rigid-rotor harmonic oscillator evaluations at the saddlepoint appear to yield an adequate description of the number of states at the inner transition state. Explicit consideration of variational effects and barrier tunneling for some sample reactions found them to be of at most modest importance as long as the reaction is rapid.[7,8] Fully coupled phase space integral based studies of the anharmonic effects in the $C_2H_4$ + OH reaction suggest that only the anharmonicities arising from the lowest frequency torsional mode(s) need to be accounted for.[7] One-dimensional corrections for such modes are easily incorporated.

Multi-reference second order perturbation theory (CASPT2)[97] provides an optimum approach for determining the rovibrational properties of the inner transition state. Such calculations combine high accuracy with suitable efficiency, particularly now that analytic CASPT2 gradients are available in electronic structure programs such as MOLPRO.[98]

Furthermore, they allow for the consideration of multiple electronic states, which can be important for cases where the radical has degenerate electronic orbitals, as in $^2\Pi$ radicals. Spin orbit interactions can also be explicitly incorporated.

The use of an augmented polarized valence triple zeta basis set, such as that of Dunning and coworkers,[99] generally provides satisfactorily converged geometries and vibrational frequencies. The related augmented double zeta basis set, aug-cc-pvdz, is also often sufficient for these purposes. However, for the latter basis set significant problems have occasionally been noted, particularly for unsaturated species.[9,100]

Coupled cluster calculations provide an alternative high level approach for performing the rovibrational analysis. When perturbative corrections for triples contributions are included (i.e., with the CCSD(T)[101] or QCISD(T)[102] methods), such analyses are typically expected to be of somewhat higher accuracy than CASPT2 based analyses. The T1 diagnostic[103] provides a useful measure of the multi-reference character to the wavefunction. The reliability of CCSD(T) calculations decreases dramatically for T1 values beyond 0.03. Unfortunately, coupled cluster calculations are considerably more intensive computationally. Typically, the difference between the CASPT2 and CCSD(T) predicted geometries and vibrational frequencies is negligible, and so the additional expense is generally not warranted. Occasionally, one is tempted to implement a CCSD(T) analysis, but with a smaller basis set. However, the errors introduced by basis set limitations are often greater than those arising from the limitations of the CASPT2 approach. The CCSD and QCISD approaches are also sometimes quite effective. However, it is not clear whether such analyses are generally an improvement over CASPT2 analyses.

The more limited geometry and vibrational analyses employed in standard *ab initio* thermodynamics schemes, such as G3,[104] while typically suitable for thermodynamic evaluations of stable species, lack the accuracy required for quantitative *a priori* kinetic predictions. In essence, kinetic predictions depend much more strongly on the vibrational analyses than do low temperature thermodynamic predictions. Furthermore, saddle point geometries are more strongly dependent on the electronic structure methodology than are equilibrium geometries.

Some limitations in such MP2 or density functional theory analyses can be accounted for by mapping out the reaction path and then using a higher level method to locate a maximum along the path. The projected frequencies along the low level reaction path can be used to obtain a variational transition state theory treatment. For low temperatures, the true variational effects are generally negligible, but the improved estimate of the high level saddlepoint (and its properties) can make a significant correction to the rate estimates.

Once the saddlepoint has been located it is useful to perform higher level estimates of its energies with methods such as CCSD(T) or QCISD(T). For such calculations, it is important to use an even larger basis, such as the aug-cc-pvqz basis. The results from calculations with triple and quadruple zeta basis sets can be extrapolated to yield improved predictions.[105] Notably, the errors arising from limitations to the triple zeta basis set can be rather large when considering second row atoms such as S (e.g., 15 kcal/mol). When CCSD(T) calculations with a quadruple zeta basis are not feasible, one can instead append basis set corrections obtained from MP2 or CASPT2 simulations. Other higher level schemes such as HEAT,[106] Wn,[107] or Gn,[108] provide alternative procedures for estimating the saddle point energy.

Notably, sample calculations with the CASPT2 method suggest that its predictions for the saddlepoint energy are remarkably accurate when the saddlepoint is submerged or nearly so. Indeed, the CASPT2 predictions appear to be even more accurate than CCSD(T) predictions. More generally, when considering energies for chemically dissimilar structures, the CCSD(T) approach is considerably more accurate than the CASPT2 approach. The efficacy of the CASPT2 approach for the inner transition state region of rapid low temperature reactions appears to be related to the relatively modest change in their bonding from the reactants to the inner transition state. Regardless, CASPT2 has become our method of choice for treating both the energies and the rovibrational properties of low temperature radical-molecule kinetics. Comparisons with multi-reference configuration interaction and QCISD(T) calculations can be used to confirm the accuracy of the CASPT2 calculations. When differences are noted, an expansion of the active space can sometimes be used to resolve the discrepancy.

#### 4.3.1.2. *Outer Transition State*

*Ab initio* electronic structure simulations are also useful when considering the long-range transition state. The direct incorporation of *ab initio* simulations into the evaluation of the long-range transition state theory capture rate according to Eq. (4.10) removes any uncertainties in the appropriateness of an analytic representation for the long-range potential. The appropriateness of the underlying *ab initio* methods can be tested by comparing the predicted and experimental electrostatic properties such as the dipole and quadrupole moments and polarizabilities. Such comparisons suggest that the aug-cc-pvdz basis is generally large enough to provide converged long-range potentials, although the aug-cc-pvtz potential is preferred.

Typically, the long-range potential is not especially sensitive to the optimized geometry. Furthermore, there is essentially no variation in the optimized geometry of the reacting fragments in the long-range region of the potential (i.e., for separations of 5 Å or greater.) Thus, relatively low level quantum chemical methods may be employed in the geometry optimizations that precede the evaluation of the long-range transition state number of states according to Eq. (4.10).

Importantly, *ab initio* simulations of the long-range transition state must employ a method that incorporates dynamic correlation. Otherwise the dispersion interaction, which is generally a dominant term in the expansion, will be neglected. Thus, one cannot employ methods such as HF, CASSCF or traditional DFT methods. (Note, however, the recent progress in developing DFT methods for treating dispersion.[109])

MP2 simulations are generally of sufficient accuracy for the long-range region, and are efficient enough for the direct sampling based evaluation of Eq. (4.10). Sample calculations suggest that CCSD(T) methods yield only minor revisions in the predicted long-range capture rate. However, orbital degeneracies in the radical can lead to wavefunction convergence problems. In such cases, it is best to employ CASPT2 methods, with state averaging of the degenerate orbital configurations. Such CASPT2 simulations are again efficient enough to incorporate in a direct sampling approach.

### 4.3.2. *Radical-Radical Systems*

Standard single reference methods, including MP2, CCSD(T), and even DFT methods such as B3LYP, are not applicable to the treatment of radical-radical reactions. The closed shell, spin-restricted, versions of these approaches have singularities precisely in the transition state region. Spin-unrestricted versions yield qualitatively correct reaction path potentials, but the energies are not quantitatively correct. The resulting errors in the predicted kinetics generally exceed a factor of two.

For radical-radical reactions, the full mode coupling and anharmonicity effects for the relative and overall rotational motions must be explicitly accounted for. We have derived a direct variable reaction coordinate transition state theory approach that appears to yield accurate rate coefficients for a number of alkyl radical reactions.[110,111] This approach is analogous to that embodied in Eq. (4.10) for the long-range transition state, but includes variational optimizations of the form of the reaction coordinate and does not make the large orbital moment of inertia assumption. A detailed description of this approach was provided in some of our recent articles.

In principle, the variational minimization of the form of the reaction coordinate, which is defined in terms of a fixed distance between pivot points for each of the reactants, allows for the treatment of the dynamical non-adiabaticities that are neglected in standard adiabatic channel models.

Indeed, the reaction coordinate optimization typically yields a reduction in the predicted rate coefficient by about a factor of two from that obtained for a center-of-mass reaction coordinate. The quantum adiabatic channel calculations are generally restricted to an assumed center-of-mass separation for the reaction coordinate.

Our applications of the variable reaction coordinate (VRC) TST approach strongly suggest that the CASPT2 method generally yields quantitatively accurate interaction energies for both the inner and outer transition state regions. In particular, careful comparisons with MRCI and full-CI calculations indicate only kinetically insignificant discrepancies in the predicted interaction energies. Furthermore, the kinetic predictions are generally in quantitative agreement with experiment. Nevertheless, we have occasionally noted discrepancies between MRCI and CASPT2 based kinetic predictions. In this instance, a study of the dependence on the number of orbitals included in the active space often resolves the discrepancy.

For saturated alkyl radical reactions, a simple 2 electron, 2 orbital active space is sufficient. For more complex radicals, such as resonantly stabilized radicals, it is important to include additional orbitals, such as the full set of $\Pi$ orbitals, in the active space.[112] In the case of electronically degenerate radicals, it is again generally necessary to employ state averaged wavefunctions.

Typically, these applications have employed the cc-pvdz basis set to examine the orientation dependence, coupled with an aug-cc-pvtz analysis of 1-dimensional corrections along the reaction path potential. Sample aug-cc-pvqz calculations suggest that the latter basis set provides an essentially converged interaction potential. However, our recent analysis of the $CH_3$ + OH reaction[113] suggests that an aug-cc-pvdz basis is preferred for the orientational sampling when hydrogen bonding interactions are present.

### 4.4. Short-Range Interactions and the Two Transition State Model

There is a wide variation in the temperature dependence of neutral reactions that are rapid at low temperature. As noted by Smith *et al.*,[72] the reactions of CN with $C_2H_2$, $NH_3$, and $C_2H_6$, illustrate the range of behavior. For $C_2H_2$, the rate coefficient is close to the long-range capture rate, and is nearly temperature independent; varying from $5 \times 10^{-10}$ to $2 \times 10^{-10}\,cm^3$ molecule$^{-1}\,s^{-1}$ as the temperature increases from 23 to 700 K. In contrast, the rate coefficient for the reaction of $C_2H_6$ with CN takes a minimum value of $2 \times 10^{-11}\,cm^3$ molecule$^{-1}\,s^{-1}$ at about 200 K, rising to $1 \times 10^{-10}\,cm^3$

molecule$^{-1}$ s$^{-1}$ as the temperature falls to 23 K or rises to 1000 K. Notably, this low temperature value is still a factor of 4 below the long-range capture rate. The temperature dependence for the reaction with $NH_3$ is between these two extremes, with a rate equal to the capture rate at low temperature, but decreasing by a factor of 20 from 20 to 200 K.

Many neutral reactions have similar discrepancies between the long-range capture rate predictions and the observations from CRESU experiments. For example, only 14 of the long-range capture rate predictions from Ref. 4 are within a factor of two of the 26 CRESU results at their lowest temperature. (Note, however, that the comparison in Ref. 4 did not consider atomic reactants, which do tend to have rates that are closer to the capture rate.) Furthermore, even where there is agreement at the lowest temperature, the CRESU results often decrease with temperature, whereas the long-range capture predictions are typically nearly constant or increasing with temperature. In every instance of discordance, the capture theory predictions exceed the experimental observations. These discrepancies are suggestive of some alternative bottleneck to the reaction.

The formation of a chemical bond requires the orientation of the reactants in order to maximize the overlap of the electronic orbitals involved in the incipient bond. The restriction in the range of reactive orientations, and corresponding decrease in the entropy, becomes increasingly severe as the separation distance decreases. Meanwhile, the strength of the chemical bond increases with decreasing separation. The interplay between these two opposing contributions typically yields a maximum in the free energy, i.e., a transition state, at separations where the chemical bonding and long-range interactions are of about the same magnitude. This generally occurs at separations of about 2–4 Å, where both interactions are on the order of 1 kcal/mol.

Importantly, the orientations required for the incipient bond are generally quite different from the optimal orientation for the long-range interactions. As a result, it is fairly common for there to be a long-range van der Waals potential minimum that precedes the inner transition state. In this case, there are clearly two transition states along the path from reactants to chemically bound complex. A schematic illustration of these two transition states is provided in Fig. 4.5. One describes the formation of the van der Waals complex from reactants. Long-range capture theory should provide a good description of this outer transition state. The other transition state describes the formation of a chemical complex from the van der Waals molecule. The properties of this inner transition state arise from a delicate balance between bonding, short-range repulsion, and long-range interactions. As a result, the accurate treatment of this inner transition state requires detailed *ab initio* simulations.

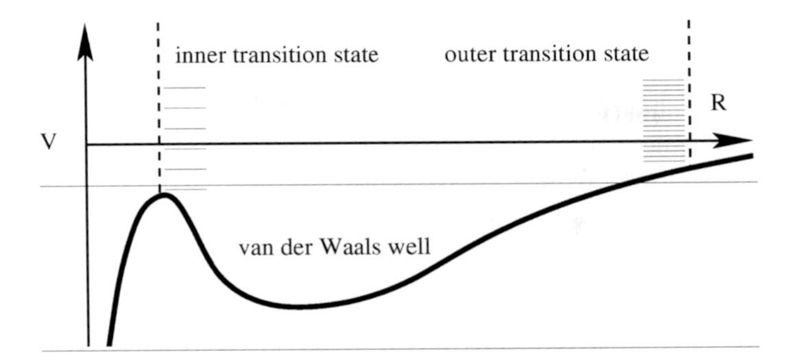


Fig. 4.5. Schematic diagram of the inner and outer transition states on a potential energy surface. The horizontal lines represent the energy levels at the two transition states.

One key result of the generally distinct nature of the inner and outer transition state regions is that long-range potential based predictions for the location of the outer transition state can underestimate the importance of the inner transition state. Such predictions only describe when the outer transition state reaches separations for which the long-range expansion is no longer appropriate. However, since its properties are distinct from those of the outer transition state, the inner transition state can provide the dominant bottleneck at much lower temperatures. Consider, for example, the limiting case where the energy of the inner transition state lies above the reactants, but there is still a well defined van der Waals complex. The inner transition state is then always the dominant bottleneck and the temperature at which the long-range transition state merges with the short-range transition state is irrelevant.

There are two factors that determine the relative importance of short-range interactions: (i) the energy of the inner transition state relative to asymptotic fragments and (ii) the entropy change at the inner transition state. The decrease in the entropy at the transition state is related to the change from free rotations of the fragments to nearly harmonic bends at the inner transition state for a number of the fragment rotational modes. Thus, the larger the number of rotational modes in the fragments, the greater the decrease in the entropy. For this reason, and due to the increased steric repulsion, the inner transition state is expected to be most important when considering the reaction of two nonlinear polyatomics and least important when considering the reaction of an atom with a diatom. The energy of the inner transition state depends on the type of interaction: ionic, radical-molecule or radical-radical. We consider these three cases separately in the following subsections.

### 4.4.1. *Radical-Molecule Reactions*

For radical-molecule reactions the formation of a bond between the radical and the molecule is coupled with the breaking of some bond in one of the reactants. The need to break a bond of the reactants prior to reaction generally leads to a well defined saddlepoint in the potential energy along the reaction path. The height of this barrier relative to reactants is a key determining factor in the rate at low temperature. If the zero-point corrected barrier is above the reactants zero-point energy, as is most often the case, then the reaction will be slow at low temperature. As noted by Herbst and coworkers,[114,115] with tunneling of low mass species such as $H_2$, the low temperature rate may be significant, but still a few orders of magnitude below the collision rate. In contrast, if the barrier is "submerged" (i.e., lies below the reactants) then the low temperature limit should be equal to the long-range capture rate.

As recently pointed out by Smith *et al.*,[72] for addition reactions of unsaturated hydrocarbons, the height of the barrier should correlate with the difference between the ionization energy (IE) of the unsaturated hydrocarbon and the electron affinity (EA) of the radical. They went on to note that simple arguments, as well as the experimental evidence, suggest that if this energy difference exceeds 8.75 eV then the barrier should be above the reactant asymptote. Conversely, if it is less than 8.75 eV then the barrier should be submerged and the reaction should be rapid at low temperature.

A subsequent combined experimental and theoretical study of the reaction of $O(^3P)$ atoms with a variety of alkenes provided further confirmation for this correlation.[9] The minimum energy path potentials for the two reactive surfaces in the $O(^3P)$ + ethylene and $O(^3P)$ + trans-butene reactions, cf. Fig. 4.6, clearly show the two transition state behavior for these reactions. For ethylene, the IE − EA difference is 9.05 eV and both barriers are significantly above the reactants' energy. In contrast, for trans-butene this difference is 7.64 eV, and the ground state surface is submerged by more than 1 kcal/mol.

When this saddlepoint is submerged the long-range transition state will be the dominant bottleneck at the lowest energies and the reaction will be rapid at low temperatures. However, the energy levels for the intermolecular motions of the fragments are more widely spaced at the inner transition state than at the outer transition state (cf., Fig. 4.5). As a result, as the excess energy increases, the number of states at the inner transition state gradually becomes smaller than the number of states at the outer transition state. Thus, the inner transition state gradually becomes the dominant bottleneck as the energy increases.

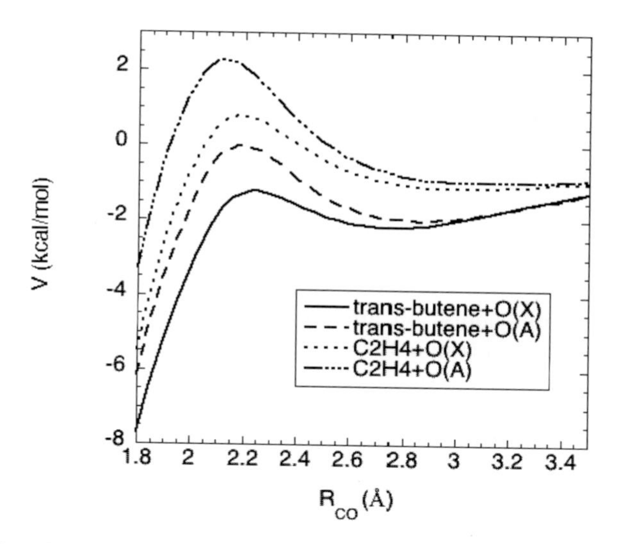


Fig. 4.6. Plot of the potential energy along the minimum energy path for the ground and first excited states of $O(^3P)$ reacting with ethylene and with trans-butene.

For some intermediate energy the numbers of states at the two transition states, $N^\dagger_{\rm inner}$ and $N^\dagger_{\rm outer}$, are roughly equivalent. In this case, the two transition states are acting in series and yield a greater reduction in the flux than would be expected from either one individually. The simple expression

$$1/N^\dagger_{\rm eff} = 1/N^\dagger_{\rm inner} + 1/N^\dagger_{\rm outer} \tag{4.23}$$

provides an accurate treatment of the net reactive flux if the dynamics is statistical and the well between the two transition states is reasonably deep. Alternatively, the unified statistical theory expression

$$1/N^\dagger_{\rm eff} = 1/N^\dagger_{\rm inner} + 1/N^\dagger_{\rm outer} - 1/N_{\rm max} \tag{4.24}$$

is more appropriate if the well is shallow enough that the flux through it, $N_{\rm max}$, is not much greater than that through the two transition states.[116] Note that Eq. (4.24) reduces to Eq. (4.23) in the limit of large $N_{\rm max}$.

A complete treatment of the kinetics through the van der Waals region should involve a master equation simulation[117] of collision induced changes in the energy. However, simplifying assumptions are appropriate in the limit of low or high pressure. The appropriateness of these limits depends on both the pressure and the van der Waals well depth. If there are enough collisions to yield thermal equilibrium populations of the van der Waals energy levels, then the high pressure limit applies and the numbers of states should be replaced with canonical partition functions. If instead there are no collisions, then the numbers of states should be evaluated for fixed total energy and total angular momentum, since those quantities are conserved between collisions. Our own sample master equation simulations suggest that, because of the small well depth for van der Waals complexes, the collisionless limit, where one employs $E$ and $J$ resolved numbers of states, is almost always appropriate.

A two transition state model for radical molecule reactions was postulated long ago by Singleton and Cvetanovic in their analysis of the reaction of $O(^3P)$ atoms with olefins.[118] Unfortunately, their analysis focused on the canonical implementation and numerous subsequent *ab initio* implementations of the two transition state model have followed their lead. The canonical implementation can yield as much as a factor of 6 overestimate of the rate coefficient for temperatures near the crossover temperature (the temperature at which the inner and outer rate coefficients are equivalent), as illustrated in Fig. 4.7 for the $OH+C_2H_4$ reaction. Notably, the $E/J$ resolved two transition state model predictions deviate significantly from the single transition state results over the full temperature range from 20 to 300 K.

An *ab initio* based variational TST study demonstrated that short-range interactions were central to the low temperature kinetics of the $CN + O_2$ radical-radical reaction[6] as suggested in an earlier study of Clary and coworkers.[43] This result implied that short-range interactions could be of

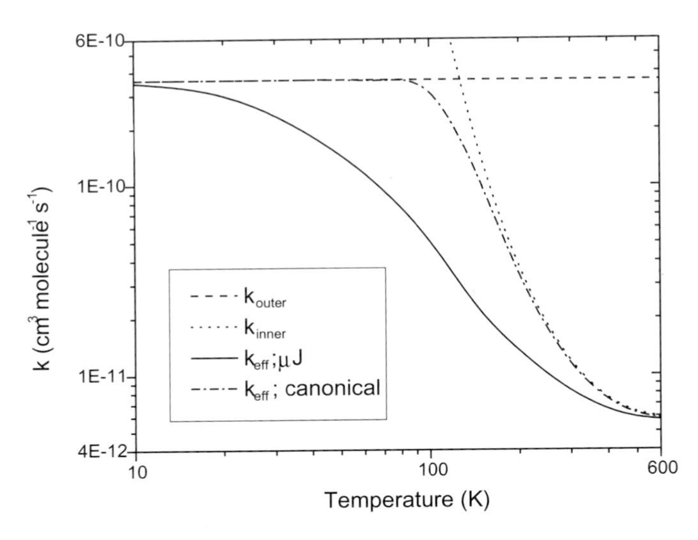


Fig. 4.7. Plot of the inner, outer, and the canonical and $E/J$ resolved implementations of the effective two transition state model predictions for the $OH + C_2H_4 \rightarrow C_2H_4OH$ reaction.

similar importance in radical-molecule reactions. Clary and coworkers suggested their importance as an explanation for the failure of their long-range potential based predictions to reproduce experiment for the $OH + HBr$,[119] $CN + NH_3$, and $CH + NH_3$ reactions.[45] Valiron and coworkers developed a simple model of rotational selectivity arising from the inner transition state that they applied to the $CN + NH_3$ reaction and to an illustrative atom-diatom system.[120,121]

Herbst and coworkers suggested the importance of short-range interactions for the $CN + C_2H_2$[3] and the $C_2H + C_2H_2$ reactions.[122] For the latter reaction, they estimated the range of validity of the long-range capture rate simply on the basis of the temperature dependence of the long-range centrifugal barriers. As discussed in the Section 4.4 overview, this estimate is unlikely to be correct, and indeed their long-range predictions, which they claim should be valid up to 50 K, were a factor of two higher than the subsequent experimental results[93] all the way down to 15 K. For the former reaction, Woon and Herbst subsequently demonstrated that even minor deviations from the long-range capture rate are not necessarily indicative of an inner transition state for the entrance channel.[5] In particular, a barely submerged exit channel transition state has a similar effect to one in the entrance channel as long as the pressure is not high enough to stabilize the molecular complex.

Both Clary and coworkers[123] and Herbst and coworkers[3] have commented on the importance of short range interactions for the reaction of $C(^3P)$ with acetylene. As Liao and Herbst pointed out, the presence of only one reactive potential surface at short-range greatly improves the agreement with experiment. Buonomo and Clary's subsequent reduced dimensionality quantum dynamics study[124] found good agreement with experiment when employing a global potential energy surface. However, this study suggested that ACCSA predictions overestimate the rate by a factor of two, due to significant non-adiabaticities in the dynamics.

The two transition state model was first applied to a rapid radical-molecule reaction at the $E/J$ resolved level by Rauk and coworkers in their study of the reaction of $CH_3$ with HBr.[125] Related two transition state models were considered by Jodkowski *et al.*, for a number of abstraction reactions with $CH_3OH$[126] and by Lin and coworkers in their study of the $OH + H_2CO$ reaction.[127] A simplified two transition state model, termed modified transition state theory, was recently introduced by Krasnoperov *et al.*[128] The latter model only considers the effect of the outer transition state on the threshold energies and so would not be expected to yield accurate low temperature rate coefficients. The other two transition state models also employ various simplified treatments, particularly for the outer transition state.

Our earliest *ab initio* implementation of the two transition state model to a barrierless reaction was to the dissociation of ketene into $^1CH_2 + CO$.[129] This reaction is an interesting case, since there is no saddle point on the minimum energy path, and yet photodissociation studies of the energy dependence of the dissociation rate and product rovibronic distributions strongly suggest an inner transition state. Our variational RRKM studies of this dissociation, which progressed with time from empirical model potentials[130] to high level *ab initio* studies,[131] provided a quantitative description for these experimental observations in terms of an inner transition state that becomes the dominant transition state at an excess energy of about $140\,cm^{-1}$. The effects of this inner transition state are significant down to $50\,cm^{-1}$, and so would affect the reverse association rate constant down to 100 K or lower.

More recently, we have implemented the two transition state model for calculations of the $SiH_2 + SiH_4$,[132] $C_2H_4 + OH$,[7] $CN + C_2H_6$,[8] $OH +$ isoprene,[133] and $O(^3P)$ + alkene reactions.[9] These implementations employ the long-range transition state theory treatment of the outer transition state and so are expected to provide an accurate treatment of the low temperature behavior. Furthermore, for some of the reactions, anharmonic and variational effects for the inner transition state were explicitly treated. For each of these reactions, the inner transition state was predicted to significantly decrease the rate constant in the low temperature regime, with all but the isoprene reaction having important effects down to 20 K.

The results from our *a priori* implementation of the two transition state model for the $CN + C_2H_6$ reaction are illustrated in Fig. 4.8. These results quantitatively reproduce the experimental observations from 25 to 1140 K.[90,134–138] Minima in rate versus temperature plots are usually indicative of the opening of new reactive pathways. However, for this reaction, the minimum in the rate constant versus temperature is simply the result of the changing dominance of the inner and outer transition states.

Our predictions for the $O(^3P)$ + alkenes reaction demonstrate the changing behavior as the saddlepoint becomes increasingly submerged. As illustrated in Fig. 4.9, for the propene case, which has a slightly positive barrier of 0.1 kcal/mol, the rate coefficient is about two orders of magnitude below the long-range capture rate. In this case, the outer transition state plays almost no role in the kinetics. In contrast, the rate constant for trans-butene, whose barrier is submerged by 0.9 kcal/mol, is an order of magnitude greater than for propene. In this case, both transition states are important all the way from a few K to 200 K.

The $O(^3P)$ + trans-butene reaction also provides an illustration of the complications that arise from our imprecise knowledge of the electronic dynamics. In particular, consider the schematic correlation diagram of the

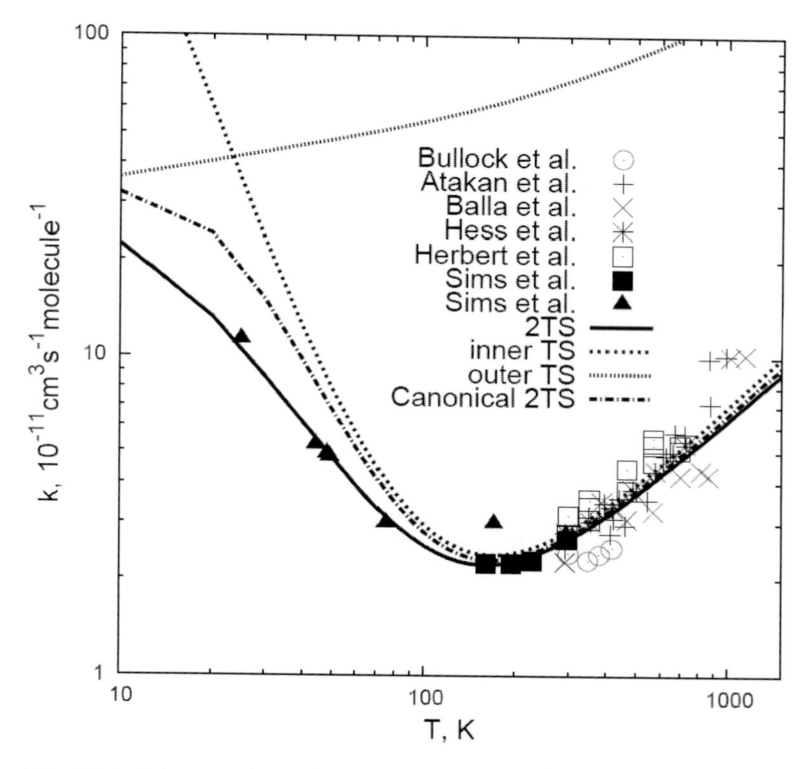


Fig. 4.8. Plot of the experimental and the long-range, short-range and two transition state model predictions for the temperature dependence of the rate constant for $CN + C_2H_6 \rightarrow HCN + C_2H_5$.

electronic states for this reaction shown in Fig. 4.10. The ground ($^3P_2$) state of the O atom is fivefold degenerate. At the inner transition state only the ground triply degenerate state (X) lies below reactants. The first excited state (A) is slightly positive relative to ground state reactants.

Thus, if the electronic dynamics is fully adiabatic (from reactants to the covalent molecular complex), one should employ an electronic degeneracy factor of 3 for both $N^{\dagger}_{\text{inner}}$ and $N^{\dagger}_{\text{outer}}$. Then, at low temperature, where the $^3P_1$ and $^3P_0$ states of the O atom are not occupied, the overall rate coefficient is predicted to be the simple long-range capture rate multiplied by 3/5. If instead, there is a statistical mixing of electronic states in the pre-reaction complex, then one should again employ an electronic degeneracy factor of 3 for $N^{\dagger}_{\text{inner}}$, but the degeneracy factor for $N^{\dagger}_{\text{outer}}$ should now be 5. In that case, the predicted low temperature rate is simply the long-range capture rate; i.e., the multiplicative factor is 5/5.

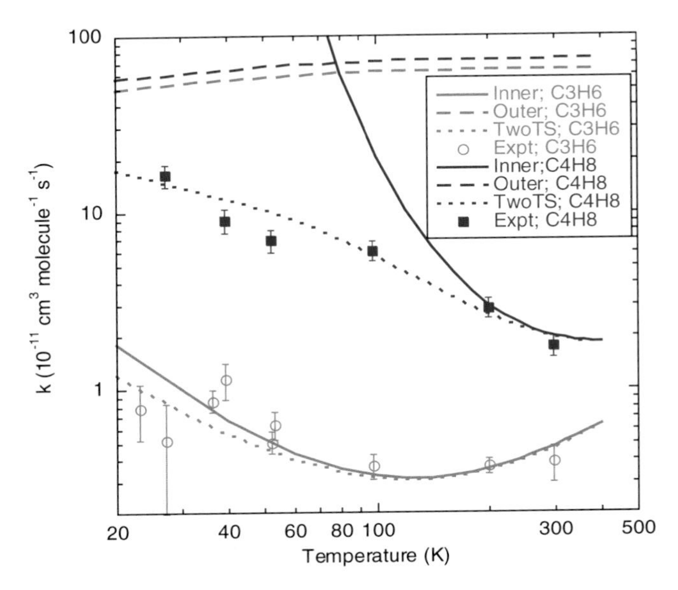


Fig. 4.9. Plot of the experimental and the long-range, short-range and two transition state model predictions for the temperature dependence of the rate constant for $O(^3P)$ + propene and $O(^3P)$ + trans-butene reactions.

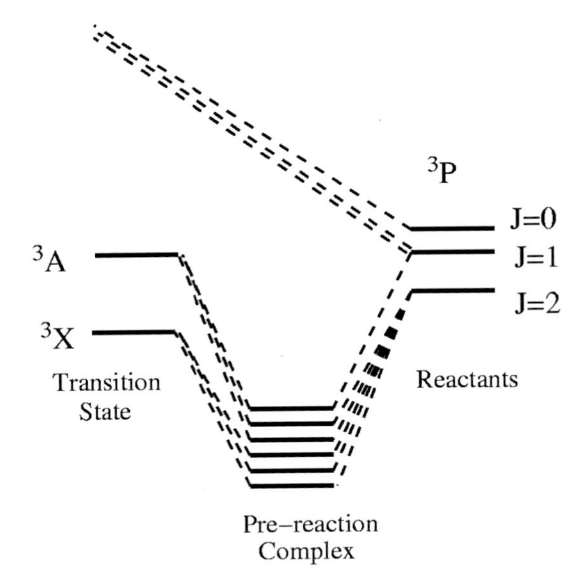


Fig. 4.10. Schematic plot of the correlation of the electronic states from reactants to the inner transition state for the $O(^3P)$ + trans-butene reaction.

The calculations from Ref. 9 employ the adiabatic electronic assumption. This assumption provides slightly improved agreement with experiment at the lowest temperatures observed (27 K). However, the distinction between the adiabatic and statistical predictions is still quite minor at 27 K, because the inner transition state still plays an important role. It would be useful to have some experimental results in the long-range capture dominated temperature regime for some system where the adiabatic statistical and electronic predictions differ.

For reactions such as $OH + C_2H_4$, the rate coefficient may differ from the long-range capture rate due to an absence of product channels with an exothermic pathway. In such simple addition reactions, the measured rate coefficient should be pressure dependent. For the $OH + C_2H_4$ reaction we have incorporated our two TS predictions in a master equation model for the pressure dependence of the kinetics.[7] The barrier height was adjusted down by 1 kcal/mol from high level *ab initio* predictions in order to reproduce the room temperature high pressure rate data. As illustrated in Fig. 4.11, the resulting predictions for the full temperature and pressure dependence were in quantitative agreement with the experimental data,[139–142] including that from Leone and coworkers in the 96 to 165 K region.

Recently, however, Blitz and coworkers have provided two additional low temperature experimental studies.[143,144] Both of these studies suggest that

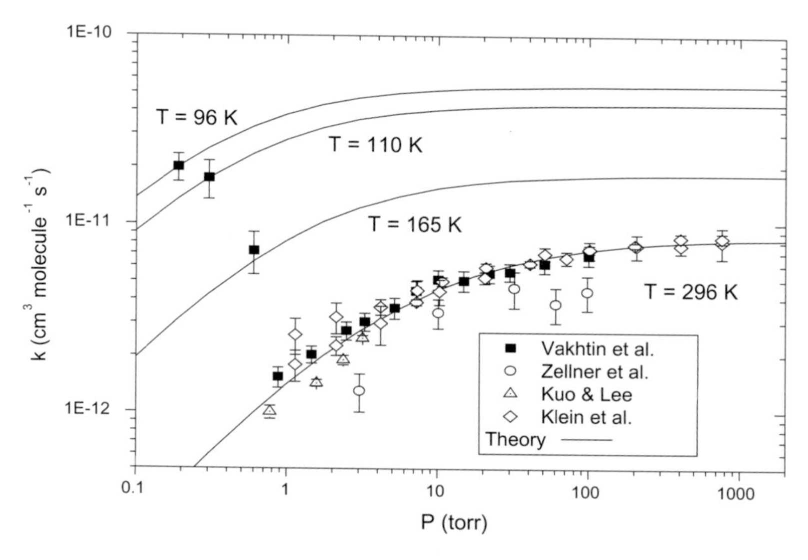


Fig. 4.11. Plot of the pressure dependence of the low temperature rate coefficients for $OH + C_2H_4 \rightarrow C_2H_4OH$.

our predictions for the high pressure limit are too high, apparently by about a factor of two to three in the 69 to 96 K region. Such a discrepancy may be indicative of a failure of the effective transition state formula, Eq. (4.23). It would be interesting to perform a dynamical study for this reaction to test the accuracy of Eqs. (4.23) and (4.24) for a given potential energy surface.

Blitz and coworkers have also modeled their data with master equation simulations employing empirical high pressure rate values. While these simulations do provide improved agreement with the full set of data, they do so by splitting the difference between theirs and Leone and coworkers experimental results. The remaining discrepancies between the fit and experiment are outside the estimated experimental error bars, suggesting some discrepancy between the various experiments. Thus, it would be useful to have further experimental studies of this reaction.

### 4.4.2. *Ion-Molecule Reactions*

Long-range complexes are ubiquitous in ion-molecule chemistry due to the increased strength of the long-range attractions. They tend to arise whenever the formation of a covalently bound complex requires structural rearrangements. With this increased attraction, the saddle points connecting the long-range ion-molecule complexes to the covalent compounds also tend to lie well below reactants. Thus, the inner transition state seldom plays an important role and, as discussed in Section 2, the long-range capture rates generally provide an accurate prediction of the kinetics for room temperature and lower. Nevertheless, if the long-range attractions are weak enough the inner transition state can be important at low temperatures even for ion-molecule reactions.

Bowers and coworkers were the first to consider two transition state models for ionic reactions. In their classic study of the decomposition of $C_4H_8^+$ cations,[145] they considered two transition state models for each of the three product channels, $C_2H_4 + C_2H_4^+$, $H + C_4H_7^+$, and $CH_3 + C_4H_5^+$. In subsequent work, they applied a related model to the dissociations of $C_4H_6^+$[146] and $C_6H_6^+$.[147] The implementations for the inner transition state were purely empirical, but did provide clear evidence for its kinetic importance.

With its low polarizability and nonpolar nature, the H atom has the weakest possible long-range interaction with ions. Furthermore, its small mass implies that the centrifugal barriers are relatively large and lie at shorter separations. Correspondingly, one expects short range interactions to be of more importance for H+ ion reactions than for other ion-molecule reactions. We have studied the H $+^3C_6H_5^+$ recombination reaction with direct CASPT2/aug-cc-pvdz VRC-TST calculations employing a 7 electron,

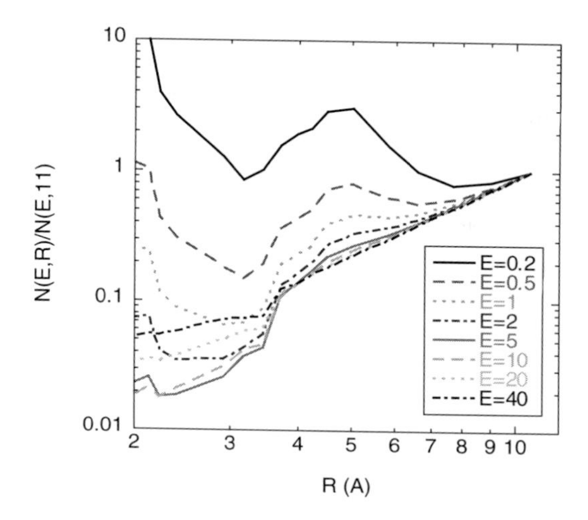


Fig. 4.12. Plot of the normalized microcanonical number of states as a function of separation distance for a range of energies (kcal/mol) for $H+^3C_6H_5^+ \rightarrow C_6H_6^+$. The numbers are normalized to their value at R = 11 Å.

8 orbital active space consisting of the H atom radical orbital, the $C_6H_5^+$ in-plane radical orbital and the $C_6H_5^+$ $\Pi$ space. At long range, R > 4 Å, the reaction coordinate was taken to be the center-of-mass separation distance. At short range, $r_{CH} < 4.2$ Å, a multifaceted dividing surface[148] consisting of fixed H to C atom separations was considered.

The radial dependence of the microcanonical number of states for the $H +^3 C_6H_5^+ \rightarrow C_6H_6^+$ reaction over a range of energies is illustrated in Fig. 4.12. At the lowest energy plotted, 0.2 kcal/mol, an inner transition state, at R ~ 3 Å, and an outer transition state, at R ~ 8 Å, are clearly visible. At $E = 0.2$ kcal/mol these bottlenecks are essentially identical, while the inner bottleneck becomes dominant (i.e., has the smaller number of states) as the energy increases. For this reaction, the inner transition state would clearly play a significant role in the low temperature kinetics.

The methyl radical has a significantly greater polarizability than the H atom, but still has no dipole. Furthermore, there are a large number of transitional modes that transform from rotations to vibrations. The radial dependence of similar direct MP2/cc-pvdz VRC-TST calculations of the microcanonical number of states for the $CH_3 + C_3H_5^+ \rightarrow C_4H_8^+$ reaction are plotted in Fig. 4.13. The qualitative behavior is similar to that for the $H+^3C_6H_5^+$ reaction. But now $N^\dagger_{\text{inner}} > N^\dagger_{\text{outer}}$ until an energy of 2 kcal/mol is reached. In this case, the low temperature kinetics will be dominated by the long-range potential, but the inner transition state should still yield a modest reduction in the rate coefficient.

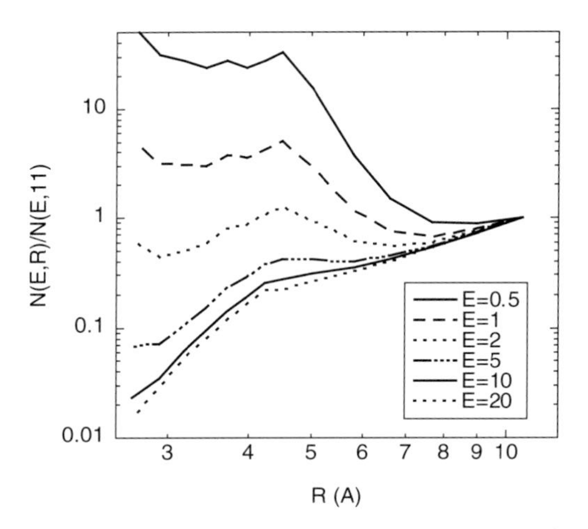


Fig. 4.13. Plot of the microcanonical number of states as a function of separation distance for a range of energies (kcal/mol) for $CH_3 + C_3H_5^+ \rightarrow C_4H_8^+$. The numbers are normalized to their value at R = 11 Å.

Ammonia provides a simple example of a molecule with a strong dipole and corresponding strong long-range interactions. However, since it is a closed shell, one might expect there to be a significant saddle point preceding its reaction. Direct MP2/cc-pvdz VRC-TST calculations for the reaction of $NH_3$ with $C_3H_7^+$ find that the inner and outer bottlenecks do not become equivalent until an energy of 8 kcal/mol. One would not expect to see any effect of short range interactions on the low temperature kinetics for this reaction.

Since most molecules do have a permanent dipole moment, the $NH_3 + C_3H_7^+$ case should be fairly typical. Unless the saddlepoint for reaction happens to lie close to or above the reactant energy, one does not expect significant effects from short-range interactions for the reaction of dipolar molecules and radicals with ions. For nonpolar radicals reacting with closed shell ions, short range interactions may be of modest significance. For closed shell ions reacting with nonpolar closed shell molecules or with H atoms, an inner transition state will generally have an effect on the low temperature kinetics. If the ion is a radical, the probability of an important inner transition state is even further reduced for each of these cases.

### 4.4.3. *Radical-Radical Reactions*

Neutral radical-radical reactions are quite different from radical-molecule reactions in that the formation of the incipient chemical bond does not require the breaking of some other covalent bond. Importantly, however, the reactants still must rotate from the optimal long-range orientations to the optimal covalent orientations and there is a strong restriction in the range of accessible angles. As a result, an inner transition state can and generally does arise. In some cases this inner transition state correlates with a submerged saddle point on the potential energy surface, while in other cases it is solely due to the reduction in entropy. Furthermore, in some cases the inner and outer transition state regions are clearly distinct, while in others the location of the transition state just gradually shifts from large separations to short separations with no discernible separation into an inner and outer transition state.

The presence or absence of a saddlepoint and the distinction between inner and outer transition states depends subtly on various factors such as (i) the presence of any dipole moments, (ii) the presence of any hydrogen bonding interactions between the two reactants, (iii) the resonance stabilization of either reactant, (iv) the relative orientation of the optimal short-range and long-range interactions, (v) the number of rotational degrees of freedom of the reactants, and (vi) the molecular size of the reactants.

A model potential based variational RRKM study of the NCNO $\rightarrow$ NC+NO dissociation provided one of the first indications of the relevance of the two transition state model to the kinetics of radical-radical reactions.[149] In this study, the dynamics on the singlet and triplet surfaces at large range was assumed to be statistical in keeping with the observed statistical distribution of the rovibronic states of the products. The resulting larger degeneracy of the long-range region yielded a rapid transition to a short-range singly electronically degenerate inner transition state with an energy of $100\,\mathrm{cm}^{-1}$.[67] By microscopic reversibility, this analysis is directly applicable to the reverse association and short-range effects should be dominant at low temperature. A subsequent experimental study of the CN+NO association yielded an estimated high pressure rate constant of $\sim 5\times 10^{-11}\,\mathrm{cm}^3$ molecule$^{-1}$ s$^{-1}$ over the 200 to 600 K range, which is in good agreement with that from the theoretical model.[150] By comparison, the long-range TST prediction for the dominant dispersion term is $\sim 7\times 10^{-11}\,\mathrm{cm}^3$ molecule$^{-1}$ s$^{-1}$ for this same range. This long-range prediction assumes electronically adiabatic dynamics with reaction on only the ground singlet state, which yields an electronic correction factor of $4/(4+4\exp(-179/\mathrm{T(K)}))$.

The $CH_3$ + H and $CH_3$ + $CH_3$ reactions are two prototypical radical-radical reactions. The latter reaction is of some interest to the modeling of low temperature planetary atmospheres.[151] For the $CH_3$ + H reaction, the long-range interactions are much weaker and the reduced mass is smaller. Both of these factors suggest that short-range interactions should be more important for the H atom case. However, for the $CH_3 + CH_3$ case, there are three more modes that transform from free rotations to bending motions

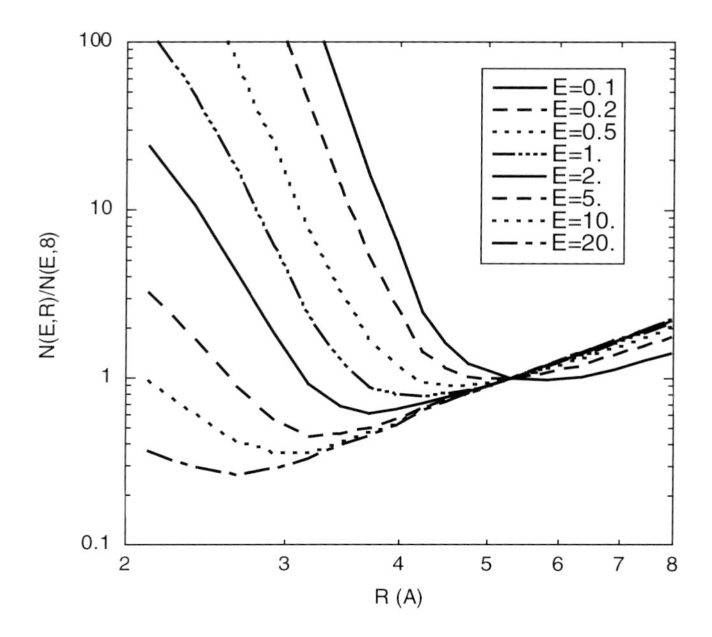


Fig. 4.14. Plot of the microcanonical number of states as a function of separation distance for a range of energies (kcal/mol) for $CH_3 + H \rightarrow CH_4$ reaction. The numbers are normalized to their value at R = 8 Å.

as the separation distance decreases. Furthermore, the additional molecular size yields more steric repulsion. These two factors yield a greater reduction in entropy, thereby suggesting that short range interactions may be more important in the $CH_3 + CH_3$ reaction.

The radial dependence of the direct CASPT2 calculated microcanonical number of states for these two reactions is plotted in Figs. 4.14 and 4.15, respectively. For the H atom case, the transition state gradually shifts from large separation to short separation, reaching a separation of 4–5 Å by an energy of 0.5 kcal/mol. There is no sign of two distinct transition state regions even if one plots the $E/J$ resolved number of states. In contrast, the $CH_3 + CH_3$ case shows some indication of distinct outer and inner transition states, but the inner transition state only becomes significant at an energy of 2 kcal/mol. Furthermore, in this case, the distinct nature of the two regions is amplified if one instead considers the $E/J$ resolved number of states.

The CASPT2/aug-cc-pvtz VRC-TST predicted rate coefficients for the $CH_3 + H$ and $CH_3 + CH_3$ reactions over the 10 to 1000 K region are plotted in Fig. 4.16. These predictions have previously been shown to be in good

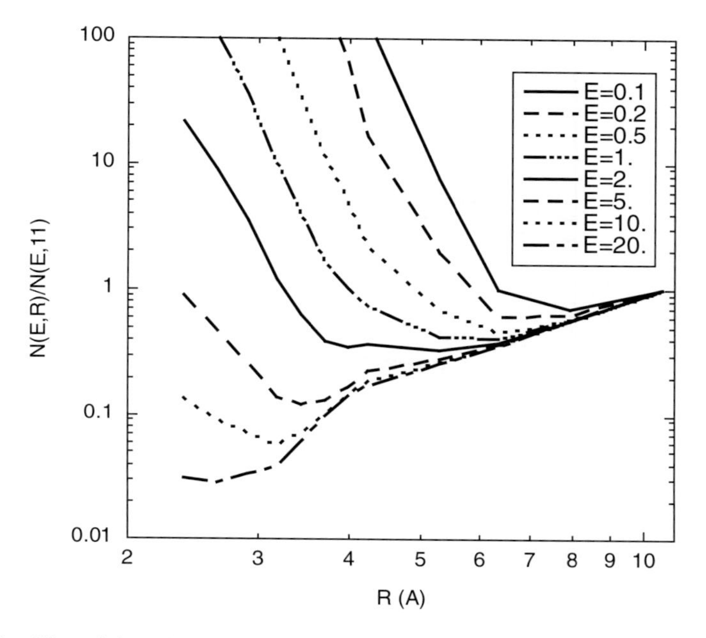


Fig. 4.15. Plot of the microcanonical number of states as a function of separation distance for a range of energies (kcal/mol) for $CH_3 + CH_3 \rightarrow C_2H_6$ reaction. The numbers are normalized to their value at R = 11 Å.

agreement with the available experimental data.[110,111] Long-range TST predictions are also plotted for a potential that includes only dispersion forces.

The $CH_3 + H$ predictions are remarkably similar to the long-range TST predictions. Results for H atom plus other larger alkyl radical reactions suggest that this similarity is solely the result of a coincidental cancelation of two opposing factors. In particular, the H+ alkyl radical reactions decrease by about a factor of two for each substitution of an H atom in $CH_3$ with a $CH_3$ group.[110] The dispersion rate for these larger radicals should be the same or larger since the polarizability is increased, while the reduced mass is essentially constant. The predicted decrease is an indication of short-range repulsions reducing the rate coefficient. This same factor in the $H + CH_3$ case is smaller. Furthermore, the increase in the attractive interactions due to bonding acts to increase the rate coefficient. Since the transition state reaches separations of about 4 Å by 1 kcal/mol, the lack of any significant deviation from the dispersion rate, indicates that the bonding and repulsion factors must cancel in the $CH_3 + H$ case.

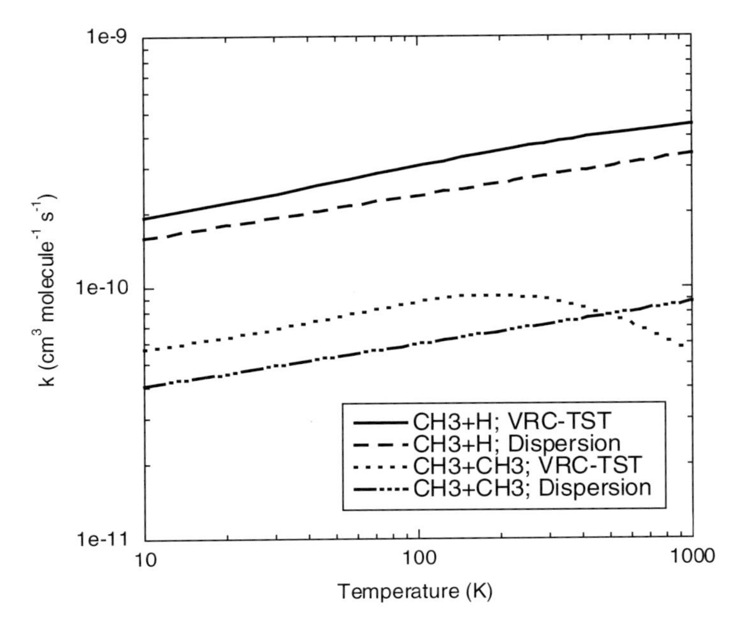


Fig. 4.16. Plot of the dispersion and VRC-TST predicted rate coefficients for $CH_3$+H → $CH_4$ and for $CH_3 + CH_3 \rightarrow C_2H_6$.

For the $CH_3$+$CH_3$ reaction, the VRC-TST predictions begin to decrease with temperature at about 300 K. Below that temperature there is again good agreement with the long-range TST predictions for the dispersion potential. For larger alkyl radicals, the rate decreases with each additional $CH_3$ group,[111] which is again in contradiction with the expectations from long-range TST. Also, the temperature at which the rate begins to deviate from the long-range TST rate is predicted to decrease with increasing size.

Resonantly stabilized radicals, which are another important class of radicals, have much weaker attractions than do saturated alkyl radicals. This reduction in the attractiveness arises from the need to break the resonance as part of the bonding process. For these radicals short-range interactions yield a reduction in the rate coefficient at remarkably low temperatures. For example, for the CN + $O_2$ reaction, the resonances in the $O_2$ molecule must be broken prior to bonding.

The CN + $O_2$ reaction is one of the few radical-radical reactions that has been studied at low temperature with CRESU experiments.[152–154] The observed rate coefficient for this reaction decreases sharply with increasing temperature above 23 K, whereas the long-range ACCSA capture rate prediction increases with temperature.[43] As illustrated in Fig. 4.17, an early

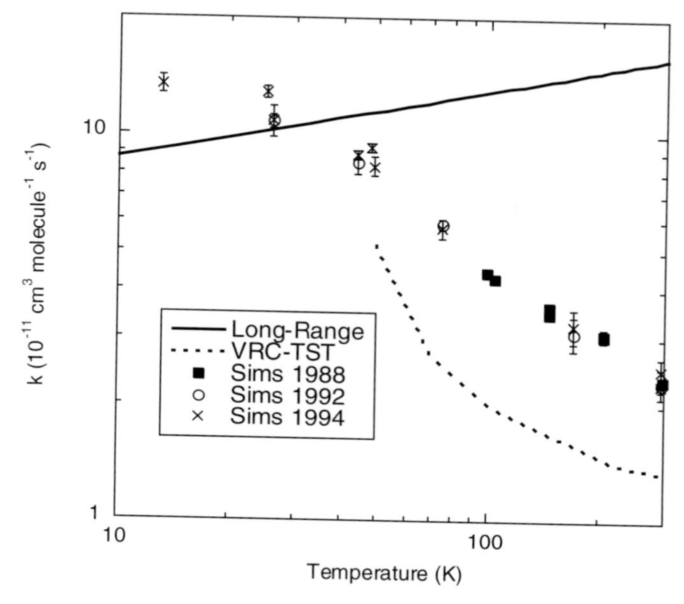


Fig. 4.17. Plot of experimental, long-range TST, and *ab intio* short-range transition theory results for the temperature dependence of the rate constant for $CN + O_2 \rightarrow$ NCOO.

*ab initio* transition state theory based implementation of VRC-TST[6] was able to describe this rapid decrease qualitatively on the basis of a short-range transition state. The discrepancies between these VRC-TST calculations and experiment can readily be explained by inadequacies in the treatment of the potential energy surface. The ACCSA predictions from Ref. 43 differ from the present long-range TST predictions, apparently by our inclusion of an electronic degeneracy factor of 1/3. This factor presumes that the reaction occurs on only the doublet electronic surface and not on the quartet surface. The underestimate at low temperatures for the long-range TST prediction suggests that either the quartet state does contribute some to the kinetics, or that the bonding and higher order interactions yield a modest increase in the long-range potential. Tan *et al.*'s study[76] of the contribution from different terms in the long-range potential, appears to have neglected the key dispersion term.

Other *ab initio* based theoretical studies of low temperature radical-radical kinetics have focused on atom-diatom reactions such as $N + OH$,[155] $O + OH$ ,[156] $Si + O_2$,[157] $C + OH$,[158] $C + NO$[80] and $O + CN$[80] reactions. The deviations from long-range predictions for these atom-diatom studies have generally been quite modest due to the small number of rotational degrees of freedom. However, these studies do emphasize yet again, the strong dependence on the contributions from the multiple nearly degenerate electronic states.

Unfortunately, there have been only a few direct studies of the electronic dynamics. In a pioneering study, Tashiro and Schinke[159] examined the quantum dynamics of the $O(^3P)+O_2$ reaction on the multiple electronic surfaces arising from the spin-orbit states of the oxygen atom. In a related study Yagi *et al.*[160] studied the quantum dynamics of the $CH_3 + O(^3P)$ reaction, but in only a two-dimensional framework. For both reactions, nonadiabatic effects were found to be of minimal importance. Nevertheless, it would be useful to examine these effects in other cases, particularly when statistical and adiabatic electronic treatments yield different estimates at low temperature.

## 4.5. Summary

In the low temperature limit the kinetics of gas phase reactions is determined by the capture rate on the long-range potential. The combination of statistical and dynamical, classical and quantum theories provides an accurate description of the long-range kinetics on a given potential energy surface. However, there are still significant uncertainties regarding the contributions from multiple degenerate electronic states. An assumption of electronically adiabatic dynamics generally yields good agreement with experiment and is the most commonly employed assumption.

Short-range interactions can yield a significant reduction in the predicted rate coefficient below that predicted by the long-range capture theory. One common effect of short-range interactions is to yield a change in the electronic degeneracy factors as only a limited number of potential energy surfaces are reactive at short-range. Another effect is to yield an inner transition state arising from a reduction in the entropy as the reactants approach one another. This inner transition state is often distinct from the outer or long-range transition state, especially for radical-molecule reactions. Two transition state models provide an effective treatment of this series of transition states. However, the accuracy of the simple models given by Eqs. (4.23) and (4.24) is still uncertain.

The short-range entropy reduction effect can become important already at 10 K, as seen in the study of the $CN+C_2H_6$ and $O(^3P)$ + alkene reactions. The accurate treatment of the short-range effects requires detailed and high level *ab initio* simulations. The CASPT2 method is a powerful approach for treating the short-range energetics in barrierless radical-radical and radical-molecule reactions. Modest adjustments in the *ab initio* predicted barrier heights can generally be used to obtain theoretical predictions that are in quantitative agreement with experiment over a wide range of temperature.

## Acknowledgements

This work is supported by the Division of Chemical Sciences, Geosciences, and Biosciences, the Office of Basic Energy Sciences, the U. S. Department of Energy. The work at Argonne (S.J.K.) was supported under DOE Contract Number DE-AC02-06CH11357. Sandia is a multi-program laboratory operated by Sandia Corporation, a Lockheed Martin Company, for the National Nuclear Security Administration under contract DE-AC04-94-AL85000.

## References

1. Smith IWM, Rowe BR. (2000) Reaction kinetics at very low temperatures: Laboratory studies and interstellar chemistry. *Acc. Chem. Res.* **33**: 261–268.
2. Smith IWM. (2006) Reactions at very low temperatures: Gas kinetics at a new frontier. *Angew. Chem. Int. Ed.* **45**: 2842–2861.
3. Liao Q, Herbst E. (1995) Capture calculations for the rates of important neutral-neutral reactions in dense interstellar clouds — $C + C_2H_2$ and $CN + C_2H_2$. *Astrophys. J.* **444**: 694–701.
4. Georgievskii Y, Klippenstein SJ. (2005) Long-range transition state theory. *J. Chem. Phys.* **122**: 194103; 1–15.
5. Woon DE, Herbst E. (1997) The rate of the reaction between CN and $C_2H_2$ at interstellar temperatures. *Astro. J.* **477**: 204–208.
6. Klippenstein SJ, Kim YW. (1993) Variational statistical study of the $CN + O_2$ reaction employing *ab-initio* determined properties for the transition state. *J. Chem. Phys.* **99**: 5790–5799.
7. Greenwald EE, North SW, Georgievskii Y, Klippenstein SJ. (2005) A two transition state model for radical-molecule reactions: A case study of the addition of OH to $C_2H_4$. *J. Phys. Chem. A.* **109**: 6031–6044.
8. Georgievskii Y, Klippenstein SJ. (2007) Strange kinetics of the $C_2H_6 + CN$ reaction explained. *J. Phys. Chem. A.* **111**: 3802–3811.
9. Sabbah H, Biennier L, Sims IR, Georgievskii Y, Klippenstein SJ, Smith IWM. (2007) Understanding reactivity at very low temperatures: The reactions of oxygen atoms with alkenes. *Science* **317**: 102–105.
10. Clary DC. (1990) Fast chemical reactions: Theory challenges experiment. *Ann. Rev. Phys. Chem.* **41**: 61–90.
11. Troe J. (1997) Recent advances in statistical adiabatic channel calculations of state specific dissociation dynamics. *Adv. Chem. Phys.* **101**: 819–851.
12. Pechukas P, Light JC. (1965) On detailed balancing and statistical theories of chemical kinetics. *J. Chem. Phys.* **42**: 3281–3291; Pechukas P, Rankin R,

Light JC. (1966) Statistical theory of chemical kinetics: Application to neutral-atom—molecule reactions. *J. Chem. Phys.* **44**: 794–805; Light JC. (1967) Statistical theory of bimolecular exchange reactions. *Faraday Discuss Chem. Soc.* **44**: 14–29.

13. Nikitin EE. (1965) *Teor. Eksp. Khim. Acad. Nauk. Ukr. SSR* **1**: 135, 428.
14. Langevin PM. (1905) *Ann. Chem. Phys.* **5**: 245; Gioumousis G, Stevenson DP. (1958) Reactions of gaseous molecule ions with gaseous molecules. V. Theory. *J. Chem. Phys.* **29**: 294–299.
15. Bass L, Su T, Bowers MT. (1978) Theory of ion-molecule collisions. Effect of anisotropy in polarizability on collision rate constant. *Int. J. Mass Spec. Ion Proc.* **28**: 389–399.
16. Clair R, McMahon TB. (1978) Orientation-dependent forces in ion-molecule reactions. Effect of anisotropic molecular polarizability on ion-molecule reaction-rates. *Int. J. Mass Spec. Ion Proc.* **28**: 365–375.
17. Hirschfelder JO, Curtiss CF, Bird RB. (1967) *Molecular Theory of Gases and Liquids.* John Wiley & Sons, New York.
18. Gorin E. (1939) Photolysis of aldehydes and ketones in the presence of iodine vapor. *J. Chem. Phys.* **7**: 256–264.
19. Nikitin EE, Troe J. (1997) Quantum and classical calculations of adiabatic and non-adiabatic capture rates for anisotropic interactions. *Ber. Bunsenges. Phys. Chem.* **101**: 445–458.
20. Benson SW. (1983) Molecular-models for recombination and disproportionation of radicals. *Can. J. Chem.* **61**: 881–887.
21. Smith GP, Golden DM. (1978) Application of RRKM theory to reactions $OH+NO_2+N_2 \rightarrow HONO_2+N_2(1)$ and $ClO+NO_2+N_2 \rightarrow ClONO_2+N_2(2)$: a modified gorin model transition state. *Int. J. Chem. Kinet.* **10**: 489–501.
22. Chesnavich WJ, Su T, Bowers MT. (1980) Collisions in a non-central field. Variational and trajectory investigation of ion-dipole capture. *J. Chem. Phys.* **72**: 2641–2655.
23. Su T, Bowers MT. (1973) Theory of ion-polar molecule collisions. Comparisons with experimental charge transfer reactions of rare gas ions to geometric isomers of difluorobenzene and dichloroethylene. *J. Chem. Phys.* **58**: 3027–3037.
24. Davidsson J, Nyman G. (1988) Extended Langevin model for rate constant calculations of exothermic atom-diatom reactions. Application to O + OH. *Chem. Phys.* **125**: 171–183.
25. Celli F, Weddle G, Ridge DP. (1980) On statistical and thermodynamic approaches to ion polar molecule collisions. *J. Chem. Phys.* **73**: 801–812.
26. Markovic N, Nordholm S. (1989) Simple estimation of thermal capture rates for ion-dipole collisions by canonical effective potential methods. *Chem. Phys.* **135**: 109–122.
27. Larregaray P, Bonnet L, Rayez JC. (2007) Mean potential phase space theory of chemical reactions. *J. Chem. Phys.* **127**: 084308; 1–4.
28. Turulski J, Niedzielski J. (1988) Implementation and limitations of the transition state theory for ion molecule systems with non-spherical dividing surfaces. *J. Chem. Soc. Faraday Trans.* **84**: 347–361.

29. Turulski J, Niedzielski J, Pezler B. (1995) The capture rate of an ion by a symmetrical top quadrupole. *Chem. Phys.* **192**: 319–323.
30. Smith SC, Troe J. (1992) Statistical modeling of ion-molecule electrostatic capture. *J. Chem. Phys.* **97**: 5451–5464.
31. Mondro SL, Vande Linde S, Hase WL. (1986) Reaction path and variational transition state theory rate constant for $Li^+ + H_2O \rightarrow Li^+(H_2O)$ association. *J. Chem. Phys.* **84**: 3783–3787.
32. Wardlaw DM, Marcus RA. (1988) On the statistical theory of unimolecular processes. *Adv. Chem. Phys.* **70**: 231–263.
33. Su T. (1988) Trajectory calculations of ion-polar molecule capture rate constants at low-temperatures. *J. Chem. Phys.* **88**: 4102–4103; **89**: 5355–5355.
34. Bhowmik PK, Su, T. (1991) Parametrization of the trajectory calculations on the ion-quadrupolar molecule collision rate constants. *J. Chem. Phys.* **94**: 6444–6445.
35. Maergoiz AI, Nikitin EE, Troe J, Ushakov VG. (1996) Classical trajectory and adiabatic channel study of the transition from adiabatic to sudden capture dynamics. I. Ion-dipole capture. *J. Chem. Phys.* **105**: 6263–6269.
36. Maergoiz AI, Nikitin EE, Troe J, Ushakov VG. (1996) Classical trajectory and adiabatic channel study of the transition from adiabatic to sudden capture dynamics. II. Ion-quadrupole capture. *J. Chem. Phys.* **105**: 6270–6276.
37. Maergoiz AI, Nikitin EE, Troe J, Ushakov VG. (1996) Classical trajectory and adiabatic channel study of the transition from adiabatic to sudden capture dynamics. III. Dipole-dipole capture. *J. Chem. Phys.* **105**: 6277–6284.
38. Maergoiz AI, Nikitin EE, Troe J, Ushakov VG. (1998) Classical trajectory and statistical adiabatic channel study of the dynamics of capture and unimolecular bond fission. IV. Valence interactions between atoms and linear rotors. *J. Chem. Phys.* **108**: 5265–5280.
39. Maergoiz AI, Nikitin EE, Troe J, Ushakov VG. (1998) Classical trajectory and statistical adiabatic channel study of the dynamics of capture and unimolecular bond fission. V. Valence interactions between two linear rotors. *J. Chem. Phys.* **108**: 9987–9998.
40. Quack M, Troe J. (1975) Complex formation in reactive and inelastic scattering. Statistical adiabatic channel model of unimolecular processes. III. *Ber. Bunsenges. Phys. Chem.* **79**: 170–183.
41. Garrett BC, Truhlar DG. (1979) Generalized transition state theory. Quantum effects for collinear reactions of hydrogen molecules and isotopically substituted hydrogen molecules. *J. Phys. Chem.* **83**: 1052–1079.
42. Clary DC. (1984) Rates of chemical reactions dominated by long-range intermolecular forces. *Mol. Phys.* **53**: 3–21.
43. Stoecklin T, Dateo CE, Clary DC. (1991) Rate constant calculations on fast diatom diatom reactions. *J. Chem. Soc. Faraday Trans.* **87**: 1667–1679.
44. Stoecklin T, Clary DC. (1992) Fast reactions between diatomic and polyatomic molecules. *J. Phys. Chem.* **96**: 7346–7351.

45. Stoecklin T, Clary DC. (1995) Fast reactions between a linear molecule and a polar symmetrical top. *Theochem. J. Mol. Struc.* **341**: 53–61.
46. Clary DC, Henshaw JP. (1987) Chemical reactions dominated by long-range intermolecular forces. *Faraday Disc.* **84**: 333–349.
47. Clary DC. (1987) Rate constants for the reactions of ions with dipolar polyatomic molecules. *J. Chem. Soc. Faraday Trans. II*, **83**: 139–148.
48. Clary DC, Smith D, Adams NG. (1985) Temperature dependence of rate coefficients for reactions of ions with dipolar molecules. *Chem. Phys. Lett.* **119**: 320–326.
49. Clary DC, Dateo CE, Smith D. (1990) Rates for the reactions of open-shell ions with molecules. *Chem. Phys. Lett.* **167**: 1–6.
50. Gonzalez AI, Clary DC. Yanez, M. (1997) Calculations of rate constants for reactions of first and second row cations. *Theor. Chem. Acc.* **98**: 33–41.
51. Dateo CE, Clary DC. (1989) Rate constant calculations on the $C^+ + HCl$ reaction. *J. Chem. Phys.* **90**: 7216–7228.
52. Graff MM, Wagner AF. (1990) Theoretical studies of fine-structure effects and long-range forces: Potential-energy surfaces and reactivity of $O(^3P)$ + $OH(^2\Pi)$. *J. Chem. Phys.* **92**: 2423–2439.
53. Clary DC, Stoecklin TS, Wickham AG. (1993) Rate constants for chemical-reactions of radicals at low-temperatures. *J. Chem. Soc. Faraday Trans.* **89**: 2185–2191.
54. Reignier D, Stoecklin T. (1999) Comparison of the spin-orbit selectivity of reaction involving atoms in a $^2P$ or $^3P$ state and a linear molecule in a $\Pi$ or $\Sigma$ state at very low temperature. *Chem. Phys. Lett.* **303**: 576–582.
55. Clary DC, Werner HJ. (1984) Quantum calculations on the rate constant for the O + OH reaction. *Chem. Phys. Lett.* **112**: 346–350.
56. Troe J. (1987) Statistical adiabatic channel model for ion-molecule capture processes. *J. Chem. Phys.* **87**: 2773–2780.
57. Maergoiz AI, Nikitin EE, Troe J. (1991) Adiabatic channel potential curves for two linear dipole rotors. I. Classification of states and numerical calculations for identical rotors. *J. Chem. Phys.* **95**: 5117–5127.
58. Ramillon M, McCarroll R. (1994) Adiabatic capture models for fast chemical reactions. *J. Chem. Phys.* **101**: 8697–8699.
59. Sakimoto K. (1982) Perturbed rotational state method for collisions between an ion and an asymmetric-top rigid rotor. *Chem. Phys.* **68**: 155–170.
60. Bates DR, Morgan WL. (1987) Adiabatic invariance treatment of hitting collisions between ions and symmetrical top dipolar molecules. *J. Chem. Phys.* **87**: 2611–2616.
61. Maergoiz AI, Nikitin EE, Troe J. (1995) Statistical adiabatic channel calculation of accurate low-temperature rate constants for the recombination of OH radicals in their ground rovibronic state. *J. Chem. Phys.* **103**: 2083–2091.
62. Vogt E, Wannier GH. (1954) Scattering of ions by polarization forces. *Phys. Rev.* **95**: 1190–1198.

63. Hessler JP. (1999) New empirical rate expression for reactions without a barrier: Analysis of the reaction of CN with $O_2$. *J. Chem. Phys.* **111**: 4068–4076.
64. Dashevskaya EI, Maergoiz AI, Troe J, Litvin I, Nikitin EE. (2003) Low-temperature behavior of capture rate constants for inverse power potentials. *J. Chem. Phys.* **118**: 7313–7320.
65. Nikitin EE, Troe J. (2005) Dynamics of ion-molecule complex formation at very low energies and temperatures. *Phys. Chem. Chem. Phys.* **7**: 1540–1551.
66. Nikitin EE, Troe J, Ushakov VG. (1995) Adiabatic and postadiabatic channel description of atom-diatom long-range half-collision dynamics: Interchannel radial coupling for $P_1$ and $P_2$ anisotropy. *J. Chem. Phys.* **102**: 4101–4111.
67. Klippenstein SJ. (1992) Variational optimizations in RRKM theory calculations for unimolecular dissociations with no reverse barrier. *J. Chem. Phys.* **96**: 367–371.
68. Olkhov RV, Smith IWM. (2007) Rate coefficients for reaction and for rotational energy transfer in collisions between CN in selected rotational levels ($X^2\Sigma^+$, v = 2, N = 0, 1, 6, 10, 15, and 20) and $C_2H_2$. *J. Chem. Phys.* **126**: 134314; 1–10.
69. Troe J. (1996) Statistical adiabatic channel model for ion-molecule capture processes. II. Analytical treatment of ion-dipole capture. *J. Chem. Phys.* **105**: 6249–6262.
70. Smith D, Adams NG, Miller TM. (1978) Laboratory study of reactions of $N^+$, $N_2^+$, $N_3^+$, $N_4^+$, $O^+$, $O_2^+$, and $NO^+$ ions with several molecules at 300 K. *J. Chem. Phys.* **69**: 308–318.
71. Marquette JB, Rowe BR, Dupeyrat G, Poissant G, Rebrion C. (1985) Ion-polar-molecule reactions: A CRESU study of $He^+$, $C^+$, $N^+$, + $H_2O$, $NH_3$ at 27, 68, and 163 K. *Chem. Phys. Lett.* **122**: 431–435.
72. Smith IWM, Sage AM, Donahue NM, Herbst E, Park IH. (2006) The temperature-dependence of rapid low temperature reactions: experiment, understanding and prediction. *Faraday Disc.* **133**: 137–156.
73. Canosa A, Sims IR, Travers D, Smith IWM, Rowe BR. (1997). Reactions of the methylidene radical with $CH_4$, $C_2H_2$, $C_2H_6$, and but-1-ene studied between 23 and 295 K with a CRESU apparatus. *Astron. Astrophys.* **323**: 644–651.
74. Berman MR, Fleming JW, Harvey AB, Lin MC. (1982) Temperature dependence of the reactions of CH radicals with unsaturated hydrocarbons. *Chem. Phys.* **73**: 27–33.
75. Butler JE, Fleming JW, Goss LP, Lin MC. (1981) Kinetics of CH radical reactions with selected molecules at room temperature. *Chem. Phys.* **56**: 355–365.
76. Tan X, Dong F, Li X. (1998) Importance of different multipole interactions in fast reactions at low temperature. *J. Phys. Chem. A.* **102**: 8169–8175.
77. Le Picard SD, Canosa A, Geppert W, Stoecklin T. (2004) Experimental and theoretical temperature dependence of the rate coefficient of the

$B(^2P_{1/2,3/2}) + O_2(X^3\Sigma_g^-)$ reaction in the [24–295 K] temperature range. *Chem. Phys. Lett.* **385**: 502–506.
78. Le Picard SD, Canosa A, Travers D, Chastaing D, Rowe BR, Stoecklin T. (1997) Experimental and theoretical kinetics for the reaction of Al with $O_2$ at temperatures between 23 and 295 K. *J. Phys. Chem. A.* **101**: 9988–9992.
79. Le Picard SD, Canosa A, Reignier D, Stoecklin T. (2002) A comparative study of the reactivity of the silicon atoms $Si(^3P_J)$ towards $O_2$ and NO molecules at very low temperature. *Phys. Chem. Chem. Phys.* **4**: 3659–3664.
80. Andersson S, Markovic N, Nyman G. (2003) Computational studies of the kinetics of C + NO and O + CN reactions. *J. Phys. Chem. A.* **107**: 5439–5447.
81. Chastaing D, Le Picard SD, Sims IR. (2000) Direct kinetic measurements on reactions of atomic carbon, $C(^3P)$, with $O_2$ and NO at temperatures down to 15 K. *J. Chem. Phys.* **112**: 8466–8469.
82. Beghin A, Stoecklin T, Rayez JC. (1995) Rate constant calculations for atom-diatom reactions involving an open-shell atom and a molecule in a $\Pi$ electronic state: Application to the $C(^3P) + NO(X^2\Pi)$ reaction. *Chem. Phys.* **195**: 259–270.
83. Geppert WD, Goulay F, Naulin C, Costes M, Canosa A, Le Picard SD, Rowe BR. (2004) Rate coefficients and integral cross-sections for the reaction of $B(^2P_J)$ atoms with acetylene. *Phys. Chem. Chem. Phys.* **6**: 566–571.
84. Canosa A, Le Picard SD, Geppert WD. (2004) Experimental kinetics study of the reaction of boron atoms, $B(^2P_J)$, with ethylene at very low temperatures (23-295 K). *J. Phys. Chem. A.* **108**: 6183–6185.
85. Chastaing D, James PL, Sims IR, Smith IWM. (1999) Neutral-neutral reactions at the temperatures of interstellar clouds: Rate coefficients for reactions of atomic carbon, $C(^3P)$, with $O_2$, $C_2H_2$, $C_2H_4$ and $C_3H_6$ down to 15 K. *Phys. Chem. Chem. Phys.* **1**: 2247–2256.
86. Chastaing D, Le Picard SD, Sims IR, Smith IWM. (2001) Rate coefficients for the reactions of $C(^3P_J)$ atoms with $C_2H_2$, $C_2H_4$, $CH_3C{\equiv}CH$ and $H_2C{=}C{=}CH_2$ at temperatures down to 15 K. *Astron. Astrophys.* **365**: 241–247.
87. Canosa A, Le Picard SD, Gougeon S, Rebrion-Rowe C, Travers D, Rowe BR. (2001) Rate coefficients for the reactions of $Si(^3P_J)$ with $C_2H_2$ and $C_2H_4$: Experimental results down to 15 K. *J. Chem. Phys.* **115**: 6495–6503.
88. Duagey N, Caubet P, Retail B, Costes M, Bergeat A, Dorthe G. (2005) Kinetic measurements on methylidyne radical reactions with several hydrocarbons at low temperatures. *Phys. Chem. Chem. Phys.* **7**: 2921–2927.
89. Goulay F, Rebrion-Rowe C, Biennier L, Le Picard SD, Canosa A, Rowe BR. (2006) Reaction of anthracene with CH radicals: An experimental study of the kinetics between 58 and 470 K. *J. Phys. Chem. A.* **110**: 3132–3137.
90. Sims IR, Queffelec JL, Defrance A, Rebrion-Rowe C, Travers D, Rowe BR, Herbert LB, Karthauser J, Smith IWM. (1993) Rate constants for the reactions of CN with hydrocarbons at low and ultra-low temperatures. *Chem. Phys. Lett.* **211**: 461–468.

91. Carty D, Le Page V, Sims IR, Smith IWM. (2001) Low temperature rate coefficients for the reactions of CN and $C_2H$ radicals with allene ($CH_2{=}C{=}CH_2$) and methyl acetylene ($CH_3C \equiv CH$). *Chem. Phys. Lett.* **344**: 310–316.
92. Opansky B, Leone SR. (1996) Rate coefficients of $C_2H$ with $C_2H_4$, $C_2H_6$, and $H_2$ from 150 to 359 K. *J. Phys. Chem.* **100**: 19904–19910.
93. Chastaing D, James PL, Sims IR, Smith IWM. (1998) Neutral-neutral reactions at the temperatures of interstellar clouds. Rate coefficients for reactions of $C_2H$ radicals with $O_2$, $C_2H_2$, $C_2H_4$, and $C_3H_6$ down to 15 K. *Faraday Disc.* **109**: 165–181.
94. Hoobler RJ, Leone SR. (1999) Low-temperature rate coefficients for reactions of the ethynyl radical ($C_2H$) with $C_3H_4$ isomers methylacetylene and allene. *J. Phys. Chem. A.* **103**: 1342–1346.
95. Vakhtin AB, Heard DE, Smith IWM, Leone SR. (2001) Kinetics of $C_2H$ radical reactions with ethene propene and 1-butene measured in a pulsed Laval nozzle apparatus at T = 103 and 296 K. *Chem. Phys.* Lett **348**: 21–26.
96. Harding LB, Klippenstein SJ, Jasper AW. (2007) *Ab initio* methods for reactive potential surfaces. *Phys. Chem. Chem. Phys.* **9**: 4055–4070.
97. Andersson K, Malqmvist PA, Roos BO. (1992) 2nd-order perturbation theory with a complete active space self consistent field reference function. *J. Chem. Phys.* **96**: 1218–1226.
98. MOLPRO is a package of programs written by Werner HJ, Knowles PJ, with contributions from Almlof J, Amos RD, Berning A, Cooper DL, Deegan MJO, Dobbyn AJ, Eckert F, Elbert ST, Hampel C, Lindh R, Lloyd AW, Meyer W, Nicklass A, Peterson K, Pitzer R, Stone AJ, Taylor PR, Mura ME, Pulay P, Schutz M, Stoll H, Thorsteinsson T.
99. Kendall RA, Dunning TH, Harrison RJ. (1992) Electron affinities of the 1st-row atoms revisited. Systematic basis sets and correlated wave functions. *J. Chem. Phys.* **96**: 6796–6806.
100. Simandiras ED, Rice JE, Lee TJ, Amos RD, Handy NC. (1988) On the necessity of f basis functions for bending frequencies. *J. Chem. Phys.* **88**: 3187–3195.
101. Raghavachari K, Trucks GW, Pople JA, Head-Gordon M. (1989) A 5th order perturbation comparison of electronic structure theories. *Chem. Phys. Lett.* **157**: 479–483.
102. Pople JA, Head-Gordon M, Raghavachari K. (1987) Quadratic configuration interaction. A general technique for determining electron correlation energies. *J. Chem. Phys.* **87**: 5968–5975.
103. Lee TJ, Taylor PR. (1989) A diagnostic for determining the quality of single reference electron correlation methods. *Int. J. Quantum. Chem.* **S23**: 199–207.
104. Curtiss LA, Raghavachari K, Redfern PC, Rassolov V, Pople JA. (1998) Gaussian-3 (G3) theory for molecules containing first and second row atoms. *J. Chem. Phys.* **109**: 7764–7776.
105. Martin JML, Uzan O. (1998) Basis set convergence in second-row compounds. The importance of core polarization functions. *Chem. Phys. Lett.* **282**: 16–24.

106. Tajti A, Szalay PG, Csaszar AG, Kallay M, Gauss J, Valeev EF, Flowers BA, Vazquez J, Stanton JF. (2004) High accuracy extrapolated *ab initio* thermochemistry. *J. Chem. Phys.* **121**: 11599–11613.
107. Karton A, Rabinovich E, Martin JML, Ruscic B. (2006) W4 theory for computational thermochemistry: In pursuit of confident sub-kJ/mol predictions. *J. Chem. Phys.* **125**: 144108: 1–17.
108. Curtiss LA, Redfern PC, Raghavachari K. (2007) Gaussian-4 theory. *J. Chem. Phys.* **126**: 084108: 1–12.
109. Grimme S, Antony J, Schwabe T, Muck-Lichtenfeld C. (2007) Density functional theory with dispersion corrections for supramolecular structures, aggregates, and complexes of bio(organic) molecules. *Org. Biomol. Chem.* **5**: 741–758.
110. Harding LB, Georgievskii Y, Klippenstein SJ. (2005) Predictive theory for hydrogen atom-hydrocarbon radical association kinetics. *J. Phys. Chem. A.* **109**: 4646–4656.
111. Klippenstein SJ, Georgievskii Y, Harding LB. (2006) Predictive theory for the association kinetics of two alkyl radicals. *Phys. Chem. Chem. Phys.* **8**: 1133–1147.
112. Harding LB, Klippenstein SJ, Georgievskii Y. (2007) On the combination reactions of hydrogen atoms with resonance stabilized hydrocarbon radicals. *J. Phys. Chem. A.* **111**: 3789–3801.
113. Jasper AW, Klippenstein SJ, Harding LB, Ruscic B. (2007) Kinetics of the reaction of methyl radical with hydroxyl radical and methanol decomposition. *J. Phys. Chem. A.* **111**: 3932–3950.
114. Herbst E, DeFrees DJ, Talbi D, Pauzat F, Koch W, McLean AD. (1991) Calculations on the rate of the ion-molecule reaction between $NH_3^+$ and $H_2$. *J. Chem. Phys.* **94**: 7842–7848.
115. Herbst E. (1994) Tunneling in the $C_2H$–$H_2$ reaction at low temperature. *Chem. Phys. Lett.* **222**: 297–301.
116. Miller WH, (1976) Unified statistical model for "complex" and "direct" reaction mechanisms. *J. Chem. Phys.* **65**: 2216–2223.
117. Miller JA, Klippenstein SJ. (2006) Master equation methods in gas phase chemical kinetcs. *J. Phys. Chem. A.* **110**: 10528–10544.
118. Singleton DL, Cvetanovic RJ. (1976) Temperature dependence of the reactions of oxygen atoms with olefins. *J. Am. Chem. Soc.* **98**: 6812–6819.
119. Clary DC, Nyman G, Hernandez R. (1994) Mode selective chemistry in the reactions of OH with HBr and HCl. *J. Chem. Phys.* **101**: 3704–3714.
120. Faure A, Rist C, Valiron P. (1999) Temperature dependence for the $CN + NH_3$ reaction under interstellar conditions: Beyond capture theories? *Astron. Astrophys.* **348**: 972–977.
121. Faure A, Wiesenfeld L, Valiron P. (2000) Temperature dependence of fast neutral-neutral reactions: A triatomic model study. *Chem. Phys.* **254**: 49–67.
122. Herbst E, Woon DE. (1997) The rate of the reaction between $C_2H$ and $C_2H_2$ at interstellar temperatures. *Astrophys. J.* **489**: 109–112.

123. Clary DC, Haider N, Husain D, Kabir M. (1994) Interstellar carbon chemistry: Reaction rates of neutral atomic carbon with organic molecules. *Astrophys. J.* **422**: 416–422.
124. Buonomo E, Clary DC. (2001). A quantum study on the reaction between $C(^3P)$ and acetylene. *J. Phys. Chem. A.* **105**: 2694–2707.
125. Chen Y, Rauk A, Tschuikow-Roux E. (1991) On the question of negative activation energies: Absolute rate constants by RRKM and G1 theory for $CH_3+HX \rightarrow CH_4+X(X = Cl, Br)$ reactions. *J. Phys. Chem.* **95**: 9900–9908.
126. Jodkowski JT, Rayez MT, Rayez JC, Berces T, Dobe S. (1998) Theoretical study of the kinetics of the hydrogen abstraction from methanol. 3. Reaction of methanol hydrogen atom, methyl, and hydroxyl radicals. *J. Phys. Chem. A.* **103**: 3750–3765.
127. Xu SC, Zhu RS, Lin MC. (2006) *Ab initio* study of the $OH+CH_2O$ reaction: The effect of the $OH–OCH_2$ complex on the H-abstraction kinetics. *Int. J. Chem. Kinet.* **38**: 322–326.
128. Krasnoperov LV, Peng J, Marshall P. (2006) Modified transition state theory and negative apparent activation energies of simple metathesis reactions: Application to the reaction $CH_3 + HBr \rightarrow CH_4 + Br$. *J. Phys. Chem. A.* **110**: 3110–3120.
129. Yu J, Klippenstein SJ. (1991) Variational calculation of the rate of dissociation of $CH_2CO$ in $^1CH_2$ and CO on an *ab initio* determined potential energy surface. *J. Phys. Chem.* **95**: 9882–9889.
130. Klippenstein SJ, Marcus RA. (1989) Application of unimolecular reaction rate theory for highly flexible transition states to the dissociation of $CH_2CO$ into $CH_2$ and CO. *J. Chem. Phys.* **91**: 2280–2292.
131. Klippenstein SJ, East ALL, Allen WD. (1996) A high level *ab initio* map and direct statistical treatment of the fragmentation of singlet ketene. *J. Chem. Phys.* **105**: 118–140.
132. Matsumoto K, Klippenstein SJ, Tonokura K, Koshi M. (2005) Channel specific rate constants relevant to the thermal decomposition of disilane. *J. Phys. Chem. A.* **109**: 4911–4920.
133. Greenwald EE, North SW, Georgievskii Y, Klippenstein SJ. (2007) A two transition state model for radical-molecule reactions: Applications to isomeric branching in the OH-isoprene reaction. *J. Phys. Chem. A.* **111**: 5582–5592.
134. Bullock GE, Cooper RE. (1972) Reactions of cyanogen radicals. Part 3. Arrhenius parameters for reactions with alkanes. *J. Chem. Soc. Faraday Trans. I* **68**: 2185–2190.
135. Hess WP, Durant JL, Tully FP. (1989) Kinetic study of the reactions of CN with ethane and propane. *J. Phys. Chem.* **93**: 6402–6407.
136. Balla RJ, Castleton KH. (1991) Absolute rate constants for the reaction of CN with $CH_4$, $C_2H_6$, and $C_3H_8$ from 292 to 1500 K using high temperature photochemistry and diode laser absorption. *J. Phys. Chem.* **95**: 8694–8701.
137. Atakan B, Wolfrum J. (1991) Kinetic studies of the reactions of CN radicals with alkanes in the temperature range between 294 and 1260 K. *Chem. Phys. Lett.* **186**: 547–552.

138. Herbert L, Smith IWM, Spencer-Smith RD. (1992) Rate constants for the elementary reactions between CN radicals and $CH_4$, $C_2H_6$, $C_2H_4$, $C_3H_6$, and $C_2H_2$ in the range 295 to 700 K. *Int. J. Chem. Kinet.* **24**: 791–802.
139. Zellner R, Lorenz K. (1984) Laser photolysis/resonance fluorescence study of the rate constants for the reactions of OH radicals with $C_2H_4$ and $C_3H_6$. *J. Phys. Chem.* **88**: 984–989.
140. Klein T, Barnes I, Becker KH, Fink EH, Zabel F. (1984) Pressure dependence of the rate constants for the reactions of $C_2H_4$ and $C_3H_6$ with OH radicals at 295 K. *J. Phys. Chem.* **88**: 5020–5025.
141. Kuo CH, Lee YP. (1991) Kinetics of the reaction $OH + C_2H_4$ in He, $N_2$, and $O_2$ at low pressure. *J. Phys. Chem.* **95**: 1253–1257.
142. Vakhtin AB, Murphy JE, Leone SR. (2003) Low temperature kinetics of reactions of OH radical with ethene, propene, and 1-butene. *J. Phys. Chem. A.* **107**: 10055–10062.
143. Cleary PA, BaezaRomero MT, Blitz MA, Heard DE, Pilling MJ, Seakins PW, Wang L. (2006) Deterimination of the temperature and pressure dependence of the reaction $OH + C_2H_4$ from 200–400 K using experimental and master equation analyses. *Phys. Chem. Chem. Phys.* **8**: 5633–5642.
144. Taylor SE, Goddard A, Blitz MA, Cleary PA, Heard DE. (2008) Pulsed Laval nozzle study of the kinetics of OH with unsaturated hydrocarbons at very low temperatures. *Phys. Chem. Chem. Phys.* **10**: 422–429.
145. Chesnavich WJ, Bass L, Su T, Bowers MT. (1981) Multiple transition states in unimolecular reactions: A transition state switching model. Application to the $C_4H_8^+$ system. *J. Chem. Phys.* **74**: 2228–2246.
146. Jarrold MF, Bass LM, Kemper PR, van Koppen PAM, Bowers MT. (1983) Unimolecular and bimolecular reactions in the $C_4H_6^+$ system. Experiment and theory. *J. Chem. Phys.* **78**: 3756–3766.
147. Jarrold MF, Wagner-Redeker W, Illies AJ, Kirchner NJ, Bowers MT. (1984) Unimolecular and bimolecular reactions in the $C_6H_6^+$ system. Experiment and theory. *Int. J. Mass Spec. Ion Proc.* **58**: 63–95.
148. Georgievskii Y, Klippenstein SJ. (2003) Transition state theory for multichannel addition reactions: Multifaceted dividing surfaces. *J. Phys. Chem. A.* **107**: 9776–9781.
149. Klippenstein SJ, Khundkar LR, Zewail AH, Marcus RA. (1988) Application of unimolecular reaction rate theory for highly flexible transition states to the dissociation of NCNO into NC and NO. *J. Chem. Phys.* **89**: 4761–4770.
150. Reznickova JG, Hippler H, Striebel F, Tevazdze L. (2000) A saturated LIF study on the high pressure limiting rate constant of the reaction $CN + NO + M \rightarrow NCNO + M$ between 200 and 600 K. *Z. Phys. Chem.* **214**: 1115–1136.
151. Smith GP. (2003) Rate theory of methyl recombination at the low temperatures and pressures of planetary atmospheres. *Chem. Phys. Lett.* **376**: 381–388.
152. Sims IR, Smith IWM. (1988) Rate constants for the radical-radical reaction between CN and $O_2$ at temperatures down to 99 K. *Chem. Phys. Lett.* **151**: 481–484.

153. Sims IR, Queffelec JL, Defrance A, Rebrion-Rowe C, Travers D, Rowe BR, Smith IWM. (1992) Ultra-low temperature kinetics of neutral-neutral reactions: The reaction of $CN + O_2$ down to 26 K. *J. Chem. Phys.* **97**: 8798–8800.
154. Sims IR, Queffelec JL, Defrance A, Rebrion-Rowe C, Travers D, Bocherel P, Rowe BR, Smith IWM. (1994) Ultralow temperature kinetics of neutral-neutral reactions. The technique and results for the reactions $CN + O_2$ down to 13 K and $CN + NH_3$ down to 25 K. *J. Chem. Phys.* **100**: 4229–4241.
155. Edvardsson D, Williams CF, Clary DC. (2006) Rate constant calculations on the $N(^4S) + OH(^2\Pi)$ reaction. *Chem. Phys. Lett.* **431**: 261–266.
156. Xu C, Xie D, Honvault P, Lin SY, Guo H. (2007) Rate constant for $OH(^2\Pi)+ O(^3P) \rightarrow H(^2S) + O_2(^3\Sigma_g^-)$ reaction on an improved ab initio potential energy surface and implications for the interstellar oxygen problem. *J. Chem. Phys.* **127**: 024304; 1–6.
157. Dayou F, Spielfiedel A. (2003) *Ab initio* calculation of the ground (1A$'$) potential energy surface and theoretical rate constant for the $Si + O_2 \rightarrow SiO + O$ reaction. *J. Chem. Phys.* **119**: 4237–4250.
158. Zanchet A, Halvick P, Rayez JC, Bussery-Honvault B, Honvault P. (2007) Cross sections and rate constants for the $C(^3P)+OH(X^2\Pi) \rightarrow CO(X^1\Sigma^+)+ H(^2S)$ reaction using a quasiclassical trajectory method. *J. Chem. Phys.* **126**: 184308; 1–6.
159. Tashiro M, Schinke R. (2003) The effect of spin-orbit coupling in complex forming $O(^3P) + O_2$ collisions. *J. Chem. Phys.* **119**: 10186–10193.
160. Yagi K, Takayanagi T, Taketsugu T, Hirao K. (2004) The effect of spin-orbit coupling on fast neutral chemical reaction $O(^3P)+CH_3 \rightarrow CH_3O$. *J. Chem. Phys.* **120**: 10395–10403.

# Chapter 5

# Molecular Spectroscopy at Low Temperatures: A High Resolution Infrared Retrospective

Scott Davis

*Vescent Photonics, 4865 E. 41st Ave, Denver, CO 80216, USA*

Feng Dong

*Los Gatos Research, 67 E. Evelyn Ave. Suite 3, Mountain View, CA 94041, USA*

David J. Nesbitt

*JILA, University of Colorado and National Institute of Standards and Technology, and Department of Chemistry and Biochemistry, University of Colorado, Boulder, Colorado 80309-0440, USA*

## Contents

## 5.1. Introduction

The environment of interstellar and interplanetary space provides both unique challenges and unique opportunities to the molecular physicist. For example, in the extremely low temperatures and pressures characterizing the interstellar medium, the relative translational energies between molecular species are very low and the mean free paths extremely long. This combination potentially enables novel species such as weakly bound molecules as well as highly reactive radicals to exist for extremely long times. This stands in contrast to our terrestrial environment, where higher atmospheric densities mean that such "exotic" molecular species are typically very short-lived, e.g., as transient intermediates during a chemical reaction or during the early stages of condensation. The short-lived or transient nature of these species makes them difficult to create and/or to study in the laboratory, which ultimately hampers prospects for identification and interpretation of their spectral signatures in outer space.

Though clearly a major challenge, the potential scientific rewards from identifying and understanding the detailed spectroscopy of such transient molecules are substantial. For example, spectral analysis of such species provides an invaluable observational window into the interstellar/interplanetary environment (e.g. as an atmospheric temperature probe of Jupiter and its moons) as well as insight into the pathways regarding chemical synthesis of molecules in the vast interstellar medium. Though there are clearly purely terrestrial and quite practical motivations for such efforts (e.g., an improved knowledge of combustion dynamics, reaction kinetics, van der Waals interactions, hydrogen bonding, condensation phenomena, etc), the prospect of observing weakly bound and/or highly reactive species in space provides additional stimulus for the material covered in this chapter.

In order to both identify and to interpret the detailed spectral signatures for weakly bound and/or radical molecules in outer space, it is of crucial importance to be able to synthesize and study these species under laboratory conditions. Fortunately, the appropriate laboratory set of terrestrially based tools do exist, based on the combination of i) adiabatic supersonic expansions and ii) high-resolution laser spectroscopy. In particular, the work described herein is based on utilizing a slit supersonic expansion to create cold ($T < 20\,K$) rotational and translational temperatures, thereby enabling the formation of weakly bound van der Waals and hydrogen bonded complexes. When coupled with a pulsed electric discharge, both highly reactive radicals and ions may be generated, again cooled down to very low temperatures. In a sense, the supersonic expansion can be used to mimic the interstellar environment, with the cold transient molecules probed by a laser beam passing through the expansion. If the laser is perchance a high resolution ($<2\,MHz$) coherent source of infrared radiation, then direct absorption spectroscopy provides both unambiguous identification of the species and, as will be discussed extensively in this chapter, a rather remarkable level of detail into both the rovibrational structure and dynamics of these weakly bound and/or highly reactive molecular transients.

By the very nature of the subject material, the topics covered in this chapter range from the relatively simple to the relatively complex. By necessity, we assume a basic knowledge of spectroscopy; our hope is to keep focused on the simple physical ideas revealed by a more diligent spectroscopic analysis, the challenging details from which the reader is spared! The corresponding structure of this chapter is as follows. First, we begin in Sec. 2 with a pedagogical introduction of intermolecular forces, potential surfaces and multidimensional "floppy" quantum mechanics, followed by a discussion in Sec. 3 of weakly bound van der Waals complexes of inert gas atoms around hydrogen halides. In Sec. 4 we follow the idea of building up a potential from pairwise additive through non pairwise additive interactions, in particular "squeezing" the high resolution spectroscopic data on (rare gas)$_n$-HX clusters for insight into sequential contributions from 2-body and 3-body forces. In Sec. 5, we introduce the more strongly hydrogen bound complexes such as HF and DF dimers, for which even with a limited number (6D) of degrees of freedom, detailed investigation of new phenomena are possible, such as intermolecular quantum tunneling (Sec. 6), vibrational predissociation (Sec. 7), permitting rigorous comparisons between theory and experiment (Sec. 8).

We next make the leap in Sec. 9 to spectroscopy of jet cooled hydrocarbon radicals, which are now quite stable with respect to dissociation but nevertheless highly reactive due to their unpaired electrons. These species play an enormously important role in terrestrial combustion chemistry; in addition, many examples of such hydrocarbon radicals have already been observed in the interstellar medium. We begin with a brief discussion of the simple yet fundamental methyl radical and halogenated methyl radicals (Sec. 10), followed by the more complex dynamics of internal rotation and hyperconjugation effects in ethyl radical (Sec. 11), as well as tunneling dynamics in cyclopropyl radical (Sec. 12) and vinyl radical (Sec. 13). Indeed, these species all prove to be quite "floppy" in their own right, executing large amplitude quantum motion that can be successfully studied through high resolution infrared spectroscopy.

As a third broad canvas of investigation, we shift to high resolution spectroscopy of jet cooled molecular ions, which represents a more recent but evolving thrust in our group. We describe two simple jet cooled ionic systems that are undoubtedly present in vast column integrated densities in the interstellar medium. First, we present in Sec. 14 spectroscopic studies of H/D isotopomeric combinations of the simplest polyatomic ion, $H_3^+$, the simplicity of which permits detailed investigation and tests of the Born Oppenheimer breakdown. Secondly, in Sec. 15, we report near IR spectroscopy of multiple isotopomers of the ubiquitous hydronium ion, $H_{3-n}D_nO^+$, the "ammonia-like" tunneling dynamics of which can be studied at high resolution and which yield information on barrier heights for inversion (see also chapter 1 by Herbst and Millar). Finally, some concluding remarks about possible future directions of such infrared studies are summarized in Sec. 16.

### 5.2. Weakly Bound Molecules: Moving Beyond the Harmonic Oscillator

We start with the simple idea of the Born Oppenheimer approximation, whereby we can discuss the ground state potential energy of a molecule as a unique function of its nuclear coordinates. A potent technique for visualizing, modeling, and ultimately understanding molecules is the intramolecular potential energy surface (PES). Stated simply, the PES is a topological map of how the potential energy changes as the constituent atoms are moved relative to one another. For diatomic molecules, which are the simplest of all molecules, the surface is only a one-dimensional function, i.e., dependent only on the distance between the two atoms. This is shown in Fig. 5.1(a) for the *ab initio* Ar-Ar van der Waals potental of Aziz.[1] The qualitative shape of this function is intuitive; when the atoms get too close to one another or "collide" the energy quickly increases, when they are far apart the energy goes to zero, and somewhere in between there is a minimum. For strongly bound and therefore long-lived terrestrial molecules (e.g., covalently bound diatoms) these minima can be deep, and consequently the atoms are "trapped" close to the bottom. The dynamics of such systems may be understood, to a fairly good approximation, as those of a simple harmonic oscillator, which is to say that least for small amplitude displacements the potential energy surface (PES) can be represented parabolically with terms up to quadratic order. The quantum eigenvalues of such a simple harmonic oscillator may be solved analytically for an arbitrary number of dimensions. Indeed, if all atoms in the molecular universe interacted solely via simple harmonic potentials, then rovibrational spectroscopy would be a much simpler field. Unfortunately, a vast array of fascinating and fundamental phenomena simply could not exist — for example, harmonically bound atoms with quadratically increasing potentials never lead to a chemical reaction.

Fortunately for chemistry, the molecular universe is filled with anharmonic interactions — i.e. bonds do break when they are stretched far enough! As a pertinent example, in weakly bound molecules (e.g. van der Waals molecules with $\mathrm{D}_0 \approx 1\,\mathrm{kJ/mol}$ as compared to $D_0 \approx 200\,\mathrm{kJ/mol}$ for covalent bonds) the vibrational wavefunctions sample regions far from the absolute minimum, and a simple harmonic oscillator is no longer sufficient to describe the dynamics. Higher order or anharmonic effects start to play a more important role. This increases the complexity and richness of the spectra and the molecular dynamics, even for a simple diatomic. As more atoms are included, this richness increases rapidly. Indeed, the complexity of highly anharmonic, long range intra- and inter-molecular forces give rise to a vast array of beautiful and important phenomena, ranging from the intricate patterns within snowflakes, the reason ice floats, key details of solvation and condensation, the promotion of chemical reaction pathways, the remarkable "lock and key" specificity of enzyme-substrate interactions, the shapes of folded proteins, to mention only a few.

Potential energy surfaces (or rather slices through them) for prototype 3 and 4 atom weakly bound molecular systems are shown in Figs. 5.1(b) and 5.1(c). As can be seen, the richness in the topology rapidly increases with the extra atoms. An analysis and understanding for both of these molecular systems: (i) $\mathrm{Ar_n}$-HX (where n is the number of Ar atoms and HX is a hydrogen halide, either HF or HCl) with Ar-HF shown in Fig. 5.1(b) and (ii) HF-HF shown in Fig. 5.1(c), provide unique insights into fundamental phenomena in chemical physics. An important motivation for this work is to build up a knowledge base for how to analyze spectroscopy from these anharmonic molecules and how to extract dynamical, structural, and ultimately potential energy surface details from the spectroscopy. In so doing we will also test our understanding of and skill at creating, either from fundamental quantum mechanics or via a semi-empirical approach, the potential energy surface itself.

The Ar-HF molecule serves as both a prototype for inert gas atoms clustering or "solvating" around a hydrogen-halide polar center and as a simple test-bed for refining and testing how we create anharmonic potential energy surfaces. Specifically, both the Ar-HF and the Ar-Ar potential energy surfaces are extremely well characterized and have been verified extensively

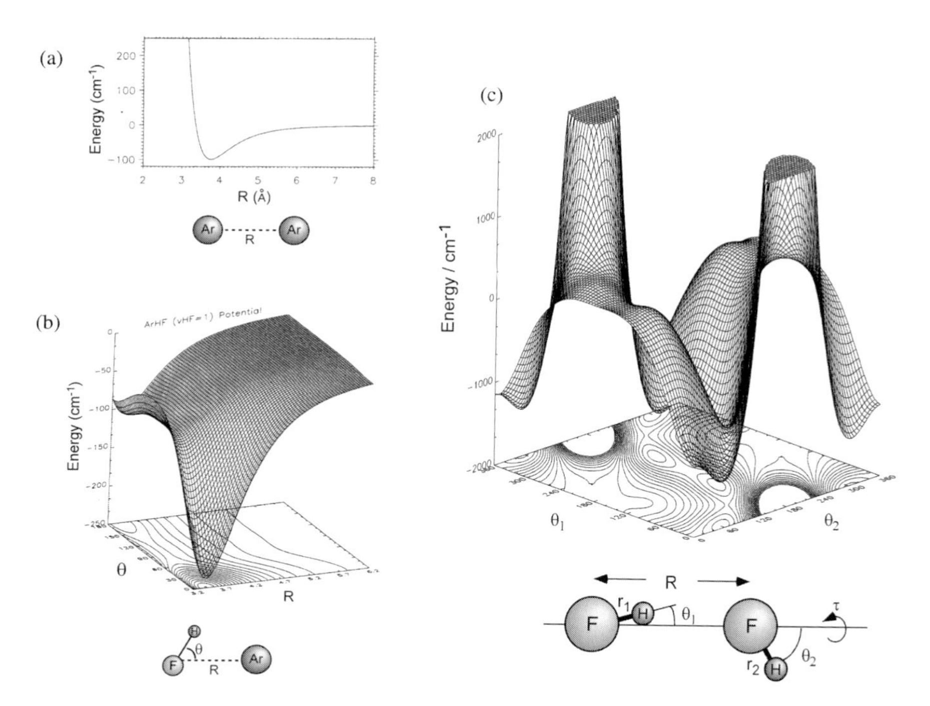


Fig. 5.1. The complex topology of weakly interacting species. (a) The 1-D Ar-Ar van der Waals potental of Aziz.[1]. (b) A surface and contour plot of the 2-D Ar-HF potential H6(4,3,2) of Hutson *et al.*[2] in the $v_{HF} = 1$ manifold. (c) A two dimensional slice through the six dimensional SQSBDE HF-HF potential surface of Quack and Suhm.[3] The six Jacobi coordinates utilized to define the four atom nuclear geometry are also displayed.

by experiment.[1,2] Therefore, the potential surfaces shown in Fig. 5.1(a) and Fig. 5.1(b) may be used to construct "pairwise additive" potential surfaces for larger $Ar_n$–HF clusters. Detailed spectroscopic data may be compared with quantum calculations on these "pairwise additive" surfaces, thereby measuring the importance (or lack thereof) of non-pairwise additive forces. It is imperative to know this as we go forward with more complicated molecules. Furthermore, van der Waals bound clusters shed light on a recurring theme throughout the cluster community: bridging the gap between isolated gas phase molecules and the condensed phase. As the number of Ar atoms is increased in the $Ar_n$-HX clusters, the hydrogen-halide molecule is being sequentally solvated one atom at a time. Analysis of the spectra sheds light on how this micro-solvation correlates with bulk phenomena.

The HF-HF hydrogen fluoride dimer, composed of a modest four nuclei and twenty electrons, represents the simplest example of a hydrogen bond. The ubiquitous importance of hydrogen bonding throughout chemistry and biology is widely acknowledged. From the multitude of crystal structures for solid water, to protein folding, to DNA base pairing, to lock and key mechanisms for enzymes, the detailed shape of the long-range, highly anisotropic hydrogen bond plays a critical role. Indeed, it is because of this richness in the potential energy surface that such physical phenomena can even happen. The prospect of understanding and ultimately predicting (e.g., from first principles *ab initio* calculations), the potential energy surface for such a prototypical hydrogen bond provides a strong impetus for such work.

Additionally, while $(HF)_2$ is the simplest of hydrogen bound molecules, in comparison to the rare gas- HX van der Waals molecules this "simplest" of hydrogen bonds is nevertheless substantially more challenging theoretically. The increased number of electrons and nuclei (3N-6 = 6 nuclear degrees of freedom) hinder both *ab initio* determination of a potential energy surface and the subsequent solutions for the vibrational Schroedinger equation on that surface. In spite of these increased difficulties, however, substantial past efforts have resulted in several trial surfaces. Shown in Fig. 5.1(c) is a two-dimensional slice through the six-dimensional surface of Quack and Suhm for $(HF)_2$ (herein referred to as SQSBDE).[3] Also indicated at the bottom of the figure are the six Jacobi coordinates utilized to define the nuclear geometry. Applying a critique of this and other trial hydrogen bonding surfaces through a comparison between spectroscopic data and calculations on those surfaces is a major thrust of our work on this molecule. Finally, in addition to serving as a prototype for hydrogen bonding, $(HF)_2$ exhibits considerable unique dynamical behavior. For example, there is novel tunneling motion between energetically equivalent configurations wherein the role of proton donor vs. proton acceptor for the two HF subunits is switched. Additionally, excitation of the high frequency HF vibration exceeds the hydrogen bond dissociation limit. Analysis of spectral lineshapes provides insight into how energy flows within and ultimately ruptures this prototype hydrogen bond.

## 5.3. Van der Waals Clusters

Though weak by comparison to covalent bonds ($D_0 \approx 200\,kJ/mol$) the strength and anisotropy of interatomic and intermolecular van der Waals bonds ($D_0 \approx 1\,kJ/mol$) play a significant role in processes such as condensation, solvation, phase transitions, and collisional energy transfer. Discerning the subtle, multidimensional, and long-ranged nature of Born Oppenheimer potential surfaces for van der Waals bound species has been the subject of considerable past effort,[4] ranging from traditional bulk transport studies to scattering experiments, which most sensitively probe the repulsive wall.[5] One particularly fruitful avenue of investigation has been high resolution, rotationally resolved, spectroscopic studies of small van der Waals clusters formed in the cold environment (<10 K) of supersonic expansions.[6–8] Notably, such spectroscopic studies of fragile van der Waals complexes provide information on the often highly anisotropic bound state region of the potential energy surface. Among the large number of molecular clusters studied, "micromatrices" of rare gas atoms clustered with a highly polar hydrogen halide chromophore have proven particularly illuminating.[2,9–24]

In our group we have performed direct absorption, near-IR spectroscopic studies on $Ar_nHF$, $Ar_nDF$ and $Ar_nHCl$ ($n = 1, 2, 3$). Due to the highly polar nature and therefore large transition dipole moment ($v = 1 \leftarrow 0$) for hydrogen halides, these clusters are especially amenable to infrared studies. Furthermore, their large permanent dipole moment enhances cluster formation, especially in the higher density environment of slit vs. pinhole expansions. In these studies, the $v = 1 \leftarrow 0$ transition in the clustered hydrogen halide chromophore has been recorded for sequentially larger clusters of HCl, HF, and DF with Ar.[12,22,25]

A sample overview spectral scan of the $Ar_nDF$ system is shown in Fig. 5.2, along with expanded regions showing the tremendous detail afforded by high-resolution spectroscopy. Each of these spectral features is experimentally measured to an absolute accuracy of <2 MHz (by determining the relative distance from a known spectral feature). This is for absorption lines that are centered at frequencies around 100 THz, meaning that the transition frequencies are known to 1 part in $10^8$. When coupled with microwave studies of the vibrational ground state, these data provide the upper quantum energy levels of the molecule to that same part in $10^8$ accuracy, a staggering level of detailed information. Furthermore, when these spectral patterns are least squares fit to known solutions to the Schroedinger equation (e.g., Watson rotational Hamiltonians), the spectra can be reproduced and, in principle, inverted to provide detailed structural information. The excellent agreement between the Watson based simulations of the spectra and the experimental data is shown in the lower spectra of Fig. 5.2. Typically, the agreement between the simulation and the data is on the order of a few MHz. Spectroscopic rotational constants, which are proportional to the vibrationally averaged inverse moments of inertia ($\propto 1/I$), provide both unambiguous assignment of cluster size and vibrationally averaged structures. The vibrationally averaged structures, as determined from spectroscopy, are shown as cartoon insets in the top panel of Fig. 5.2. Additionally, we may also discern how these structures change upon vibrational and rotational excitation, i.e., we learn how stiff these

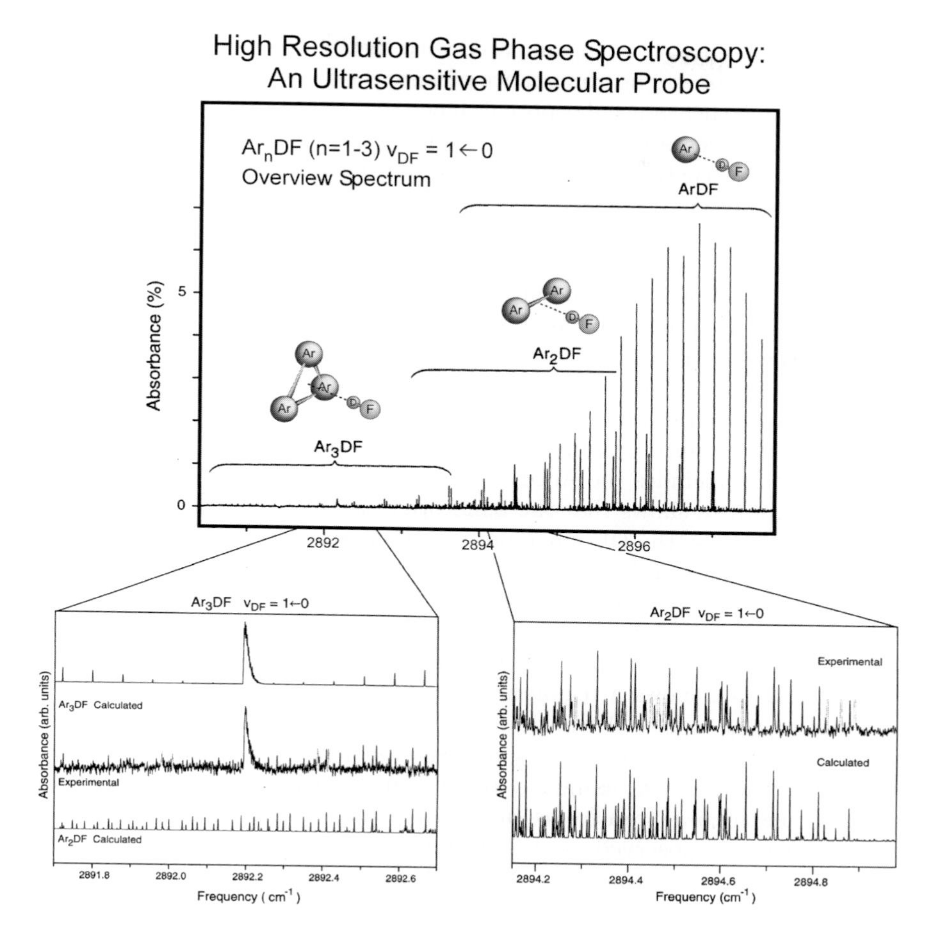


Fig. 5.2. High resolution spectroscopy of gas phase molecular species provides unprecedented detail into the subtle molecular physics. Top panel: Experimental spectrum near the $Ar_3DF$ $v_{DF} = 1 \leftarrow 0$ origin, and spectral simulations of $Ar_2DF$ and $Ar_3DF$ calculated at a rotational temperature of 8.5 K. The shaded peaks that appear as doublets in the experimental spectrum at 2892.6 and 2892.2 cm$^{-1}$ are $P(13)$ and $P(14)$, respectively, of the $v_1$, $K = 1 \leftarrow 0$ transition due to "contaminant" $(DF)_2$. All of the structure can be attributed to transitions of either $Ar_2DF$, $Ar_3DF$ or $(DF)_2$, and there is no evidence for transitions attributable to $Ar_nDF$ complexes with $n > 3$. Lower right panel: Experimental and calculated spectra for $Ar_2DF$ near the band origin. The latter is generated using the spectral constants obtained from a least squares fit of the experimental data. The shaded peaks in the experimental spectrum are transitions of ArDF and $(DF)_2$, and have been suppressed for clarity. Lower left panel: Experimental spectrum near the $Ar_3DF$ $v_{DF} = 1 \leftarrow 0$ origin.

molecules are. Finally, when compared with theory we can use this detailed information to assess the importance of many-body terms and to assess our ability to construct molecular potential energy surfaces. In order to provide a feel for the level of detail with which such spectra are analyzed, and to illustrate how structural and dynamical information is extracted from the analysis, we present a few pertinent detailed examples. Specifically, we will show how analysis of the $Ar_n$-HX cluster spectra sheds light on both the importance of many-body terms in the potential energy surface and the process of solvation, one atom at a time.

Before structural or dynamic information may be extracted from the spectra, the spectra itself must be carefully recorded and analyzed. In so doing, the high-resolution spectroscopist must pay extra attention; here the devil is very much in the details, or perhaps more accurately, seemingly in the noise. For example, as already mentioned, a section of the $v_{DF} = 1 \leftarrow 0$ spectrum that includes the ArDF $P$ branch is shown in the upper panel of Fig. 5.2. Analysis shows that the band origin is at $2897.6593(2)\,cm^{-1}$, i.e., $8.9656\,cm^{-1}$ below the DF monomer rotationless origin of $2906.6609\,cm^{-1}$. The larger clusters of Ar with DF were first observed as "contaminants" in the spectrum of the ArHF/DF dimer. This is not noise, but rather upon closer analysis this clutter was positively identified as rotationally resolved spectra of the larger $Ar_nDF$ clusters. The expanded spectra to the lower right in Fig. 5.2 shows a region around P(18) of ArDF; the shaded peaks are due to $v_{DF} = 1 \leftarrow 0$ transitions of ArDF and $\nu_1$, $K = 1 \leftarrow 0$ transitions of $(DF)_2$, and have been suppressed for graphical clarity. The dense spectral structure was assigned to the $v_{DF} = 1 \leftarrow 0$ band of $Ar_2DF$. Among the high-$J$ transitions of the $Ar_2DF$ $P$ branch, a pronounced, broad feature is observed near $2892.2\,cm^{-1}$ (see the lower left expanded spectra of Fig. 5.2), which is ascribed to the $Ar_3DF$ $Q$ branch. From the minutia of this spectral detail a wealth of fundamental information may be extracted.

## 5.4. "Squeezing" the Spectra, One Atom at a Time: Solvation and Many-Body Terms

Although the kinetics and thermodynamics of solvation at the macroscopic level are reasonably well characterized, a more microscopic understanding at the quantum state level of detail has remained elusive. At the most basic level, potential energy surfaces capable of reproducing solute-solvent and solvent-solvent interactions to spectroscopic accuracy are extremely difficult to obtain. Second, even with such pair potentials, there is an incomplete knowledge about the importance and magnitude of nonpairwise additive (i.e. many body) terms in the full potentials. Finally, there are fundamental computational limitations for solving the full multidimensional dynamics even for a simple diatom solute with a few solvent molecules/atoms, which renders most problems intractable. Attacking the problem of solvation and associated pairwise and nonpairwise additive terms by building up sequentially from smaller clusters of weakly bound species offers obvious advantages, though it is still a considerable theoretical challenge to pursue at the quantum state level for systems with even a few solvent species.

For weakly bound systems in which the requisite experimental data are available, semi-empirical methods have been quite successful at generating highly accurate potential surfaces. Traditionally, results from crossed molecular beam studies are used for modeling the repulsive wall, and bulk measurements such as virial coefficients, viscosity, thermal conductivity, and pressure broadening coefficients[5] are used for the attractive region. Unfortunately, the bulk properties often involve substantial thermal averaging over quantum states, collision angles, and collision energies, which precludes a quantum state specific modeling of the interaction. Furthermore, nonpairwise additive contributions are often inextricably folded into the pair potentials extracted from these measurements. Although the use of these temperature and density dependent "effective" pair potentials greatly reduces the computational demands of liquid simulations, relatively little insight is gained from their form and the numerical values cannot be extended easily even to closely related systems.

Alternatively, high resolution spectroscopy of weakly bound clusters is a powerful method for probing the structure and dynamics of small molecular aggregates, especially in the bound region ($E < D_0$). For a number of prototypical systems it has been possible to characterize a significant portion of the potential through the rotationally resolved study of intermolecular bending and stretching states. In addition, by exploiting the low temperatures and densities of supersonic expansions for cluster formation, these complexes can be formed and studied in an environment nearly free from external perturbations. The study of binary complexes, or dimers, has been an area of especially active interest, both experimentally and theoretically. This is appropriate, since a detailed understanding of these simplest, i.e., dimer, interactions is a necessary prerequisite before proceeding to larger oligomers. The fruits of numerous experimental and theoretical studies are accurate potential energy surfaces for prototypical systems such as HF dimer,[3] and complexes of rare gases with HF,[2] HCl,[19,20] $H_2$,[26] $H_2O$,[27] and $NH_3$,[28] which are sufficiently accurate to use in the theoretical modeling of the structure and dynamics of larger complexes.

Investigations of van der Waals trimers and larger oligomers at rotational resolution are now possible as a result of impressive increases in experimental sensitivity in the microwave,[15–18,29–32] far-IR,[33–36] near-IR,[21,37–39] and visible[40–44] regions of the spectrum. For the complexes in which the corresponding pair potentials are known accurately, structural and dynamical properties can be compared to predictions based on calculations on pairwise additive potential energy surfaces, as first demonstrated on $Ar_2HCl$ clusters by Saykally, Hutson and coworkers.[33–35,45] Deviations between the observed and calculated values then highlight contributions due to nonpairwise additive, or three-body interactions. Fig. 5.3 shows a graphical representation of non-pairwise additive contributions to the potential energy surface.

Although the study of three body terms has a long history for atomic systems (see Ref. 46 for a review), little experimental data exists on the precise form or magnitude of these terms for molecular systems. Consequently, the studies discussed here provide the first quantitative probe of these nonpairwise additive terms, a crucial step towards describing larger solute-solvent interactions. Among the numerous weakly bound dimers that have been characterized experimentally, the rare gas-hydrogen halide complexes have arguably been studied in the most detail. For example, complexes of the rare gases He-Xe with HF/DF, HCl/DCl, and HBr/DBr have been studied in the microwave,[47–52] far- IR,[53,54] and near-IR.[54–56] Within our group, ArHF/DF has emerged as the system about which the most data are available. The experimental data[14,49,50,53,55,57–62] sample a

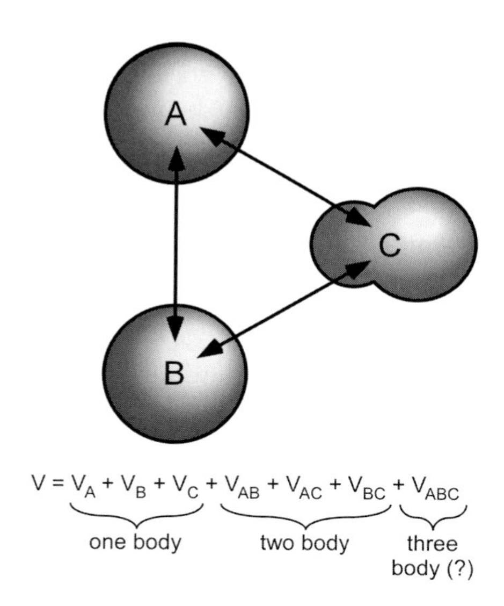


Fig. 5.3. Schematic of the multibody contribution to the total potential in a three body molecular cluster.

significant region of the full three-dimensional potential energy surface for the lowest four vibrational levels of the ground electronic state. Hutson has constructed a potential energy surface[2] from the ArHF/DF data that is a function of all three internal degrees of freedom, i.e., the intermolecular bending and radial coordinates and the intramolecular vibrational coordinate (through the diatom mass reduced vibrational quantum number $\eta = \{v + 1/2\}/\sqrt{\mu_{HF/DF}}$). This potential, designated the H6(4,3,2), reproduces the data to near spectroscopic accuracy,[2] as well as properties such as line shape parameters[63] for HF in a bath of Ar and rotationally inelastic scattering cross sections.[64] Additionally, the diatomic Ar-Ar system has served as a prototype for understanding interatomic van der Waals forces, and its 1-D potential energy curve is well known from crossed beam and UV spectroscopic studies.[65]

With the pair potentials well determined, $Ar_nHF/DF$ provides a particularly good system for studying the transformation to a more fully solvated complex. Additionally, rotationally resolved spectra for $Ar_nHF/DF$ ($n = 1-4$) were already reported in Fourier transform microwave studies by Gutowsky and coworkers,[15–18,29] and infrared results have been reported for $Ar_nHF$ ($n = 1 - 4$) by McIlroy *et al.*[21]

Furthermore, comparison between theory and experiment for the equilibrium geometries of the complexes (shown for $n = 1 - 3$ in Fig. 5.2) and vibrational redshifts[11,22,66] provide critical tests of theoretical potentials. Furthermore, the spectroscopic data implicitly include the effects of nonpairwise interactions, and comparison with theoretical calculations using accurate pairwise potentials permits the quantitative determination of their sign and magnitude.

Ernesti and Hutson[9,11] have presented calculations for $Ar_2HF/DF$ in which the five low frequency intermolecular degrees of freedom are treated explicitly. The rotational constants calculated at the pairwise-additive level of approximation are in good agreement with, but consistently larger than, the experimental values (see Table 5.1), indicating that the pairwise additive approximation leads to an interaction energy that is too attractive. The pair potentials reproduce the rotational constants of the respective dimers to better that 0.1%, suggesting that the disagreement in the $Ar_2HF/DF$ rotational constants is not attributable to deficiencies in the pair potentials.

These results suggest that three-body terms that are repulsive at the equilibrium geometry must be included to bring the calculated rotational constants in line with experiment. Hutson and coworkers[9,11,45] have investigated the effect of several three-body terms on reducing the discrepancies with experiment. Specifically, the inclusion of terms accounting for (i) the anisotropic Axilrod-Teller triple dipole (DDD) interaction, and (ii) the

Table 5.1. Comparison of the $Ar_2HF$ and $Ar_2DF$ rotational constants with those predicted by Ernesti and Hutson.[9,11] The numbers in parentheses are the errors between the calculations and experimental constants

| | Experiment | Pairwise | +3-Body |
|---|---|---|---|
| ArHF | | | |
| B″ | 0.10226 | 0.10228 (+0.02%) | |
| B′ | 0.10262 | 0.10262 (+0.00%) | |
| $Ar_2HF$ | | | |
| A″ | 0.11929 | 0.11993 (+0.05%) | 0.11987 (+0.5%) |
| B″ | 0.05801 | 0.05889 (+1.5%) | 0.05801 (+0.0%) |
| C″ | 0.03873 | 0.03911 (+1.0%) | 0.03870 (−0.1%) |
| A′ | 0.11937 | 0.11988 (+0.4%) | 0.11987 (+0.4%) |
| B′ | 0.05812 | 0.05910 (+1.7%) | 0.05813 (+0.0%) |
| C′ | 0.03879 | 0.03921 (+1.1%) | 0.03876 (−0.1%) |
| ArDF | | | |
| B″ | 0.10140 | 0.10122 (−0.18%) | |
| B′ | 0.10169 | 0.10150 (−0.19%) | |
| $Ar_2DF$ | | | |
| A″ | 0.11697 | 0.11805 (+0.92%) | 0.11798 (+0.86%) |
| B″ | 0.05818 | 0.05941 (+2.11%) | 0.05824 (+0.10%) |
| C″ | 0.03854 | 0.03913 (+1.53%) | 0.03636 (−5.66%) |

interaction between the dipole moments induced on the argon atoms by the permanent dipole moment of the HF/DF, does not change the predicted constants significantly. However, Hutson has identified a third term which, when coupled with the above two terms, does change the constants in the direction necessary to improve agreement with experiment. This term arises from the interaction of the permanent dipole moment of the DF/HF with the quadrupole formed on the $Ar_2$ due to overlap distortion effects.[9,45] This interaction, when included with the DDD and induced dipole-induced dipole terms, improves the agreement between the calculated and experimental rotational constants as shown in Table 5.1.

The identification of a three-body term that is linked directly to the presence of a molecular species, and thus has no counterpart in atomic systems, provides intriguing new insight as such terms are expected to be important for all related molecular systems. As can be seen from Table 5.1, however, the agreement is still not at the level of the uncertainty of the pair potentials, signifying that further refinements to the three body terms are still necessary. Since the present data sample limited regions of the intermolecular potentials of the complexes, efforts to refine the form of these three body terms are best made in conjunction with data from the intermolecular bending states of $Ar_2DF$ and $Ar_2HF$,[13] which sample a more substantial region of the potential surface. Such theoretical efforts are presented elsewhere.[10]

In addition to providing insight into multi-body terms, van der Waals bound clusters help bridge the gap between isolated gas phase molecules and the condensed phase. With computational limitations aside, it is a common belief that the underlying physics for an isolated molecule, a partially solvated molecule (cluster), and a fully solvated molecule (bulk) should all be the same (see top of Fig. 5.4). While this is difficult to test in practice, studies on the "middle group" of molecular clusters provide some insight. One relatively simple experimental handle on this problem is the shift in the vibrational frequency of the chromophore with successive addition of "solvation" atoms (from zero to infinity).

The complexation-induced frequency shifts for $Ar_nDF/HF$ allow the characterization of the small shifts in monomer vibrational energy upon the incremental addition of "solvent" atoms. For large values of $n$, the

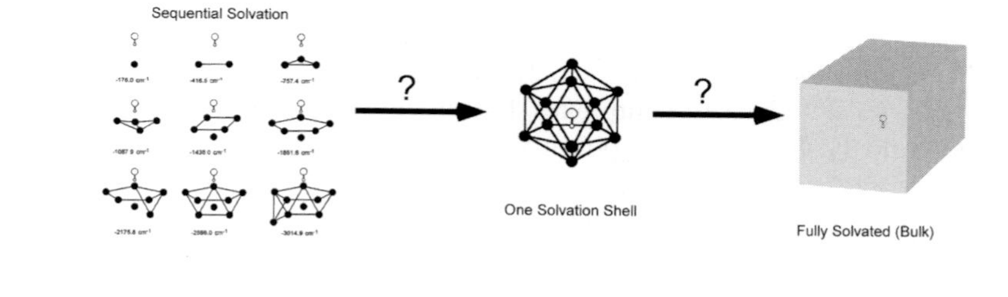


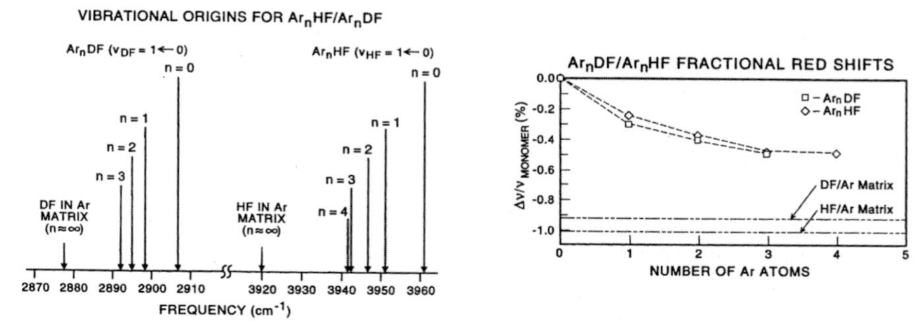


Fig. 5.4. Top: A graphic representation of the role played by $Ar_n$-HX clusters in the transition from isolated gas phase molecules and the condensed phase. Bottom Left: Vibrational orgins for complexes of DF/HF and argon as a function of $n$, the number of Ar atoms in the complex. The corresponding vibrational origins of DF and HF observed in argon matrices are also shown. Bottom Right: Fractional vibrational redshifts for $Ar_nHF$ and $Ar_nDF$ as a function of $n$. The redshifts for $n = 3$ are nearly 50% of the matrix values, in which the DF/HF is surrounded by 12 nearest neighbors in the first solvation shell.

shifts might be expected to approach asymptotically the values observed in condensed phase environments. The Monte Carlo calculations of Lewerenz[66] indicate that the wavefunctions for the Ar atoms in the complexes are highly localized, with the complexes best described as a DF/HF pointing towards an argon "microcrystal". Consequently, a natural point of comparison for the vibrational frequency shifts is the corresponding shift observed in an Ar matrix. The incremental perturbations to the monomer stretching potential can be expected to decrease as successive solvation shells are filled, and thus should be most dramatic for the first solvation shell, i.e., precisely the region sampled by the $Ar_nDF$ complexes.

The vibrational origins of $Ar_nDF$ and $Ar_nHF$ complexes are listed in Table 5.2, along with the corresponding shifts observed in an Ar matrix. The redshift of 8.696 $cm^{-1}$ for ArDF reflects a 0.30% decrease in the DF stretching frequency. While this change is small compared to the DF vibrational frequency, it correlates with a 7.5% increase in binding energy $D_0$.[67] For $Ar_2DF$, the second Ar leads to an incremental redshift of 2.981 $cm^{-1}$, i.e., only $\approx 1/3$ the shift of the first atom. This behavior suggests a strong orientation dependence to the shifts, and can be contrasted with the vibrational redshifts for $Ar_nCO_2$ ($n = 1, 2$) and the electronic blueshifts for $I_2$–(rare gas)$_n$ ($n = 1, 2, \ldots 6$) complexes, for which the shifts are linear with $n$. For $CO_2$ and $I_2$, however, the potential energy surface has multiple equivalent minima in which the rare gas atoms can reside, and the shifts upon sequential atom addition are approximately the same until all the equivalent minima are filled. For DF/HF, there is only one minimum correponding to the lowest energy orientation; hence with two or more solvent atoms each must compete for the preferred spot collinear with the molecular axis, such that the overall interaction per atom is reduced.

The $Ar_3DF$ origin is 2.777 $cm^{-1}$ below that of $Ar_2DF$, leading to a total redshift of 14.454 $cm^{-1}$. The 2.777 $cm^{-1}$ incremental redshift for $Ar_3DF$ is slightly less than the 2.981 $cm^{-1}$ shift of $Ar_2DF$. This nonlinear but monotonic increase in frequency shifts for the $Ar_nDF$/HF complexes is illustrated schematically in the lower left of Fig. 5.4. Interestingly, the redshifts for $Ar_3DF$ and $Ar_3HF$ are almost halfway to the values observed in Ar matrices. Furthermore, as suggested by the minor incremental redshift observed between the $C_{3v}$ $Ar_3HF$ and $Ar_4HF$ structures (which differ nominally by an Ar in the second shell), the frequency shifts appear to be dominated by the first solvation shell.[21,68] Thus, the frequency shifts which accompany the addition of the remaining nine argon atoms (predominantly around the F atom) to complete the first solvation shell must be of the same magnitude as the first three, further emphasizing the strong angular anisotropy and positional dependence of the redshifts.

Table 5.2. Experimental vibrational redshifts for DF and HF with sequential addition of argon "solvent" atoms. Also shown are redshifts calculated using diffusion quantum Monte Carlo techniques from Ref. 66 and bound state variational calculations by Ernesti and Hutson from Refs. 9,11. The two columns reflect the values calculated within the approximation of pairwise additivity, and including the corrective three-body terms as described more fully in the text.

| | Vibrational Origin | Redshift $cm^{-1}$ | Incremental Reshift, $cm^{-1}$ | QDMC[-a] | | Variational[-b] | |
|---|---|---|---|---|---|---|---|
| | | | | Pairwise | +3-Body | Pairwise | +3-Body |
| DF | 2906.6609[c] | | | | | | |
| ArDF | 2897.9653[d] | −8.696 | −8.696 | −9.3(3) (+6.9%) | | −8.694 (−0.02%) | |
| $Ar_2DF$ | 2894.9844 | −11.677 | −2.981 | −13.1(4) (+12.2%) | −12.6(4) (+7.9%) | −12.06 (+3.3%) | −11.46 (−1.9%) |
| $Ar_3DF$ | 2892.2065 | −14.454 | −2.777 | −14.9(6) (+3.1%) | −15.0(6) (+3.8%) | | |
| Ar Matrix | 2876.87[e] | −29.79 | | | | | |
| HF | 3961.4229[f] | | | | | | |
| ArHF | 3951.7688[g] | −9.654 | −9.654 | −9.8(4) (+1.5%) | | −9.655 (+0.01%) | |
| $Ar_2HF$ | 3946.5919[h] | −14.827 | −5.173 | −15.6(4) (+5.2%) | −16.1(5) (+8.6%) | −15.354 (+3.6%) | −14.577 (−1.7%) |
| $Ar_3HF$ | 3942.1634[h] | −19.260 | −4.429 | −22.3(6) (+15.8%) | −21.8(6) (+13.2%) | | |
| $Ar_4HF$ | 3941.7260[h] | −19.697 | −0.437 | −20.6(8) (+4.6%) | −21.0(7) (+6.6%) | | |
| Ar Matrix | 3920.05[e] | −41.37 | | | | | |

[a]Ref. 66, [b]Ref. 9,11, [c]Ref. 70, [d]Ref. 70, [e]Ref. 71, [f]Ref. 72, [g]Ref. 61, [h]Ref. 21.

A comparison of the incremental redshifts for $Ar_nDF$ and $Ar_nHF$ for a given $n$ requires that the isotopic dependence of solvation induced frequency shifts be accounted for. Buckingham[69] has presented a model based on first- and second-order perturbation theory which predicts that $\Delta\nu_{\text{redshift}}/\nu_{\text{monomer}}$ is independent of isotopic composition. Thus for a given $n$, the model predicts:

$$\frac{\Delta v_{Ar_n DF}}{v_{DF}} = \frac{\Delta v_{Ar_n HF}}{v_{HF}}$$

Figure 5.4 shows a plot of $\Delta\nu_{\text{redshift}}/\nu_{\text{monomer}}$ vs $n$ for $Ar_nDF$ and $Ar_nHF$. The ratio is indeed nearly the same for both the HF and DF complexes, indicating that the isotopic dependence of the redshifts for these systems is well described by the above model. The slightly larger fractional shifts for the DF complexes most likely reflect the reduced zero-point bending motion of the DF complexes, which keeps the DF more localized about the intermolecular axis than the HF, leading to an enhanced redshift interaction. It is interesting to note that the fractional redshifts in the matrix limit are nearly 10% smaller for DF than HF; thus the approach to the matrix value with number of Ar atoms is indeed faster for the deuterated species.

Calculations of the vibrational redshifts for the $Ar_nDF/HF$ complexes have been reported by Lewerenz[66] for $Ar_nDF/HF$ ($n = 1 - 4$). Within the pairwise-additive approximation, the vibrational redshifts calculated for $v_{DF/HF} = 1$ of $Ar_nDF$ and $Ar_nHF$ complexes (listed in Table 5.2) show good agreement with, but systematically overpredict, the experimental values. Since the redshifts correlate with the increase in binding energy upon vibrational excitation, it is apparent that the nonpairwise-additive terms necessary to correct the predicted redshifts must lead to a binding energy smaller than that predicted by pairwise-additivity. This is consistent with results from the preceding section, which outlined the need for repulsive three-body terms in the interaction potential. Inclusion of the isotropic triple dipole term discussed above does not improve the agreement with experiment, as shown in Table 5.2. In fact, within the quoted error limits the values are largely unchanged, indicating that the DDD term does not have a significant effect on the vibrational redshift.

The calculations of Ernesti and Hutson[11] allow an even more rigorous determination to be made of three body effects on the vibrational redshifts for $Ar_2DF$ and $Ar_2HF$. The redshifts predicted for $Ar_2DF$ and $Ar_2HF$ using pairwise additive surfaces are $12.06\,\text{cm}^{-1}$ and $15.35\,\text{cm}^{-1}$, respectively, which are $0.38\,\text{cm}^{-1}$ and $0.53\,\text{cm}^{-1}$ larger than the experimental values. The sign of the discrepancy is again consistent with the neglect of repulsive, three-body terms. When the three-body terms discussed in the preceding section are included, the predicted values are $11.46\,\mathrm{cm}^{-1}$ and $14.78\,\mathrm{cm}^{-1}$ for $Ar_2DF$ and $Ar_2HF$, respectively, reducing the discrepancy between the experimental results by approximately a factor of two. Refinements to this analysis should achieve improved agreement with experiment, and consequently a more detailed characterization of nonpairwise additive contributions to the full potential energy surface.

## 5.5. "Plucking" a Hydrogen Bond: Vibrational Dynamics on the Simplest Hydrogen Bonded Interaction Potential

Hydrogen bonds play a key role in a wide array of important scientific phenomena, ranging from the mechanism of DNA base pairing, to dynamics of protein folding, to the unique structural differences between ice and water. However, a quantitative understanding of hydrogen bonding at the level of detailed potential energy surfaces remains an outstanding experimental and theoretical challenge, even for the smallest of molecular systems. As the simplest 4-atom prototype of a hydrogen bonded cluster, the hydrogen fluoride dimer $(HF)_2$ and its $(DF)_2$ isotopomer provide the logical "benchmark" system for quantitatively refining our intuitions on hydrogen bonding phenomenon. Indeed, despite their small size, these systems display a richness of dynamical behavior that already pushes both theory and experiment to the current state-of-the art limits. In our laboratory[11,25,73,74] we have undertaken a series of studies directed toward providing detailed high resolution IR spectroscopic data on vibrational motion in both $(HF)_2$ and $(DF)_2$, in order to elucidate the potential surface and dynamics for this simplest of hydrogen bonds.

Shown in Fig. 5.5 are the $3N-6 = 6$ normal modes, which naturally divide into "high frequency" (a few thousand $\mathrm{cm}^{-1}$) *intra*molecular and "low frequency" (a few hundred $\mathrm{cm}^{-1}$) *inter*molecular vibrations. This disparity of time scales for intra- vs. inter-molecular vibrations is largely responsible for surprisingly slow predissociation lifetimes, i.e., $10^5$–$10^6$ fold longer than the vibrational period of the HF or DF subunits. Additionally, the rates for vibrational predissociation are highly mode specific,[74–77] demonstrating a nonstatistical dependence on internal energy.[78,79] Furthermore, the combination of light H/D atoms and sufficient molecular symmetry result in quantum mechanical tunneling between equivalent configurations,[80–83] which can be observed spectroscopically. This considerable dynamic range of behavior makes these systems ideal for rigorously testing state-of-the art *ab initio* and semi-empirical energy surfaces for hydrogen bonding.

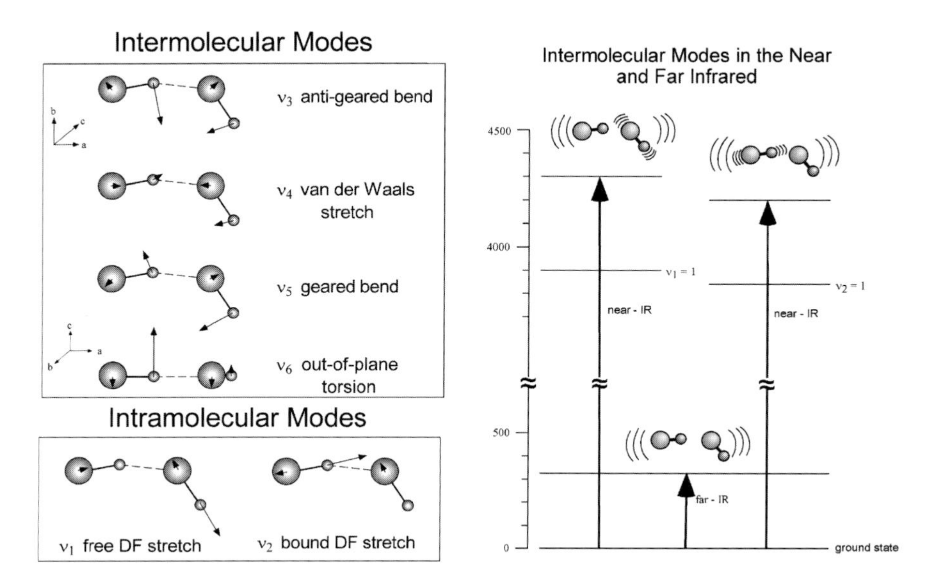


Fig. 5.5. Normal modes of the 4 atom hydrogen bonded $(DF)_2$. The 6 vibrational degrees of freedom naturally separate into high frequency (intramolecular) and low frequency (intermolecular) modes. The intermolecular modes correspond to stretches and bends of the hydrogen bond, with typical frequencies of a few hundred $\mathrm{cm}^{-1}$s. The intramolecular modes, however, correspond to stretches of the covalent DF bonds and therefore absorb near the $\approx 3000\,\mathrm{cm}^{-1}$ free DF frequency. Shown at right is a schematic diagram illustrating how the far-IR intermolecular modes may be observed in the near-IR via combination bands built on the high frequency intramolecular vibrations.

As shown on the right in Fig. 5.5 the *intermolecular* modes of $(HF)_2$ or $(DF)_2$ can be accessed in the *near-IR* as *combination bands* built upon the high frequency $v_1$ and $v_2$ *intramolecular* vibrations. Accessing the intermolecular modes via combination band excitation has the distinct advantage of shifting the absorption from the far to near-IR, where laser methods have been developed to provide continuous, single mode tuning over a large (i.e. several thousand $\mathrm{cm}^{-1}$) region of the spectrum. Furthermore, the additional *intramolecular* excitation allows vibrational predissociation dynamics to be studied, as well as its dependence on *intermolecular* mode. Finally, combination band data may be used to extract intermolecular vibrational frequencies, as well as the dependence of tunneling rates, vibrationally averaged "structures", etc. upon incremental intermolecular excitation. Of particular importance to further refinement of the hydrogen bond potential, such high resolution data provide opportunities for direct, rigorous comparison with full 6-D quantum calculations on trial energy surfaces. Of particular relevance to our work is the analytical potential surface (SQSBDE) by Quack and Suhm,[3] which was fitted to the points of Karpfen *et al.*,[84] but also empirically modified to reflect the microwave, far- and near-IR spectroscopic and bond dissociation data[85,86] available at that time. Indeed, this potential has served as a benchmark surface for full quantum calculations in numerous contexts[87–95] and has proven reasonably consistent with much of the experimental results. As will be demonstrated herein, the SQSBDE potential does a reasonably good job at reproducing spectroscopic properties of the intermolecular vibrations; however, there are significant discrepancies that this study also reveals.

## 5.6. Quantum Mechanical Tunneling

In addition to serving as a prototype for hydrogen bonding, $(HF)_2$ exhibits considerable unique dynamical behavior. Perhaps foremost among these is the novel tunneling between energetically equivalent configurations. Shown in Fig. 5.6 is an alternative 2-D slice through the 6-D Quack and Suhm SQSBDE potential surface, where the plotted coordinates are the Jacobi intermolecular distance $R$ and a geared bending motion, i.e., the motion which most directly connects the two equivalent minima (see Fig. 5.1 for a definition of the Jacobi coordinates). Though energetically identical, the two configurations interchange the role of proton donor vs. proton acceptor for the two HF subunits. In the presence of vibrational excitation in either of the HF monomers, this tunneling motion is also accompanied by a vibrational energy transfer.

The quantum mechanical tunneling results in a splitting of the otherwise degenerate energy levels. This manifests itself in the spectroscopy as a splitting or doubling of the spectral features. The rate of this tunneling motion is experimentally determined from the magnitude of this splitting, and therefore presents one compelling avenue of deriving dynamical information. An example of this spectroscopic splitting is shown in the lower right of Fig. 5.6, where two separate band origins for the $v_2 + v_6$ combination band are clearly visible. An analysis of how this splitting depends on both intra- and intermolecular excitation sheds light on the details of the PES and ultimately allows us to extract the magnitude of the tunneling potential energy barrier.

Intermolecular excitation is predicted to increase the donor-acceptor tunneling rate in a strongly mode specific fashion. A large body of theoretical work [3,77,89–91,93,94,96–102] has been directed at describing the HF dimer energy level splittings and the full-6D calculations for the ground HF-stretch vibrational states predict that both the $v_4$ "van der Waals stretch" and $v_5$ "geared bend" should enhance the tunneling rate by factors of 2 and 17, respectively. The geared bend intermolecular mode is predicted to exhibit the larger effect since this mode correlates strongly with the tunneling coordinate over a $C_{2h}$ transition state. Quantitative predictions for the tunneling splittings in states that include one quantum of intramolecular excitation (i.e., which are measured in the present combination band study) are not yet available. However, one would similarly anticipate a strong preferential enhancement of tunneling rates for eigenfunctions with significant components of geared bend vs. van der Waals stretch motion.

For purposes of comparison, we can attempt to isolate this effect of intermolecular excitation on tunneling by examination of the ratio of the observed combination state splitting to the corresponding $v_1$ or $v_2$ excited state values for a given $K$ level. Examination of the $v_1$ data reveals that both intermolecular modes enhance the tunneling rate dramatically, leading to excited state splittings on the order of $2\,\mathrm{cm}^{-1}$, i.e., between 8- to 13-fold larger than the corresponding $0.2\,\mathrm{cm}^{-1}$ tunneling splittings in $v_1$ or $v_2$. However, although the sign and magnitude of these ratios are in good qualitative agreement with theoretical predictions, there is far less mode selectivity observed than anticipated. Specifically, the observed increase in the tunneling splitting for $v_4$ is *greater* than predicted while the observed increase for $v_5$ excitation is *less* than predicted. This is consistent with a strong mixing of "van der Waals stretch" and "geared bend" modes.

The degree of mode mixing in $v_1 + v_4$ and $v_1 + v_5$ can be quantified via the following simple two state analysis. From a two state perspective, we consider tunneling in two zero order "van der Waals stretch" and "geared bend" states, with values $\Delta v_{\mathrm{tun\ vdw}}$ and $\Delta v_{\mathrm{tun\ geared}}$, respectively. As a reasonable first estimate, these uncoupled, zero order states would be quite similar to those predicted from the 4D quantum mechanical (QM) calculations,[94] which yield a mode specific enhancement ratio of $\Delta v_{\mathrm{tun\ geared}}/\Delta v_{\mathrm{tun\ vdw}} = 7.2$. A plot of the 4D QM wavefunctions is shown in the upper right of Fig. 5.6. As discussed in more detail elsewhere, if we accept from 4D QM predictions[94] that $\Delta v_{\mathrm{tun\ vdw}}$ is enhanced 2.31-fold and $\Delta v_{\mathrm{tun\ geared}}$ 16.6-fold from the tunneling splitting measured in $v_1$, the experimentally observed tunneling splittings correspond to a rather significant (i.e., 30%–70%) mixing of the zero order geared bend and van der Waals stretch character in the actual $v_4$ and $v_5$ intermolecular modes.

We can take this analysis one step further. In the limit of strong mixing of these low frequency intermolecular degrees of freedom, one might anticipate that tunneling rates become independent of mode and depend only on total intermolecular energy and some effective 1D tunneling coordinate. This suggests that the energy dependence of the tunneling rates for the two lowest modes can be characterized by a simple 1D tunneling model and

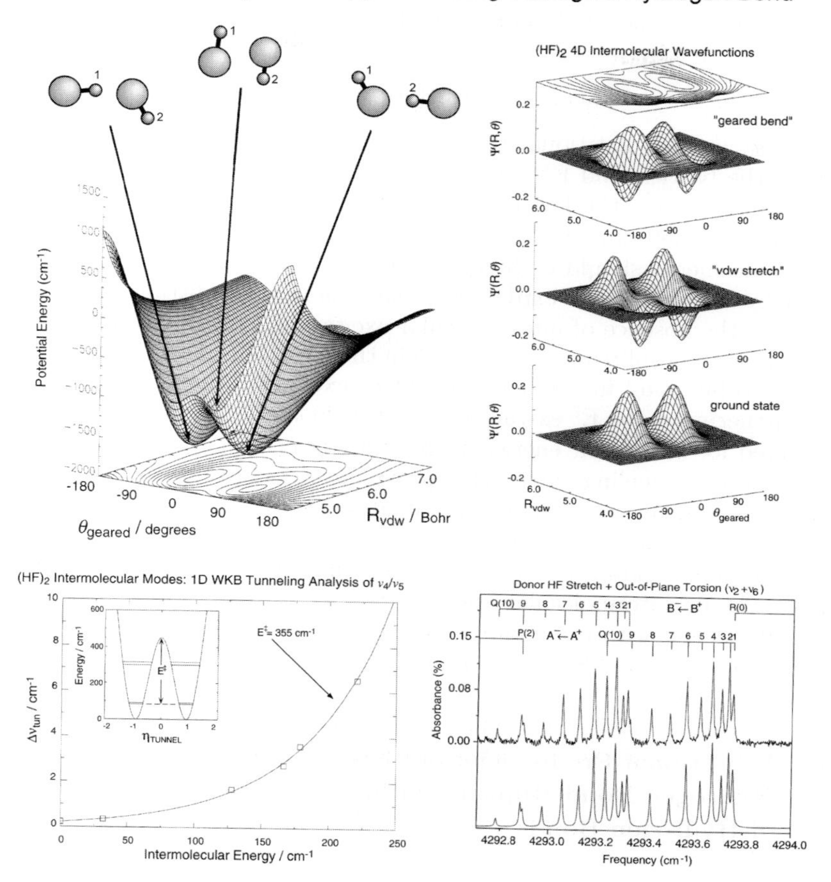


Fig. 5.6. Upper Left: A two dimensional slice through the SQSBDE surface displaying the two isoenergetic minima connected by interconversion tunneling. Lower Left: Plot of the observed tunneling splittings versus intermolecular energy along with a WKB least squares fit. The inset shows a cartoon of the effective 1D tunneling potential used in the WKB analysis along with a definition of the barrier height, $E^{\ddagger}$. The effective 1D barrier height ($E^{\ddagger}$) is $355(2)\,cm^{-1}$ where the reported uncertainty reflects the fit, not the simplifying approximations of a 1D tunneling model. Upper Right: Wavefunctions ($\Psi(R_{vdw}, \theta_{geared})$) for the ground state of $(HF)_2$ and the first two excited intermolecular states from 4D calculations on the SQSBDE potential energy surface. Only the lower tunneling component is shown for each state. $R_{vdw}$ is the van der Waals stretching coordinate (Bohr) and $\theta_{geared}$ is a perfectly geared bending coordinate (degrees) that interconnects the two tunneling states (see text for details). A contour plot of the SQSBDE potential is included at the top of the figure; the energy contours are at equally spaced $200\,cm^{-1}$ intervals in the attractive part of the potential. Lower Right: Example spectra showing the clear splitting or doubling of the spectral features due to the interconversion tunneling.

a WKB analysis of the tunneling splittings. For a 1D parabolic barrier to interconversion, the semiclassical WKB approximation predicts tunneling splittings given by:

$$\Delta\nu_{\text{tun}} = \frac{v_0}{\pi}\exp\{-\beta[\text{E}^{\ddagger} - \text{E}_{\text{inter}}]\}$$

where $v_0$ is the vibrational frequency with which the tunneling species strikes the barrier, and E‡ is the height of the barrier from the zero point level (see inset in the lower left of Fig. 5.6). If an effective 1D model is a reasonable approximation, a plot of the tunneling splittings versus intermolecular energy should be fit by a single exponential, with the pre-factor given by $\Delta\nu_{\text{tun}} = v_0/\pi\exp\{-\beta\text{E}^{\ddagger}\}$, i.e., tunneling splitting observed in $v1/v2$ in the absence of intermolecular excitation. Such a plot of the measured tunneling splittings is presented in the lower left of Fig. 5.6 indicating a remarkably good fit for both $v_4$ and $v_5$ excitation built on either $v_1$ or $v_2$. Furthermore, the $\text{E}^{\ddagger}$ extracted from this fit is $355(2)\,\text{cm}^{-1}$, which when corrected for zero point energy in the geared bend coordinate suggests an empirical 1D tunneling barrier of approximately $438(5)\,\text{cm}^{-1}$. This is in reasonable agreement with the initial estimates of the barrier height ($400\,\text{cm}^{-1}$) in HF-stretch excited states of Pine *et al.*[77] From the parameters determined in the fit and the known ground state tunneling splitting, the barrier height in the ground intramolecular states is predicted to be $366\,\text{cm}^{-1}$, again in good agreement with the minimum energy tunneling path on the SQSBDE surface of Quack and Suhm ($350\,\text{cm}^{-1}$).

## 5.7. Vibrational Predissociation: Energy Flow and Hydrogen Bond Rupture within the Simplest of Hydrogen Bonds

A second arena for dynamical investigations arises from the disparity of vibrational frequencies, or timescales, within such a hydrogen bonded molecule. Specifically, the vibrational modes fall into two distinct classes; *intra*molecular and *inter*molecular. The *intra*molecular vibrations correspond to the high frequency ($\approx 4000\,\text{cm}^{-1}$) covalent stretches of the two HF subunits, while the *inter*molecular modes correspond to the lower frequency vibrations ($\approx 400\,\text{cm}^{-1}$) between the two hydrogen bound diatoms. The validity of an adiabatic separation between these two "classes" of vibrations (such as was demonstrated in the Ar-HX systems) is one of the questions these studies address. A striking example of where a separation between *intra* and *inter*molecular degrees of freedom breaks down is provided by vibrational predissociation. As indicated in Fig. 5.7, excitation of a $4000\,\text{cm}^{-1}$ HF stretch provides ample energy to rupture the weak *inter*molecular hydrogen bond. The timescale for this to occur is experimentally measurable through the magnitude of excess broadening in spectral transitions ($\Delta\nu = 1/(2\pi\tau_{\text{prediss.}})$). The combination of the high resolution infrared laser and the sub-Doppler slit supersonic source utilized in these experiments permits the detection of otherwise unobservable broadening, and consequently detection of vibrational predissociation with timescales ranging from approximately 200 ps to 70 ns.

The predissociation lifetimes are determined from a Voigt deconvolution of the HF dimer lineshapes. The resulting Lorentzian component from such an analysis can be ascribed completely to predissociation broadening ($\Delta\nu_{\text{pd}}$), since the factors such as pressure and power broadening are negligible for the greatly suppressed collision frequencies in a supersonic jet and typical ($\leq 10\,\mu$W) laser power levels.[103] The Gaussian component for each transition arises from residual Doppler broadening in the planar expansion, and is determined either from (i) an unconstrained least squares fit (floating both Gaussian and Lorentzian components) to selected strong transitions in the $(HF)_2$ band or (ii) by independent fits to ArHF transitions present as "impurities" in the jet for which predissociative broadening is immeasurably small.[59,60,62] High resolution scans (7.5 MHz step size, signal averaging 4–6 slit jet pulses) over transitions for several $J$ levels in a given band are fit separately. No statistically significant $J$ dependence is found in all the bands studied.

Even in states with *inter*molecular excitation, the predissociation rate depends predominately on the *intra*molecular mode. This is demonstrated in Fig. 5.7 where typical lineshapes for transitions in the strong $v_1 + v_5$ and $v_2 + v_5$ combination bands are shown. The greater than 10-fold difference in the predissociation rates for the nearly isoenergetic $v_2 + v_5$ and $v_1 + v_5$ states is another clear demonstration of the *non-statistical* nature[78] of the predissociation dynamics in $(HF)_2$. Similarly, though too weak to extract quantitative linewidth information, the linewidths for $v_2 + v_4$ excitation are definitely broader ($\sim$300 MHz) than $v_1 + v_4$ (25(5) MHz and 40(8) MHz). Thus, predissociation from $v_2 + v_4$ appears to be much *faster* than from the corresponding $v_1 + v_4$ upper states, again by roughly an order of magnitude.

Though the effect of *intermolecular* excitation on the predissociation lifetimes is less pronounced than the *intramolecular* dependence noted above, it is still quite significant. In the $v_1$ supported states, where the transitions are sufficiently narrow that small fractional changes are easier to detect, both $v_4$ and $v_5$ excitation results in a $\approx$4-fold *increase* in the predissociation broadening. For the $v_2$ supported state, the effect of $v_5$ excitation is in the opposite direction, i.e., $v_2 + v_5$ intermolecular excitation leads to a *decrease* in the predissociation rate compared with the $v_2$ fundamental.

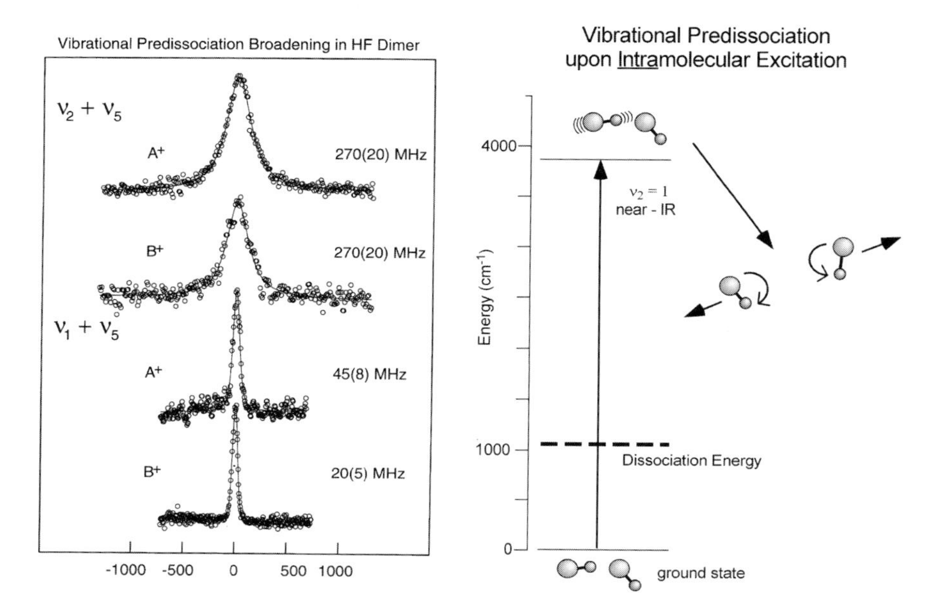


Fig. 5.7. Left: Four sample rovibrational lineshapes for the $K = 0 \leftarrow 0$ transitions of the $v_2 + v_5$ (top) and $v_1 + v_5$ (bottom) $(HF)_2$ combination bands. The circles represent the experimental data; the lines are the simulated Voigt profiles with homogeneous (lifetime) contributions to the broadening indicated in MHz. The mode specificity observed for the $v_1/v_2$ intramolecular fundamentals is largely reiterated in the combination bands. Right: A graph of the relative dissociation and *intra*molecular vibrational energies for $(HF)_2$.

These opposing trends suggest the following simple picture. In the *absence* of intermolecular excitation, the frequencies of the $v_2$ and $v_1$ fundamentals differ by $62.8\,\text{cm}^{-1}$, with corresponding predissociation rates that differ by factors of 34 to 50, for the different tunneling states, respectively. In the presence of intermolecular excitation, however, the two intramolecular modes are closer in energy; specifically the energy difference between the $v_2 + v_5$ and $v_1 + v_5$ origins is reduced by 20% to only $50.7\,\text{cm}^{-1}$. Therefore, one way to view the effect of intermolecular excitation is that it *weakens* the hydrogen bond and *decreases* the splitting between the intramolecular vibrations and thereby mixes more "donor" stretch character into the $v_1$ mode and "acceptor" stretch character into the $v_2$ mode, respectively. Thus, $v_1$ states in the presence of intermolecular excitation mix in a small amount of $v_2$ character and predissociate *more quickly*, while conversely, $v_2$ states mix in $v_1$ character and predissociate *more slowly*. Indeed, due to the large 34–50 fold difference in the predissociation rates of the zero order $v_1$ and $v_2$ states, only a relatively small (10%) amount of mixing would be necessary to produce the observed changes for the $v_4$, $v_5$ combination states. This trend is also supported by spectroscopic studies of the $v_6$ (out-of-plane torsion) and $v_3$ (anti-geared bend) combination bands, the details of which are presented elsewhere.[25]

Interestingly, the predissociation rates for $v_1 + v_4$ and $v_1 + v_5$ from a given tunneling level are equal to within experimental uncertainty. Furthermore, even the 1.5 fractional increase in predissociation rates *between* states in the $v_1$ fundamental, is quantitatively echoed in each of the $v_1 + v_4$ and $v_1 + v_5$ combination bands. This could in principle be due to a fortuitous match between predissociative enhancement for van der Waals stretch and geared bend excitation, though this conjecture is not supported by Fermi Golden Rule calculations[93] on the SQSBDE potential surface. On the other hand, these observations are consistent with the simple picture of strong state mixing between the zero order stretch-bend states, as supported by other spectroscopic data.

One final comment on the predissociation dynamics from combination states is in order. In previous work, Bohac and Miller[104] measured the state resolved photofragment distributions produced following $v_1 + v_4$ and $v_1 + v_5$ $(K = 0)$ excitation. The $v_1 + v_5$ excited fragments indicated a slightly hotter rotational distribution, whereas the $v_1 + v_4$ state led to higher translational recoil, i.e., qualitatively consistent with a zero order "geared bend" and "van der Waals stretch" description of the intermolecular modes. These differences in the HF fragment distributions were interpreted by the authors as evidence for *weak* stretch-bend coupling in the photodissociation dynamics. On the other hand, the weight of spectroscopic evidence discussed in this paper indicates significant mixing of "van der Waals stretch" and "geared bend" degrees of freedom in the HF-stretch excited states. The conclusions from these two studies need not be inconsistent, however. Specifically, vibrational predissociation in $(HF)_2$ is a "rare" event, occurring on a time scale of tens of thousands of monomer vibrations.[75,105] Therefore from a transition state theory perspective,[106] the predissociation dynamics may be dominated by poor overlap of the initial wavefunction to some critical intermolecular configuration, which could be substantially different from the geometries sampled by single quantum excitation of the low frequency intermolecular modes. Indeed, this picture would be at least qualitatively consistent with the experimental observation of a relatively modest dependence of vibrational predissociation rate on intermolecular excitation. However, the coupling between angular and radial degrees of freedom could be much weaker as the inter/intramolecular coordinates approach this critical configuration, which would therefore influence the degree of state mixing inferred from a photofragment distribution.

## 5.8. Intermolecular Energies: A Critical Test of the $(HF)_2$ Hydrogen Bond Potential Energy Surface

One important aspect of this work has been to provide a rigorous experimental platform for testing trial potential energy surfaces. The low frequency *inter*molecular modes that we have studied represent highly anharmonic, large-amplitude, low frequency stretches and librations of the hydrogen bond itself and are therefore highly sensitive to the detailed shape of the potential surface. A comparison of these values to theoretical calculations is therefore extremely valuable for critiquing and refining trial surfaces. The near-IR combination band data provide intermolecular frequencies that reflect the potential surface adiabatically averaged over $v_1/v_2$ intramolecular excitation. As presented in detail elsewhere,[25] there is a strong correlation between *intra*molecular redshifts and *inter*molecular frequencies for $(HF)_2$ combination bands. Specifically, in each case where an intermolecular mode is observed in combination with both $v_1$ and $v_2$, the $v_2$ supported intermolecular levels are higher in energy than the corresponding $v_1$ supported levels, and by an amount proportional to the intramolecular red shift. Consequently, plots of the combination band derived *inter*molecular frequencies as a function of the parent *intra*molecular redshift are found to be nearly linear. Extrapolation to zero redshift can therefore provide empirical estimates of *inter*molecular frequencies in the absence of intramolecular excitation, which prove remarkably accurate when compared with available far-IR measurements in $(HF)_2$.[25,73] The same extrapolation procedure, therefore, is applied to all of the $(HF)_2$ and $(DF)_2$ combination band results. For the two intermolecular modes ($v_3$ and $v_6$) that are not observed on both intramolecular modes, the slope employed for the extrapolation is obtained by performing a single parameter fit to the $v_4$ and $v_5$ data, as discussed elsewhere.[25] It is worth noting that the magnitude of these intramolecular induced shifts is relatively small (<7%), but necessary to include in any rigorous comparison between experiment and theory.

The resulting data are graphically represented in Fig. 5.8, which plots the extrapolated values for both $(HF)_2$ and $(DF)_2$, along with theoretical predictions from full 6-D calculations by Zhang *et al.*,[92] Each of the intermolecular modes exhibits a significant (magnitude >20%) frequency shift upon isotopic substitution. For the two highest frequency

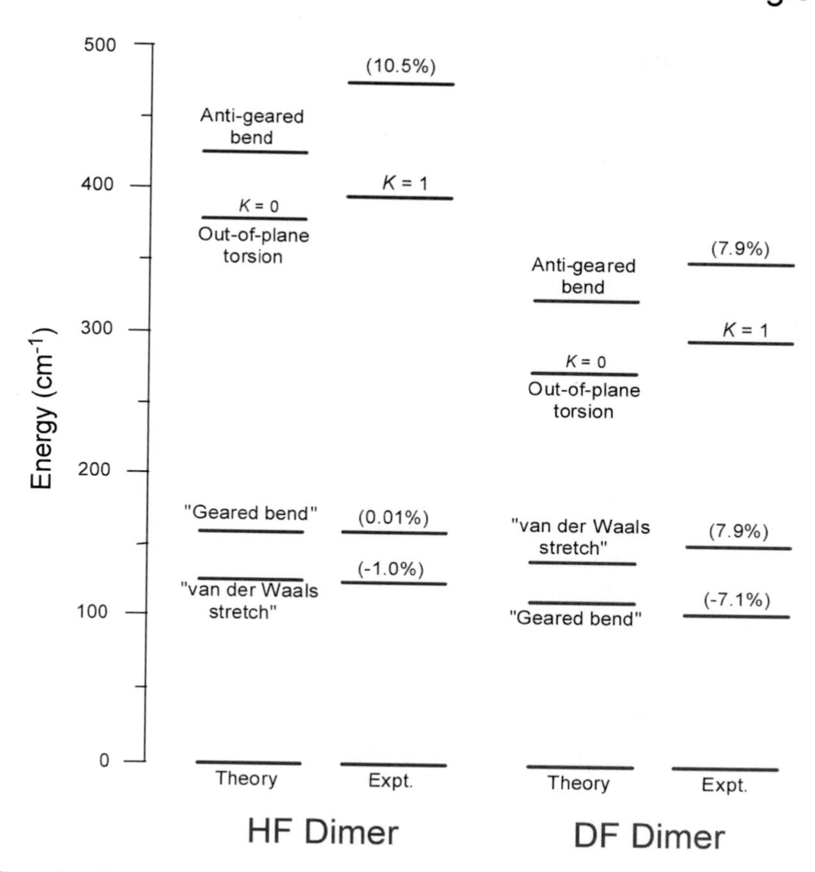


Fig. 5.8. Both experimental and theoretical values for all the intermolecular energies of both $(HF)_2$ and $(DF)_2$. The theoretical frequencies are determined from 6D QM calculations on the SQSBDE surface. Percent discrepancies between experiment and theory are given above the experimental values. For the two lowest energy modes ("geared bend" and "van der Waals stretch") note the considerably worse agreement between experiment and theory in DF vs. HF dimer. Also note the surprisingly large (23%) and positive shift for the $v_4$ "van der Waals stretch" upon deuteration.

vibrations, the anti-geared bend and the out-of-plane torsion, the observed isotope dependence is close to expectations. Specifically, these modes involve predominantly librational motion of the H/D atom; the observed frequency shifts ($\Delta\nu/\nu \approx -26\%$) are in qualitative agreement with the $\Delta\nu/\nu \approx -29\%$ prediction from reduced mass considerations for librational motion. Deviating slightly more from expected H/D librational shifts is the $v_5$ "geared bend", whose frequency decreases $\Delta\nu/\nu \approx -34\%$ upon deuteration. The most striking deviation from expected isotopic shifts, however, comes from the $v_4$ "van der Waals stretch". Simple modeling of $v_4$ as a pure van der Waals stretch motion indicates deuteration should result in a relatively small ($\Delta\nu/\nu \approx -2\%$) and *negative* isotope shift. The observed isotopic shift, however, for the $v_4$ vibration is rather large ($\Delta\nu/\nu \approx -3\%$) and *positive*. The source of this discrepancy could be the bend-stretch coupling in *either* the $(HF)_2$ or $(DF)_2$ isotopomer, which would seriously limit the predictive reliability of any 1-D model of the isotope shifts. Indeed, as previously noted, the $(HF)_2$ experimental data provide strong spectroscopic evidence that the van der Waals stretch and geared bend degrees of freedom are highly mixed.[25]

The intermolecular frequencies in Fig. 5.8 provide an excellent opportunity for comparison with theoretical predictions of Zhang and coworkers[93] on the SQSBDE surface. These calculations incorporate all 6 degrees of freedom, i.e., they allow for coupling between the high frequency intramolecular and low frequency intermolecular modes. As demonstrated in Fig. 5.8, the agreement between experiment and theory for $v_4$ and $v_5$ in $(HF)_2$ is remarkably good, though given the inability of the surface to reproduce the more demanding tests such as tunneling splittings,[91] vibrational predissociation,[94,95,107] etc., this may be somewhat fortuitous. The agreement between experiment and theory for the $v_3$ antigeared bend of $(HF)_2$, on the other hand, is rather poor (+10.5%). For the $v_6$ out-of-plane torsion, only $K = 1$ experimental values and $K = 0$ theoretical predictions exist; thus a truly rigorous comparison is not possible without knowledge of the upper $A$ constant. However, the extrapolated $K = 1 - K = 0(\text{theory}) = 16(8)\,\text{cm}^{-1}$ difference in the upper state agrees within uncertainty to the $A = 24(3)$ cm$^{-1}$ value for $(HF)_2$ excited in the $v_6$ fundamental.[3,108] Additionally, far-IR measurements[109,110] of the $v_6$ vibrational origin for a series of $K \geq 1$ values have been extrapolated[3] to $K = 0$, yielding an intermolecular frequency in agreement with the 6-D calculations within experimental uncertainty. Therefore, with the exception of the $v_3$ anti-geared bend, $(HF)_2$ intermolecular frequencies calculated on the SQSBDE surface are in excellent agreement with experiment.

Considerably more limited success between experiment and theory is achieved for the $(DF)_2$ intermolecular energies. Specifically, agreement between theory and experiment for the $v_4/v_5$ "van der Waals stretch" and "geared bend" modes (which for the $(HF)_2$ values were right on) is now only within 7–8%. Notably, however, the switching of the $v_4/v_5$ energetic ordering upon deuteration is correctly predicted. The discrepancy for the $v_3$ anti-geared bend mode is also only within 8%, though this reflects an even slightly better agreement with experiment than observed for the corresponding mode in $(HF)_2$. Finally, the $v_6$ out-of-plane torsion values for experiment ($K = 1$) and theory ($K = 0$) differ by 23 $cm^{-1}$; the corresponding experimental $A$ value for $v_6$ excited $(DF)_2$ is not known, but the ground state value (16.1945(2) $cm^{-1}$)[74,77] suggest that the SQSBDE theoretical prediction is too low by 2–3%. Therefore, in contrast to $(HF)_2$, where the SQSBDE surface impressively predicted all but the $v_3$ intermolecular frequency, the $(DF)_2$ results indicate significant disagreement for all 4 of the intermolecular modes, and suggest that theoretical work still remains on fully reproducing and understanding the shape of a hydrogen bond.

## 5.9. Slit Discharge Methods: Spectroscopy of Jet Cooled Hydrocarbon Radicals

As highly reactive transient intermediates, radicals play a key role in a vast array of fundamental gas and condensed phase processes ranging from atmospheric chemistry to industrial synthesis in the petroleum and polymer industry.[111–113] Even at extremely cold temperature, nearly barrierless chemical reactions can proceed at gas kinetic rates (see chapter 2 by Canosa *et al.*). However, since the concentrations under steady state conditions can be vanishingly small, this enhanced reactivity also makes them experimentally challenging for detailed spectroscopic study under high-resolution gas-phase conditions. Although conventional electrical discharge can maintain relatively high densities of radicals/ions, the rovibrational temperatures can be inconveniently high due to the high-energy electron impact ionization and/or dissociation, i.e. 300–500 K even with cryogenically cooled discharge walls. Therefore, the population in any given quantum state sampled by a high resolution probe laser will be significantly diluted and the spectra will be substantially congested, as in the top panel in Fig. 5.9 where the simulated high-resolution spectra of ethyl radical at 500 K is shown.

A significant improvement of situating the discharge in the throat of a pin-hole supersonic expansion, pioneered by Engelking and coworkers,[114] yields appreciable densities of rotationally cold radicals (as low as 15 K) at the expansion orifice and has proved quite useful for a variety of emission and LIF studies. However, as these pinhole sources offer short path lengths and have densities that drop off quadratically with distance downstream, they are generally insufficient to use with direct absorption techniques. Furthermore, the spectral widths observed in a pinhole expansion are typically limited by angular spread of the uncollimated free jet, which when

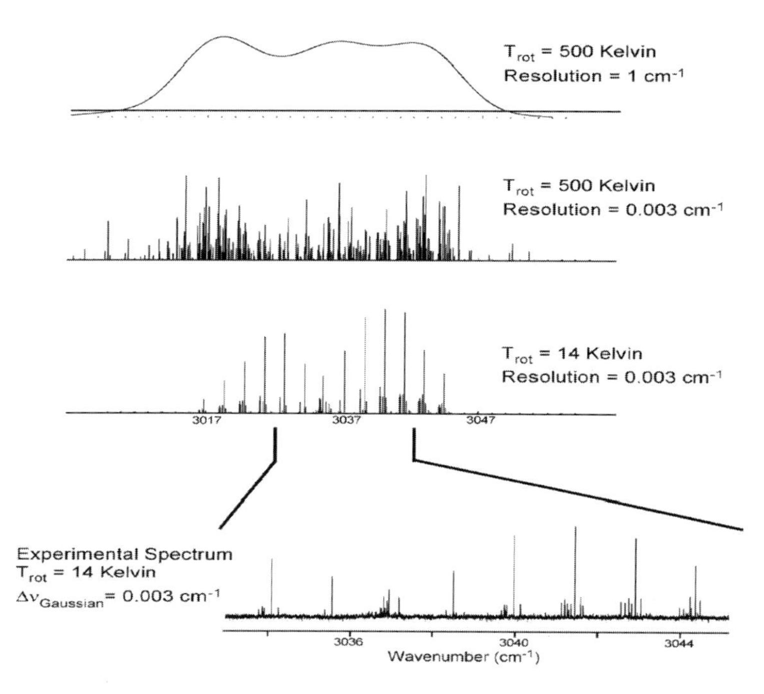


Fig. 5.9. Simulation of the infrared spectrum of the ethyl radical at two different temperatures and two different resolutions. Two advantages of low rotational temperature are evident: a less congested spectrum and more population per low energy state.

viewed with high resolution methods adds significantly to the experimental Doppler profile.

In recent years, there has been progress toward building shaped discharge expansions for matching the ion or radical density appropriately to a long pass direct absorption geometry. Foster *et al.*,[115] have used cw discharges across metal knife edges defining a slit expansion, and succeeded in detecting radicals such as $NH_2$ with direct diode laser absorption. A multiple pin extension of the original Engelking method has been used by Meerts *et al.*,[116] to generate a cw Corona slit discharge, which was able to detect the $NO^+$ and $N_2H^+$ molecular ions with modest signal to noise (S/N) ratios. Efforts by Amano and coworkers[117] with a combined hollow cathode/cw slit discharge have succeeded in detecting several ions such as $H_3^+$, $N_2H^+$, and $H_3O^+$ in the near-IR. A dramatic improvement in sensitivity has been reported for a cw expansion excited by a single 1.5 kV

Corona pin behind a slit nozzle.[117] This work demonstrated S/N greater than 200:1 on single rovibrational lines in $H_3^+$, though with rotational temperatures varying between 33K and 93K and spectral linewidths of order 300 MHz.

Over the past decade, our group has developed pulsed slit supersonic discharge sources[118,119] to permit high resolution spectroscopic access via direct IR laser absorption methods to radicals[120–127] and molecular ions[128–132] at unprecedented densities. The specific advantages of the slit jet discharge for study of radicals are manifold (see Fig. 5.10). First of all, the plasma discharge is localized ($\approx 1\,\text{mm}$) behind the slit expansion jaws; this minimizes gas contact to a few $\mu$s, which is too short for radical-radical or three body radical recombination processes to represent significant loss pathways. Secondly, the nascent hot radicals are rapidly cooled into low quantum states by supersonic expansion ($T_{rot} \approx 10 - 20\,\text{K}$). This represents a crucial improvement in reducing spectral congestion (bottom panel

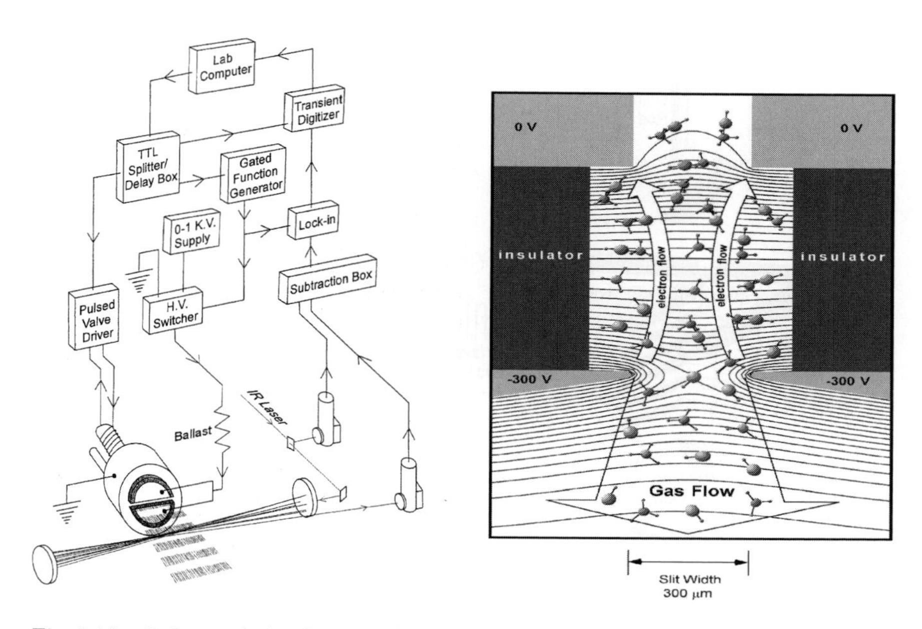


Fig. 5.10. Left panel: A schematic drawing of gated high frequency modulated slit-jet discharge and the lock-in detection. Right panel: Detailed cross sectional view of the slit discharge radical source. In such a negative bias configuration, the light electrons travel upstream of the expansion and the heavy cations continue along downstream. Highly localized discharge is confined between two electrodes separated at 1mm distance. This translates into a $\approx 1\,\mu$s transit time, which is too short to permit subsequent radical-radical chemistry.

in Fig. 5.9) and thereby greatly extending the range of radical complexity that can be experimentally tackled via high resolution IR laser methods. Thirdly, the 1-D expansion geometry results in a collimated velocity distribution along the slit axis, permitting sub-Doppler frequency resolution in an unskimmed jet (typical linewidth of 75–100 MHz). This permits investigation of rotational fine/hyperfine structure in the radical spectra, and also enhances peak absorption by another 20-fold over traditional Doppler limited methods. Additional advances in detection sensitivity have become possible via high frequency slit discharge modulation methods (left panel in Fig. 5.11).[119] In this configuration, high voltage digital switches modulate the discharge bias at up to 100 KHz, yielding "ribbons" of modulated jet cooled radical density in the laser probe multipass region (typical absorption paths of 80 cm). Most importantly, strong discharge confinement behind the slit orifice and efficient rovibrational cooling in the slit expansion greatly minimizes interference due to modulation of precursor absorption, which permits the desired radical absorption signals to be sensitively extracted

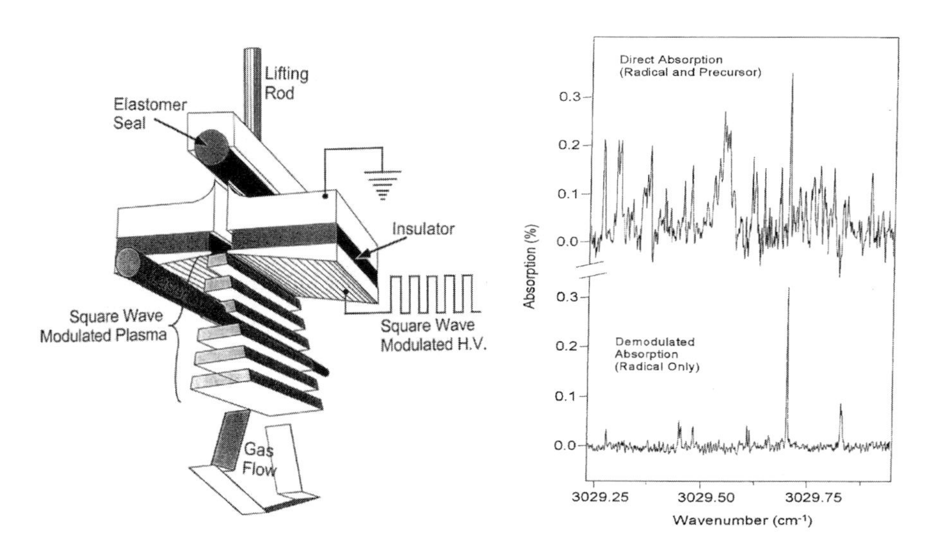


Fig. 5.11. Left panel: Schematic of the pulsed valve discharge modulation source. A 100 kHz square wave discharge ($\approx 1$ kV, 1A) is strongly confined in the 1 mm$\times$300 $\mu$m $\times$ 4 cm region behind the slit expansion jaws. This yields 'slabs' of spatially modulated jet-cooled radicals and molecular ions for detection via direct IR laser absorption and lock-in detection methods. Right panel: Upper trace: direct absorption spectra with conventional discharge modulation in a cw discharge source but without lock-in detection; Lower trace: spectra with frequency modulated discharge and lock-in detection, revealing substantial elimination of radical precursor and improves the absorption sensitivity to the near shot-noise level.

via lock-in detection. Right panel in Fig. 5.11 demonstrates the significant difference between the spectra with and without concentration modulation. In favorable systems, radical concentrations of $\approx 10^{14}$–$10^{15}\,\mathrm{cm}^{-3}$ at the slit orifice can be obtained, with the dissociation efficiency more than 10%. Such slit discharge source capabilities open up an enormous range of radical systems for spectroscopic investigation. From a terrestrial perspective, open-shell hydrocarbon species are particularly important due to their kinetic role in fossil fuel combustion as well as in the processing of biomass emission in the troposphere. This has stimulated our efforts toward obtaining high-resolution infrared spectra for several hydrocarbon radicals in the CH stretch region for the first time, explicitly including methyl ($CH_3$),[120] ethyl ($CH_3$-$CH_2$),[122,126] vinyl($CH_2 = CH$),[127] allyl ($CH_2$-CH-$CH_2$),[121] cyclopropyl,[123] and halogenated methyl radicals ($FCH_2$ and $ClCH_2$).[124,125] More interestingly, intramolecular dynamics — spin-rotation, internal rotation, hyperconjugation, large amplitude tunneling dynamics and even intra-molecular vibrational-redistribution (IVR) have been revealed from the detailed spectroscopic studies.

## 5.10. Methyl and Halogenated Methyl Radicals

Methyl radical, the smallest alkyl radical, is an oblate top molecule with a symmetric, anharmonic potential in the out-of-plane bending coordinate and a planar vibrationally averaged structure which leads to no permanent dipole moment for the molecule and therefore no pure rotational transition moment for spectroscopic access in the microwave region. There have been numerous electron spin resonance (ESR) studies of methyl radical.[133,134] However, the requirement of trapping the radicals in a matrix makes the perturbing effects of the matrix difficult to characterize. Despite the lack of a permanent dipole, methyl radical exhibits nonvanishing dipole moment derivatives with respect to vibrational motion, which allows high-resolution gas phase spectroscopic studies of the strong $\nu_2$ out-of-plane umbrella mode as well as the much weaker $\nu_3$ asymmetric CH stretch mode.

The rotational energy levels for both ground and $\nu_3 = 1$ vibrationally excited states of $CH_3$ are shown schematically in Fig. 5.12. At slit expansion temperature of $\sim$25 K, essentially all of the population collapses into the two lowest spin allowed rotational states. Therefore, a total of 5 transitions are allowed from these two states, which are also shown in Fig. 5.12, i.e., the middle panel shows the one $I = 3/2$ transition and right most panel shows all four possible $I = 1/2$ transitions. At sub Doppler resolution, each rovibrational transition displays additional structure, both varying the number of peaks (up to four distinct features for the $(1,0) \leftarrow (1,1)$ transition) as

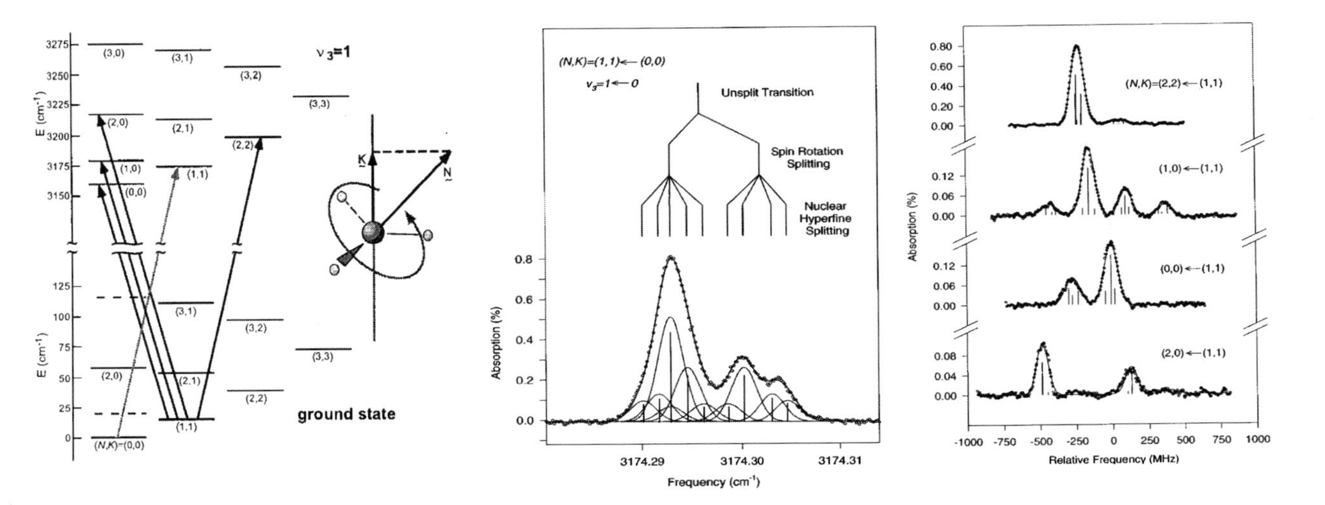


Fig. 5.12. Left panel: Schematic of the rotational energy levels for both the ground and $v_3 = 1$ vibrationally excited state of $CH_3$. At supersonic jet temperatures of ≈25 K, essentially all of the population collapses into the two lowest spin allowed rotational states. The five dipole allowed transitions from these two states are indicated above. Middle panel: Five transitions for both that from $I = 1/2$ (right panel) and those from $I = 3/2$ (middle panel) observed in the slit jet expansion. The shape of the absorption profile is dictated by both spin–rotation and nuclear hyperfine splitting. The magnitude of the splittings and the positions of each sub transition are shown in the tree above. A stick spectrum is plotted below, along with the Gaussian from each sub transition plotted in grey.

well as the widths of these peaks. The resolution of each transition into multiple lines is due predominantly to spin-rotation splittings, whereas the variation in spectral widths predominantly reflects nuclear hyperfine interactions. This can be quantitatively simulated when the total Hamiltonian includes both spin-rotation and nuclear hyperfine effects,

$$H = H_{rot} + H_{sr} + H_{hf},$$

where $H_{rot}$ is the symmetric top rotational Hamiltonian with centrifugal distortion corrections, $H_{sr}$ is the spin-rotation interaction Hamiltonian, and $H_{hf}$ is the hyperfine term.

The Fermi contact interaction in methyl radical is a measure of the unpaired spin density at the hydrogen nuclei, which has been determined by ESR as an unsigned value. The sign of the Fermi contact parameter ($a_f$) in such a prototype $\pi$ radical with spin density predominantly located on a p-like orbital on the C atom arises from two competing effects. The first is due to an anomalously soft out-of-plane bend potential, which results in large amplitude "umbrella" bending motion even at the zero point level. This motion bends the H atoms out of plane and into overlap with the p-radical orbital, contributing to a positive sign of $a_f$. The second is through spin polarization of the C-H bond molecular orbitals. As initially proposed by McConnel,[135] the unpaired electron spin on the carbon can weakly polarize the spin density across each C-H bond via the Pauli exclusion principle. Specifically, the C-H bond electron density near the C atom is preferentially triplet coupled with the unpaired p orbital to generate positive spin density and therefore minimize spatial overlap and electrostatic repulsion. Since the electron pair in the filled CH molecular orbital is singlet coupled, this positive spin polarization localized near the C atom leads to a net negative density at the hydrogen nuclei, and thus predicts a negative sign for $a_f$. The key point is that these two mechanisms result in opposite and therefore competing spin densities at the hydrogen nuclei, which in turn predicts opposite signs for the Fermi contact term. As seen from Fig. 5.13, the experimental spectra unambiguously confirm the negative sign of Fermi contact, and therefore the predominance of spin polarization. Interestingly, such high resolution near-IR measurements represent the first *direct* experimental verification of the predominance of spin polarization effects, due to the lack of a $CH_3$ permanent dipole moment and therefore its inaccessibility in the microwave regime.

In similar studies of substituted methyl radicals, fine and hyperfine structures have also been observed for both $FCH_2$ and $ClCH_2$ radical, as illustrated by sample data for the $2_{11} \leftarrow 2_{02}$ transition in the $FCH_2$ radical (Fig. 5.14). The progression from panel (a) to panel (c) reflects

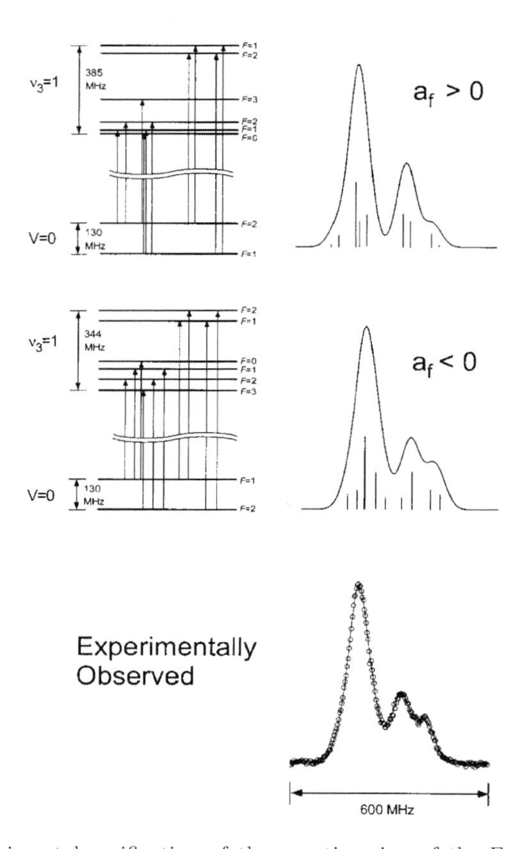


Fig. 5.13. Experimental verification of the negative sign of the Fermi contact interaction. The top of the figure displays the detailed nuclear hyperfine levels for both positive and negative values of $a_f$, along with the predicted absorption profiles for the current sub-Doppler experimental resolution. Shown at the bottom is the observed profile, clearly indicating quantitative agreement for $a_f < 0$. This experimental verification of the absolute sign of $a_f$ for $CH_3$ confirms the importance of spin polarization vs large amplitude bending motion in this simplest of $\pi$ radicals.

systematic inclusion of spin-rotation (fine structure) and nuclear interactions (hyperfine structure) on the spectral line shapes. The simulated spectra match experiment only when both contributions are included in the Hamiltonian. Furthermore, in the halogenated methyl radical systems, the halogen substitution breaks the local mode degeneracy in the $CH_3$ group, thereby localizing the $CH_2$ vibrations into symmetric and antisymmetric motions. Both symmetric and antisymmetric $CH_2$ stretches have been observed in $FCH_2$ (with the intensity ratio of 1.8:1), whereas only the symmetric stretch has been observed in $ClCH_2$ (which indicates an

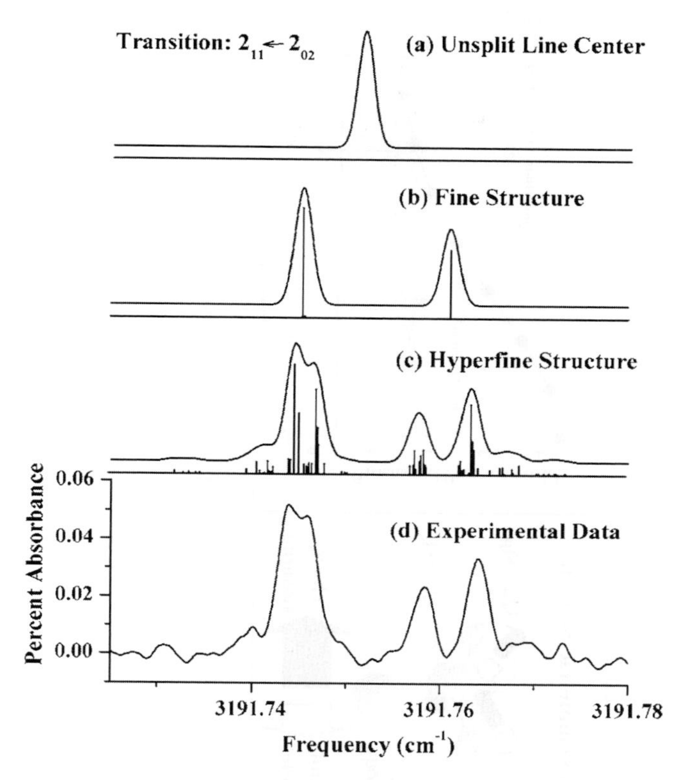


Fig. 5.14. Sample data and contour predictions for the $2_{11} \leftarrow 2_{02}$ transition, systematically including fine and hyperfine terms in the Hamiltonian. (a) Asymmetric top only. (b) Asymmetric top plus spin rotation. (c) Asymmetric top, spin rotation, and hyperfine terms. (d) Experimental data.

intensity ratio of over 30:1 based on the experimental sensitivity). Experimental and *ab initio* predicted intensity ratios are in good agreement, but quite different from a simple geometric "bond-dipole" CH intensity ratio of $I_{sym}/I_{asym} \approx 1 : 3$. This clearly indicates that electronegative atom substitution can lead to a severe breakdown of the bond-dipole model in describing stretch intensities in $CH_2X$ radicals.

Close inspection of the *ab initio* results reveals that this breakdown arises from larger (and in fact opposing) dipole derivatives due to flow of atomic charge densities under normal mode $CH_2$ stretch displacements. Based on a simple Mullikan charge description, the calculations (figuratively shown in the top panel of Fig. 5.15) indicate a significant decrease ($-\delta_{qH}$) in H atom positive charge densities upon symmetric extension of the C–H bond. This decrease is accompanied by a corresponding increase

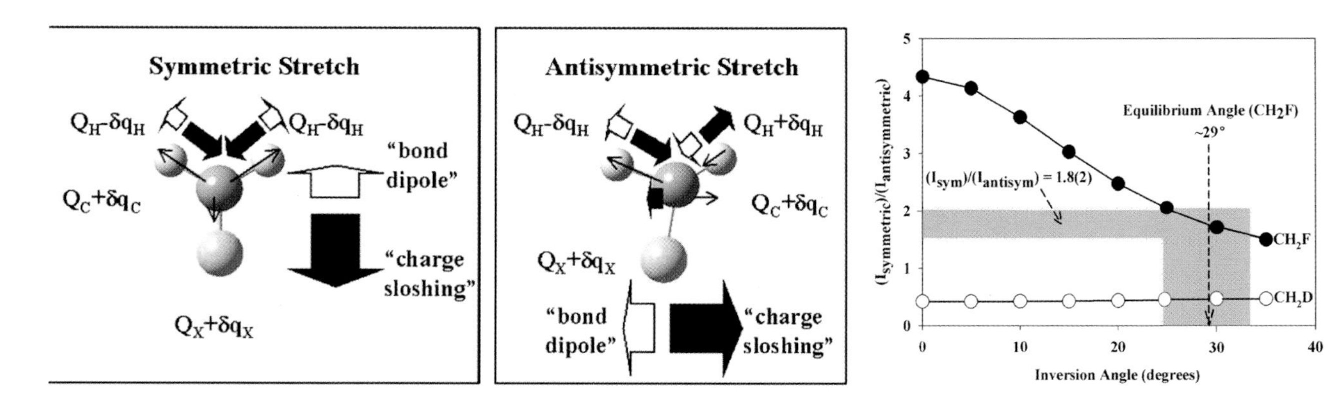


Fig. 5.15. Left and middle panels: Cartoon illustrating the competition between "bond dipole" and "charge sloshing" contributions to symmetric and antisymmetric CH stretch intensities in $CH_2X$. Right panel: Calculated intensity ratios for symmetric and antisymmetric $CH_2$ stretches in $CH_2F$, highlighting a remarkably strong dependence on inversion angle. Note the essentially constant ratio predicted for $CH_2D$. Though large amplitude averaging is clearly important, the experimental intensity ratio of 1.8(2):1 for $CH_2F$ is strongly suggestive of a nonplanar equilibrium structure.

($+\delta_{qC}$ and $+\delta_{qX}$) in both carbon and halogen charge densities (i.e., to less negative values). This "charge-sloshing" results in net flow of positive charge *away* from the C–H bond displacement, yielding transition dipole moment contributions in the *opposite* direction from bond dipole expectations. Indeed, detailed estimates for the $CH_2$ symmetric stretch in $CH_2F$ show that the magnitude of this charge-sloshing moment is in excess of the conventional bond-dipole contribution, thus actually *reversing* the direction of the overall transition dipole.

Similarly for the antisymmetric $CH_2$ stretch, the hydrogen atom charge densities decrease ($-\delta_{qH}$) and increase ($+\delta_{qH}$) upon extension and compression of the C–H bonds, respectively. Both carbon and fluorine atom charge densities increase ($+\delta_{qC}$ and $+\delta_{qX}$) with antisymmetric CH extension, though the magnitudes of $\delta_{qC}$ and $\delta_{qX}$ are far smaller since the H atom charge redistributions are of opposing sign. Once again, the charge-sloshing component of the transition moment more than cancels that of the bond dipole contribution and indeed reverses the sign of the overall transition dipole. Interestingly, since the net CH stretch transition moment reflects a competition between these two contributions, a strong dependence of these vibrational band strengths on the nature of the halogen substitution is anticipated, in agreement with the 1.8:1 and >30 : 1 ratios observed in the $FCH_2$ and $ClCH_2$ radicals, respectively.

A more dramatic piece of structural evidence can also be derived from a closer investigation of such an anomalous ratio for $FCH_2$, as shown in the right-hand panel of Fig. 5.15. At the planar $C_{2v}$ configuration, the predicted intensity ratio of 4.3:1 is significantly larger than experimental results. However, at CCSD(T) theoretically predicted equilibrium geometry of $\theta = 29°$, the intensity ratio is in good agreement with experiment. This provides strong support for a *nonplanar* equilibrium structure of $FCH_2$, though undoubtedly with a large amplitude bending wavefunction sampling both wells. This would be consistent with previous spectroscopic studies[136,137] which suggested a vibrationally averaged planar geometry, but concluded that the presence of a small barrier at the planar configuration could not be ruled out.

## 5.11. Ethyl Radical: Internal Rotation and Hyperconjugation

Ethyl radical is a particularly important benchmark species in hydrocarbon combustion. It is the simplest hydrocarbon radical species with a single C–C bond and therefore serves as a prototype for large-amplitude internal rotation dynamics in open shell systems. It is also the smallest alkyl species with a CH bond in $\beta$ position with respect to the radical center and thus serves as a potential window into hyperconjugation effects. Therefore, a great deal of effort both experimentally and theoretically has been put into this simple radical over the past three decades. The first rotationally resolved spectra in the ethyl radical was reported in the $CH_2$ bending region by Sears *et al.*[138] based on flash photolysis diode laser studies. However, due to limited coverage of the diode laser source, the complete assignment and analysis of these spectra required the rotationally resolved spectra from our laboratories of jet-cooled ethyl radical in the CH stretch region.

*Ab initio* calculations indicate that the ethyl radical behaves qualitatively like a methyl radical with one H atom replaced by a methyl substitution, i.e., with an essentially near planar radical equilibrium geometry and a very soft out of plane $CH_2$ bending coordinate. Thus, excitation of the symmetric $CH_2$ stretch vibration should be accompanied by a transition dipole moment nearly parallel to the C–C bond. Because the relatively heavy C–C framework dominates the moments of inertia for ethyl radical, this bond essentially coincides with the B axis of a near prolate top. Furthermore, H atoms ($I = 1/2$) on both the methyl and methylene groups are feasibly interchangeable by a combination of rotation around the C–C bond and out of plane bending of the methylene moiety with 4:2 and 3:1 statistical weights. As clearly developed by Sears *et al.*[138], this internal motion leads to four nuclear spin symmetries (A$'$,A$''$,E$'$,E$''$) with nuclear spin weights of 12:4:6:2, of which collisional conversion between different nuclear-spin states is negligibly slow on the 10 $\mu$s time scale of jet expansion. Thus, at $T_{rot} \approx 20\,K$, the molecules are essentially cooled down to the lowest state within each nuclear-spin manifold in a ratio of ≈12:4:6:2. Fully one-half of the population is in the A$'$ symmetry manifold with approximate quantum numbers $k = m = 0$, where k is the projection of total angular momentum (J) on the molecule fixed frame and m is the internal rotor angular momentum for relative rotation between methylene and methyl groups. As the dominant features in the spectrum, one therefore predicts parallel ($k = 0 \leftarrow 0$) and perpendicular ($|k| = 1 \leftarrow 0$) bands out of A$'$ ground internal rotor state (m = 0), with somewhat weaker transitions from the other spin manifolds. Indeed, under jet-cooled conditions, the spectra are dominated by the most populated A$'$ manifold, as shown in Fig. 5.16.

The top two panels show the sample spectra of $CH_2$ symmetric and asymmetric stretches. Least-squares fits of transition frequencies to a near prolate top model Hamiltonian reproduce the data within the 7 MHz experimental uncertainty and provide rotational constants for both ground and vibrationally excited states. Structural analysis based on the measured B and C rotational constants imply a C–C bond distance of 1.49 Å, which

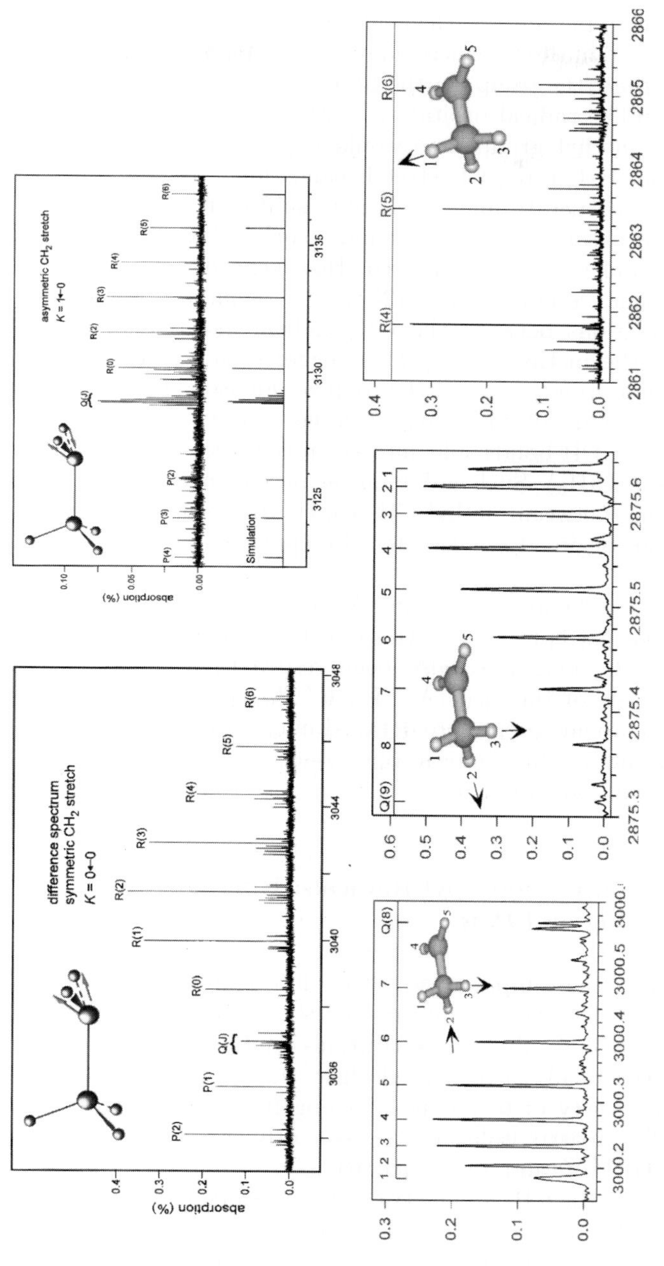


Fig. 5.16. Sample data from 5 CH stretch vibrational bands in ethyl radical: top 2 panels — symmetric and asymmetric $CH_2$ stretch; bottom 3 panels — 3 C-H stretches from methyl group.

is consistent with partial (≈15%) double bond character for the ethyl radical carbon frame and in excellent agreement with theoretical predictions. The bottom three panels show the sample spectra of three CH stretches for the $CH_3$ group. In the absence of hyperconjugative coupling between the p-like radical orbital on methylene breaking the 3-fold symmetry of the methyl group, one would expect symmetric (A-type) and antisymmetric (B/C-type) CH stretch bands for $CH_3$, with the B/C-type bands nearly degenerate and much blue shifted from the A-type band. Indeed, we do observe an A-type band , correlating with the symmetric C-H stretch, and very far to the red. However, the splitting between B- and C-type bands is not small (≈125 $cm^{-1}$), signaling the presence of significant interactions between the $CH_2$ radical moiety and the opposing CH bond on the methyl group. This suggests an improved zeroth-order vibrational description of methyl group as an isolated CH stretch (e.g. C-$H_1$), strongly red-shifted by hyperconjugation, with localized vibrations in the remaining CH bonds split into symmetric and asymmetric stretches (e.g. C-$H_2$ and C-$H_3$). Such a dynamical picture highlights a remarkably strong coupling between methyl CH-stretch vibrations and C–C torsional geometry and begins to elucidate discrepancies with previous matrix observations.

In addition to the five bands from A′ nuclear spin symmetry, transitions from other nuclear spin symmetries due to low barrier internal rotation around the C–C bond have also been resolved, as shown in Fig. 5.17. Detailed analysis of the rich (A″, E′, E″: $k, m \neq 0$) internal rotor fine structure will provide even more detailed information about the hyperconjugative coupling in this simplest open-shell prototype of large amplitude internal rotation dynamics.

## 5.12. Allyl and Cyclopropyl Radicals: Ring Opening, Tunneling and IVR

Allyl radical is the simplest radical exhibiting resonance delocalization and one of the key radical intermediates in organic chemistry through mechanisms such as allylic substitutions and allylic isomerizations. It has been theoretically investigated since the 1950s, and experimentally studied by a variety of techniques ranging from ESR[133,139] and electron diffraction[140] to matrix isolation,[141] resonantly enhanced multiphoton ionization (REMPI)[142] and resonance Raman.[143] The first high-resolution infrared study of the $CH_2$ symmetric wagging mode of the allyl radical was performed by Hirota and co-workers in 1992.[144] However, due to room temperature Doppler broadening and diode laser frequency noise, the excess

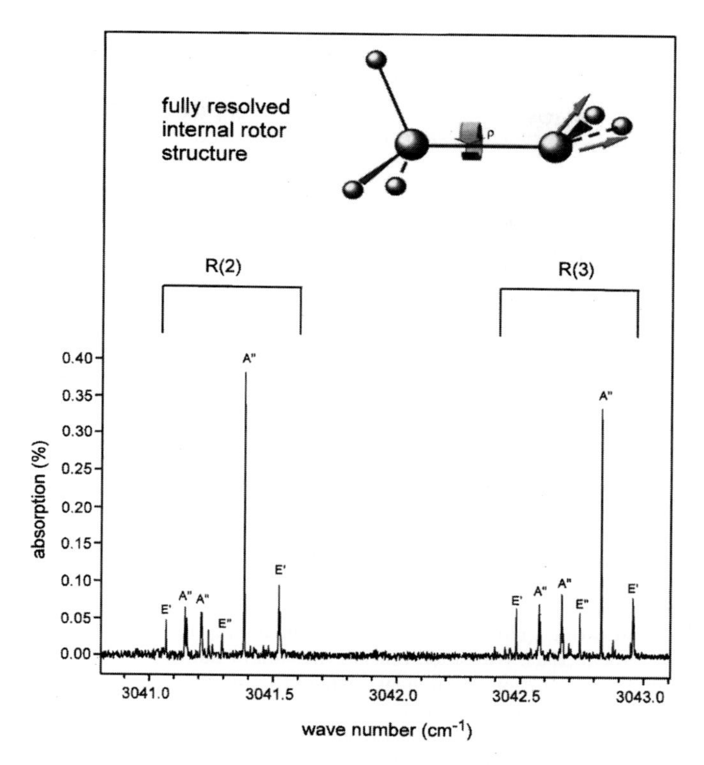


Fig. 5.17. A detailed section of the symmetric $CH_2$ stretch spectrum illustrating additional spectral structure arising from transitions out of excited internal rotor levels. The lettered labeling on the additional transitions indicates the nuclear spin symmetry, which in turn impacts the intensity of the different features.

line widths observed in this $\nu_{11}$ band were simply attributed to unresolved splitting from spin-rotation interaction.

Two new bands — in-phase ($\nu_1$) and out-of-phase ($\nu_{13}$) antisymmetric $CH_2$ stretching vibrations of allyl radical have been obtained in the slit jet discharge spectrometer, as the sample spectra shown in the top panel of Fig. 5.18. The data have been successfully analyzed with a Watson asymmetric rotor Hamiltonian, yielding precise band origins and rotational constants for both bands. The high quality of least squares fits to ground state combination differences indicates that the rotational level structure in the lower state is well behaved, while the reduced quality of fits to the vibrational transitions, on the other hand, suggest the presence of Coriolis mediated rotational perturbations in the upper state. Due to sub-Doppler resolution ($\Delta\nu \approx 70\,\text{MHz}$) in the slit jet expansion,

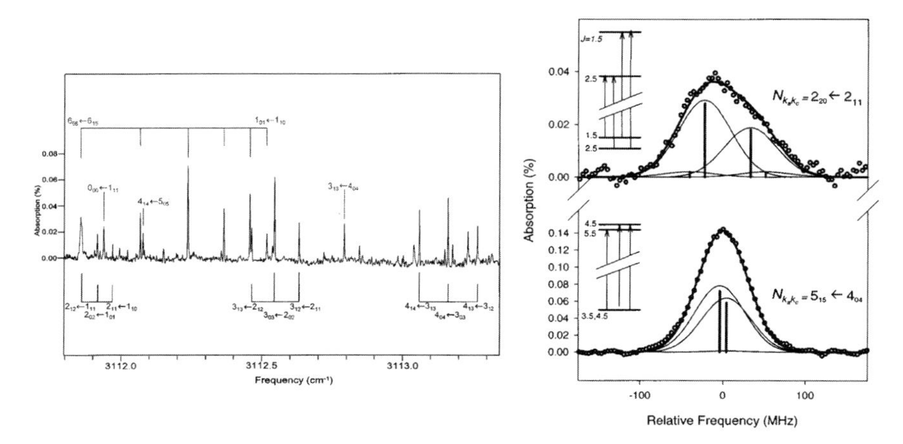

Fig. 5.18. Left panel: Sample section of the allyl radical spectrum illustrating transitions in an overlapping region for both out-of-phase ($\nu_1$, labeled above) and in-phase ($\nu_{13}$, labeled below) $CH_2$ asymmetric stretch bands. The expected 10:6 intensity alternation due to nuclear spin statistics is clearly illustrated by the $K_a = 0 \leftarrow 1$, Q(J) branch series in $\nu_1$ , while the tight clusters of $\Delta K_a = 0$ transitions in the R(J) branch for the $\nu_{13}$ band reflect the near prolate symmetric top behavior of allyl radical. Right panel: Sample data indicating quantum state excess broadening due to spin–rotation interactions in allyl radical. Each transition reflects the sum of multiple overlapping Gaussians from individual spin–rotation components, where the relative intensities and frequencies are determined from matrix diagonalization of a combined rigid asymmetric top and spin rotation Hamiltonian. Note the substantially larger broadening in the $2_{20} \leftarrow 2_{11}$ transition versus the $5_{15} \leftarrow 4_{04}$ transition. This reflects the greater splittings observed in higher $K_a$ states as well as for states with $J \approx K_a$, due to the predominance of spin–rotation effects due for rotation around the A-axis.

quantum-state-dependent excess broadening of the rovibrational transitions is also observed (right panel in Fig. 5.18), which can be ascribed to spin–rotation interactions. Based on a least squares analysis of the high-resolution line shapes, the data are consistent with a spin rotation constant of $\varepsilon_{aa} \approx -67$ MHz.

The cyclopropyl radical is particularly interesting from a dynamical perspective, since it differs from allyl only by the additional closure of the C–C bond. Thus, cyclopropyl is strictly a "resonance" excited state of allyl, with the two species representing "entrance" and "exit" channels on a potential surface corresponding to unimolecular ring opening. Indeed, the CCC ring in cyclopropyl radical leads to energetically unfavorable bond angles, providing a benchmark for hydrocarbon strain effects in open shell systems small enough for both experimental and theoretical studies. The radical center in cyclopropyl is predicted to have a symmetric double minimum with respect to CH bending perpendicular to the carbon plane (top panel

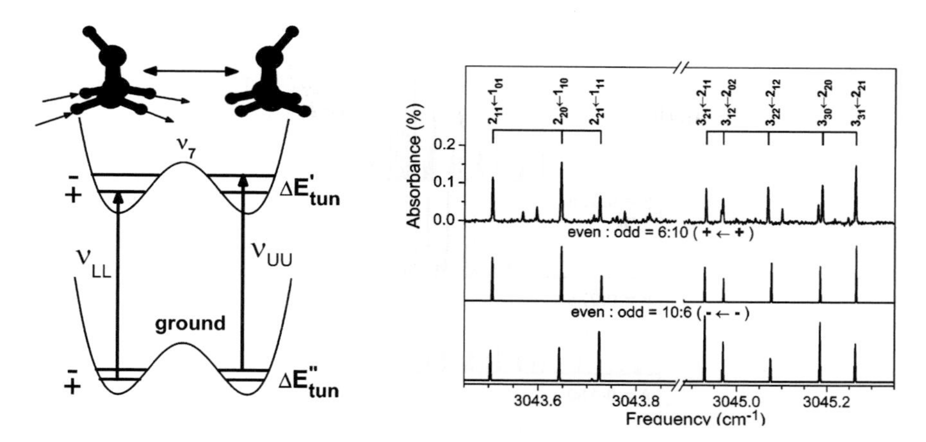


Fig. 5.19. Left panel: Illustration of double-minimum potentials for the ground and excited states ($\nu_7$) of cyclopropyl radical. According to symmetry considerations and nuclear spin statistics, only two transitions with nonzero oscillator strength are allowed, i.e., c-type bands for $\nu_7^+ \leftarrow 0^+$ and $\nu_7^- \leftarrow 0^-$, which give only the difference of the ground- and excited-state tunneling splittings. Right panel: Comparison of experimental spectra with predicted spectra for different nuclear spin weights (6:10 and 10:6 for the upper and lower lines, respectively). The band can be unambiguously assigned to $\nu_7^+ \leftarrow 0^+$, i.e. the nuclear spin ratio is 6:10 for $K_a + K_c$ = even:odd.

of Fig. 5.19). Though classically inaccessible from the ground state at the planar HCCC transition-state configuration, this double-well potential nevertheless results in rapid quantum tunneling from one minimum to the other. This tunneling represents an elementary prototype for stereochemistry around a hydrocarbon radical center, the retention or inversion of which for asymmetrically substituted cyclopropyl species is critical in *chiral* organic and pharmaceutical synthesis pathways.

High-resolution infrared spectra of jet-cooled cyclopropyl radical have been obtained in our laboratories for the first time, specifically sampling the in-phase antisymmetric $CH_2$ stretch ($v_7$) vibration. In addition to yielding the first precise gas-phase structural information, the spectra reveal quantum level doubling into lower (+) and upper (−) states due to tunneling of the lone $\alpha$-CH with respect to the CCC plane. The bands clearly reveal intensity alternation due to H atom nuclear spin statistics (6:10 and 10:6 for even:odd $K_a + K_c$ in lower (+) and upper (−) tunneling levels, respectively) consistent with $C_{2v}$ symmetry of the cyclopropyl-tunneling transition state, as the sample data shown in Fig. 5.19. The two *ground-state* tunneling levels fit extremely well to a rigid asymmetric rotor Hamiltonian, but there is clear evidence for both local and global state mixing in the vibrationally *excited* $\nu_7$ tunneling levels. In particular, the upper (−)

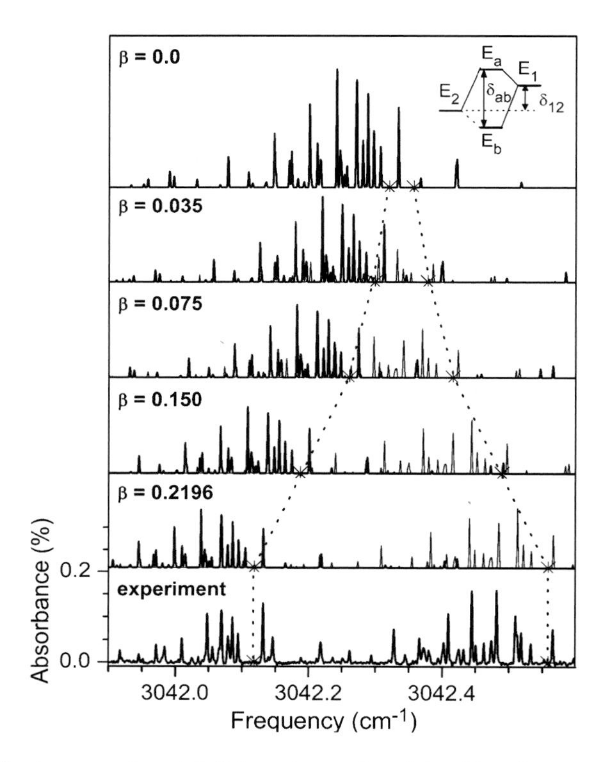


Fig. 5.20. The two experimental Q-branches assigned to $\nu_7^- \leftarrow 0^-$ can only be quantitatively reproduced by adding an anharmonic coupling matrix element ($\beta$). The low-frequency band with more intensity (band b, while $I_b : I_a \approx 0.54 : 0.46$) is assumed to be the "bright" state in the fitting. As $\beta$ increases, the "dark" state "borrows" more intensity from the "bright" state, and the separation between the two shifted band origins increases (as the dotted lines shown between the different panels).

tunneling component of the $\nu_7$ state is split by anharmonic coupling with a nearly isoenergetic dark state, which thereby acquires oscillator strength via intensity sharing with the bright state (Fig. 5.20).

With the reasonable assumption of thermal equilibrium between the $0^+$ and $0^-$ states, tunneling splittings for cyclopropyl radical are estimated to be $3.2 \pm 0.3\,\text{cm}^{-1}$ and $4.9 \pm 0.3\,\text{cm}^{-1}$ in the ground and $\nu_7$ excited states, respectively. This analysis indicates ground state stereoracemization of the $\alpha$-CH radical center to be a very fast process [$k \approx 2.0(4) \times 10^{11}\,\text{s}^{-1}$], with the increase in tunneling rate upon $CH_2$ in-phase asymmetric stretch excitation. Modeling of the ground-state-tunneling splittings with high level *ab initio* 1D potentials indicates an improved barrier height of $V_0 = 1115 \pm 35\,\text{cm}^{-1}$ for $\alpha$-CH inversion through the cyclopropyl CCC plane.

## 5.13. Vinyl Radical: Tunneling and H-Atom "Roaming"

The vinyl radical ($H_2C{=}CH$) is the simplest open-shell olefinic hydrocarbon radical, and its importance as an intermediate in combustion as well as in the low temperature chemistry of planetary atmospheres has prompted many studies, both experimental and theoretical. The reaction of vinyl radical with oxygen/hydrogen plays an important role both in saturated and unsaturated hydrocarbon fuel/oxygen flames, and in the formation of acetylene in the flame which is probably the source of the polycyclic aromatic hydrocarbons (PAH) and soot. Although the first observation of vinyl radical was made through mass spectrometry, electron spin resonance (ESR) studies[133,145] indicated that the structure of the vinyl radical is essentially that of the ethylene molecule with one hydrogen removed and that the unpaired electron occupies a $sp^2$ hybrid orbital. This radical center also permits a relatively low energy in plane pathway for motion of the lone $\alpha$-CH bond from one side to the other, which therefore raises the interesting possibility of large amplitude tunneling dynamics, or alternatively, intramolecular H atom transfer, that would be apparent in high resolution studies. Indeed, Kanamori *et al.*[146] reported the first high-resolution infrared spectroscopy of the $CH_2$ wagging mode of the vinyl radical near $895\,\mathrm{cm}^{-1}$, with a 3:1 (1:3) nuclear spin statistical alternation for $K_a$ = even:odd levels in the ground and excited tunneling states, respectively. This is to be expected due to the two H atoms ($I = 1/2$) in the $CH_2$ moiety, which become equivalent under high resolution detection conditions by feasible tunneling of the $\alpha$-CH across a $C_{2v}$ transition state geometry. Recent studies[147] of pure rotational excitation as well as combined rotational-tunneling transitions provide a nearly complete description of the ground vibrational state and definitively establish the magnitude of the tunneling splitting to be $0.5428\,\mathrm{cm}^{-1}$.

To date, however, no high-resolution infrared studies in the CH stretch region have been reported, though such studies would provide important benchmarks for state-of-the-art theoretical predictions as well as facilitate further laser kinetic studies of vinyl radical under real time combustion conditions. The absence of such studies reflects the requisite level of experimental sensitivity; the CH stretch band strengths are predicted to be down by more than 30 fold with respect to the $CH_2$ wagging mode, based on density functional theory (DFT) calculations with a respectably large basis set (B3LYP/$6-311++$G(3df,3pd)). Though weak, vinyl radical still offers sufficient oscillator strength to detect in the slit jet discharge spectrometer with quite respectable S/N.

Sample data for first high-resolution IR spectra of jet-cooled vinyl radical in the C–H stretch region (specifically sampling the symmetric $CH_2$

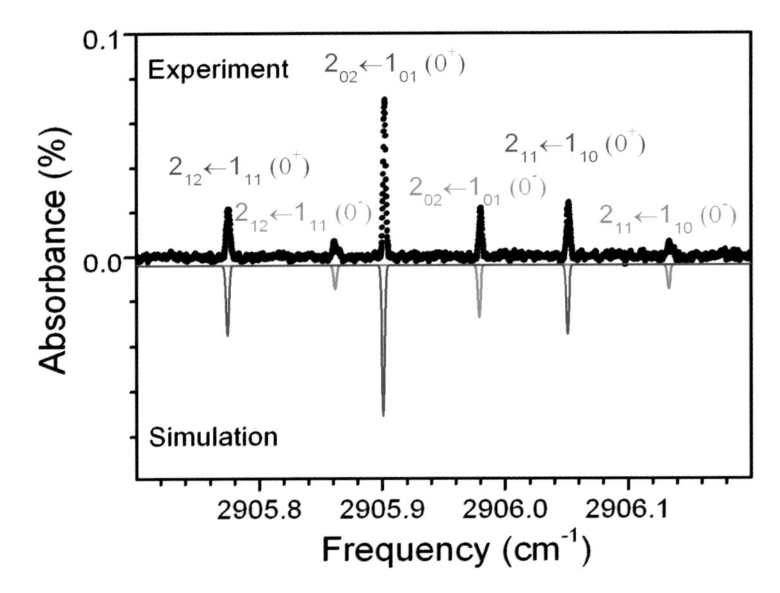


Fig. 5.21. Sample experimental (upward going) and simulated (downward going) spectral regions for vinyl radical $CH_2$ symmetric stretch absorptions out of ground ($0^+ \leftarrow 0^+$) and first excited ($0^- \leftarrow 0^-$) tunneling manifolds. Note the absorbance sensitivity of $\approx 5 \times 10^{-5}$, which corresponds to $<10^8$ vinyl radicals/$cm^3$/quantum state.

stretch) are shown in Fig. 5.21. Two A-type bands have been observed around $2900\,cm^{-1}$, which can be unambiguously assigned by ground state combination differences to the $0^+ \leftarrow 0^+$ and $0^- \leftarrow 0^-$ $CH_2$ stretch transitions out of the lower ($0^+$) and upper ($0^-$) tunneling levels. Tunneling state symmetries have been assigned by the *decrease* in vibrationally adiabatic barrier height expected for $CH_2$ symmetric stretch excitation, which translates into an *increase* in tunneling splittings and vibrational band origins ($\Delta\nu \approx 0.071\,cm^{-1}$) for lower-lower and upper-upper A-type tunneling selection rules. Of special dynamical relevance is the anomalous nuclear spin statistics observed in the $0^+ \leftarrow 0^+$ and $0^- \leftarrow 0^-$ fundamental tunneling bands, which clearly reveal a 1:1 intensity ratio for $K_a$ = even:odd states populated in the expansion, instead of 3:1 intensity alternation. On the other hand, symmetry analysis indicates that a 1:1 ratio of ortho($I = 1$):para ($I = 0$) states would be consistent with feasible exchange of *all three* H atoms in vinyl radical, with the $2^3 = 8$ nuclear spin states transforming as a $\Gamma = 4A_1^{'} + 2E^{'}$ irreducible representation. This would suggest the presence of a large amplitude pathway for labile H atom exchange in vinyl radical, arising from vibrationally excited states formed by C–Br bond cleavage of vinyl bromide in the discharge.

A key point of this argument is that the 1:1 ratio of ortho ($I = 3/2$): para ($I = 1/2$) nuclear spin statistics for feasible exchange of the 3 H atoms is established in the discharge region and then frozen into the resulting expansion. The presence of such large amplitude H atom exchange dynamics in forming vinyl radical is strongly reminiscent of the "roaming" H atom behavior noted by Suits, Bowman and coworkers in near threshold studies of formaldehyde photolysis,[148,149] as well as studies of acetaldehyde photolysis of Kable and Houston.[150]

It is particularly interesting that alternate generation of vinyl radical in these same $0^+$ and $0^-$ states via 193 nm photolysis of vinyl bromide yields the "normal" 3:1 nuclear spin intensity alternation associated with feasible exchange of the 2 H atoms on the $CH_2$ group, as first verified by Kanamori *et al.*,[146] This suggests that there may be multiple dynamical pathways for formation of vinyl, depending on degree of internal vibrational excitation left in the nascent radical species by either photolytic bond cleavage or dissociative electron attachment processes. Detailed exploration of such intriguing dynamical processes would clearly be accelerated by development of a high level *ab initio* vinyl potential energy surfaces in full dimensionality, which we hope the current work will stimulate.

## 5.14. Jet-Cooled Molecular Ions: Breakdown of the Born-Oppenheimer Approximation in $H_2D^+/D_2H^+$

There have been pioneering achievements by Saykally,[151] Oka[152] and others based on high resolution direct absorption laser spectroscopy of molecular ions in discharge cells. Although our studies have primarily focused on jet cooled radicals, the slit jet discharge can also generate exceptionally high concentrations of jet cooled molecular ions, at sufficient densities to make direct IR laser absorption spectroscopy possible with high S/N. For example, direct absorption on the $^rR_0(1)$ line of $H_3^+$ with frequency modulated discharge methods yields signal to rms noise of $\approx 8500 : 1$ due to $2 \times 10^{12}$ jet cooled ions/cm$^3$ in the laser probe region. This is already comparable to total ion densities achieved in velocity modulation discharges under rotationally much hotter conditions. Therefore, several molecular ions have also been studied in detail.

Protonated molecular hydrogen, $H_3^+$, is arguably the most fundamental molecular ion, playing a central role in a wide range of chemistry, physics, and especially astrochemistry (see chapter 1 by Herbst and Millar). At the theoretical level, this simplest of polyatomic molecules (i.e., 3 H nuclei and 2 electrons) has served as a benchmark system for development of high accuracy *ab initio* methods, especially as a prototype of treatment of 2e$^-$ three-center bonding. Exact rovibrational solution of $H_3^+$ and isotopomers as a full five-particle system remains an unsolved challenge. Though a reasonably accurate description of 2-electron bonding in $H_3^+$ and isotopomers is available from conventional electronic structure calculations, the level of effort required to match experimental observation to spectroscopic accuracy is far from trivial. Specifically, large amplitude motion of the H/D nuclei leads to considerable complications and challenges in the nuclear dynamics for this light three-atom system. Indeed, Tennyson and co-workers [153] have clearly shown that a large contribution to error in the *ab initio* determination of $H_3^+$ rotation-vibration state energies arises from the breakdown of the fundamental Born-Oppenheimer approximation, i.e., sufficiently rapid electronic motion on the time scale of nuclear displacements to justify the concept of a potential energy surface (PES). In fact, the magnitude of such effects for $H_3^+$ and isotopomers is on the order of several $cm^{-1}$ even for fundamental vibration/rotation levels; this is already $10^3$-fold in excess of high-resolution Doppler limited spectroscopic uncertainties and predicted to increase further with overtone excitation. The existence of such non-Born-Oppenheimer effects undermines the simple yet virtually universal picture of nuclear dynamics occurring on a well defined, isotopically invariant potential surface. Detailed overtone studies of asymmetrically substituted isotopomers such as $H_2D^+$ and $D_2H^+$ are likely to prove especially enlightening, since these ions require additional non-Born-Oppenheimer corrections[153,154] that are absent in the more symmetric $H_3^+$ and $D_3^+$.

Nevertheless, progress in these directions has been extremely promising based on approximate three particle variational calculations with suitably modified potentials and effective non-Born–Oppenheimer kinetic energy terms in the Hamiltonian.[155] Indeed, predictions based on this *ab initio* procedure now reproduce existing experimental data for $H_3^+$ to nearly spectroscopic levels of precision, i.e., within a few hundredths of a $cm^{-1}$. By virtue of the additional non-Born-Oppenheimer terms in the Hamiltonian, a particularly stringent test is provided by the asymmetrically substituted isotopomers, which prior to these studies had not been observed beyond the fundamental manifold.

This motivates high-resolution overtone studies of the $H_2D^+$ and $D_2H^+$ species in the slit jet discharge spectrometer. State-of-the-art non-Born-Oppenheimer theoretical predictions are used simultaneously to expedite high resolution searches for overtone stretching/bending rovibrational $2\nu_2$, $2\nu_3$, and $\nu_2 + \nu_3$ transitions in both $H_2D^+$ and $D_2H^+$. Measured line positions and intensities are then compared with theoretical predictions, permitting a rigorous evaluation of such non-Born-Oppenheimer models at high internal energies and as a systematic function of isotopomeric composition.

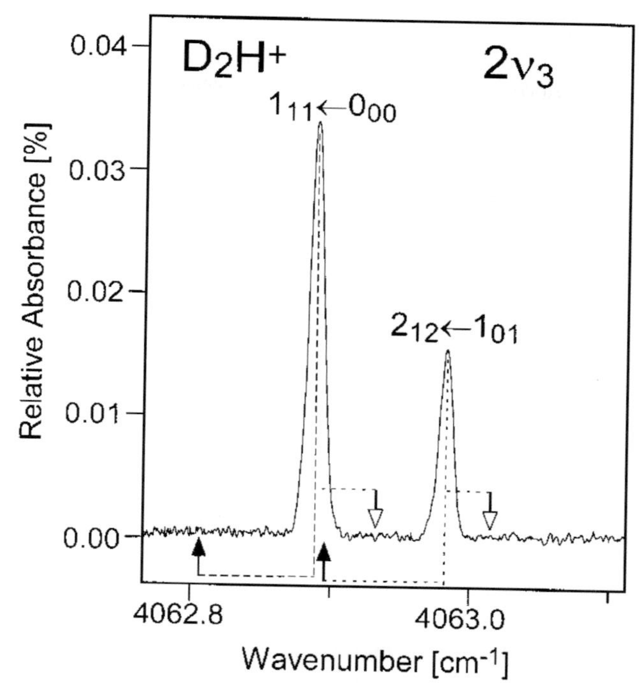


Fig. 5.22. An experimental scan over two lines corresponding to $2\nu_3$ $1_{11} \leftarrow 0_{00}$ and $2_{12} \leftarrow 1_{01}$ transitions in $D_2H^+$ ion. Full and open arrows indicate theoretical predictions based on semi-empirical and *ab initio* non-Born-Oppenheimer model treatments, respectively.

The data indicate an overall remarkable level of agreement between theory and experimental results, with maximal discrepancies of only $\approx 0.2\,\mathrm{cm}^{-1}$ for the semi-empirical and less than $0.1\,\mathrm{cm}^{-1}$ for the *ab initio* predictions. In fact, the majority of the predicted frequencies are within $0.04\ \mathrm{cm}^{-1}$ of the experimental data. By way of example, sample data and theoretical predictions are shown in Fig. 5.22 for two overtone transitions in the $2\nu_3$ manifold of $D_2H^+$, where the arrows below (above) the zero absorption frequency axis refer to the semi-empirical (*ab initio*) set of predictions, respectively. Indeed, the second set of fully *ab initio* predictions show roughly an additional order of magnitude increase in accuracy, with a typical prediction error of $\approx 0.05\,\mathrm{cm}^{-1}$ and in line with previous studies. Although still significantly outside the range of experimental uncertainties ($\approx 0.001\ \mathrm{cm}^{-1}$), these results should be compared with intrinsic non-Born-Oppenheimer correction terms on the order of several $\mathrm{cm}^{-1}$.[156] This confirms that the non-Born-Oppenheimer models employed are achieving near spectroscopic levels of accuracy, even for overtone transitions in the asymmetrically substituted isotopomers. The information gained in these experiments can be used to further refine theoretical models of non-Born-Oppenheimer effects.

## 5.15. Protonated Water: Isotopically Mediated Tunneling Dynamics in Hydronium Ion

As the strongest acid possible in an aqueous environment, hydronium ($H_3O^+$), is arguably the most important ion in all of chemistry and biology. Consequently,it has been the focus of extensive experimental and theoretical attention. Also, models of isotopically substituted deuterium chemistry play a very important role in interpreting the chemical and physical processes occurring in interstellar clouds (see chapter 1 by Herbst and Millar). As a consequence, probes of the total deuterium/hydrogen ratio provide excellent quantitative insight into cosmochemistry in the galaxies, for which the spectroscopic study of deuterated isotopomers in the laboratory are an essential prerequisite. Furthermore, from a dynamics perspective, such 4-atom systems represent the current state-of-the-art challenge for theory and experiment, due to unusually facile quantum motion along the umbrella inversion coordinate. The systematic mass-dependent tunneling dynamics of $H_3O^+$, $H_2DO^+$, $HD_2O^+$, and $D_3O^+$ represent a systematic "tuning" of effective tunneling mass between fully protonated and deuterated species. The dramatic effects on tunneling splittings can be used to map out the intramolecular bending energy surface for this prototypically important species and further provide stringent tests of the full 6D potential energy surface. For example, in $H_2DO^+$, this "simple" H/D substitution localizes the remaining OD stretch, rotates the symmetry axis of the planar transition state by 90°, makes hybrid OH stretch bands IR allowed, and therefore permits the upper and lower state tunneling splittings to be obtained from a single rovibrational spectrum, as illustrated in the top panel of Fig. 5.23.

The fundamental challenge in isotopomer studies for such a light, anharmonic and rapidly tunneling hydride is that the spectra extend over hundreds of $cm^{-1}$, even at low slit jet discharge temperatures. This represents a major scanning effort at high resolution. Furthermore, the large amplitude, anharmonic and likely Coriolis perturbed nature of these vibrations requires precise four line combination differences ($\approx$10 MHz) for definitive spectral assignment, opportunities for which are limited at low jet temperatures and greatly aided by reliable predictions of band origins. Of crucial importance to these studies, therefore, are the recent theoretical advances by

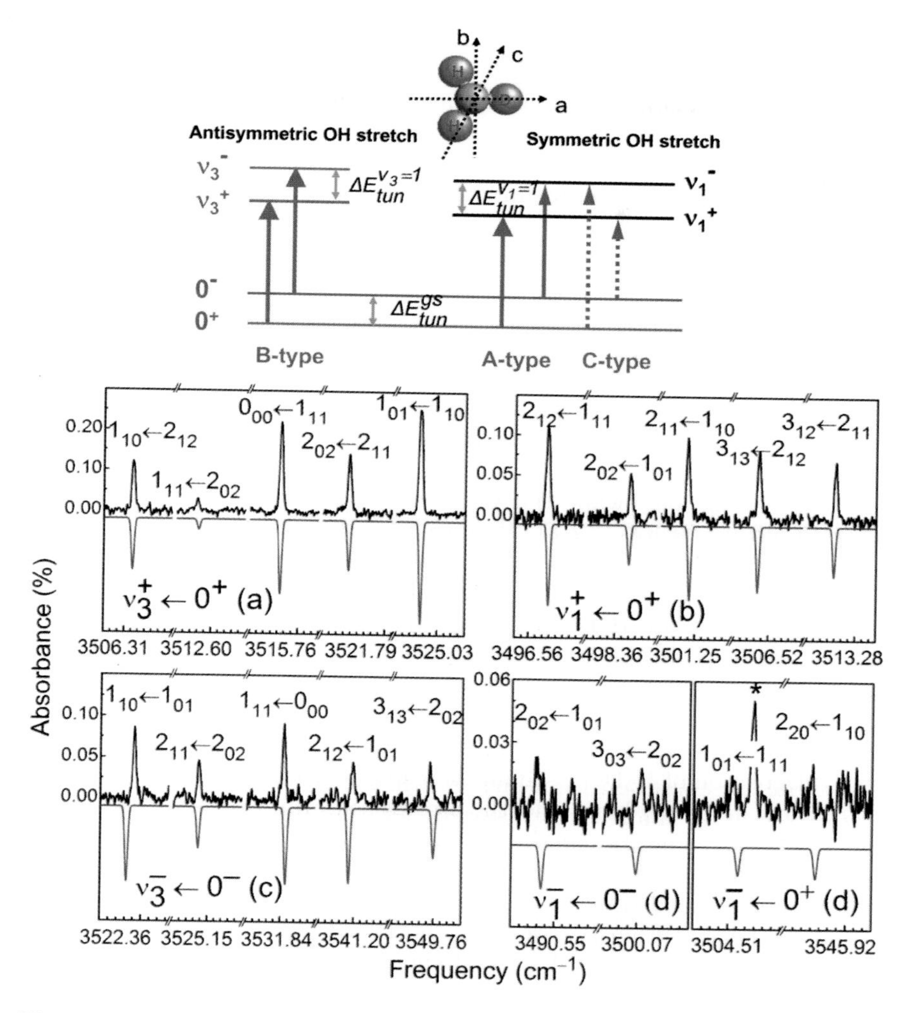


Fig. 5.23. Top panel: Antisymmetric ($\nu_3$) and symmetric ($\nu_1$) OH stretch transitions in singly deuterated $H_2DO^+$. In addition to the four allowed tunneling bands (dark lines), two weaker tunneling bands in the $\nu_1$ manifold ($\nu_1^+ \leftarrow 0^-$ and $\nu_1^- \leftarrow 0^+$) acquire oscillator strength due to the lowering of the transition state symmetry from $D_{3h}$ to $C_{2v}$. In conjunction with the allowed transitions, these additional bands cross the tunneling gap and permit spectroscopic determination of tunneling splittings in both ground and excited states. Bottom panel: Sample data for the five bands in $\nu_1$ and $\nu_3$ of $H_2DO^+$, along with the simulated spectra, assigned by four-line combination differences and nuclear spin statistics. The small discrepancies between experimental and simulated transitions are real and reflect weak local perturbations and/or breakdown in the rigid rotor Hamiltonian due to large amplitude motion. The line marked with an asterisk is the $1_{10} \leftarrow 1_{01}$ rotational transition in $\nu_3^+ \leftarrow 0^+$ band due to $HD_2O^+$ "contaminant."

Bowman *et al.*,[157] and Halonen *et al.*, [158] based on high level ab initio development of the full 6D potential surface for $H_3O^+$. First principles agreement with experimentally known band origins and tunneling splittings for $H_3O^+$ and $D_3O^+$ looks extremely promising ($\approx$14 cm$^{-1}$ and $\approx$1.8 cm$^{-1}$ for the band origins deviation, $\approx$11 cm$^{-1}$ and 0.8 cm$^{-1}$ deviation for the tunneling splittings, respectively), with similar accuracies anticipated for mixed isotopomer predictions.

Even with concentration modulation which effectively discriminates the ions from the neutrals, the interference due to other molecular ions can also be a big concern since the predicted signal-to-noise at the peak absorbance is only $\approx$25. Therefore, a thoughtful experimental strategy has to be designed in order to obtain all of the high-resolution spectra for both $H_2DO^+$ and $HD_2O^+$. $HD_2O^+$ is clearly the best first choice, since there is only one OH stretch vibration and it can also be generated relatively purely in a $D_2O/H_2$ mixture discharge (no $H_2DO^+$ will be produced on the time scale of jet discharge expansion), while $H_2DO^+$ can be studied as the second target after all of the features of $HD_2O^+$ have been assigned.

Based on a simple dipole bond model at the equilibrium geometry and 40 K thermal population of excited tunneling states, the six bands of OH stretches in $H_2DO^+$ are predicted to exhibit intensity ratios for $\langle\nu_3^+|\mu_b|0^+\rangle$, $\langle\nu_3^-|\mu_b|0^-\rangle$, $\langle\nu_1^+|\mu_a|0^+\rangle$, $\langle\nu_1^-|\mu_a|0^-\rangle$, $\langle\nu_1^-|\mu_c|0^+\rangle$ and $\langle\nu_1^+|\mu_c|0^-\rangle$ of 100:20:30:7:10:2. Assuming a similar signal-to-noise ratio (S/N) for the strongest line as observed in $HD_2O^+$, the peak S/N values for each of these bands are estimated to be ~25, 5, 8, 2, 3, and 1, respectively. Detection of $H_2DO^+$ should therefore be achievable for both B-type transitions in the $\nu_3$ band and the stronger A-type transition out of the lower ($0^+$) tunneling state in the $\nu_1$ band. Detection of the remaining bands (at least two of which would be necessary to infer tunneling splittings) is also feasible, provided one has sufficiently accurate predictions for vibrational frequencies from high-level *ab initio* calculations to guide a high-resolution spectral search. Fig. 5.23 shows sample spectra for all of the five vibrational sub-bands of $H_2DO^+$ observed in both $\nu_3$ and $\nu_1$ bands. Three allowed bands ($\nu_3^+ \leftarrow 0^+$, $\nu_3^- \leftarrow 0^-$, and $\nu_1^+ \leftarrow 0^+$) have been observed to an experimental precision of ~0.0003 cm$^{-1}$, where +/− refer to wave function symmetry with respect to the planar tunneling transition state.

Equally crucial to spectroscopic determination of the tunneling splittings is the successful detection and assignment of two weaker bands $\nu_1^- \leftarrow 0^+$ and $\nu_1^- \leftarrow 0^-$, the first of which is forbidden in $H_3O^+$ and $D_3O^+$ but becomes weakly allowed in the partially deuterated species as a result of tunneling symmetry reduction from $D_{3h}$ to $C_{2v}$.

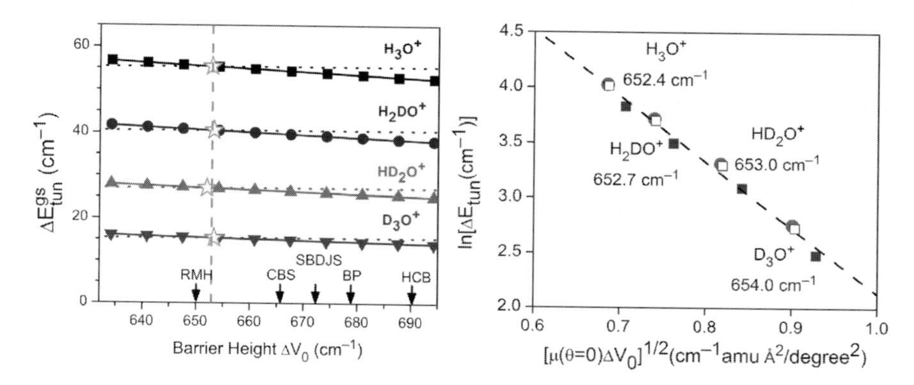


Fig. 5.24. Left panel: Tunneling splittings ($\Delta E^{gs}_{tun}$) for the four isotopically labeled hydronium ions in the ground state vs. scaled inversion barrier height ($\Delta V_0$). By matching the calculated splittings (solid symbol) to the observed tunneling splittings (dotted lines), the barrier height is experimentally estimated to be 652.9(6) $cm^{-1}$ (dashed line). For comparison, previous reports of *ab initio* and semiempirical barrier heights are also shown (arrows).[157–160] Right panel: Semiclassical WKB tunneling analysis for hydronium ion isotopomers, in excellent agreement with the predicted $\ln\{\Delta E_{tun}\} = A - \beta\{\mu(\theta = 0)\Delta V_0\}^{1/2}$ scaling. Open diamonds represent a WKB guided interpolation between RMH's (circle) and HCB's (square) predictions.[157,158] This alternative procedure yields a barrier height of $\Delta V_0 = 653.0(7)\,cm^{-1}$, strongly confirming the $\Delta V_0 = 652.9(6)\,cm^{-1}$ reported value derived from 1D vibrationally adiabatic analysis.

Similar to $H_3O^+$ and $D_3O^+$, the tunneling splittings of $HD_2O^+$ and $H_2DO^+$ in the excited state are decreased substantially more than those in the ground state. This can be explained both qualitatively and quantitatively by the evolution from $sp^2$ bonding in the transition state geometry to $sp^3$ bonding through the large inversion angle. Previous theoretical predictions of vibrational frequencies and tunneling splittings for the full series of isotopomers prove to be in remarkably good agreement with experiment (left panel in Fig. 5.24), though systematic discrepancies still exist. Given the reasonable vibrationally adiabatic approximation, an experimentally based estimated inversion barrier height of $\Delta V_0 = 652.9(6)\,cm^{-1}$ can be extracted according to the large amplitude Hamiltonian dynamical analysis of Rush and Wiberg.[161] The quantitative accuracy of this estimate is further confirmed by combining these experimental results with the full 6D calculation results of Halönen[158] and Bowman[157], exploiting semiclassical WKB ideas to construct a simple but robust interpolation metric (right panel in Fig. 5.24). The results from this WKB analysis yield a barrier height of $\Delta V_0 = 653.0(7)\,cm^{-1}$, in good agreement with vibrationally adiabatic analysis, and suggesting a precision in the full 6D Born-Oppenheimer barrier height of less than a few $cm^{-1}$.

## 5.16. Looking Toward the Future

The combination of high resolution tunable IR lasers, slit jet cooled supersonic expansions, quantum shot noise limited absorbance detection and pulsed discharge sources has provided a general and remarkably powerful tool for spectroscopic study of cold clusters as well as highly reactive radicals and molecular ions. Yet for all that has been done, this seems just the beginning, particularly for the more recent studies of radicals and molecular ions.

As one ongoing example, we are making significant efforts toward understanding the high resolution spectroscopy of protonated methane, $CH_5^+$, which has been speculated as an extremely abundant molecular ion in the interstellar medium.[131,132] This system has long represented a major challenge both experimentally and theoretically, due to "fluxional" large amplitude quantum motion over 120 equivalent minima on the potential energy surface. This is an extremely challenging system, which has been the focus of intense theoretical and experimental effort over the last several decades.[162–165] What makes this species both intriguing and elusive for spectroscopists and dynamicists alike is the extreme quantum nature of its intramolecular motion, which precludes any conventional notion of "molecular structure." The minimum energy structure ($C_s$ symmetry) is in fact best characterized as $CH_3^+$ strongly bound to an $H_2$ molecule, but with relatively low "torsion" (30 $cm^{-1}$) and "flip" (300 $cm^{-1}$) barriers between 120 equivalent minima leading to facile zero point permutation of all 5 protons in a highly fluxional, pseudo-rotational motion. This large amplitude motion has made $CH_5^+$ exceedingly difficult to analyze by high-resolution spectroscopy under conventional discharge conditions. Indeed, velocity modulation spectra thought to be largely from $CH_5^+$ in the CH stretch region have been obtained in heroic efforts by Oka and coworkers,[164] but which have still eluded rovibrational assignment.

A recent breakthrough for us has been to obtain the first high resolution $CH_5^+$ spectra under *jet-cooled* conditions in the slit discharge spectrometer.[131,132] The $CH_5^+$ is cleanly formed by 50 KHz modulated proton transfer from $H_3^+$ to $CH_4$ in the 1 mm discharge region, and promptly cooled by supersonic expansion prior to synchronous detection by direct IR absorption in the CH stretch region. The $CH_5^+$ spectral carrier identity has been unambiguously determined by chemical titration methods and preliminary rotational assignments have been made from precise 4-line combination differences. By way of theoretical support, collaborative quantum/classical calculational efforts with Bowman and McCoy[131,166,167] have been performed using full-12D *ab initio* potential energy and dipole moment surfaces for $CH_5^+$. The resulting spectral predictions indicate excellent qualitative agreement with the high resolution jet-cooled CH stretch spectra and argue for a tentative assignment of the band origin to an asymmetric CH stretching motion of the transient $CH_3^+$ subunit. However, clearly a great deal of theoretical and experimental effort will be necessary for complete rotational assignment and resolution of this long standing spectroscopic puzzle.

## Acknowledgements

This work has been supported by a grant from the Department of Energy. Additional funding from the National Science Foundation has been crucial in development of the IR laser spectrometer, as well as support from Air Force Office of Scientific Research for refinement of the pulsed slit supersonic expansion sources. Finally, we would also like to thank the many Nesbitt group members, past and present, who have contributed to the work in this chapter, with special gratitude to Dr. Chris Lovejoy, Dr. Andrew McIlroy, Dr. John T. Farrell, Dr. Erin S. Whitney, Dr. Chandra Savage, Professor Michal Farnik, and Professor David Anderson. We have all enjoyed the joys and challenges of high resolution spectroscopy, hopefully some small fraction of which has been captured herein.

## References

1. Aziz RA. (1993) *J. Chem. Phys.* **99**: 4518.
2. Hutson JM. (1992) *J. Chem. Phys.* **96**: 6752.
3. Quack M, Suhm MA. (1991) *J. Chem. Phys.* **95**: 28.
4. Beswick JA, Jortner J. (1981) *Adv. Chem. Phys.* **47**: 363.
5. Maitland GC, Rigby M, Smith BE, Wakeham W. (1987) *Intermolecular Forces.* Clarendon, Oxford.
6. Bemish RJ, Wu M, Miller RE. (1994) *Faraday Disc.* **97**: 57.
7. Nesbitt DJ. (1988) *Chem. Rev.* **88**: 843.
8. Nesbitt DJ. (1994) *Ann. Rev. Phys. Chem.* **45**: 367.
9. Ernesti A, Hutson J. (1995) *Phys. Rev.* **51**: 239.
10. Ernesti A, Hutson J. (1997) *J. Chem. Phys.* **106**: 6288.
11. Ernesti A, Hutson JM. (1994) *Faraday Disc. Chem. Soc.* **97**: 119.
12. Farrell JT, Davis S, Nesbitt DJ. (1995) *J. Chem. Phys.* **103**: 2395.
13. Farrell JT, Nesbitt DJ. (1996) *J. Chem. Phys.*
14. Farrell JT, Sneh O, McIlroy A, Knight AEW, Nesbitt DJ. (1992) *J. Chem. Phys.* **97**: 7967.
15. Gutowsky HS, Chuang C, Klots TD, Emilsson T, Ruoff RS, Krause KR. (1988) *J. Chem. Phys.* **88**: 2919.
16. Gutowsky HS, Klots TD, Chuang C, Keen JD, Schmuttenmaer CA, Emilsson T. (1985) *J. Am. Chem. Soc.* **107**: 7174.

17. Gutowsky HS, Klots TD, Chuang C, Schmuttenmaer CA, Emilsson T. (1985) *J. Chem. Phys.* **83**: 4817.
18. Gutowsky HS, Klots TD, Chuang C, Schmuttenmaer CA, Emilsson T. (1987) *J. Chem. Phys.* **86**: 569.
19. Hutson JM. (1989) *J. Chem. Phys.* **91**: 4448.
20. Hutson JM. (1992) *J. Chem. Phys.* **96**: 4237.
21. McIlroy A, Lascola R, Lovejoy CM, Nesbitt DJ. (1991) *J. Chem. Phys.* **95**.
22. McIlroy A, Nesbitt DJ. (1992) *J. Chem. Phys.* **97**: 6044.
23. Nesbitt DJ. (1994) *Reaction Dynamics in Clusters and Condensed Phases.* Kluwer Academic Publishers, Netherlands.
24. Suhm MA, Nesbitt DJ. (1995) *Chem. Soc. Rev.* **24**: 45.
25. Anderson DT, Davis S, Nesbitt DJ. (1996) *J. Chem. Phys.* **105**: 4488.
26. Roy RJL, Hutson JM. (1987) *J. Chem. Phys.* **86**: 837.
27. Cohen R, Saykally C, Saykally RJ. (1993) *J. Chem. Phys.* **98**: 6007.
28. Schmuttenmaer CA, Cohen RC, Saykally RJ. (1994) *J. Chem. Phys.* **101**: 146.
29. Gutowsky HS, Klots TD, Chuang C, Keen JD, Schmuttenmaer CA, Emilsson T. (1987) *J. Am. Chem. Soc.* **109**: 5633.
30. Gutowsky HS, Klots TD, Dykstra CE. (1990) *J. Chem. Phys.* **93**: 6216.
31. Ruoff RS, Emilsson T, Klots TD, Chuang C. (1988) *J. Chem. Phys.* **89**: 138.
32. Xu Y, L GC, Connelly JP, Howard BJ. (1993) *J. Chem. Phys.* **98**: 2735.
33. Elrod MJ, Loeser JG, Saykally RJ. (1993) *J. Chem. Phys.* **98**: 5352.
34. Elrod MJ, Steyert DW, Saykally RJ. (1991) *J. Chem. Phys.* **95**: 3182.
35. Elrod MJ, Steyert DW, Saykally RJ. (1991) *J. Chem. Phys.* **94**: 58.
36. Pugliano N, Saykally RJ. (1992) *Science* **257**: 1937.
37. Fraser GT, Pine AS, Lafferty WJ, Miller RE. (1987) *J. Chem. Phys.* **87**: 1502.
38. Jucks KW, Miller RE. (1988) *J. Chem. Phys.* **88**: 2196.
39. Suhm MA, Farrell JT, Ashworth S, Nesbitt DJ. (1993) *J. Chem. Phys.* **98**: 5985.
40. Drobits JC, Lester MI. (1987) *J. Chem. Phys.* **86**: 1662.
41. Hair SR, Cline JI, Bieler CR, Janda KC. (1989) *J. Chem. Phys.* **90**: 2935.
42. Kenny JE, Johnson KE, Sharfin W, Levy DH. (1980) *J. Chem. Phys.* **72**: 1109.
43. Levy DH. (1981) *Adv. Chem. Phys.* **47**: 323.
44. Sands WD, Beiler CR, Janda KC. (1991) *J. Chem. Phys.* **95**: 729.
45. Cooper A, Hutson J. (1993) *J. Chem. Phys.* **98**: 5337.
46. Elrod MJ, Saykally RJ. (1994) *Chem. Rev.* **94**: 1975.
47. Baiocchi FA, Dixon TA, Joyner CH, Klemperer W. (1981) *J. Chem. Phys.* **75**: 2041.
48. Buxton LW, Campbell EJ, Keenan MR, Balle TJ, Flygare WH. (1981) *Chem. Phys.* **54**: 173.
49. Harris SJ, Novick SE, Klemperer W. (1974) *J. Chem. Phys.* **60**: 3208.
50. Keenan MR, Buxton LW, Campbell EJ, Flygare WH. (1981) *J. Chem. Phys.* **74**: 2133.

51. Keenan MR, CEJ, Balle TJ, Buxton LW, Minton TK, Soper PD, Flygare WH. (1980) *J. Chem. Phys.* **72**: 3070.
52. Novick SE, Davies P, Harris SJ, Klemperer W. (1973) *J. Chem. Phys.* **59**: 2273.
53. Dvorak MA, Reeve SW, Burns WA, Grushow A, Leopold KR. (1991) *Chem. Phys. Lett.* **185**: 399.
54. Lovejoy CM, Nesbitt DJ. (1990) *J. Chem. Phys.* **93**: 5387.
55. Fraser GT, Pine AS. (1986) *J. Chem. Phys.* **85**: 2502.
56. Howard BJ, Pine AS. (1985) *Chem. Phys. Lett.* **122**: 1.
57. Chang H-C, Klemperer W. (1993) *J. Chem. Phys.* **98**: 9266.
58. Dixon TA, Hoyner CH, Baiocchi FA, Klemperer W. (1981) *J. Chem. Phys.* **74**: 6539.
59. Huang ZS, Jucks KW, Miller RE. (1986) *J. Chem. Phys.* **85**: 6905.
60. Lovejoy CM, Hutson J, Nesbitt DJ. (1992) *J. Chem. Phys.* **97**: 8009.
61. Lovejoy CM, Nesbitt DJ. (1989) *J. Chem. Phys.* **91**: 2790.
62. Lovejoy CM, Schuder MD, Nesbitt DJ. (1986) *J. Chem. Phys.* **85**: 4890.
63. Green S, Hutson J. (1994) *J. Chem. Phys.* **100**.
64. Chapman WB, Weida MJ, Nesbitt DJ. (1997) *J. Chem. Phys.* **106**: 2248.
65. Aziz RA. (1989) *Inert Gases.* Springer Verlag, Berlin.
66. Lewerenz M. (1996) *J. Chem. Phys.* **104**: 1028.
67. Howard BJ. (1987) *Structure and Dynamics of Weakly Bound Molecular Complexes.* Reidel, Dordrecht.
68. Liu S, Bacic Z, Moskowitz JW, Schmidt KE. (1994) *J. Chem. Phys.* **100**: 7166.
69. Buckingham AD. (1960) *Trans. Faraday Soc.* **56**: 753.
70. Lovejoy CM. (1991) *Intramolecular Dynamics of Van der Waals Complexes as Elucidated by Infrared Spectroscopy in Slit Supersonic Expansions.* University of Colorado, Boulder.
71. Anderson DT, Winn JS. (1994) *Chem. Phys.* **189**: 171.
72. Guelachvili G. (1976) *Opt. Comm.* **19**: 150.
73. Anderson DT, Davis S, Nesbitt DJ. (1996) *J. Chem. Phys.* **104**: 6225.
74. Davis S, Anderson DT, Nesbitt DJ. (1996) *J. Chem. Phys.* **104**: 8197.
75. Block PA, Miller RE. (1994) *Chem. Phys. Lett.* **226**: 317.
76. Deleon RL, Muenter JS. (1984) *J. Chem. Phys.* **80**: 6092.
77. Pine AS, Lafferty WJ, Howard BJ. (1984) *J. Chem. Phys.* **81**: 2939.
78. Bohac EJ, Marshall MD, Miller RE. (1992) *J. Chem. Phys.* **96**: 6681.
79. Nesbitt DJ. (1991) *Mode Selective Chemistry (proceedings, 24th Jerusalem Symposium).*
80. Fraser GT. (1991) *Int. Rev. Phys. Chem.* **10**: 189.
81. Loeser JG, Schmuttenmaer CA, Cohen R, Elrod MJ, Steyert DW, Bumgarner RE, Blake GA. (1992) *J. Chem. Phys.* **97**: 4727.
82. Schuder MD, Nelson JDD, Nesbitt DJ. (1993) *J. Chem. Phys.* **99**: 5045.
83. Schuder MD, Lovejoy CM, Lascola R, Nesbitt DJ. (1993) *J. Chem. Phys.* **99**: 4346.
84. Kofranek M, Lischka H, Karpfen A. (1988) *J. Chem. Phys.* **121**: 137.
85. Dayton DC, Jucks KW, Miller RE. (1989) *J. Chem. Phys.* **121**: 137.

86. Miller RE. (1990) *Acc. Chem. Res.* **23**: 10.
87. Gregory JK, Clary DC. (1995) *Chem. Phys. Lett.* **237**: 39.
88. Jensen P, Bunker PR, Karpfen A, Kofranek M, Lischka H. (1990) *J. Chem. Phys.* **93**: 6266.
89. Necoechea WC, Truhlar DG. (1994) *Chem. Phys. Lett.* **224**: 297.
90. Necoechea WC, Truhlar DG. (1994) *Chem. Phys. Lett.* **231**: 125.
91. Quack M, Suhm MA. (1995) *Chem. Phys. Lett.* **234**: 71.
92. Zhang DH, Wu Q, Zhang JZH. (1995). *J. Chem. Phys.* **102**: 124.
93. Zhang DH, Wu Q, Zhang JZH, Dirke Mv, Bacic Z. (1995) *J. Chem. Phys.* **102**: 2315.
94. Zhang DH, Zhang JZH. (1993) *J. Chem. Phys.* **99**: 6624.
95. Zhang DH, Zhang JZH. (1993) *J. Chem. Phys.* **98**: 5978.
96. Althorpe SC, Clary DC, Bunker PR. (1991) *Chem. Phys. Lett.* **187**: 345.
97. Bunker PR, Carrington JT, Gomez PC, Marshall MD, Kofranek M, Lischka H, Karpen A. (1989) *J. Chem. Phys.* **91**: 5154.
98. Hancock GC, Truhlar DG. (1989) *J. Chem. Phys.* **90**: 3498.
99. Marshall MD, Jensen P, Bunker PR. (1991) *Chem. Phys. Lett.* **176**: 255.
100. Mills IM. (1984) *J. Chem. Phys.* **88**: 532.
101. Quack M, Suhm MA. (1991) *Chem. Phys. Lett.* **183**: 187.
102. Sun H, Watts RO. (1990) *J. Chem. Phys.* **92**: 603.
103. Lovejoy CM, Nesbitt DJ. (1987) *J. Chem. Phys.* **86**: 3151.
104. Bohac EJ, Miller RE. (1993) *J. Chem. Phys.* **99**: 1537.
105. Fraser GT. (1989) *J. Chem. Phys.* **90**: 2097.
106. Schinke R. (1993) *Photodissociation Dynamics.* Press Syndicate of the University of Cambridge, Cambridge.
107. Dirke Mv, Bacic Z, Zhang DH, Zhang JZH. (1995) *J. Chem. Phys.* **102**: 4382.
108. Puttkamer Kv, Quack M, Suhm MA. (1989) *Infrared Phys.* **29**: 535.
109. Puttkamer Kv, Quack M. (1987) *Mol. Phys.* **62**: 1047.
110. Puttkamer Kv, Quack M, Suhm MA. (1988) *Mol. Phys.* **65**: 1025.
111. Miller JA, Kee RJ, Westbrook CK. (1990) *Ann. Rev. Phys. Chem.* **41**: 345.
112. Olah GA, Molnar A. (1995) *Hydrocarbon Chemistry.* John Wiley & Sons, Inc., New York.
113. Patten TE, Matyjaszewski K. (1998) *Adv. Mat.* **10**: 901.
114. Engelking PC. (1986) *Rev. Sci. Instr.* **57**.
115. Comer KR, Foster SC. (1993) *Chem. Phys. Lett.* **202**: 216.
116. Hilpert G, Linnartz H, Havenith M, ter Meulen JJ, Meerts WL. (1994) *Chem. Phys. Lett.* **219**: 384.
117. Xu Y, Fukushima M, Amano v, McKellar ARW. (1995) *Chem. Phys. Lett.* **242**: 126.
118. Anderson DT, Davis S, Zwier TS, Nesbitt DJ. (1996) *Chem. Phys. Lett.* **258**: 207.
119. Davis S, Farnik M, Uy D, Nesbitt DJ. (2001) *Chem. Phys. Lett.* **344**: 23.
120. Davis S, Anderson DT, Duxbury G, Nesbitt DJ. (1997) *J. Chem. Phys.* **107**: 5661.
121. Uy D, Davis S, Nesbitt DJ. (1998) *J. Chem. Phys.* **109**: 7793.

122. Davis S, Uy D, Nesbitt DJ. (2000) *J. Chem. Phys.* **112**: 1823.
123. Dong F, Davis S, Nesbitt DJ. (2006) *J. Phys. Chem. A* **110**: 3059.
124. Whitney ES, Feng D, Nesbitt DJ. (2006) *J. Chem. Phys.* **125**: 054304.
125. Whitney ES, Häber T, Schuder MD, Blair AC, Nesbitt DJ. (2006) *J. Chem. Phys.* **125**.
126. Häber T, Blair AC, Nesbitt DJ, Schuder MD. (2006) *J. Chem. Phys.* **124**: 054316.
127. Dong F, Roberts M, Nesbitt DJ. (2007).
128. Farnik M, Davis S, Kostin MA, Polyansky OL, Tennyson J, Nesbitt DJ. (2002) *J. Chem. Phys.* **116**: 6146.
129. Dong F, Uy D, Davis S, Child M, Nesbitt DJ. (2005) *J. Chem. Phys.* **122**: 224301.
130. Dong F, Nesbitt DJ. (2006) *J. Chem. Phys.* **125**: 144311.
131. Huang XC, McCoy AB, Bowman JM, Johnson LM, Savage C, Dong F, Nesbitt DJ. (2006) *Science* **311**: 60.
132. Savage C, Dong F, Nesbitt DJ. (2007) manuscript in preparation.
133. Fessenden RW, Schuler RH. (1963) *J. Chem. Phys.* **39**: 2147.
134. Jackel GS, Gordy W. (1968) *Physical Review* **176**: 443.
135. McConnel HM. (1955) *J. Chem. Phys.* **24**: 764.
136. Endo Y, Hamada C, Saito S, Hirota E. (1983) *J. Chem. Phys.* **79**: 1605.
137. Yamada C, Hirota E. (1986) *J. Mol. Spectrosc.* **116**: 101.
138. Sears TJ, Johnson PM, Jin P, Oatis S. (1996) *J. Chem. Phys.* **104**: 781.
139. McManus HJ, Fessenden RW, Chipman DM. (1988) *J. Phys. Chem.* **92**: 3778.
140. Vajda E, Tremmel J, Rozsondai B, Hargittai I, Mal'tsev AK, Kagramanov N, Nefedov OM. (1986) *J. Am. Chem. Soc.* **108**: 4352.
141. Holtzhauer K, Cometta-Morini C, Oth JFM. (1990) *J. Phys. Org. Chem.* **3**: 219.
142. Minsek DW, Chen P. (1993) *J. Phys. Chem.* **97**: 13375.
143. Liu X, Getty JD, Kelly PB. (1993) *J. Chem. Phys.* **99**: 1522.
144. Hirota E, Yamada C, Okunishi M. (1992) *J. Chem. Phys.* **97**: 2963.
145. Kasai PH. (1972) *J. Am. Chem. Soc.* **94**: 5950.
146. Kanamori H, Endo Y, Hirota E. (1990) *J. Chem. Phys.* **92**: 197.
147. Tanaka K, Toshimitsu M, Harada K, Tanaka T. (2004) *J. Chem. Phys.* **120**: 3604.
148. Lahankar SA, Chambreau SD, Townsend D, Suits F, Farnum J, Zhang XB, Bowman JM, Suits AG. (2006) *J. Chem. Phys.* **125**.
149. Townsend D, Lahankar SA, Lee SK, Chambreau SD, Suits AG, Zhang X, Rheinecker J, Harding LB, Bowman JM. (2004) *Science* **306**: 1158.
150. Houston PL, Kable SH. (2006) *Proc. Natl. Acad. Sci. U.S.A.* **103**: 16079.
151. Saykally RJ. (1988) *Science* **239**: 157.
152. Oka T. (1980) *Phys. Rev. Lett.* **45**: 531.
153. Dinelli BM, Polyansky OL, Tennyson J. (1995) *J. Chem. Phys.* **103**: 10433.
154. Polyansky OL, Dinelli BM, Lesueur CR, Tennyson J. (1995) *J. Chem. Phys.* **102**: 9322.
155. Polyansky OL, Tennyson J. (1999) *J. Chem. Phys.* **110**: 5056.

156. Dinelli BM, Polyansky OL, Tennyson J. (1995) *J. Chem. Phys.* **103**: 10433.
157. Huang XC, Carter S, Bowman JM. (2002) *J. Phys. Chem. B* **106**: 8182.
158. Rajamäki T, Miani A, Halonen L. (2003) *J. Chem. Phys.* **118**: 10929.
159. Botschwina P. (1986) *J. Chem. Phys.* **84**: 6523.
160. Sears TJ, Bunker PR, Davies PB, Johnson SA, Spirko V. (1985) *J. Chem. Phys.* **83**: 2676.
161. Rush DJ, Wiberg KB. (1997) *J. Phys. Chem. A* **101**: 3143.
162. Boo DW, Lee YT. (1993) *Chem. Phys. Lett.* **211**: 358.
163. Bunker PR. (1996) *J. Mol. Spec.* **176**: 297.
164. White ET, Tang J, Oka T. (1999) *Science* **284**: 135.
165. Asvany O, Kumar P, Redlich B, Hegemann I, Schlemmer S, Marx D. (2005) *Science* **309**: 1219.
166. Huang XC, Johnson LM, Bowman JM, McCoy AB. (2006) *J. Am. Chem. Soc.* **128**: 3478.
167. McCoy AB, Braams BJ, Brown A, Huang XC, Jin Z, Bowman JM. (2004) *J. Phys. Chem. A* **108**: 4991.

# Chapter 6

# The Production and Study of Ultra-Cold Molecular Ions

D. Gerlich

*Faculty of Natural Science, Technical University,*
*09107 Chemnitz, Germany*
*gerlich@physik.tu-chemnitz.de*

## Contents



## 6.1. Introduction

### 6.1.1. *Cold Molecules*

The late 20th century has seen significant experimental advances in cooling and trapping neutral atoms culminating in the successful formation of

Bose-Einstein condensates. Following the exciting progress with cold atoms the experimental challenge became to prepare ultracold molecules. Unfortunately the laser-cooling techniques applied to atoms are not directly transferable to molecules, since these methods rely upon cyclic absorption and spontaneous emission within an almost pure two-level system. The multilevel structure of the electronic states of molecules and the manifold of allowed transitions does not allow for simple iterative excitation with one single laser line.

With the invention of optical dipole traps and optical lattices for storing cold atoms, the association of two atoms opened up a field called *ultracold molecules* (see Chapter 9 by van de Meerakker *et al.*). Stimulating the association of two colliding atoms by a laser, cold molecules have been produced in atomic traps. In most cases, however, these molecules are just homonuclear diatomics made from alkaline metal atoms which can easily achieve temperatures in the micro-Kelvin range due to the sub-Doppler cooling methods. Some efforts to extend these association techniques to heteronuclear molecules have been successful. However, besides being restricted to a few specific atoms, there is the problem that these molecules are only translationally cold. They have been formed from two atoms, which have been nearly at rest before they attracted each other; however the result is a more or less excited molecule. In order to get cold molecules for applications in chemical processes, it is not sufficient to lower the translational motion, one has to freeze also the internal degrees of freedom. Therefore it is necessary to distinguish between *slow* and *cold*. For some diatom systems, specific laser based strategies have been developed to transfer the slow molecules to low-lying vibrational rotational states, may be even to the ground state.

As described in other chapters of this book there are other strategies to produce ultracold *neutral* molecules, e.g. to decelerate polar molecules in pulsed electrostatic fields (see Chapter 9). Combining this method with suitable traps one can create and study ensembles of the molecules. Experimental advances have made it possible to cool selected molecules in well-defined states to temperatures below 1 K and to trap them. Besides deceleration buffer gas cooling in cold $^3$He environments has also been utilized for loading magnetic traps with paramagnetic molecules.[1] Although cooling and manipulating neutral molecules have become the subject of intense studies in recent years and although many research groups have contributed to this field, the activities with neutrals are still restricted to a small number of specific molecules. It is obvious that one needs more general strategies for contributing to the new interdisciplinary field of molecular matter at very low temperatures ($T < 1$ K). In order to understand *ultracold chemistry,* e.g. needed for modeling dense interstellar clouds or for deposition of ultracold molecules on surfaces, one has to study the quantum-mechanical details of the cold chemical systems including large polyatomic molecules, clusters, and cold nanoparticles. With charged particles and various cooling schemes, this is possible.

In contrast to neutrals, charged objects are much easier to trap because of their strong interactions with electromagnetic fields. As a consequence a wide variety of experimental set-ups for confining ions and charged particles has been developed ranging from small ion trap devices to large storage rings. One of the obvious advantages of deep potential wells is that it is possible to first trap externally created ions and to cool them afterwards. A nice review of cooling methods in traps has been given by Itano *et al.*[2] As can be seen from that summary and as discussed below, the main driving force was and still is to increase the precision and accuracy in spectroscopy and metrology. Besides laser based cooling methods there are other quite general schemes to cool a finite ensemble of charged particles such as evaporation, optical pumping or chemically removing energetic ions. In addition there are interesting methods such as resistive cooling or active-feedback cooling.[3] These methods make use of the currents induced by the motions of the charges in suitable electrodes. Note that, via the same coupling, parasitic voltages also can heat the motion of trapped ions.

Based on extremely sensitive detection schemes, sophisticated experiments on single objects became possible. Interesting results include the observation of quantum jumps in isolated atomic ions, ultrahigh resolution mass spectrometry on clusters or unstable nuclei, and extremely long time studies on single nanoparticles.[4] Confinement of just one charge in a trap avoids space charge problems, which have to be accounted for if an ensemble of ions is to be cooled to low temperatures. An alternative strategy which has been discovered many decades ago and which has gained more and more attention in the last decade, is to work in a regime where space charge effects dominate.[5] Progress in laser cooling trapped ions has made it possible to lower the *temperature* so far (typically a few mK; see below concerning the definition of *temperature*) that spatially ordered structures form. These so-called Coulomb crystals of ions have been observed in Penning and Paul traps.[6] In such crystalline arrangements, which can be stored for hours, single specific ions can be addressed and manipulated. Strings of well-localized ions have become very attractive physical systems for studying multi-particle entanglement as well as for quantum computer research. For cooling of molecular ions, the subject of this chapter, the process of sympathetic cooling is of significant importance. As will be discussed in detail below, the motion of almost any molecular ion can be reduced below 100 mK in ion traps through Coulomb interactions with laser-cooled atomic ions.

### 6.1.2. *Interaction with Radiation*

A general problem in creating cold molecules is whether or how efficient the internal degrees of freedom couple to the cooling process. In the case of sympathetically cooled molecular ions, the vibrational or rotational motion is largely unaffected by the Coulomb interactions as explained below.[7,8] If molecules are not exposed to another interaction, long enough trapping finally equilibrates internal excitation with the black body radiation penetrating the ion cloud. An example is the rotational temperature of the polar molecules CH, $CH^+$, and CN in dense cold interstellar clouds. As discussed in the introduction of Chapter 3, the observed population of the rotational states indicates that rotation is in equilibrium with the 2.7 K temperature of the cosmic background radiation. As a consequence of the permanent interaction with the black body radiation, innovative instruments, which aim at reaching temperatures of 1 K or below, either need an efficient cooling mechanism or the ion environment has to be operated at temperatures of a few K. With small traps this is rather easy to realize while operating a storage ring at low temperatures is a challenge.[9]

Long time storage in a well-controlled environment provides ideal conditions for studying the interaction of the trapped objects with electromagnetic radiation. Several of the fundamental aspects of light-atom interactions such as spectroscopy at the limits and applications to novel frequency standards are meanwhile also extended to molecules, e.g. $HD^+$.[10] One of the obvious conditions for obtaining high resolution is to reduce the Doppler effect. Orienting the micro-motion in rf traps transverse to the laser beam reduces the first-order Doppler effect, while elimination of the second-order Doppler effect requires getting the ions at rest. In the case of rf-based traps this means that one has to reduce the oscillatory motion. One general solution for that is to work with multi-electrode traps. One nice related illustration is a lamp-based $Hg^+$ frequency standard which uses a linear rf 12-pole for ion trapping.[11] Due to the improvement of the kinetic energy distribution in this trap, the second order Doppler fractional frequency shift has been reduced to $2.4 \times 10^{-13}$ with a stability of $<1 \times 10^{-16}$.[11] If one works, however, with one single charged object, the harmonic potential of a quadrupole trap is the ideal solution for cooling the translational motion to the limits of the uncertainty principle. The extreme localization in space in such a system ($<1\,\mu$m) even makes it possible to detect one single ion in absorption.[12]

Traps are often used for accumulating ions, which are not so easy to produce, e.g. isotopes in nuclear physics applications. In the case of molecular ions this means that one may use also rather inefficient ways to create or to prepare them in specific states. So far photoionization schemes, which allow molecular ions in specific states to be obtained, have been used rather seldom for creating a trapped ensemble. Using photons with an energy just at the ionization limit of the neutral, ions can easily be formed in their ground state. There are several multi- or multiple-photon based schemes for preparing ions in specific states. The state selective preparation of $H_2^+$ and $CO^+$ via resonance enhanced multi-photon ionization is mentioned in Chapter 3 (Sec. 3.5.5). Another strategy for preparing ions in specific states is to use optical pumping schemes including coherent manipulation of internal degrees of freedom. A partly destructive method, similar to evaporation of fast atoms, is to remove ions in unwanted states, e.g. by using photo-induced processes followed by a chemical reaction or by photofragmentation. Since the trapped ensemble is finite, the state population of the remaining ions becomes a hole, provided it is isolated. The most efficient and universal method which is available today is to use first buffer gas cooling and to start with excitation via electromagnetic waves from a few low lying states. Work is in progress in several laboratories to transfer an ensemble of ions to their ground state in collisions with He. An interesting question is whether such a sample, e.g. $CO^+$, can be used to perform infrared or micro-wave absorption spectroscopy.

### 6.1.3. *Cooling Complex Systems*

For systems that are more complex than diatomic ions, buffer gas cooling is today the only general way to relax all degrees of freedom. Already simple molecular ions with a few atoms can have very complex energy levels because of internal isomerization, pseudo-rotation, tunneling, etc. In such situation, standard methods of spectroscopy often fail and it helps if one can record spectra of ions which have been thermalized to well-defined low temperatures. The need of thermalizing polyatomic ions to a given temperature has been emphasized by Wang *et al.*[13] who utilize high resolution photoelectron spectroscopy for studying anions. In their experiment vibrational cooling of $C_{60}^-$ anions has been achieved via collisions with a cold buffer gas (70 K) in a Paul trap, which has been attached to a cold head. Relative to spectra taken at room temperature, vibrational hot bands have been completely eliminated, yielding well-resolved vibrational structures and an accurate electron affinity for neutral $C_{60}$. Many more interesting results have been achieved within the past two years including spectra from several cold singly and doubly charged fullerenes.[14] As discussed below and in Chapter 3 of this book, an rf quadrupole trap can be used rather successfully for collisional relaxation instead of a higher order multipole, provided that the ions are much heavier than the neutral buffer gas; however, a Paul trap should not be used if one intends to cool the ions to the limit and if translational motion plays a role. A few selected examples for cooling complex molecular ions, e.g. protonated bio-relevant molecules are discussed below.

In recent years a lot of attention has been paid to the cation $CH_5^+$. Its potential energy surface is very flat near the various minima and the hydrogen atoms can change their location on a sub-ns time scale. This leads to very complex spectra especially if one does not cool the highly fluxional ion. As described by Davis *et al.*, in Chapter 5, molecular ions such as protonated methane can be formed and cooled in supersonic jets. In this way high-resolution absorption spectra of $CH_5^+$ have been recorded in the 3000 $cm^{-1}$ range.[15] Also a temperature variable 22-pole ion trap has been used successfully in combination with a free electron laser to record infrared spectra of $CH_5^+$.[16] In this experiment the applied chemical probing scheme was based on laser induced proton transfer to carbon dioxide requiring the trap to be operated at and above 110 K. Both experiments have provided important spectral information; however, it is not yet sufficient to predict rotational spectra of $CH_5^+$ with the accuracy, required for detecting this important ion in space. It has been pointed out by Bunker *et al.*[17] that the mm-wave spectrum of $CH_5^+$ is strongly affected by the large amplitude motion of the H atoms. From their simulated spectra, calculated at 300 and 77 K it is obvious that it is mandatory to cool the ions to much lower temperatures, especially if one intends to detect the $J = 1 \leftarrow J = 0$ absorption spectrum which is predicted to be in the 220–235 GHz region. In principle, there are experimental techniques available today to cool complex ions such as $CH_5^+$ with buffer gas to 1 K or even below. The basics of these techniques are mentioned below.

### 6.1.4. *Preview*

The following text concentrates on methods to cool molecular ions which are confined in traps. Concerning the use of inhomogeneous high frequency fields for guiding and trapping ions, this chapter is closely related to Chapter 3 in the center of which were experimental studies of collisions at low translational energies while here internal energies play a role. After a general introduction describing methods to prepare, detect and manipulate charged molecules in a trap, recent advances in laser based cooling methods are reviewed. Special emphasis is given to sympathetic cooling since this method can be applied to all kind of ions, as well as molecules. The main part deals with buffer gas cooling, i.e., inelastic collisions with a cold or slow non-reactive buffer gas, which is presently the most general scheme to cool all degrees of freedom of a complex molecule, a cluster or a nanoparticle. Selected instruments and results illustrate the wide range of applications of cold ion traps ranging from fine structure states of atomic ions via di- and tri-atomic ions to complex molecules of astrochemical and biological interest. So far, buffer gas cooling in ion traps has been limited to typically 10 K; however, a new experimental strategy is presented which will allow trapped ions to be cooled to temperatures below 1 K.

## 6.2. Cooling Ions in Traps

### 6.2.1. *Trapping, Probing, and Detecting*

As already mentioned in Chapter 3, charged particles are easy to manipulate with electric and magnetic fields. Therefore, in comparison to neutral molecules, it is much simpler to confine positive or negative ions in specific regions of space. However, ions are also very sensitive to stray fields and space charge, making it simultaneously problematic to keep them cold. A variety of trapping devices has been described in the literature and been summarized in several review articles, see for example Refs. 2, 3, and 18–20. Many of the basic ideas have been developed already in the 1950s[21] and 1960s as can be seen from a still very attractive paper written by Dehmelt[3] or from the overview given by W. Paul in his Nobel lecture.[22] In Penning and ICR traps (ion cyclotron resonance) confinement is achieved by a static magnetic field. These devices are well-suited for high resolution mass spectrometry and related studies; however, they cannot be used efficiently for buffer gas cooling since they belong to the class of *dynamic* traps where the confining force is based on the motion of the ion. The same obviously holds for storage rings or other electrostatic traps in which ions are moving on quasi-periodic orbits with rather high velocities. In contrast, traps that use inhomogeneous time varying fields in the adiabatic regime, have real three-dimensional potential minima and, therefore, can be used for buffer gas cooling. Many aspects of the technique and the basic theory, especially the limits of the effective potential approximation, have been explained in Chapter 3 with emphasis on applications in low energy collision dynamics. In this chapter the cooling process itself is discussed and *in situ* applications of the cold stored objects, especially spectroscopic applications, are described.

Dehmelt[3] has stated in his article on *Radiofrequency spectroscopy of stored ions* that keeping a system at rest in space and free from any outside perturbation is of limited value unless one has dedicated methods for preparing, manipulating, and analyzing the stored objects. Interesting schemes for creating oriented, aligned or state-selected atomic ions have been discussed already before laser methods became integrated in such experiments. As described in Chapter 3, many of today's applications of traps use the flexibility of external ion sources for filling the trap and also external detectors for counting each of the extracted ions. Mass selection is achieved by quadrupole or magnetic mass spectrometers. For mass analysis of a trapped ion cloud, the combination of pulsed ejection and time-of-flight mass spectrometry is ideal; however, the method is not yet fully developed as discussed in the conclusion of Chapter 3. There are also several non-destructive detection schemes, which leave the stored objects in the trap. Commercially established is the detection of the image current induced by the periodic motion of trapped particles in pick-up electrodes. The commonly used high resolution Fourier transform ion cyclotron resonance mass spectrometry (FTICR-MS) has been extended recently to an electrostatic trap, the Orbitrap.

Other methods to obtain information on the distribution of mass, or of mass to charge ratio in the case of multiply charged objects, are based on resonant excitation of periodic motions in magnetic traps or in suitable harmonic potentials, superimposing oscillatory electric fields. In the case of an rf quadrupole, this method is called *notch filter*, if the excitation is used to eject specific ions. The method also can be applied to a linear multipole trap if one creates a harmonic potential along the axis with a set of ring electrodes (see Fig. 3.13 of Chapter 3). A significant advantage of such an arrangement is that the ions of a specific mass can be ejected towards the detector. As discussed below, mass selective *in situ* excitation also can be monitored via optical detection. For molecular ions, more information is desirable besides the mass over charge ratio. There are a variety of specific methods, e.g. for distinguishing different isomers or for analyzing the internal population of various vibrational or electronic states. Rather general is the method of chemically probing the trapped ions by exposing them to a pulse of suitable molecular reactants passing as collimated beam through the trap. As discussed in Chapter 3, atoms or radicals have also been used. Especially advantageous are chemical processes in which only the reaction products leave the trap while the prepared cold ion cloud remains stored, i.e., the product ions must be formed with additional kinetic energy in the laboratory frame. An example is the resonant proton transfer between a fast molecule and its protonated pendant. The probing of $CH_5^+$ with a beam of methane is mentioned below.

The combination of ion traps with lasers has opened up a wide range of innovative experiments. In addition to the information obtained from detecting reemitted or stimulated photons, there are many laser based schemes for probing or manipulating a trapped ensemble including optical pumping, laser induced dissociation, or light stimulated chemical reactions. Lasers are utilized for state specific creation of ions in the trap, e.g. via resonance-enhanced multi-photon dissociation. It is important to note that, for testing cold ensembles, the Doppler profile of a transition provides information on translational motion. As summarized in what follows and illustrated with selected results, the majority of laser applications is still in high resolution spectroscopy, metrology, and fundamental quantum-electrodynamics; however, applying the sensitive optical manipulation and detection schemes to molecular ions allows for unique contributions to the emerging field of ultracold chemistry.

### 6.2.2. *Laser Based Detection and Cooling*

The methods of laser induced fluorescence (LIF) and laser based cooling has entered the field of ion trapping in the late 1970s as briefly mentioned in the Nobel lecture of W. Paul.[22] The high sensitivity of various optical detection schemes has allowed individual trapped ions to be monitored over long times, resulting in interesting new experimental strategies. For example, information about the lifetime of metastable states has been deduced from *not seeing* the stored ion, i.e., from the disappearance of the LIF signal, a process called shelving.[2] A detailed understanding of the structure and dynamics of the stored system allows unique tests with photons including chemical shelving in the case of the reversible formation of a weakly bound molecule with a cold, inert buffer gas (see below the He–$N_2^+$ example). Permanent state-specific perturbation via optical pumping of a trapped low temperature ensemble with an infrared laser can avoid or reduce clustering with the cooling gas. The ability to monitor ions, be it a single one or a few thousands, in good isolation from the outside world has led to many improved spectroscopic measurements, and exquisite tests of QED, including the observation of quantum jumps, as mentioned above.

While laser cooling of a free particle is based on the Doppler effect and uses red shifted photons, for trapped ions one makes use of the fact that the translational motion is quantized by the boundary conditions. In a harmonic potential the oscillatory motion adds equidistant sidebands to optical transitions. Fig. 6.1 shows results from an interesting experiment[23] where sideband cooling has been applied to one single $Hg^+$ ion which has been confined in an rf quadrupole trap. The effective potential was set such that the eigenfrequency of the secular motion was $\omega/2\pi = 2.96$ MHz. Using a resolved sideband transition, based on the electric quadrupole transition $^2S_{1/2}$–$^2D_{5/2}$, the single mercury ion has been transferred almost completely into the ground state of its confining potential. Comparison of the absorption strengths of the upper and lower sideband (anti-Stokes and Stokes line) has confirmed that the ion was in the lowest state, $n_v = 0$, for 95% of the time. This corresponds to a time averaged temperature of

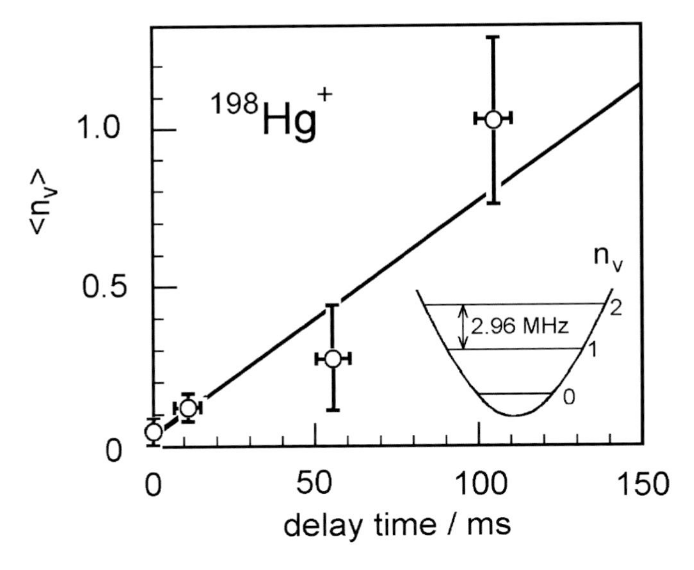


Fig. 6.1. Laser cooling towards the zero-point energy of the motion using three lasers.[23] The quantum number $n_v$ characterizes the quantized motion of a single $^{198}Hg^+$ ion in the harmonic effective potential of the confining rf quadrupole trap (eigenfrequency $\omega/2\pi = 2.96\,\text{MHz}$, $\hbar\omega = 12\,\text{neV}$). Using side band laser transitions $\Delta n_v = -1$ transitions can be pumped. Analysis of side band resolved spectra reveals that the system can be cooled so far that it occupies the ground state $n_v = 0$ for 95% of the time, corresponding to a temperature of $47\,\mu\text{K}$. As soon as the cooling laser is switched off, the ion motion is heated with a rate of $\langle dn_v/dt \rangle = 6/\text{s}$. After 2 s it reaches a mean value of $\langle n_v \rangle = 12$ corresponding to $T \sim 1.7\,\text{mK}$.

$T = 47\,\mu\text{K}$. In order to maintain such a low translational temperature, the ion has to be exposed continuously to the radiation field. Switching off the cooling laser leads to a heating rate, $\langle dn_v/dt \rangle$, of 6 vibrational quanta per sec, see Fig. 6.1. After 2s the mean expectation value of the vibrational quantum number is already $\langle n_v \rangle = 12$. This corresponds to a translational temperature of $T_T = 1.7\,\text{mK}$ of the 3-dimensional oscillator. For more details of this fundamental experiment see Ref. 23.

Besides sensitive and spectrally resolved detection of the light emitted from trapped ions, progress in the development of imaging devices (e.g. intensified CCD cameras) made it possible to record pictures of the trapped ensemble with high spatial resolution. This has allowed many tests to be performed on so-called Coulomb crystals. Trapped ions that are cooled with laser light below a translational temperature $T_T$, arrange themselves in spatial structures. A detailed analysis of regular arrays of a few laser-cooled $Hg^+$ ions which have been observed in an rf quadrupole trap has been reported by Wineland *et al.*[6] The ratio of Coulomb potential energy per ion to the thermal translational energy, $k_B T_T$, has been estimated to be 120 in these clusters. The spectroscopy of such a pseudo-molecule is unique in that one can probe individual atoms of the structure separately due to $\mu$m distances. Order-chaos transitions in trapped ion clouds have been discussed for example by Blümel *et al.*[24] Fig. 6.2 shows two typical pictures of two trapped $Ba^+$ ions taken by Hoffnagle *et al.*[25] If laser cooling reduces the kinetic energy of the ion pair sufficiently they form a diatomic molecule in the trap. As can be seen from the left panel, the repulsion between the two positive charges, together with the confining force imposed by the trap, keeps them at an equilibrium distance of about 6 $\mu$m. Sideband transitions allow vibrational transitions of such molecules to be probed. Reducing the cooling efficiency or heating the trapped ion pair leads to a phase transition: the molecule dissociates and the axial and radial motions of the two ions become uncoupled.

By storing and cooling a cloud of many particles of the same sign of charge in a trap, large ion Coulomb crystals are formed and are the basis of many interesting experiments. Such an one-component plasma has been

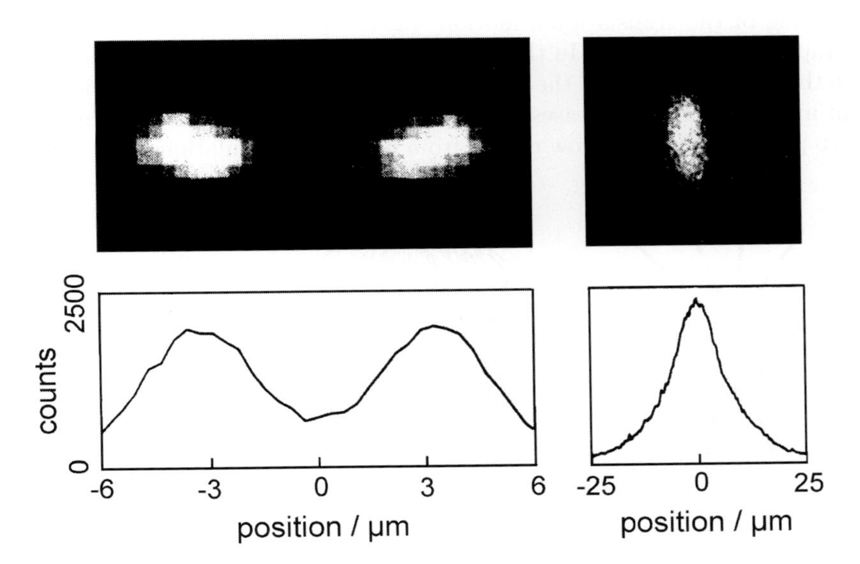


Fig. 6.2. Order (left side) and chaos (right side, note the difference in scale) of a pair of laser-cooled $Ba^+$ ions which are confined in a Paul trap.[25] Laser cooling leads to the formation of a bound diatomic Coulomb molecule. At slightly higher temperatures, the bound is broken and the two ions move independently in the trap. The spatial confinement within a few tens of $\mu$m is determined by their mean energy and the effective potential.

first seen in a quadrupole trap for microscopic particles.[21] There have been also activities to obtain regular structures of ions in storage rings; however, the cooling was obviously not efficient enough to reduce the average kinetic energy per ion sufficiently (<mK in the moving frame) below the repulsive Coulomb energy for obtaining the transition from the liquid to the solid state phase. Many experimental and theoretical studies to understand the structure and the dynamics of ion crystals have been performed during the last few years; see for example Refs. 5, 26, and 27. The resulting structures, ranging from strings of a few ions via planar arrangements to complex multi-shell compositions, depend on the number of ions and the boundary conditions imposed by the trapping field. Today most experiments use linear quadrupole traps, which are closed in the axial direction by electrostatic barriers. It can be foreseen that higher order multipole traps will be used soon.

A few selected properties of Coulomb crystals are illustrated in Fig. 6.3. The left part shows a photo from the pioneering experiment performed by Wuerker *et al.*[21] An ensemble of 32 charged aluminum particles having a diameter of a few $\mu$m was stored in the effective potential of a three dimensional quadrupole trap operated with an ac voltage of some hundred Hz. Cooling of the translational motion was achieved by collisions with room temperature buffer gas. In this example the ions were confined in the plane of the ring electrode of the Paul trap leading to a radial oscillation with an amplitude that increases with the distance from the center. The other two panels of Fig. 6.3 show results from numerical simulations of a 1000-ion

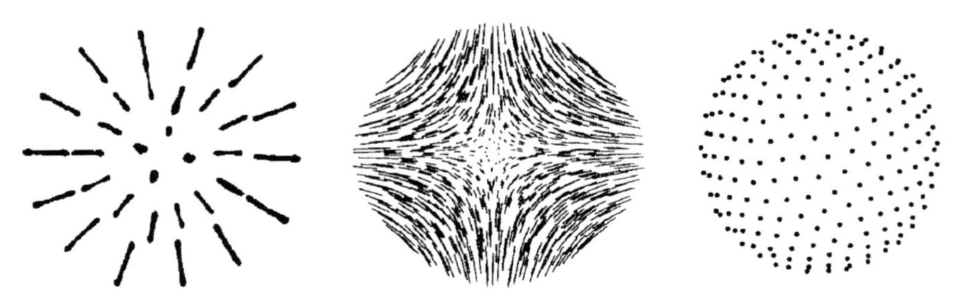

Fig. 6.3. Motions and crystalline arrangements of trapped charged particles. The left figure is a photo of a cluster of 32 charged micro-particles stored in a Paul trap.[21] The two other figures are results from a numerical simulation of a 1000-ion Coulomb crystal confined in a linear quadrupole trap.[28] In the left and the center panel the micro-motion of the particles in the oscillatory electric field can be seen. The amplitude increases proportionally to the distance from the center. The time-averaged positions plotted in the right part for a selected sample shows that the ions remain well-localized. This is the basis for defining an effective translational temperature of the ion cluster by subtracting the periodic oscillation from the overall motion.

Coulomb crystal confined in a linear quadrupole trap published by Schiffer *et al.*[28] As in the picture on the left, the lines in the middle image indicate the micro-motion. Other than in the left picture, here the ions are stored in a linear quadrupole and the image shows the motion in a plane orthogonal to the axis of the rod system. As a consequence the direction of the rf field induced micro-motion is radial in front of the rods and tangential in between them. More details on the micro-motion can be found in Chapter 3 (see for example Fig. 3.1). The time-averaged positions plotted in the right part for a selected sample of ions shows that they remain well-localized. As discussed already in Chapter 3 and in more detail below, this observation is the basis for defining an effective translational temperature of the ion cluster.

### 6.2.3. *Sympathetic Cooling in Traps*

The early applications of ion Coulomb crystals have been restricted to single atomic species suitable for laser cooling; however, it is also possible to replace any single ion in the ordered structure by another one, including molecular ions. The strong coupling via the Coulomb repulsion, a process called sympathetic cooling, efficiently freezes the translation of all ions embedded in the crystal including those which do not interact directly with the cooling laser. There are some restrictions concerning the mass over charge ratio but a wide range can be covered if one uses for cooling an ion with a suitable mass. There are several atomic ions available with masses between $^{9}Be^{+}$ and $^{198}Hg^{+}$. Many applications have been reported including bi-crystals consisting of two different ions both of which can be imaged, isotopically mixed crystals created via isotope selective resonance-enhanced two-photon ionization or a cluster of 1000 laser cooled $Mg^{+}$ ions in which 95% have been converted into $MgH^{+}$ in reactions with $H_2$. The unique versatility of the method of sympathetic cooling has been demonstrated recently with two specific ions confined in a linear Paul trap, one $^{27}Al^{+}$ and one $^{9}Be^{+}$.[29] The beryllium ion provides laser cooling and is used simultaneously for monitoring of the $^{1}S_{0} \rightarrow {}^{3}P_{0}$ transition in $^{27}Al^{+}$. In this arrangement the frequency of this clock transition could be determined with a fractional uncertainty of $5 \times 10^{-15}$.

Under ultrahigh vacuum conditions, ion clusters can last for several hours since loss or chemical modification is only caused by a collision with background gas. Such long storage times of sympathetically cooled clusters and a nearly 100% detection efficiency of individual ion events within that structure makes the technique very versatile. Therefore, recent years have seen many activities on two or more ion species crystals and new groups enter this field. Usually only one species in the crystal is laser cooled and monitored while the other ones are dark, i.e., not seen by the detector; however, they are fixed at specific locations of the crystal. Various ion species usually segregate in the trap, one of the reasons being that the effective potential is weaker for higher masses. This means that detection and identification relies on the observation of changes in the spatial structure of the crystal. In order to understand and characterize the complex three-dimensional structures of the non-neutral plasmas, numerical simulations of the motion of up to several thousand species confined in a multi-species ion crystal have become routine.[27,28,30] Such calculations are useful for extracting information from the observed CCD images. They also help to develop and quantify new schemes for identifying particles newly formed in the crystal. Very efficient is, for example, the nondestructive identification of species via resonant excitation with an oscillating electric field resulting in high mass-to-charge resolution. The secular frequencies of ions in the effective harmonic potential of the quadrupole trap are shifted and broadened by the Coulomb interaction.

Of special interest for the present chapter is the possibility that complex ions can be integrated into Coulomb crystals. The first detection of charged molecules in a laser-cooled $Mg^+$ ensemble has been reported by Baba and Waki.[31] Subsequent applications of forming and confining translationally cold molecular ions include $MgH^+/MgD^+$,[32] $CaO^+$, $BeH^+$, $HD^+$, and others. For a recent summary see the publications by Roth *et al.*[27,33] These authors have created large samples of ultracold (<20 mK) molecular ions including triatomic ones, e.g. $N_2H^+$, $H_3^+$, and deuterated variants by sympathetic cooling and crystallization via laser-cooled $Be^+$ ions in a linear radiofrequency trap. Probably the largest ion, which has been so far integrated into a Coulomb crystal, was a protonated organic molecule with a mass of 410 u. The ions have been generated in an external electrospray ionization source and transferred and cooled to the sub-K range by Coulomb interaction with laser-cooled barium ions.[34] In principle, these techniques allow a large variety of charged objects to be sympathetically cooled, including ions of much higher mass, such as protonated proteins. There is no doubt that these ultracold ions are interesting targets for a variety of high-precision measurements in fundamental physics. However it is not yet clear how to combine sympathetic cooling with other efficient schemes for really cooling both the translational as well the internal degrees of freedom, e.g. for applications in ultracold chemistry, for state-specific preparation of ions, or for absorption spectroscopy on internally frozen molecules.

The problem of the *real* kinetic energy of the *ultra-cold* ions has been discussed already in Chapter 3 and is illustrated in Fig. 6.3. Briefly, the effective temperature is just a practical parameter which describes surprisingly well the ordering process in a Coulomb crystal.[28] Its definition is based on the small random deviations of the ion trajectories from the coherent local periodic motion imposed by the oscillating electric field. Apparently there is only an extremely small coupling between this thermal energy and the orders of magnitude higher kinetic energy contained in the rf driven motion. So far no dependence of the ion temperature on the distance from the center line has been observed. As emphasized in Chapter 3, collisions with such ultra-cold ions are rather hot, since they occur with the full momentary relative velocity. One solution is to restrict collisional studies to strings of ions localized close to the axis of a linear multipole trap.

An experimental problem with *ultra-cold* molecular ions created in a crystal, which has not yet been solved, is how to control their internal energy. As already indicated in the introduction, the Coulomb interaction couples efficiently to the translational degrees of freedom while the internal degrees of freedom apparently are almost completely unaffected by the laser cooled environment. Without any other interaction the population of vibrational and rotational states is mainly determined by the production mechanism of the charged molecules and the following decay processes. Ionization with energetic electrons or formation via an exothermic reaction can lead to highly excited ions, while selective photoionization may be suitable for creating internally cold ions. If there is no other cooling process, after a long time molecular ions may come into equilibrium with the black-body radiation provided the coupling is efficient enough.

A special test case for internal relaxation is the cooling process of diatomic polar molecules. To investigate the efficiency of rotational relaxation, Bertelsen *et al.*[35] used $MgH^+$ ions which have been sympathetically cooled to a translational temperature below 0.1 K. Information on the population of specific rotational states has been obtained using rotational resonance enhanced multiphoton dissociation. Although the $MgH^+$ ion has a rather large permanent dipole moment of 3.6 Debye, no significant cooling of rotation via the laser cooled atomic ions could be found, indicating that the charge-dipole interaction is very weak. A similar experiment has been performed with $HD^+$ ions which have been created by electron bombardment and which have been sympathetically cooled with laser-cooled $Be^+$ ions.[8] Also in this case destructive laser induced fragmentation has been applied to determine the rotational population of the trapped ions. While the translational temperatures were in the mK range, the effective rotational temperature has been found to be close to room temperature. The observed separation of rotation and translation led Koelemeij *et al.*[8] to propose that such a system may be used as a thermometer for the ambient blackbody radiation.

### 6.2.4. *Buffer Gas Cooling*

In contrast to the specific process of laser cooling, which just reduces the translational energy, interactions with a cold buffer gas is a true cooling method. It can provide thermalization of all degrees of freedom through elastic and inelastic collisions. Gas friction has been used in the above mentioned trapping experiment for cooling the randomly moving aluminum particles so far (in this example 300 K) that they arrange themselves into a Coulomb crystal.[21] Cooling of molecules in supersonic beams is very efficient and can cool them to temperatures of a few Kelvin. Buffer gas cooling also has been applied successfully for loading paramagnetic atoms and molecules into a magnetic trap. In such experiments $^3$He buffer gas is employed to reach temperatures below 1 K since the depth of the trap is rather shallow. Once loading has taken place the buffer gas is removed by cryopumping. While in this application the velocity of the injected atoms or radicals have to be slowed down sufficiently that the weak magnetic forces store them, ion traps have much deeper potential wells and are much easier to fill. Cooling to room temperature of ions produced by electron bombardment has been shown to be quite efficient in so-called storage ion sources, which also use rf fields for confining the ions. As described in Ref. 18 and in Chapter 3, the first liquid nitrogen cooled rf trap became operational two decades ago. Meanwhile there are several instruments that rely on buffer gas cooling in high order multipole traps. As already described in Chapter 3 (see Fig. 3.14) one can use long interaction times or an intense short gas pulse for thermalization.

Buffer gas cooling is very general, the only condition is that the ion is capable of surviving multiple collisions. It already has been observed in the case of weakly bound clusters that cooling of translation leads partly to dissociation. As will be discussed below the reverse mechanism can also play a role, clustering with the atoms or molecules used for cooling. Due to different efficiencies in removing specific forms of energy stored initially in the molecular ions, it may take many collisions before a real thermal ensemble is reached; however, the time can be varied from ms to min and the neutral's number density from $10^7$ to $10^{16}\,\mathrm{cm}^{-3}$. In most ion trap experiments, He, n-$H_2$ or p-$H_2$ are injected into the trap effusively; however, there are also new initiatives to use very slow beams for cooling, e.g. also H atoms.

In an ideal situation both the trapped ions and the cooling neutrals reach a thermal equilibrium at a common temperature $T$ which is usually defined by the surrounding walls. Under such conditions the distribution of the relative velocity, $f(g)$ (definition see Eq. (3.10) of Chapter 3 and also Ref. 18) is given by

$$f(g) = f_M(g;\mu,T) = (4\pi^{1/2})(\mu/2kT)^{3/2} g^2 \exp\left(-\frac{\mu}{2kT} g^2\right), \qquad (6.1)$$

where $\mu$ is the reduced mass and $k$ the Boltzmann constant. In practice the translational temperature of trapped ions often deviates from the buffer gas temperature. If in such a situation the motion of the ions and the buffer gas can be approximated by two Maxwellians with different temperatures, $T_1$ and $T_2$, Eq. (6.1) still holds if one uses for $T$ the mass weighted collision temperature

$$T = (m_1 T_2 + m_2 T_1)/(m_1 + m_2). \tag{6.2}$$

After a sufficient number of collisions, it is the collision temperature $T$, which determines the internal temperatures of stored ions, and not its translational temperature $T_1$. This means that one can cool efficiently internal degrees of ions even if they are translationally hot, provided one takes a light buffer gas. For example, the internal temperature of $C_{60}^-$ ions stored in a quadrupole trap by Wang *et al.*[13] could be cooled to $T = 15.5\,\mathrm{K}$ with He at $T_2 = 10\,\mathrm{K}$, even if the ions had a kinetic energy distribution corresponding to $T_1 = 1000\,\mathrm{K}$. Another example illustrating the importance of Eq. (6.2) is shown in Fig. 6.4. In order to test the new low temperature ion spectrometer in Basel, $N_2O^+$ ions have been used in which predissociative rovibronic states are accessible through the absorption of a single UV photon.[36] The ions have been cooled with 10 K He buffer gas. In Fig. 6.4 the measured spectrum, i.e., the number of $NO^+$ fragments measured as a function of the wavelength, can be compared with three spectra simulated for rotational temperatures of 15, 25 and 35 K. The molecular constants used and more details of the experiment can be found in Ref. 36. In principle, it is also possible to assess the translational temperature of the $N_2O^+$ ions through the Doppler broadening of the rovibronic lines. However, in the present case the bandwidth of the OPO radiation ($\sim 0.5\,\mathrm{cm}^{-1}$) was much greater than the Doppler width ($0.017\,\mathrm{cm}^{-1}$ at 25 K). On the other hand it also cannot be excluded that the kinetic energy of the $N_2O^+$ ions is much higher. Assuming that the rotational temperature $T_{rot} \sim 25\,\mathrm{K}$ is equal to the collision temperature $T$ in Eq. (6.2) and using $m_1 = 44\,\mathrm{u}$, $m_2 = 4\,\mathrm{u}$ and $T_2 = 10\,\mathrm{K}$ one obtains for the translational temperature $T_1 = 190\,\mathrm{K}$. Possible reasons for heating the translational energy of the ions are discussed below.

Another test for the efficiency of buffer gas cooling is the study of the growth of weakly bound cluster ions until collision induced dissociation leads to a stationary state. The decreasing binding energy of ion clusters with increasing complexity allows one to use such equilibrium distributions for determining the trapping conditions. The dynamics of the growth and destruction of hydrogen cluster ions from $H_5^+$ to $H_{23}^+$ with para and normal hydrogen at 10 K has been reported by Paul *et al.*[37] Figure 6.5 shows two typical mass spectra of cluster equilibria in the trap at a temperature of

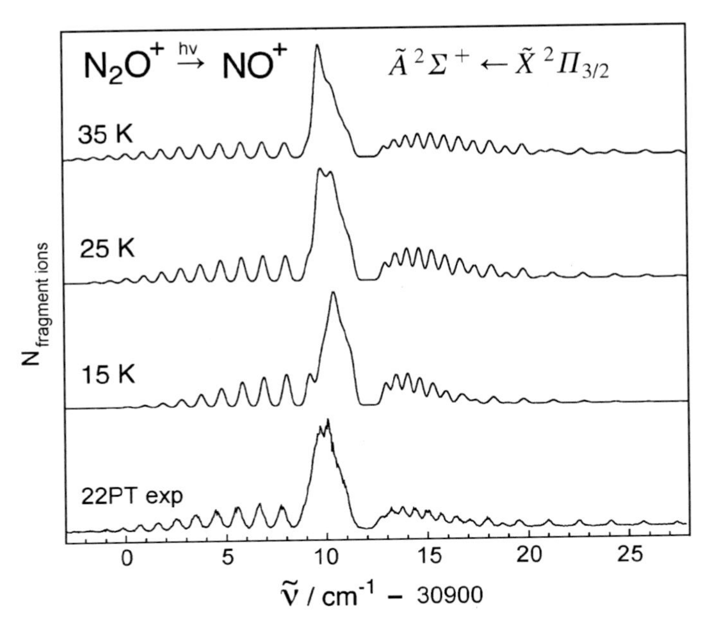


Fig. 6.4. Rotational population as ion thermometer. The photofragmentation spectrum of $N_2O^+$ (lowest plot) has been measured in a recently completed machine which has been developed in Basel for ion spectroscopy:[36] The ions have been cooled in collisions with 10 K He buffer gas. Comparison of the rotationally resolved intensities with the three simulated spectra (15, 25, and 35 K) indicates that the rotational temperature is about 25 K. As discussed in the text this indicates that there must be some parasitic heating.

10 K. $H_5^+$ ions have been injected into n-$H_2$ (upper panel) or p-$H_2$ (lower panel) and stored for 9.9 s at a hydrogen number density of $1.2 \times 10^{14}$ cm$^{-3}$. Since there is only a small chance of double deuteration, the odd masses correspond to $H_n^+$ while the even ones are $DH_{n-1}^+$. In both cases the maximum of the distribution is n = 19. With normal hydrogen one obtains a few more n = 21 clusters, while para hydrogen apparently favors the formation of deuterated clusters in collisions with the traces of HD in the hydrogen ($3 \times 10^{-4}$). This is an indication, that nuclear spin restrictions influence the thermodynamic equilibrium. Ortho-hydrogen, i.e., $H_2$ in the first rotational state at 10 K ($E_{rot}$ = 14.4 meV), also plays a specific role in heating the ensemble and there is the additional possibility that rotation leads to a dynamic stabilization of an attached $H_2$. A general conclusion is that the calibration of the size of hydrogen clusters as a thermometer is complicated by traces of o-$H_2$. A related problem, which has been discussed elsewhere, is the process of deuterium enrichment in $H_3^+$ ions.[38]

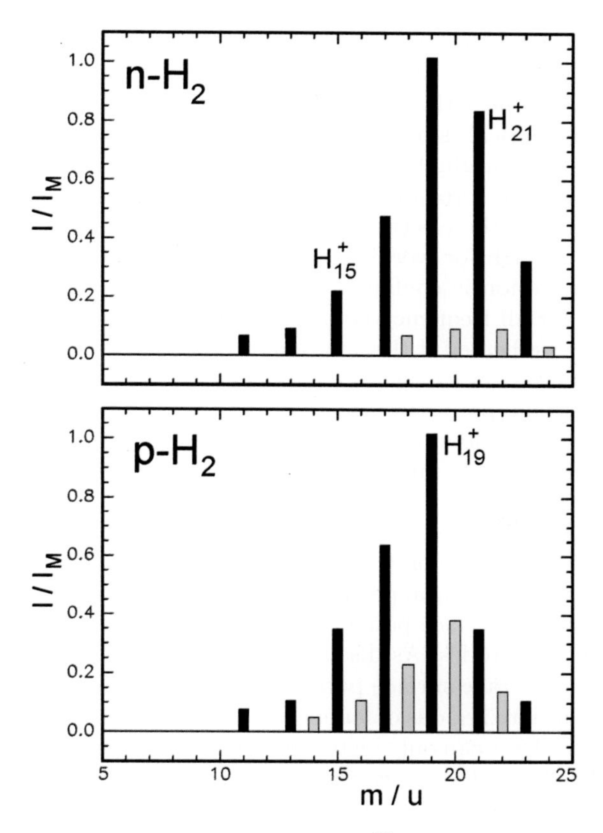


Fig. 6.5. Hydrogen cluster ions as thermometer.[37] The competition between cluster growth and fragmentation leads to a stationary equilibrium distribution, for typical densities after a few seconds. The plotted mass spectra ($H_n^+$ in black and $DH_{n-1}^+$ in yellow) have been measured at 10 K with normal and para-hydrogen at a number density of $10^{14}$ cm$^{-3}$. In both cases the maximum of the distribution is n = 19; however, in n-$H_2$ more $H_{21}^+$ is formed while p-$H_2$ favors deuteration.

### 6.2.5. *Heating of Stored Ions*

Since the first application of multi-electrode traps, there have been several attempts to measure directly the actual kinetic energy distribution of buffer gas cooled ions using narrow bandwidth diode-lasers. In an early test with $N_2^+$ (see below), good agreement was found between the nominal and the translational temperatures between 300 K and 50 K.[39] There is also one recent example where the temperature derived from the Doppler profile was in good agreement with the 10 K of the walls of the trap.[40] On the other hand several recent publications report that ions trapped in a

22-pole trap have higher kinetic energies than expected from the buffer gas temperature.[36,41,42] It is certainly necessary to localize and eliminate the mechanisms which heat the translational motion of stored ions.

Detailed numerical simulations have shown that the time-varying external force has a negligible effect on the energy distributions of a buffer gas cooled ensemble, provided some conditions are fulfilled.[18] The basic requirement is that the electric field $\boldsymbol{E}_0(\boldsymbol{r},t)$ (see Eq. (3.1) in Chapter 3) oscillates fast enough for working in the adiabatic limit, at least in those regions where the ion is confined. Note, however, that this condition has to be fulfilled for all frequency components of the time dependent electric field. Ions may be heated if the sinusoidal signal from the rf generator has some superimposed low frequency components. Another explanation for ion acceleration is that parasitic low frequency electric fields may penetrate into the trapped ion cloud from the entrance or exit electrode. One fundamental problem with rf traps is that rf-mediated heating always occurs if collisions take place in high field regions. This effect is significantly reduced if one uses traps with wide, nearly field-free regions and if one avoids space charge effects. In practice only a few thousand ion per $cm^3$ are used.

The biggest experimental problem which may hinder ion cooling is most probably caused by surface patch effects, i.e., local distortions of the work function of the electrodes. As discussed by Gerlich,[18] local potential deviations from the average surface potential can be $\pm 100\,mV$ or higher. Such loci can pull ions into regions of large rf fields causing permanent heating. In practice, patch effects can only be reduced by cleaning the surfaces. Another strategy, which has not yet been tested, is to use superconducting surfaces. An important diagnostic tool is based on the external correction electrodes, surrounding the trap (see Fig. 3.13 of Chapter 3). Since this tool is of central importance for reaching low temperatures with buffer gas cooling, Fig. 6.6 shows the calculated influence of a ring electrode, surrounding an octopole. Using, for example, an external voltage $U_{ext} = +1\,V$, the center line is raised by $2.5\,mV$. Such well-controlled local changes of the potential can be used to push ions away from critical regions. The barriers are also very practical for determining the homogeneity of the potential with time of flight methods (see Fig. 6.2 of Ref. 43). Other applications include the axial confinement of the ion cloud with very accurate barrier heights. In this way one also can eliminate influences from the entrance or exit electrode.

What are the consequences if a storage device is not operated with parameters which have been called in Ref. 18 *safe operating conditions*? In a recent publication, Mikosch *et al.*[44] discussed some experience they made in the "unsafe" region, where the adiabaticity parameter $\eta > 0.3$ (see Eq. (3.3) in Chapter 3). They observed the evaporation of buffer gas thermalized $Cl^-$ anions out of a multipole rf ion trap. Working at conditions where the

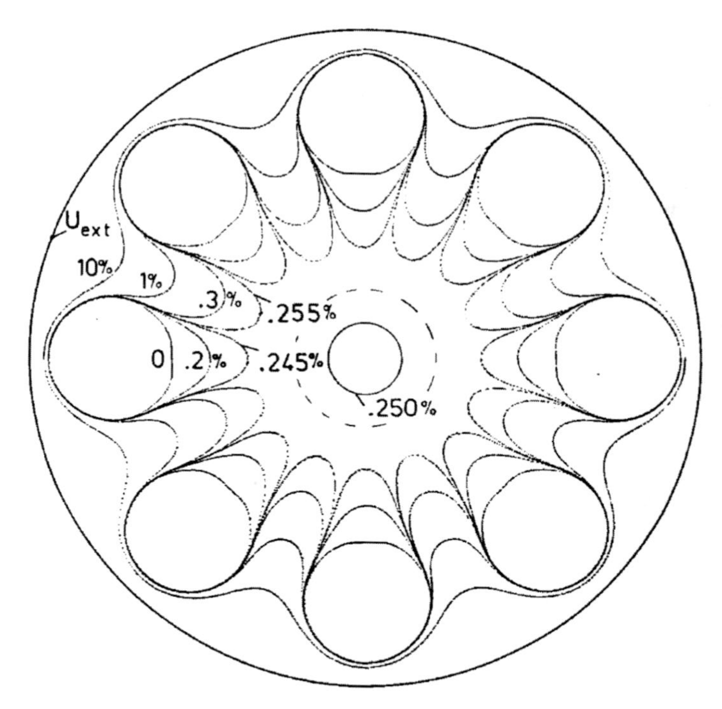


Fig. 6.6. One of the crucial problems in rf traps are patch effects on surfaces. For test and correction purposes several thin ring electrodes are surrounding the linear multipole (see Fig. 3.13 of Chapter 3). In the present example an octopole with a dc potential of 0 V is surrounded by a cylinder the potential of which has been set to $U_{ext}$. The contour lines which are given in % of $U_{ext}$, show the resulting electrostatic potential. At 1 V the potential of the center line is shifted by 2.5 mV. For more details see Ref. 18.

adiabatic approximation is not anymore fulfilled, energy was transferred from the rf field to the ion motion. This resulted in a more or less controllable temperature dependent loss of ions explained with a parameter called *trap depth*. The interplay between rf based heating and confining makes this definition rather questionable. A more suitable method for determining the actual height of the effective potential is based on the use of a dc difference applied between the rods (see Figs. 12 and 37 of Ref. 18). If one really intends to measure the ion temperature via an evaporation rate it is much better to use an axial electrostatic barriers mentioned above. As can be estimated from Fig. 6.6, this method can be calibrated with an accuracy better than 0.1 mV. The combination of buffer gas cooling and collision induced heating in an "unsafe" rf region is — most probably — not suited for quantitative measurements. For low temperature applications it is recommended to confine ions in a region where the adiabaticity parameter $\eta$ is smaller than 0.1.

### 6.2.6. *Specific Instruments*

Many applications prove that ion traps are versatile tools for preparing, confining, and cooling molecular ions. A selection of instruments and their applications in studying low energy collisions has been presented in Chapter 3. In specific experiments, e.g. in some of the Coulomb crystals, the ions are produced and characterized *in situ* while in other instruments, the ion trap is integrated in a complex arrangement containing additional external ion sources, molecular beams and various detectors for ions and photons. In some applications, a low temperature ion trap is operated just as a source for cold ions. One example is the installation of a cryogenic 22-pole ion trap on a high-voltage platform for preparing cold $H_3^+$ and for injecting them with high kinetic energies into the Test Storage Ring.[45] The excellent phase space of the cold ion cloud has allowed a transfer efficiency of almost 50% to be reached. Effects of molecular rotation and evidence for nuclear spin effects have been reported for dissociative recombination of this astrophysically important ion with electrons.[46] In another experiment a 22-pole ion trap has been used in combination with an electrospray source for cooling and injecting bunches of protonated biomolecules into an electrostatic ion storage ring.[47] There the time dependence of the fragmentation of protonated amino acids and peptides after absorption of a 266 nm photon has been measured. Another example of ion cooling in an rf trap is the photo detachment experiment mentioned in the Introduction to this chapter.[13] In this instrument a Paul trap has been used successfully as a source for cold $C_{60}^-$ ions. As discussed above the favorable mass ratio between light buffer gas and heavy ions allows the use of such a trap. In general, however, a linear 2n-pole with $n > 2$ is better suited because the stability parameter $\eta$ is r-dependent, i.e., the adiabatic approximation becomes better if the ions get colder. In addition a linear electrode arrangement makes injection and extraction of ions much easier.

The first 22-pole trap based apparatus has been described in detail elsewhere.[19] Most of the early applications concentrated on the study of low temperature ion-molecule reactions (see Chapter 3). Of course in all these applications it is mandatory to have control over, or to get information on, the actual temperature of the stored species. Especially flexible are methods based on laser induced processes as discussed above. Selected laser induced reaction (LIR) schemes which can been performed with trapping machines have been summarized recently by Schlemmer *et al.*[48] The sensitive method provides spectral information and rate coefficients for several competing processes including reactions of the excited species, collisional relaxation, and radiative decay. Sophisticated LIR applications use today several pulsed gas inlets or pulsed effusive or supersonic beams (see Fig. 3.13 of Chapter 3). The first 22-pole trap based instrument for studying cold anions has been completed recently.[44] As a first application, absolute photodetachment cross sections of cooled $OH^-$ ions have been determined.[49] As discussed in the results section below, variation of the temperature will provide state specific photodetachment cross sections.

Using a 22-pole trap and cryogenic cooling with helium gas, a special instrument has been developed in Basel for measuring electronic transitions to bound exited states for large cations of relevance to astrochemistry.[36] For directly comparing electronic spectra with astronomical observations, it is mandatory in the case of large molecules to cool their vibrational and rotational temperatures to those found in the interstellar medium. Previous studies have employed pulsed molecular beam methods to produce cold polyatomic cations. While these methods have proven useful in rotationally cooling the created species, spectral congestion is still present due to the fact that vibrational modes are often not fully relaxed. First test measurements on $N_2O^+$ have been mentioned above (see Fig. 6.4). Further applications of the new instrument are mentioned below. A general strategy of this instrument is to use two or multiple photon fragmentation for detecting the absorption of the first photon. A rather important observation made with this machine is, that polyatomic ions can become transparent to VUV radiation if they are cooled to low temperatures. This results in very low background two-color spectra since the VUV photon only leads to fragmentation if the first resonantly absorbed photon has heated the ion.

Another 22-pole based instrument recently became operational in Lausanne aimed at recording electronic and vibrational spectra of cold closed-shell biomolecular ions.[50] Representative of such types of machines, Fig. 6.7 shows the Lausanne apparatus, a tandem quadrupole mass spectrometer the central element of which is the rf ion trap cooled to less than 10 K. The ions of interest are produced in the gas-phase by electrospray, mass-selected in a quadrupole, and then injected into the trap where they are cooled via collisions with cold helium. After irradiating the ions with IR and/or UV laser pulses, the content of the trap is ejected and sent through an analyzing quadrupole before being detected. Spectra are generated by monitoring the appearance of a particular fragment ion mass as a function of the laser wave number.

An ion trapping apparatus for sub-K cooling of ions is close to completion in Chemnitz. It is based on the simple idea to take for buffer gas cooling only the slow tail from a thermal beam of cold neutrals. The important parts of this instrument, which is based on the apparatus shown

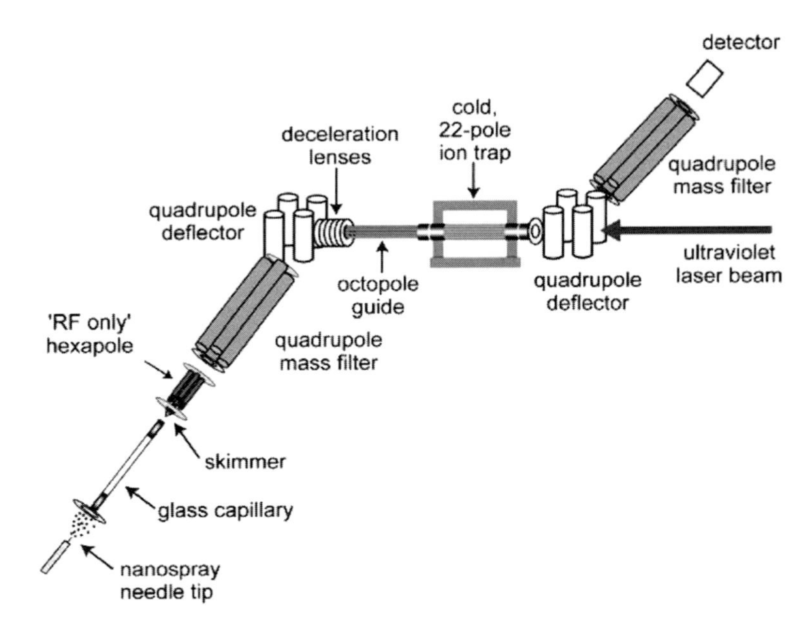


Fig. 6.7. The Lausanne photofragment-spectrometer for measuring spectra of cold biomolecular ions as well as their clusters with solvent molecules.[50] In the center of the tandem quadrupole mass spectrometer is a 22-pole ion trap which can be cooled to temperatures below 10 K. The ions of interest are produced by an electrospray source, mass-selected in a quadrupole, and then injected into the trap where they are cooled via collisions with cold helium. After irradiating the ions with IR and/or UV laser pulses, the contents of the trap are ejected and sent through an analyzing quadrupole before being detected. Spectra are generated by monitoring the appearance of a particular fragment ion mass as a function of the laser wave length.

in Fig. 3.15 of Chapter 3, are shown in Fig. 6.8. The ions are pre-cooled in the ion trap in the usual way, i.e., with a pulse of He buffer gas. A special cold head (Oerlikon Leybold 4.2 GM), in principle, allows temperatures as low as 3.5 K to be obtained; so far 5.8 K has been reached, indicating an overall heat load of 2 W. The very low temperature in the surrounding of the trap is also important for cryo-pumping and for reducing the blackbody radiation. Further cooling of stored ions is achieved in collisions with a pulsed beam of slow atoms ($^3He$, $^4He$, may be also H) or molecules ($H_2$, HD, $D_2$).

The construction of a special pulsed effusive beam source, which is based on a two-stage cryocooler (Sumitomo SRDK-101E-A11C), is indicated schematically on the left side of Fig. 6.8. An efficient method for getting a cold gas pulse is to evaporate adsorbed gas inside a cold tube with a short voltage pulse from a thin substrate that can be cooled down

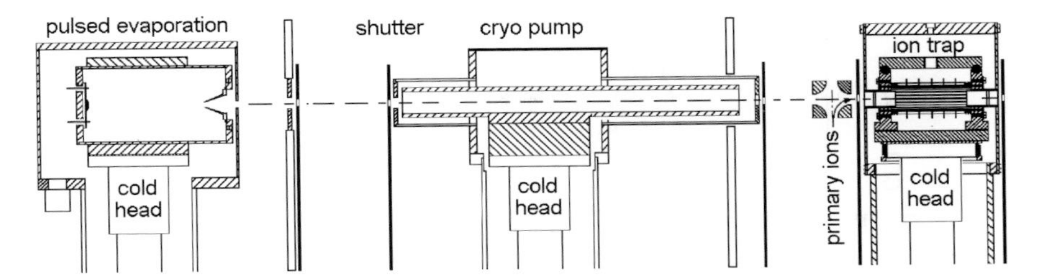


Fig. 6.8. Ion trapping apparatus for sub-K cooling of ions. A pulsed, cold effusive beam is produced via desorption of helium, hydrogen, or deuterium from a cold 3.6–8 K surface. Using fast shutters with sub-ms switching time and operating at a repetition frequency of typically 100 Hz, only the slow part of the Maxwellian distribution is allowed to pass through the 22-pole trap. Large differential pumping capacity is provided using several turbomolecular pumps and efficient cryo-pumping. In addition it is planned to separate the various chambers with shutters operating synchronously.

to 3.5 K. Of basic importance for the method is the integration of UHV compatible fast shutters which are able to operate with ms (better sub-ms) switching time and at repetition frequencies of some hundred Hz. Magnetically driven shutters have been tested; however they are not yet fast and reliable enough. Other ideas to solve this technical problem are based on piezo actors. Rotating wheels or tuning fork choppers also can be used; however, they are less flexible. A special role has the shutter at the entrance of the ion trap. It is synchronized with the pulsed beam and delayed such that only the slow part of the Maxwellian distribution passes through the ion cloud. As indicated in Fig. 6.8 shutters are also used for improving the separation of the various vacuum chambers, which are differentially pumped by magnetically suspended turbo pumps and by cryopumping.

Although molecular beams are well-described in the literature, most applications concentrate on high Mach numbers and high densities and not so much on the slow velocity tail of an effusive beam. In order to test the idea, the beam source shown in Fig. 3.15 of Chapter 3 has been utilized in combination with a mechanical chopper and a universal detector. The arrangement, which has been optimized for forming a cold effusive beam of H atoms, is described in detail in Ref. 51. In Fig. 6.9 the measured time of flight distribution of an effusive $D_2$ beam, which has been produced with the accommodator set at 10 K, is compared with a 10 K Maxwell-Boltzmann distribution (solid line). The flight path was 37 cm. The differences at long flight times indicate that the long channel of the accommodator already leads to some acceleration of the slow $D_2$ molecules towards the mean velocity. Avoiding the long channel and restricting the gas flow to a few mbar l/s will lead to a rather unperturbed cold tail of an effusive beam.

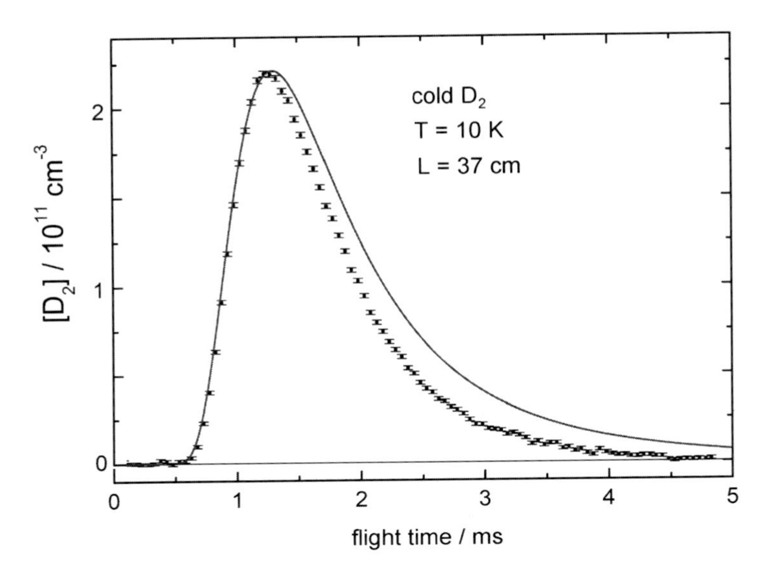


Fig. 6.9. Measured time of flight distribution of an effusive beam of $D_2$ created with the apparatus shown in Fig. 3.15 of Chapter 3. The flight path was 37 cm. Comparison of the experimental distribution with a Maxwell-Boltzmann distribution (solid line) indicates that the long channel of the accommodator leads to some acceleration of the slow molecules towards the mean velocity.

Based on the geometry shown in Fig. 6.8 and assuming a source — trap distance of 23 cm, the resulting number densities have been estimated which can be reached in the trap with the cut-off beam. A typical result is shown in Fig. 6.10 for a 5 K He beam with a flow of 0.5 mbar l/s. Cutting away half of the beam leads already to a velocity distribution with a mean velocity corresponding to 1 K. The reduction of the time averaged number density due to the duty cycle (pulse length $<1$ ms, frequency typical 100 Hz) may require cooling times of several seconds, in specific cases minutes; however this is no problem with the rf trapping technique. One very important advantage of cooling ions in a linear rf multipole with a slow beam of neutrals is that the collisions are restricted to the nearly field free region of the trap.

## 6.3. Selected Results

### 6.3.1. *Fine Structure Relaxation in Atomic Ions*

One important question in many low energy collisions is the role of fine structure energy. It is therefore important to create ions in specific states

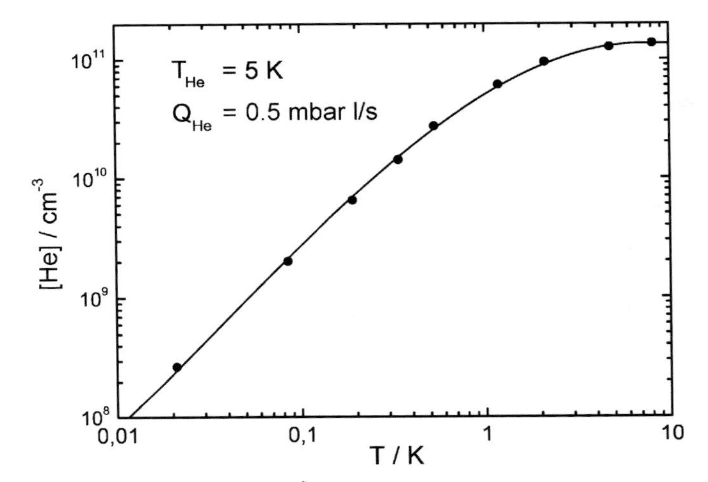


Fig. 6.10. Calculated mean temperature and peak density of an effusive beam in the trap, the fast tail of which has been cut off. Estimates show that for many ions the collisional cooling rate coefficients are big enough to obtain ions with internal temperatures below 1 K using the arrangement shown in Fig. 6.8.

or to cool them from an initially high temperature population to a low temperature equilibrium. In the following, typical test procedures and some problems are discussed with reference to the two atoms $Ar^+$ and $N^+$.

Due to spin-orbit interaction the electronic ground state of $Ar^+$ splits into two fine structure states, the $^2P_{3/2}$ ground state and the 0.178 eV higher lying $^2P_{1/2}$ state. In a fully thermalized ensemble already 99.95% of the ions are in the ground state at room temperature, while at 20 K the probability to find an ion in the excited state is not measurable ($<10^{-45}$). This simple two state system has been selected as an example to illustrate the method of buffer gas relaxation and for briefly mentioning the experimental method of chemically probing specific states of a trapped ensemble of ions with a pulsed supersonic beam of reactant gas. Here the fact is used that, in comparison to the ground state, charge transfer from $H_2$ to $Ar^+$ is significantly faster for the $^2P_{1/2}$ excited state ($k_{1/2} \sim 7.5 \times k_{3/2}$ at a collision energy of 0.1 eV, see Ref. 52 for more details).

There have been many experimental studies on fine structure changing collisions with argon ions, most of which have been performed at collision energies high in comparison to the excitation energy of the $^2P_{1/2}$ state. With the exception of the following unpublished data, which are from the PhD thesis of E. Haufler[52] there have been so far no experimental studies determining the fine structure relaxation

$$Ar^+(^2P_{1/2}) + He \rightarrow Ar^+(^2P_{3/2}) + He \tag{6.3}$$

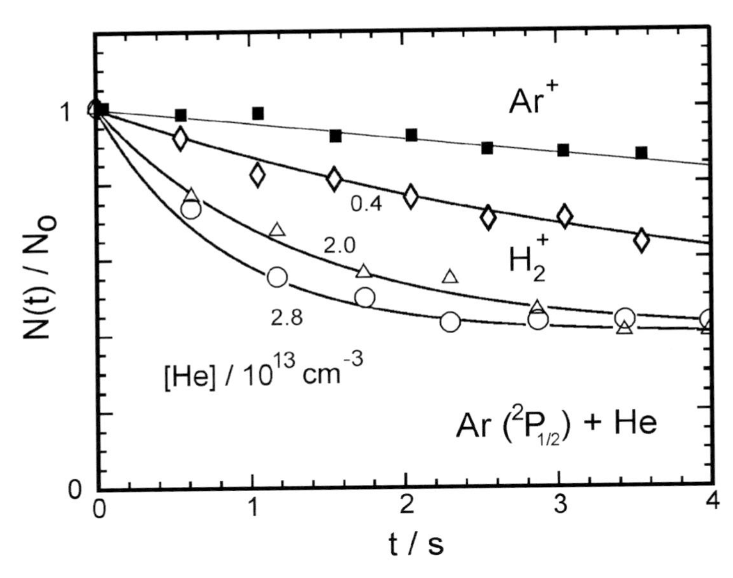


Fig. 6.11. Collisional and radiative thermalization of the 2-level system $Ar^+(^2P_{1/2})$ and $Ar^+(^2P_{3/2})$, observed via chemical probing with a delayed pulsed beam of $H_2$. The solid squares show the slow decay of the total number of $Ar^+$ ions in the trap due to reaction with traces of hydrogen. The open symbols show the number of $H_2^+$ charge transfer products for the indicated three different He number densities.

at low temperatures. Figure 6.11 shows some typical data measured at 20 K with a 22-pole ion trap. The experimental set-up was similar to that shown in Fig. 3.13 of Chapter 3. $Ar^+$ ions have been created by bombardment of argon with energetic electrons in an external ion source. It is rather safe to assume that the two states have been initially populated with their statistical high temperature ratio $^2P_{1/2} : {}^2P_{3/2} = 1 : 2$. These ions are injected into the trap containing He buffer gas at different number densities in the range of $10^{13}\,cm^{-3}$ (see Fig. 6.11) and stored there for times varying between 0 and 4 s. After this time a pulse of hydrogen passes the ion cloud and a few $H_2^+$ ions are formed by electron transfer. The total number of $H_2^+$ products decreases as a function of time. Besides a minor loss of $Ar^+$ ions due to reactions with hydrogen background gas this is mainly due to fine structure relaxation, i.e., due to the lower reactivity of the $Ar^+(^2P_{3/2})$ ground state ions.

The lines in Fig. 6.11 are fits accounting for the various competing mechanisms. A detailed analysis of the kinetic system reveals that the final experimental result, the relaxation rate coefficient $k^*$, is independent of the initial population and of the ratio of the two rate coefficients for charge

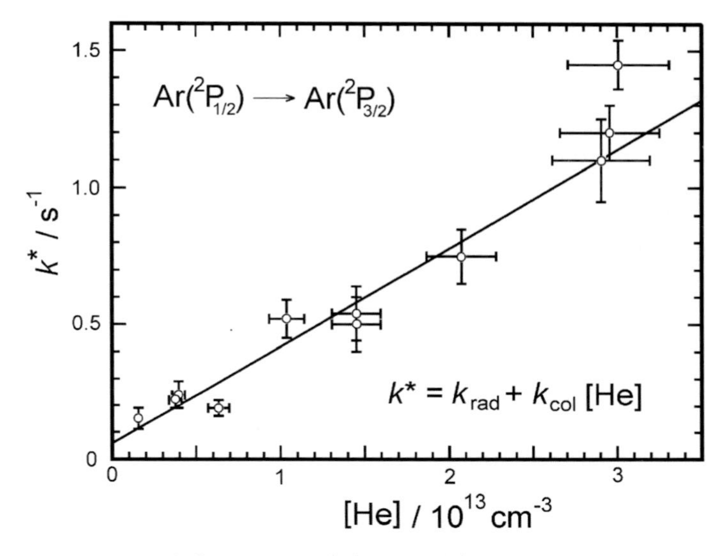


Fig. 6.12. Effective $Ar^+(^2P_{1/2}) \rightarrow Ar^+(^2P_{3/2})$ relaxation rate $k^*$, measured through the fine structure dependent $Ar^+ + H_2$ charge transfer for various He densities. The parameters of the the linear fit are given in the text, for [He] $\rightarrow$ 0 one obtains the radiative rate $k_r = 1/\tau_r = (0.06 \pm 0.04)\,s^{-1}$.

transfer.[52] A collection of measured data is plotted in Fig. 6.12 as a function of the helium number density [He]. This dependence has been fitted with

$$k^* = k_{rad} + k_{col}[He]. \tag{6.4}$$

It is obvious from the very small relaxation rate coefficient, $k_{col} = (3.6 \pm 0.8) \times 10^{-14}\,cm^3\,s^{-1}$, that one has to take very large He densities for getting a thermalized sample of $Ar^+$ with He as buffer gas. The deduced radiative relaxation rate $k_{rad} = (0.06 \pm 0.04)\,s^{-1}$, corresponding to a radiative lifetime of 17 s, is in good agreement with a lifetime of 19.0 s calculated for the magnetic dipole transition.

A second example of fine structure relaxation at low temperatures is the three state system $N^+(^3P_J)$. Here spin-orbit coupling leads to the three states $^3P_0$ (0 meV), $^3P_1$ (6.1 meV), $^3P_2$ (16.2 meV). The numbers in brackets give the excitation energy relative to the $^3P_0$ ground state. In the following the question is raised how one can prepare a nearly pure $N^+(^3P_0)$ ensemble in collision with He or $H_2$. Note that at 20 K an ensemble at equilibrium would still contain 8% of $^3P_1$. The experiment was based on the idea that hydrogen is an ideal quencher since the low temperature relaxation process

$$N(^3P_J)^+ + H_2 \rightarrow N(^3P_{J'})^+ + H_2 \tag{6.5}$$

with $J' < J$ is supported by the weakly endothermic reaction

$$N(^3P_J)^+ + H_2 \rightarrow NH^+ + H. \tag{6.6}$$

Some remarks concerning the reaction of $N^+(^3P_J)$ with $H_2$ have been made in Chapter 3, more details can be found in Ref. 53 and references therein. Using reaction rate coefficients from phase space theory in which it has been assumed that translation, rotation and fine structure energy are equivalent, the competition between reactions 6.5 and 6.6 has been modeled. A few examples are plotted in Fig. 4 of Ref. 53. For the unknown relaxation rate coefficients $k(^3P_J \rightarrow {}^3P_{J-1}) = 10^{-n}\,cm^3/s$ has been assumed with $n = 9$, 10 or 11. None of the calculated curves could reproduce the data measured in a 22-pole trap.

Fig. 6.13 shows the decline of the relative $N^+$ intensity as a function of storage time due to the $NH^+$ formation. The measurements have been performed with $[n\text{-}H_2] = 3 \times 10^{12}\,cm^{-3}$ at a trap temperature of 15 K. The $N^+$ ions, created externally by electron bombardment, must include species in excited states. Most probably one starts with the high temperature statistical population of the fine-structure states, i.e., $^3P_0 : {}^3P_1 : {}^3P_2 = 1:3:5$. Simulations with different assumptions concerning relaxation show that the competition between relaxation and reaction of $N^+$ ions always leads to a curved decay of primary ions. The perfect mono-exponential decay, in Fig. 6.13 extending over four orders of magnitude, is a strong indication that all three $N^+$ fine structure states react with the same rate coefficient. This, on the other hand, impedes the observation of relaxation processes. Note that the rate coefficient, here $k = 5.0 \times 10^{-11}\,cm^3/s$, strongly depends on the ortho — to para ratio of the hydrogen gas. For critical tests, experiments with state selectively prepared $N^+(^3P)$ reactants are needed or another probing gas than hydrogen has to be found.

### 6.3.2. *Rotation of Diatomic Molecules*

One of the diatomic molecular ions very often used for testing low temperature environments is $N_2^+$, most probably because it can be easily probed via exciting the A state with cheap laser-diodes. This allows one to determine the rotational and translational temperature of stored $N_2^+$ ions with the method of laser induced charge transfer. The slightly endothermic (179 meV) charge transfer with Ar is used in the following way. In a first step photons induce the transition

$$N_2^+(X, v = 0) + hv \rightarrow N_2^+(A, v = 2). \tag{6.7}$$

The excited ions undergo radiative decay within $\mu s$; however, due to the Franck-Condon-factors, the majority of the $N_2^+(X)$ formed is vibrationally

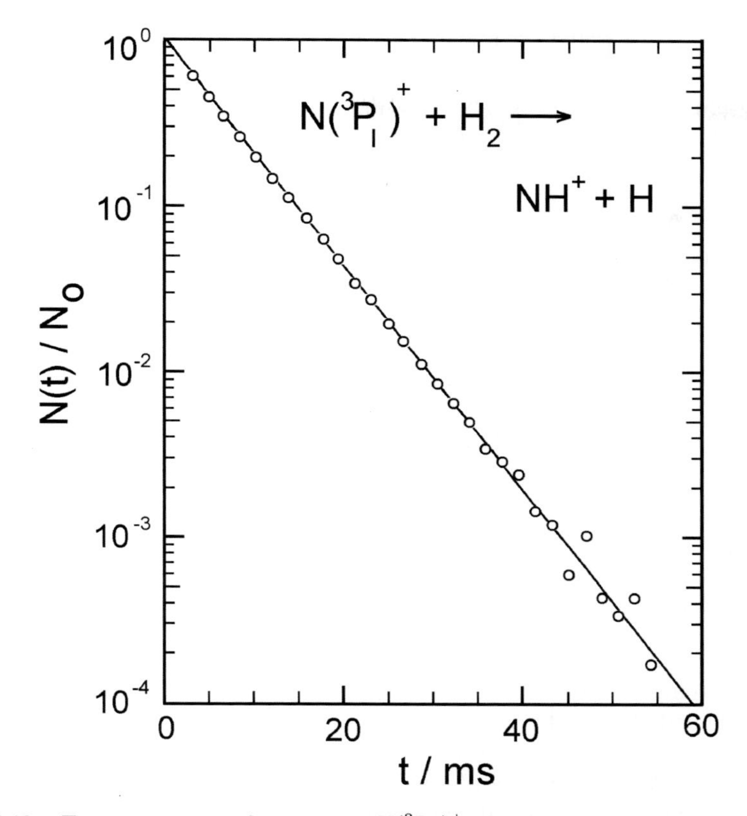


Fig. 6.13. Test measurement for preparing $N(^3P_0)^+$ground state ions via inelastic collisions (fine structure relaxation) or chemical quenching at 15 K in a 22-pole trap. $N^+(^3P_I)$ ions which have been produced by electron bombardment with the high temperature population of the three fine structure states ($^3P_0$: $^3P_1$: $^3P_2 = 1 : 3 : 5$) are stored in normal hydrogen at a number density of $3 \times 10^{12}\,cm^{-3}$. The mono-exponential decay over 4 orders of magnitude indicates that the reaction rate coefficient for forming $NH^+$ is independent on the finestructure state. For more details see text, Chapter 3 and Ref. 53.

excited. Therefore, the ions have sufficient internal energy to react with Ar via the charge transfer process

$$N_2^+(X, v > 0) + Ar \rightarrow Ar^+ + N_2. \tag{6.8}$$

Fig. 6.14 shows an early measurement of the translational temperature, derived from the Doppler profile, as a function of time after turning on the cold head.[39] The light source was a 10 mW single-mode cw laser-diode (Sharp LTO24 MDO, 783–787 nm). The line width was better than 250 MHz as determined with an etalon (free spectral range 2 GHz, finesse 200). This derived translational temperature is in good overall agreement with the

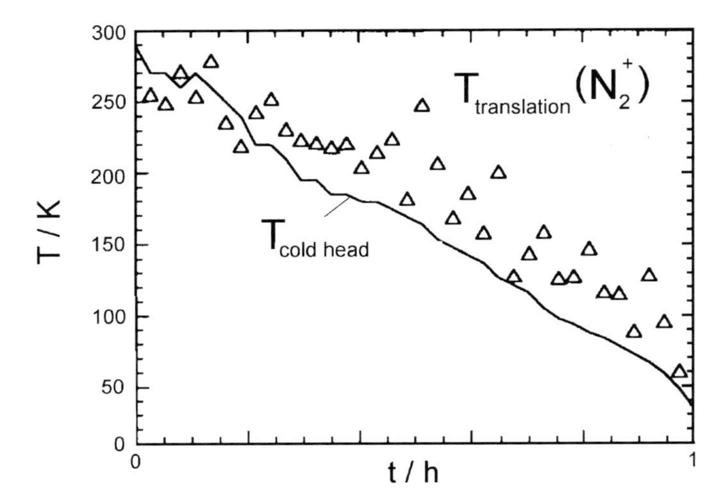


Fig. 6.14. $N_2^+$ cooling as a function of time.[39] The translational temperature of the ions which follows the temperature of the cold head, has been determined via Doppler resolved laser induced reaction. A detailed discussion of the laser induced charge transfer $N_2^+$ + Ar can be found in Ref. 54.

nominal temperature that has been measured with a thermistor directly at the cold head. The slight deviation can be partly explained by a temperature gradient caused by heat conduction or by uncertainties of the temperature calibration. Also rf heating of the ions may have been partly responsible for the deviation in this early experiment. Many more details including the determination of the rotational temperature of the stored ions can be found in the publication by Schlemmer *et al.*[54]

Filling the trap with Ar target gas limits the temperature by the onset of condensation. At 35 K the vapor pressure of Ar is $10^{-4}$ mbar. One method for using reaction (6.8) for probing $N_2^+$ also at very low temperatures is to use a skimmed Ar beam instead of the effusive gas inlet. Although straightforward with a set-up such as shown in Fig. 3.13 of Chapter 3, corresponding experiments have not yet been reported. Another method is based on the use of He buffer gas and the formation of He–$N_2^+$, a weakly bound (12.5 meV) van der Waals cluster. Ionic clusters of nitrogen and helium have been investigated in detail by Bieske *et al.*[55] For various clusters $He_n$–$N_2^+$ spectra have been reported between 390 and 392 nm, very close to the $N_2^+$ (B←X) band origin. The absorption of one of the near UV photon is detected via dissociation of the cluster ion. A third method to detect stored $N_2^+$ ions is LIF. There have been some successful activities; however, only large ensembles of $N_2^+$ ions have been seen via LIF.

So far there has been no experimental detection of a single molecular ion such as $N_2^+$ via LIF, although conservative estimates reveal that this should be possible with today's techniques at a few K where only the ground rotational state is populated under thermal conditions. For efficient LIF-detection, the cycle (i) laser excitation, (ii) photon emission has to be closed with (iii) relaxation. This means, the multi-level system has to be coupled efficiently to a low temperature bath, the method of choice being He buffer gas at high densities. Such a single molecular ion experiment has a lot of interesting applications. For example the sequence excitation, fluorescence, and collisional relaxation can be followed time resolved. Other interesting applications are the observations of the formation of the He–$N_2^+$ cluster (reactive shelving) or the detection of the unlikely event that the homonuclear molecule changes its nuclear spin in a collision with He.

Cold, trapped $HD^+$-ions are ideal objects for direct spectroscopic tests of quantum-electrodynamics, relativistic corrections in molecules, or for determining fundamental constants such as the electron-proton mass ratio. It is also of interest for many applications since it has a dipole moment. The potential of localizing trapped ions in Coulomb crystals has been demonstrated recently with spectroscopic studies on $HD^+$ ions with sub-MHz accuracy.[10,56] The experiment has been performed with 150 $HD^+$ ions which have been stored in a linear rf quadrupole trap and sympathetically cooled by 2000 laser-cooled $Be^+$ ions. IR excitation of several rovibrational infrared transitions has been detected via selective photodissociation of the vibrationally excited ions. The resonant absorption of a 1.4 $\mu$m photon induces an overtone transition into the vibrational state $v = 4$. The population of the $v = 4$ state so formed is probed via dissociation of the ion with a 266 nm photon leading to a loss of the ions from the trap. Due to different Franck-Condon factors, the absorption of the UV photon from the $v = 4$ level is orders of magnitude larger than that from $v = 0$.

The relative accuracy of the transition frequencies, which can be achieved within a sympathetically cooled ensemble, is in principle better than with buffer gas cooled ions or with fast ion beams. Molecular dynamics simulations of the motion of the trapped $HD^+$ ions lead to an estimate of a Doppler broadening of only 10 MHz under ideal conditions, whereas a thermal 10 K ion ensemble leads to a Doppler width (FWHM) of 280 MHz. In reality the reported line broadening was 40 MHz indicating an effective temperature of 0.2 K.[56] Note that this temperature describes only the axial motion of the $HD^+$ ions, the radial motion does not lead to a first-order Doppler effect. The explanation given by the authors is based on a non-linear coupling between the axial and radial motion of the ions via Coulomb interaction. In a more detailed analysis, Koelemeij *et al.*[10] distinguish between translational (secular) temperature of 50 mK resulting in 20 MHz Doppler broadening and an axial micro-motion caused by field imperfections. In the case of $HD^+$ cooled by $^9Be^+$, the ions of interest are confined close to the centerline of the linear trap because of their lower mass. Nonetheless one has to be aware that, already 10 $\mu$m away from the trap axis, the radial micro-motion energy corresponds to 0.5 K.

It can be foreseen that, in the near future, the method of sympathetic cooling is going to lead to outstanding new measurements. For extending the method several improvements are possible. In the $HD^+$ example the detection is based on a combination of mass-specific excitation and laser based destructive of the ions. If one finds non-destructive detection schemes it will become possible to work with a small number of ions arranged along the axis of a linear multipole where the micro-motion is close to zero. A nice example for permanent monitoring is the $^{27}Al^+$–$^9Be^+$ pair confined in a linear Paul trap.[29] An extension of the method towards larger ions and spectroscopic applications will require the internal degrees of freedom of the ions to be cooled. One obvious method is to transfer the linear multipole trap into a cryogenic environment.

As third diatomic example, which illustrates the need for cooling ions, as well as demonstrating the potential of low temperature ion traps, is the negative ion $OH^-$. The absolute cross section for the photodetachment process

$$OH^- + h\nu \rightarrow OH + e^-, \tag{6.9}$$

has been measured in the 22-pole based machine mentioned above at a rotational and translational temperature of 170 K. This machine, which is described in detail in Ref. 49, has been developed for investigating fundamental questions concerning negative ions, such as reaction rates or lifetimes. As light source a He-Ne laser has been used. The photon energy (1.96 eV) is sufficient to expel the electron from $OH^-$ (electron affinity 1.8 V). Figure 6.15 shows the loss of $OH^-$ ions with and without laser radiation present. Absolute cross sections are obtained from the laser induced decay constant. Using a tomography scan of the photodetachment laser through the trapped ion cloud, the derived cross section is independent on assumptions concerning the radial ion distribution and thus features a small systematic uncertainty. The tomography also yields the column density of the $OH^-$ anions in the 22-pole ion trap in good agreement with the distribution expected from a trapping potential with a large field free region bound by steep walls.

Photodetachment of electrons from negative ions is a fundamental destruction process in chemical environments such as the upper atmosphere or interstellar space. In comparison with previous results, the measured cross sections seems to indicate that the cross section is proportional to

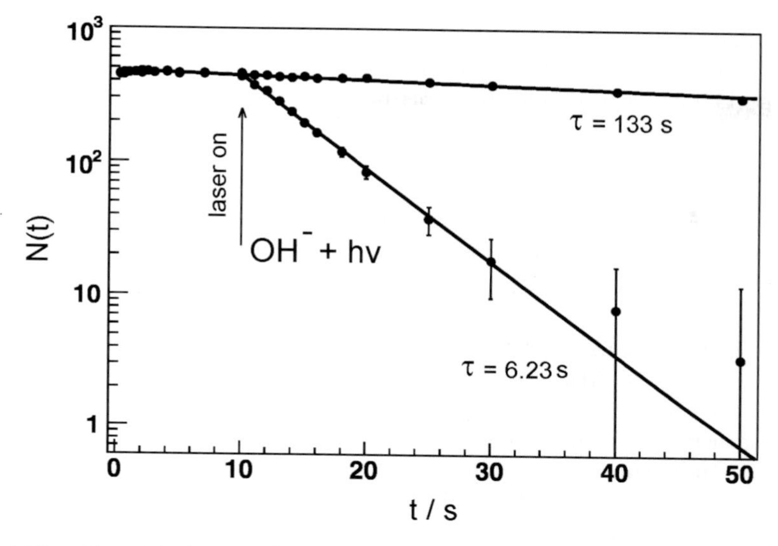


Fig. 6.15. Photodetachment of electrons from $OH^-$ trapped at 180 K in a 22-pole.[49] The small loss of ions (time constant 133 s) is significantly increased, if a He-Ne laser is switched on (here at 10 s). The photon energy (1.96 eV) is sufficient to detach the electron from the anion $OH^-$ (electron affinity 1.8 eV). The solid lines are exponential fits. Measurements performed at various temperatures of the trap allow state specific rate coefficients to be extracted.

(2J + 1), where J is the rotational quantum number of $OH^-$. Since such a strong dependence would have a significant influence on the photodestruction of this cation in the atmosphere or in interstellar space, the measurements reported by Trippel *et al.*[49] need to be extended toward lower temperatures.

### 6.3.3. *Preparing Ultracold $H_3^+$*

$H_3^+$ ions and deuterated variants play an important role in interstellar chemistry. Various aspects of the importance of these ions at low temperatures have been recently discussed by Oka,[57] Gerlich *et al.*[38] and Asvany *et al.*[42] There are many motivations for studying this simplest triatomic ion at low temperatures including ultrahigh resolution spectroscopy, the role of nuclear spin in collisions, fermionic and bosonic behavior in chemical reactions, or the relevance of the ion for dense interstellar clouds.

One fundamental collision process is the interaction of $H_3^+$ with slow electrons. There have been several efforts in recent years to measure the dissociative recombination rate coefficients for state selected ions. One of the approaches, installed at the ion storage ring in Heidelberg (see above), is to use a 22-pole trap in which the $H_3^+$ ions are relaxed at different temperatures before injection into the TSR, using either normal- or para-hydrogen or also helium. The diagnostic tools for extracting information about the state populations have been reviewed by Krekel *et al.*[58] As can be seen in Fig. 6.16 there are only a few rotational states populated in $H_3^+$ at low temperatures. In order to determine their actual population the method of laser induced chemical probing has been used by Mikosch *et al.*[41] The detection scheme is based on the reaction

$$H_3^+ + Ar \rightarrow ArH^+ + H_2 \tag{6.10}$$

which is very slow for ions in the vibrational ground state since it is endothermic with $\Delta H_0 = 0.57\,eV$. Laser excitation to the third vibrational bending level provides enough energy for reaction (6.10) to proceed. The used transitions are indicated in Fig. 6.16. As already mentioned above in the $N_2^+$ study, the use of argon for chemical probing limits the lowest temperature because of condensation. In the present case the trap was operated at 55 K. The absorption profiles, recorded as gain in $ArH^+$ as a function of the laser wavelength can be found in Fig. 3 of Ref. 41. The product signal count rate was typically 0.1 ions per trap filling.

From the so obtained spectra, the frequencies of the three transitions have been determined with experimental errors of only $\pm 0.017\,cm^{-1}$, increasing the accuracy by about a factor of 4 in comparison with earlier results. The relative intensities and the widths of the three line profiles

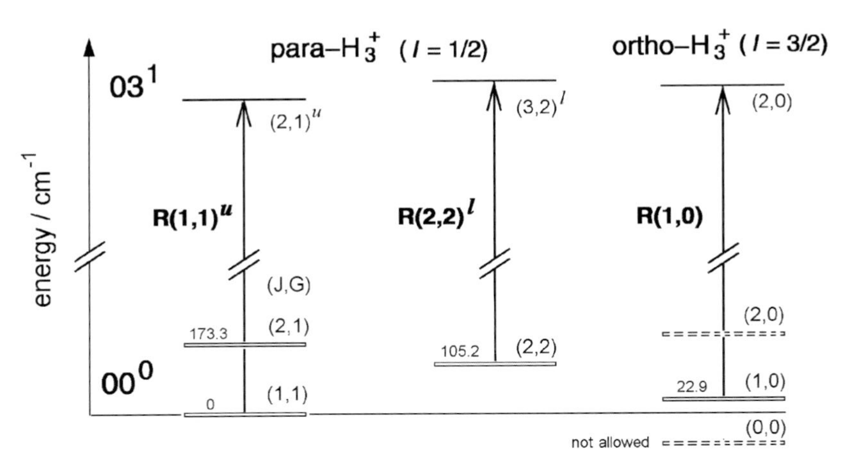


Fig. 6.16. Energy levels of $H_3^+$ used for chemical probing.[41] The population of the three lowest rotational states of $H_3^+$ has been determined via overtone transitions to the third vibrational bending level $(0,3^1)$ followed by a chemical reaction with Ar. The states $(J, G) = (0, 0)$ and (2,0) are forbidden.

provide three important experimental parameters, the translational temperature (170 K), the rotational temperature of the para-$H_3^+$ ensemble (150 K) and the ortho to para ratio (~1 : 1). The large Doppler widths of 470 MHz is unexpected. It leads to the conclusion that the ions have been heated by some of the effects mentioned above. In addition the rotational energy of o-$H_2$ may play an important role in heating.[38]

A very important process involving $H_3^+$ is the H-D exchange reaction

$$H_3^+ + HD \leftrightarrow H_2D^+ + H_2. \tag{6.11}$$

State specific rate coefficients in both directions are needed to fully understand the process of deuteration in low-temperature interstellar clouds. So far it is generally assumed that, at low temperatures, the rules of equilibrium thermodynamics adequately describe the outcome of a bimolecular proton-deuteron-exchange reaction. In a thermal equilibrium where all states are coupled efficiently enough, the equilibrium constant for reaction (6.11) is proportional to exp(231.8 K/T) (see Fig. 1 of Ref. 38). In reality, however, the actual equilibrium constants obtained in ion traps are orders of magnitude lower due to traces of o-$H_2$. As discussed by Gerlich *et al.*[38] this is not so much caused by the average energy contributed by a few $H_2(J = 1)$ molecules but more by nuclear spin restrictions which allow for accumulating non-thermal amounts of energy in the $H_2D^+$ ions. In order to confirm these predictions, it is necessary to measure precisely the population of $H_2D^+$ at very low temperatures and for very clean para-$H_2$. One possible method is to induce reaction (6.11) in the backwards direction with a laser.

Recently the LIR-trapping machine in Köln[48] has been used to make the first steps towards the determination of the state populations under the conditions described in the previous paragraph. Laser induced D-H exchange in collisions of $H_2D^+$ and $D_2H^+$ with $H_2$ has been used to obtain high-resolution IR spectra.[42] The nominal temperature was 17 K; however, as already discussed above, some unexplained heating effects lead to a translational temperature of 27 K. In addition just normal-hydrogen has been used. Despite these *hot* conditions nice spectra could be recorded. In total, line positions of 27 new overtone and combination transitions have been detected. Comparison to high accuracy *ab initio* calculations revealed some minor mode-dependent differences. As reported by Asvany *et al.*[42] populations have not yet been extracted from these measurements. Future activities in this direction must reach temperatures below 10 K, use pure para-$H_2$, and test the influence of traces of ortho-$H_2$. It may also be better to use a pulsed beam of Ar or $O_2$ for *in situ* probing of the state populations.

For studying nuclear spin restrictions in great detail and for obtaining a $H_3^+$ ion ensemble in a real thermal equilibrium, it is planned to use the exchange reaction

$$H_3^+ + H \leftrightarrow H_3^+ + H. \tag{6.12}$$

for thermalization. At a temperature of 5 K the (1,1) ground state (see Fig. 6.16) is already populated with 99.7% probability; lowering the temperature to 4 K reduces the population of the competing $22.9\,cm^{-1}$ higher lying (1,0) state to $5\times10^{-4}$. An instrument similar to that shown in Fig. 3.15 in Chapter 3 is going to be used for such studies, including measurements of rate coefficients for ortho-para transitions or reaction of $H_3^+$ with D.[51] A very cold H atom beam is prepared by operating the accommodator at the lowest temperature possible and using weak magnetic guiding fields for transferring slow hydrogen atoms into the 22-pole trap. Similar to Fig. 6.8 cryopumping removes efficiently warmer background gas. The ideal solution would be to trap cold H atoms in the region of the 22-pole trap. The idea of such a nested trap is not new and it is technically possible; however, it requires significant modifications of the trapping region. The use of superconducting electrodes for creating both the electric rf field and the magnetic field may allow a rather simple set-up to be constructed.

### 6.3.4. *Electronic Spectra of Ions of Astrophysical Interest*

A special ion trap based spectrometer for recording electronic spectra of large ions at low vibrational and rotational temperatures has been constructed in Basel.[36] It has been described above together with first results, e.g. the single-photo-fragmentation of $N_2O^+$ (see Fig. 6.4). These rotationally resolved spectra are demonstrating the power and the limits of such an instrument. A variety of additional interesting results have been reported recently.[59–61] As an example, Fig. 6.17 shows in the upper part a spectrum of the 2,4-hexadiyne-cation isomer of $C_6H_6^+$. As described in detail in Ref. 59, the ions were vibrationally and rotationally relaxed to around 30 K by collisions with cryogenically cooled helium in a 22-pole trap. The fragments are dominantly $C_6H_5^+$ with minor amounts of $C_6H_4^+$ and $C_4H_4^+$. Measured breakdown curves show that the yield of these two ions remains constant, around 20% and 5%, respectively. Comparison of the spectrum recorded with cold ions with the room temperature spectrum reveals that the origin band at 486 nm becomes very narrow and that there is no evidence of hot bands in the upper spectrum. The analysis of the well resolved vibrational pattern and of the rotational K-structure indicates that the vibrational and rotational degrees of freedom are equilibrated to around 30 K.

Another application of the instrument in Basel has been to study for the first time electronic spectra of cold protonated polyacetylenes $HC_{2n}H_2^+$

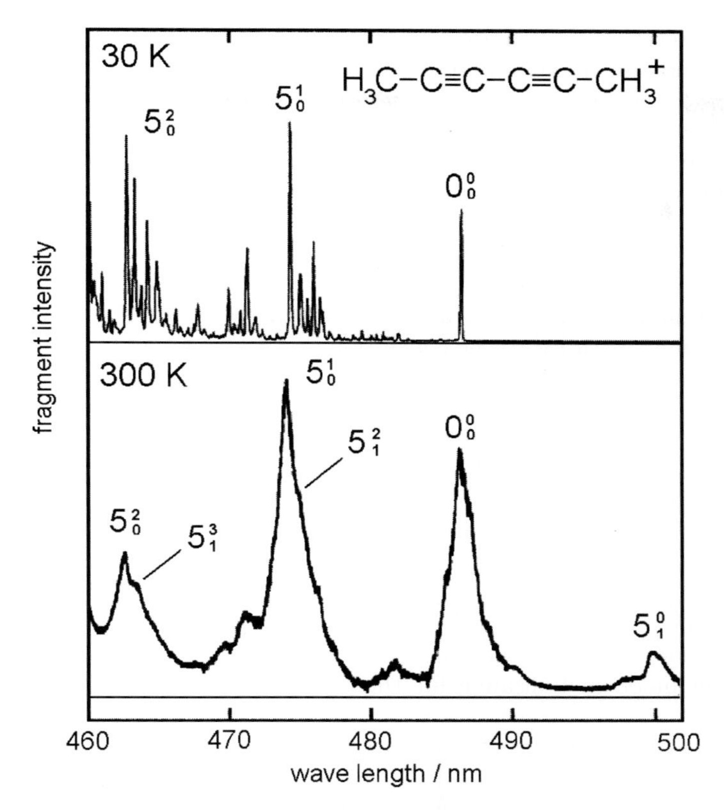


Fig. 6.17. One photon dissociation spectrum of the 2,4-hexadiyne-cation isomer of $C_6H_6^+$ in the Basel 22-pole ion trap. In the upper panel, the vibrational and rotational degrees of freedom of the ions were cooled to ~30 K in collisions with helium. Comparison with the 300 K spectrum (lower panel) reveals that the origin band at 486 nm becomes very narrow and that there is no evidence of hot bands. For more details see Ref. 59.

$(n = 3, 4)$ in the gas phase.[60] In this case, the detection of the spectra was based on photodissociation with two lasers. A tunable laser promotes the cold ions to an electronically excited state. A subsequent UV photon is then used to initiate fragmentation of the excited ions. For optimum signals the two laser beams must be overlapped both in time and space. Cooling dynamics of excited ions has been observed by varying the delay between the pump and the probe lasers while monitoring the intensity of the fragment ions. These protonated systems were chosen due to their astrophysical significance; however, no correspondence between experimental bands in the electronic spectra and features listed in the diffuse interstellar band (DIB) literature has been found.

In a recent paper, Dzhonson *et al.*[61] have reported the gas-phase origin bands of different polyacetylene cation chains. Also in this application a two-color two-photon pump-probe scheme was applied to dissociate the polyatomic species in the low temperature ion trap. The observed bands are not rotationally resolved because the rotational constants vary from 0.15 cm$^{-1}$ for $HC_4H^+$ to less than 0.01 cm$^{-1}$ for species larger than $HC_{10}H^+$. Figure 6.18 shows the rotational profile for the A $^2\Pi_g$–$X^2\Pi_u$ transition of $HC_6H^+$. The laser resolution was limited to 0.3 cm$^{-1}$. In addition, lifetime broadening might occur as a result of intramolecular processes. Simulating this spectrum with known spectroscopic constants and the laser line width led to the conclusion that the ions were cooled to 30 K. Interest in these species stems from the observation of hydrocarbons in combustion and interstellar environments. Since microwave spectra of these linear cations are not available due to their centrosymmetric nature, electronic spectroscopy offers a means of identification in diffuse clouds. Also for the polyacetylene cation chains no distinct matches between experimental and observed spectral features could be found.

### 6.3.5. *Spectroscopy of Cold Biological Molecules*

A very successful installation of a temperature variable 22-pole trap has been made in Lausanne for the investigation of biologically related

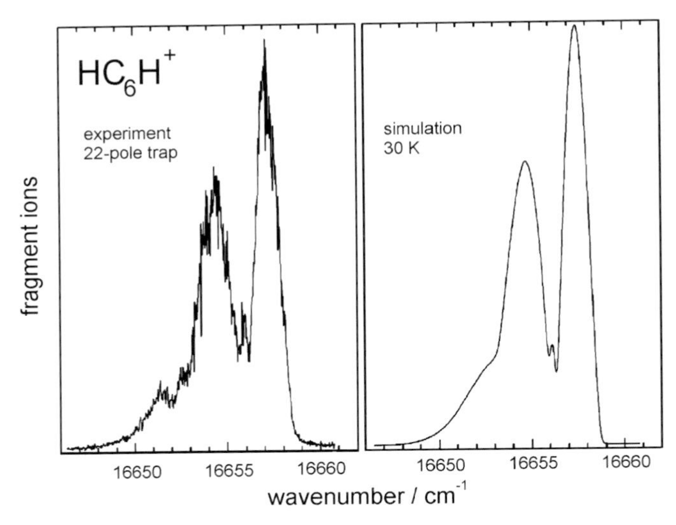


Fig. 6.18. Rotational profile for the $A^2\Pi_g$-$X^2\Pi_u$ transition of $HC_6H^+$ measured in the Basel 22-pole trap spectrometer.[61] The comparison with the simulated spectrum shows that a rotational temperature of 30 K has been obtained. For the simulation a line width of 0.3 cm$^{-1}$ was used.

molecules in the gas phase. It is important to study ionic species since biological molecules are charged in solution, and the presence of a nearby charge can strongly influence their photophysics. The aim, to use spectroscopic tools as a probe of the structure and dynamics of molecular architectures of biological interest, significantly profits from cooling the ions to very low temperatures. The instrument that is shown in Fig. 6.7 uses electrospray ionization (ESI) as a soft ionization technique to bring large thermally labile, bare molecular ions into the gas phase directly from solution. There is almost no limit to the size of molecules that can be transferred into the gas phase and stored in the trap. As a first demonstration of the potential of this innovative tool, cold protonated tyrosine, $TyrH^+$, and tryptophan, $TrpH^+$, have been studied by Boyarkin *et al.*[50]

In Fig. 6.19 one can compare electronic photofragmentation spectra of $TyrH^+$ measured at room temperature (upper panel) and of ions stored in

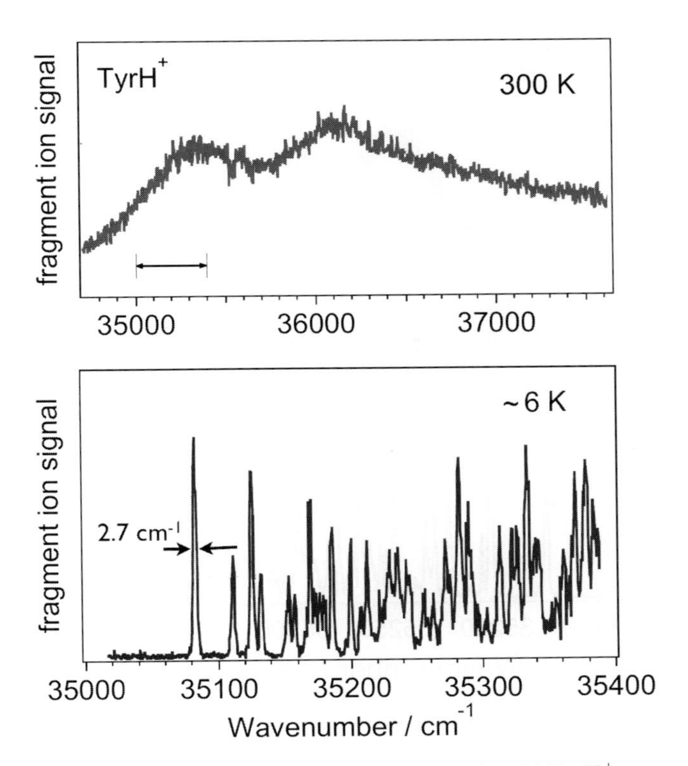


Fig. 6.19. Electronic spectra of room temperature and 6 K cold $TyrH^+$ measured in the Lausanne 22-pole ion spectrometer.[50] The lower spectrum shows only the small section indicated by the arrow in the upper part. The product fragment corresponding to m/z 136 u was monitored in each case.

a 22-pole ion trap, the walls of which have been cooled to 6 K.[50] In both cases fragment ions with m/z = 136 u/e have been recorded as a function of the laser wavenumber. The absence of hot-bands in the lower electronic spectrum demonstrates that the trapped ions attain a vibrational temperature of ∼10 K. Features in the spectra arising from distinct conformations of the molecules clearly appear. Using an IR-UV double-resonance technique different conformers have been identified via their vibrational spectra.

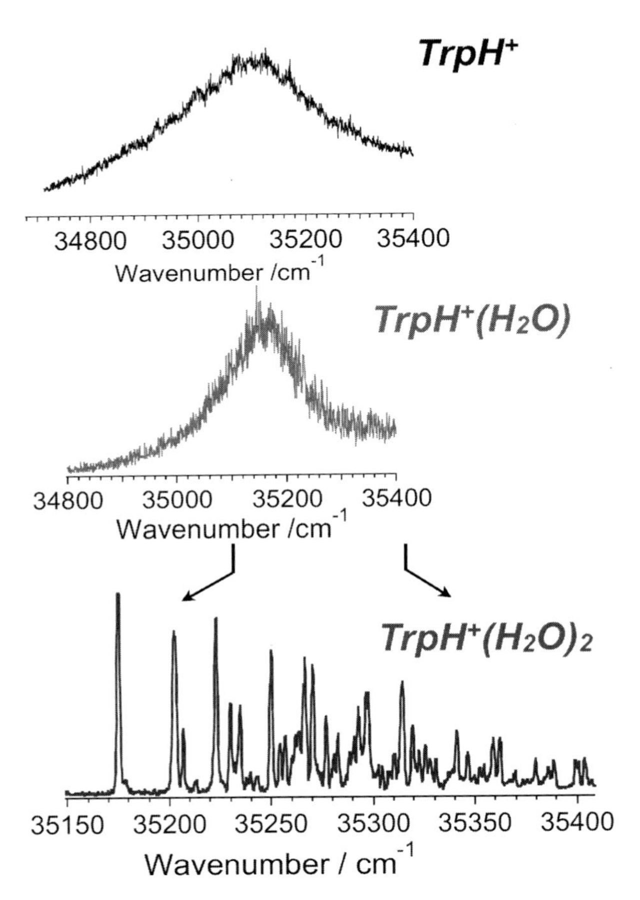


Fig. 6.20. Electronic photofragment action spectrum of protonated tryptophan with 0, 1, and 2 attached water molecules. The three spectra have been measured at 10 K using the Lausanne 22-pole ion spectrometer.[62] The broad overall band observed for the bare, protonated tryptophan (top panel) becomes slightly narrower when one water molecule is added. Attaching a second one leads to sharp lines indicating a significant reduction of the decay rate of the excited state.

The second example is tryptophan, which is widely used as a test case for protein structures and dynamics. Mercier *et al.*[62] used the low temperature trap to cool $TrpH^+$ to 10 K and to study its spectrum with a controlled number of water molecules. The results presented in Fig. 6.20 reveal a very surprising microsolvation effect for $TrpH^+–(H_2O)_n$ going with n from 0 to 2. Even at low temperatures, the bare $TrpH^+$ exhibits a broad electronic spectrum, implying ultrafast, nonradiative decay of the excited state. Comparison of the two upper spectra reveals that the band becomes slightly narrower when one water molecule is attached. The addition of a second water molecule apparently leads to a substantial lengthening of the excited state lifetime and a fully vibrationally resolved electronic spectrum is obtained. Quantum chemical calculations gave important hints as to how the interaction with the water molecules destabilizes the photodissociative states and increases the lifetime of the excited intermediate state.[62]

## 6.4. Conclusions

There are many innovative ways to explore the fundamental principles that govern low temperature chemistry. This chapter has concentrated on ions. All the discussed experimental approaches to produce and to study cold molecular ions are based on trapping. It is obvious that the long storage times which can be achieved under extremely good vacuum conditions, allow for experiments with unprecedented sensitivity and accuracy. Besides ultrahigh resolution spectroscopy one can observe very rare processes such as radiative association, spin changing collisions or spontaneous isomerization. There are interesting applications of the trapped ions in quantum optics, frequency metrology and quantum information processing.

Some experimental strategies are based on permanent interrogation of one single ion, other technique use a small ensemble of ions to reduce space charge effects, while several new approaches are based on a strongly interacting plasma. Many interesting experiments have been made with sympathetic cooling and there are several proposals to use laser-based schemes for manipulating the population of the internal states.[7] However, there is presently no general method to get really ultra-cold molecular ions for chemical applications, besides coupling to a low temperature black body radiation field. This is certainly sufficient for some applications, especially if one makes use of non-destructive detection schemes and extremely long storage times. For simple molecular systems such as $HD^+$ or $H_3^+$ such a set-up will allow unique experiments, especially if one arranges these ions in a chain along the axis of a linear multipole. One fascinating possibility for non-destructive detection of molecular ions is the application of an ion in the neighborhood. It can be foreseen that one finds a suitable atomic transition for distinguishing between a rotating or non-rotating polar ion in the neighborhood of the sensing ion.

Many remarks have been made in this chapter concerning the meaning of cold, ultracold, slow etc. It is by no means straightforward to define thermodynamic quantities such as temperature or equilibrium constant in a strongly correlated Coulomb system where the cooling ions heavily oscillate in space and spend quite some time in the electronically excited state. The concepts of temperature and equilibrium are not well defined in systems of particles that are confined by time-varying external forces. Indeed, the adiabatic approximation described in Chapter 3 helps to transfer such a system into a conservative one; however, there are limitations. Very serious, from a practical and technical point of view are the problems of parasitic heating of ions via "noise" and local deviations from an ideal equipotential surface imposed by the conductors the ion traps need. Working at low temperatures and introducing superconducting materials may reduce surface patch effects, will create less heat, and certainly can be used to improve the quality of the resonance circuit needed for the creation of the rf field. Simultaneously, parasitic oscillations can be reduced.

The methods described to obtain cold molecular ions are far from producing an ionic ensemble in the quantum degenerate regime. However, the spectra recently recorded for large molecules of biological and astrochemical relevance have shown that there is a lot to do between 300 and 10 K in order to learn more about their structure and dynamics. A "polyatomic" test case is the $CH_5^+$ ion which has been mentioned in the introduction. It is the mm-wave spectrum that will most directly provide the experimental information needed to understand the large amplitude motion of the molecule. With sufficiently cold ions one can attempt to detect the $J = 1 \leftarrow 0$ spectrum which is predicted to be in the region 220–235 GHz. At 3 K already more than 99.5% are in the rotational ground state (mean rotational constant $B = 3.7\,cm^{-1}$). An alternative approach would be to use a supersonic beam of methane for distinguishing between different states via the large distance proton transfer reaction: $CH_4 + CH_5^+(J, K) \rightarrow CH_4H^+ + CH_4$.[63]

Finally the temperature regime below 1 K for ion chemistry is within our reach. Chopped very slow beams of non-reactive coolers appear particularly promising at the frontier of cooling of all degrees of freedom in the laboratory frame. Combination of ion trapping in high order rf fields with cold neutrals confined in magnetic fields are an ideal solution for specific experiments. Attractive, and technically possible, is the combination of a trap for H atoms with a multi-electrode rf trap; however, there have not yet been any attempts to deploy such an apparatus. The sub-K range opens up many interesting experiments. One example is the formation and study of a deuterated variant of the $H_4^+$ ion confined in the shallow $H_3^+$–H van der Waals potential. Efficient buffer gas cooling may create ensembles of ions where a significant fraction is always in the ground state. With the sensitivity of modern GHz and THz techniques this may allow cold, well localized molecular ions, such as $CO^+$ or $H_3^+$, to be studied via absorption spectroscopy.

## References

1. Doyle JM, Friedrich B, Kim J, Patterson D. (1995) Buffer-gas loading of atoms and molecules into a magnetic trap. *Phys. Rev. A.* **52**: R2515–R2518.
2. Itano WM, Bergquist JC, Bollinger JJ, Wineland DJ.(1995) Cooling methods in ion traps. *Physica. Scripta.* **T59**: 106–120.
3. Dehmelt HG. (1967) Radiofrequency spectroscopy of stored ions I: Storage. *Advan. At. Mol. Phys.* **3**: 53–72.
4. Schlemmer S, Illemann J, Wellert S, Gerlich D. (2001) Non-destructive, high resolution and absolute mass determination of single, charged particles in a 3D quadrupole trap. *J. Appl. Phys.* **90**: 5410–5418.
5. Drewsen M, Jensen I, Lindballe J, Nissen N, Martinussen R, Mortensen A, Staanum P, Voigt D. (2003) Ion Coulomb crystals: A tool for studying ion processes. *Int. J. Mass. Spec.* **229**: 83–91.
6. Wineland DJ, Bergquist JC, Itano WM, Bollinger JJ, Manney CH. (1987) Atomic-ion coulomb clusters in an ion trap. *Phys. Rev. Lett.* **59**: 2935–2938.
7. Vogelius IS, Madsen LB, Drewsen M. (2006) Rotational cooling of molecular ions through laser-induced coupling to the collective modes of a two-ion Coulomb crystal. *J. Phys. B: At. Mol. Opt.* **39**: 1267–1280.
8. Koelemeij JCJ, Roth B, Schiller S. (2007) Blackbody thermometry with cold molecular ions and application to ion-based frequency standards. *Phys. Rev. A.* **76**: 023413-1–023413-6.
9. Zajfman D, Wolf A, D Schwalm, Orlov DA, Grieser M, von Hahn R, Welsch CP, Crespo Lopez-Urrutia JR, Schröter CD, Urbain X, Ullrich J. (2005) Physics with colder molecular ions: The Heidelberg Cryogenic Storage Ring CSR. *J. Physics* **4**: 296–299.
10. Koelemeij JCJ, Roth B, Wicht A, Ernsting I, Schiller S. (2007) Vibrational spectroscopy of $HD^+$ with 2-ppb accuracy. *Phys. Rev. Lett.* **98**: 173002-1–173002-4.
11. Prestage JD, Tjoelker RL, Maleki L. (2000) Mercury-ion clock based on linear multi-pole ion trap. *In Proc. 2000 IEEE International Frequency Control Symposium*, p. 706.
12. Wineland DJ, Itano WM, Bergquist JC. (1987) Absorption spectroscopy at the limit: Detection of a single atom. *Opt. Lett.* **12**: 389–391.

13. Wang X-B, Woo H-K, Wang L-S. (2005) Vibrational cooling in a cold ion trap: Vibrationally resolved photoelectron spectroscopy of cold $C_{60}^-$ Anions. *J. Chem. Phys.* **123**: 051106-1–051106-4.
14. Wang XB, Woo HK, Yang J, Kappes MM, Wang LS. (2007) Photoelectron spectroscopy of singly and doubly charged higher fullerenes at low temperatures: $C_{76}^-, C_{78}^-, C_{84}^-$ and $C_{76}^{2-} C_{78}^{2-} C_{84}^{2-}$. *J. Phys. Chem. C.* **111**: 17684–17689.
15. Huang X, McCoy AB, Bowman JM, Johnson LM, Savage C, Dong F, Nesbitt DJ. (2006) Quantum deconstruction of the infrared spectrum of $CH_5^+$. *Science* **311**: 60–63.
16. Asvany O, Kumar P, Redlich B, Hegemann I, Schlemmer S, Marx D. (2005) Understanding the infrared spectrum of bare $CH_5^+$. *Science* **309:** 1219–1222.
17. Bunker PR, Ostojic B, Yurchenko S. (2004) A theoretical study of the millimeterwave spectrum of $CH_5^+$. *J. Molec. Struc.* **695–696**: 253–261.
18. Gerlich D. (1992) Inhomogeneous electrical radio frequency fields: A versatile tool for the study of processes with slow ions. *Adv. in Chem. Phys.* **LXXXII**: 1–176.
19. Gerlich D. (1995) Ion-neutral collisions in a 22-pole trap at very low energies. *Physica Scripta* **59**: 256–263.
20. Dunn G. (1995) Ion-electron and ion-neutral collisions in ion traps *Physica Scripta* **59**: 249–255.
21. Wuerker RF, Shelton H, Langmuir RV. (1959) Electrodynamic containment of charged particles. *J. Appl. Phys.* **30**: 342–349.
22. Paul W. (1990) Electromagnetic traps for charged and neutral particles. *Rev. Mod. Phys.* **62**: 531–540.
23. Diedrich F, Bergquist JC, Itano WM, Winelandg DJ. (1989) Laser cooling to the zero-point energy of motion. *Phys. Rev. Lett.* **62**: 403–406.
24. Blümel R, Chen JM, Peik E, Quint W, Schleich W, Shen YR, Walther H. (1988) Phase transitions of stored laser-cooled ions. *Nature* **334**: 309–313.
25. Hoffnagle J, DeVoe RG, Reyna L, Brewer RG. (1988) Order-chaos transition of two trapped ions. *Phys. Rev. Lett.* **61**: 255–258.
26. Bowe P, Hornekaer L, Brodersen C, Drewsen M, Hangst JS. (1999) Sympathetic crystallization of trapped ions. *Phys. Rev. Lett.* **82**: 2071–2074.
27. Roth B, Blythe P, Schiller S. (2007) Motional resonance coupling in cold multispecies Coulomb crystals. *Phys. Rev. A.* **75**: 023402-1–023402-8.
28. Schiffer JP, Drewsen M, Hangst JS, Hornekær L. (2000) Temperature, ordering, and equilibrium with time-dependent confining forces. *PNAS* **97**: 10697–10700.
29. Rosenband T, Schmidt PO, Hume DB, Itano WM, Fortier TM, Stalnaker JE, Kim K, Diddams SA, Koelemeij JCJ, Bergquist JC, Wineland DJ. (2007) Observation of the$^1S_0 \rightarrow {}^3P_0$ clock transition in$^{27}Al^+$. *Phys. Rev. Lett.* **98**: 220801-1–220801-4.
30. Zhang CB, Offenberg D, Roth B, Wilson MA, Schiller S. (2007) Molecular dynamics simulation of cold single- and multi-species ion ensembles in a linear Paul trap. *Phys. Rev. A.* **76**: 012719-1–012719-13.
31. Baba T, Waki I. (1996) Cooling and mass analysis of molecules using laser-cooled atoms. *Jpn. J. Appl. Phys.* **35**: L1134–L1137.

32. Molhave K, Drewsen M. (2000) Formation of translationally cold $MgH^+$ and $MgD^+$ molecules in an ion trap. *Phys. Rev. A.* **62**: 11401–11405.
33. Roth B, Blythe P, Daerr H, Patacchini L, Schiller S. (2006) Production of ultracold diatomic and triatomic molecular ions of spectroscopic and astrophysical interest. *J. Phys.B* **39**: 1241–1258.
34. Ostendorf A, Zhang CB, Wilson MA, Offenberg D, Roth B, Schiller S. (2006) Sympathetic cooling of complex molecular ions to millikelvin temperatures. *Phys. Rev. Lett.* **97**: 243005-1–243005-4.
35. Bertelsen A, Jørgensen S, Drewsen M. (2006) The rotational temperature of polar molecular ions in Coulomb crystals. *J. Phys. B: At. Mol. Opt. Phys.* **39**: L83–L89.
36. Dzhonson A, Gerlich D, Bieske EJ, Maier JP. (2006) Apparatus for the study of electronic spectra of collisionally cooled cations: Para-dichlorobenzene. *J. Mol. Struc.* **795**: 93–97.
37. Paul W, Lücke B, Schlemmer S, Gerlich D. (1995) On the dynamics of the reaction of positive hydrogen cluster ions ($H_5^+$ to $H_{23}^+$) with para and normal hydrogen at 10 K. *Int. J. Mass Spectrom. Ion Proc.* **150**: 373–387.
38. Gerlich D, Windisch F, Hlavenka P, Plašil R, Glosik J. (2006) Dynamical constraints and nuclear spin caused restrictions in $H_m D_n^+$ collision systems and deuterated variants. *Phil. Trans. R. Soc. Lond. A.* **364**: 3007–3034.
39. Paul W, Gerlich D. (1993) Temperature of trapped ions mesured with spectroscopic methods. In eds. T. Anderson *et al.* ICPEAC 1993, Book of Contributed Papers pp. 807–808.
40. Glosik J, Hlavenka P, Plašil R, Windisch F, Gerlich D. (2006) Action spectroscopy of $H_3^+$ using overtone excitation. *Phil. Trans. R. Soc. Lond. A.* **364**: 2931–2942.
41. Mikosch J, Kreckel H, Wester R, Plasil R, Glosik J, Gerlich D, Schwalm D, Wolf A. (2004) Action spectroscopy and temperature diagnostics of $H_3^+$ by chemical probing. *J. Chem. Phys.* **121**: 11030–11037.
42. Asvany O, Hugo E, Müller F, Kühnemann F, Schiller S, Tennyson J, Schlemmer S. (2007) Overtone spectroscopy of $H_2D^+$ and $D_2H^+$ using laser induced reactions. *J. Chem. Phys.* **127**: 154317-1–154317-11.
43. Gerlich D. (2003) Molecular ions and nanoparticles in rf and ac traps. *Hyperfine Interactions* **146–147**: 293–306.
44. Mikosch J, Frühling U, Trippel S, Schwalm D, Weidemüller M. Wester R. (2007) Evaporation of buffer-gas-thermalized anions out of a multipole rf ion trap. *Phys. Rev. Lett.* **98**: 223001-1–223001-4.
45. Wolf A, Krekel H, Lammich L, Strasser D, Mikosch J, Glosik J, Plasil R, Altevogt S, Andrianarijaona V, Buhr H, Hoffmann J, Lestinsky M, Nevo I, Novotny S, Orlov DA, Pedersen HB, Terekhov AS, Toker J, Wester R, Gerlich D, Schwalm D, Zajfman D. (2006) Effects of molecular rotation in low-energy electron collisions of $H_3^+$. *Phil. Trans. R. Soc. Lond. A.* **364**: 2981–2997.
46. Kreckel H, Motsch M, Mikosch J, Gloslk J, Plasil R, Altevogt S, Andrianarijaona V, Buhr H, Hoffmann J, Lammich L, Lestinsky M, Nevo I, Novotny S,

Orlov DA, Pedersen HB, Sprenger F, Terekhov AS, Toker J, Wester R, Gerlich D, Schwalm D, Wolf A, and Zajfman D. (2005) High-resolution dissociative recombination of cold $H_3^+$ and first evidence for nuclear spin effects. *Phys. Rev. Lett.* **95**: 263201-1–263201-4.
47. Andersen JU, Cederquist H, Forster JS, Huber BA, Hvelplund P, Jensen J, Liu B, Manil B, Maunoury L, Brøndsted Nielsen S, Pedersen UV, Rangama J, Schmidt HT, Tomita S, Zettergren H. (2004) Photodissociation of protonated amino acids and peptides in an ion storage ring. Deter-mination of Arrhenius parameters in the high-temperature limit. *Phys. Chem. Chem. Phys.* **6**: 2676–2681.
48. Schlemmer S, Asvany O. (2005) Laser induced reactions in a 22-pole ion trap. *J. of Phys. Conf. Ser.* **4**: 134–141.
49. Trippel S, Mikosch J, Berhane R, Otto R, Weidemüller M, Wester R. (2006) Photodetachment of cold $OH^-$ in a multipole ion trap. *Phys. Rev. Lett.* **97**: 193003-1–193003-4.
50. Boyarkin OV, Mercier SR, Kamariotis A, Rizzo TR. (2006) Electronic spectroscopy of cold, protonated Tryptophan and Tyrosine. *J. Am. Chem. Soc.* **128**: 2816–2817.
51. Borodi G, Luca A, Mogo C, Smith M, Gerlich D. (2007) On the combination of a low energy hydrogen atom beam with a cold multipole ion trap. *to be submitted to Rev. Sci. Instr.*
52. Haufler E. (1996) Niederenergetische Elektronen- und Protonen-Transferprozesse, Experimente zur Stoßdynamik in Hochfrequenz-Ionenführungsfeldern. *Doktorarbeit TU Chemnitz.*
53. Tosi P, Dmitriev O, Bassi D, Wick O, Gerlich D. (1994) Experimental observation of the energy threshold in the ion-molecule reaction $N^+ + D_2 \rightarrow ND^+ + D$ *J. Chem. Phys.* **6**: 4300–4307.
54. Schlemmer S, Kuhn T, Lescop E, Gerlich D. (1999) Laser excited $N_2^+$ in a 22-Pole trap, Experimental studies of rotational relaxation processes. *Int. J. Mass. Spec.* **187**: 589–602.
55. Bieske EJ, Soliva AM, Friedmann A, Maier JP. (1992) Electronic spectra of $N_2^+$–$(He)_n$ ($n = 1, 2, 3$). *J. Chem. Phys.* **96**: 28–34.
56. Roth B, Koelemeij JCJ, Daerr H, Schiller S. (2006) Rovibrational spectroscopy of trapped molecular hydrogen ions at millikelvin temperatures. *Phys. Rev. A.* **74**: 040501-1–040501-4.
57. Oka T. (2006) Introductory remarks. *Phil. Trans. R. Soc. A.* **364**: 2847–2853.
58. Kreckel H, Mikosch J, Wester R, Glosik J, Plasil R, Motsch M, Gerlich D, Schwalm D, Zajfman D, Wolf A. (2005) Towards state selective measurements of the $H_3^+$ dissociative recombination rate coefficient. *J. Physics.* **4**: 126–133.
59. Dhzonson A, Maier JP. (2006) Electronic absorption spectra of cold organic cations: 2,4-Hexadiyne. *Int. J. Mass. Spec.* **255–256**: 139–143.
60. Dzhonson A, Jochnowitz EB, Kim E, Maier JP. (2007) Electronic absorption spectra of the protonated polyacetylenes $HC_{2n}H_2 +$ ($n = 3, 4$) in the gas phase. *J. Chem. Phys.* **126**: 044301-1–044301-5.
61. Dzhonson A, Jochnowitz EB, Maier JP. (2007) Electronic gas-phase spectra of larger polyacetylene cations. *J. Phys. Chem. A* **111**: 1887–1890.

62. Mercier SR, Boyarkin OV, Kamariotis A, Guglielmi M, Tavernelli I, Cascella M, Rothlisberger U, Rizzo TR. (2006) Microsolvation effects on the excited-state dynamics of protonated Tryptophan. *J. Am. Chem. Soc.* **128**: 16938–16943.
63. Gerlich D. (2005) Probing the structure of $CH_5^+$ ions and deuterated variants via collisions. *Phys. Chem. Chem. Phys.* **7**: 1583–1591.

# Chapter 7

# Chemical Dynamics Inside Superfluid Helium Nanodroplets at 0.37 K

Alkwin Slenczka

*Institut für Physikalische und Theoretische Chemie der Universität Regensburg, Universitätsstraße 31, 93053 Regensburg, Germany, Alkwin.Slenczka@chemie.uni-regensburg.de*

J. Peter Toennies

*Max Planck Institut für Dynamik und Selbstorganisation, Bunsenstraße 10, 37073 Göttingen, Germany, jtoenni@gwdg.de*

## Contents

## 7.1. Introduction

Since the early 90's when the first reports of rotationally resolved IR spectra of a chromophore molecule inside helium nanodroplets were published more than 1000 publications have appeared describing a plethora of new unusual molecular phenomena. These developments derive from the many unique properties of liquid helium most of which have been well known for more than 70 years. But it is only recently that beams of freely levitated nano-sized droplets were used as a cryomatrix which has made it possible to take full advantage of these properties. Perhaps most important, helium is the only substance which remains liquid down to temperatures of zero Kelvin since it solidifies only under applied pressures of 25 bar. This property, which ultimately derives from the very weakly attractive van der Waals interaction between helium atoms, indicates that helium droplets are the only purely liquid clusters. The weak interaction between helium atoms is a direct consequence of the stability of the helium atom with only two electrons in a closed 1s shell. The stability is also responsible for the extremely high energy of $E = 19.82\,\mathrm{eV}$ of the first electronically excited state and the huge ionization potential of 24.5 eV, as well as the unusually low polarizability of $\alpha = 0.21\,\dot{A}^3$ (compared even to the H atom with $\alpha = 0.67\,\dot{A}^3$). Thus bulk helium is transparent down to wavelengths of 59.14 nm making it an ideal "window" over a wide spectral range.

The stability of the He atom is also responsible for an exceedingly weak van der Waals interaction with all other atoms and molecules. Not surprisingly the interaction potential of helium with most molecules are among the weakest of all atom-molecule partners. The van der Waals potential for interactions with alkali and alkali earth atoms are so weak that they cannot support a bound state. This explains why these are the only known surfaces which are not wetted by liquid helium. In connection with the nanodroplet experiments the interaction potentials also explain why virtually all atoms and molecules with the exception of the alkali and alkaline earth atoms are heliophilic, implying that they prefer to be submerged inside the liquid, whereas the alkali and alkaline earth atoms are heliophobic and remain on the surface.

The weak van der Waals potential between He atoms and the bosonic nature of $^4$He also are basic for understanding why $^4$He is the only bulk superfluid below $T_c = 2.18\,K$ at 0.05 bar. The rare isotope $^3$He, a fermion, on the other hand, only becomes superfluid at a three orders of magnitude lower temperature. There are many well known *macroscopic* manifestations of superfluidity such as (i) flow without resistance (ii) a vanishing viscosity (iii) the ability to "creep" out of vessels against the forces of gravity (iv) the fountain effect which is driven by a type of Maxwell demon which separates the superfluid from the normal fluid components and (v) an enormous thermal conductivity which is 30 times greater than that of copper. Table 7.1 compares some properties of liquid argon (also a cryomatrix) with those of helium in the normal and in the superfluid state.

With all these attractive properties it is indeed surprising that the full potential of helium as a cryogenic matrix for spectroscopy was not put into practice until recently. Two problems had to be overcome. First the use of bulk liquid helium requires a cryostat which allows for access through optical windows. More serious, however, is the high mobility of

Table 7.1. Comparison of some physical properties of normal liquid helium and superfluid helium with those of liquid Argon

| Property | Liquid Ar | Normal$^4$ He $T > 2.18\,K$ | Superfluid$^4$ He $T < 2.18\,K$ |
|---|---|---|---|
| Triple point | 83.8 K 0.678 bar | — | — |
| Density [g/cm$^3$] | 1.39 (87.3 K) | 0.146 (4.22 K) | 0.149 (1.0 K) |
| Viscosity [poise] | $2.35 \cdot 10^{-5}$ (85 K) | $30.0 \cdot 10^{-6}$ (4.2 K) | $\leq 10^{-11}$ (T < 2 K) |
| Heat conductivity [cal(cm · s · K)$^{-1}$] | $31.1 \cdot 10^{-5}$ (85 K, 1 bar) | $7.2 \cdot 10^{-5}$ (4.2 K) | > 238 (T < 2 K) |
| Heat of vaporization [K][a] | 76 | 9.69 | 7.2 (T = 0 K) |
| Lowest Electronically Excited State [eV] | 11.55 | 19.82 | |
| Ionization Potential [eV] | 15.76 | 24.6 | |
| Atom-Atom Potential well depth [K][a] | 143.4 | 11.0 | |
| Atom-Atom Equilibrium Distance [Å] | 3.76 | 2.97 | |

[a]Here and elsewhere in this review energies will be frequently divided by the Boltzmann constant and given in units of Kelvin.

foreign species in bulk helium. Thus once injected into the liquid the foreign molecules quickly coagulate to form clusters which either float to the surface or fall to the bottom of the container. Up to now it has only been possible to evaporate certain metal atoms by laser ablation inside the liquid and probe the atoms by a delayed laser pulse.

Both of the above problems are overcome by using an isolated liquid helium droplet as a type of nanocryomatrix. Droplets are easily formed by expansion of the cold gas through a small nozzle into vacuum. By evaporation the droplets rapidly cool to very low temperatures of 0.37 K for $^4$He and 0.15 K for $^3$He droplets. Thus the $^4$He droplets are cold enough to be superfluid whereas the $^3$He droplets are not. Being liquid the droplets readily pick up and absorb into their interior all heliophilic species which they encounter in their path. Thus each of the droplets serves as an individual potential cage for trapping a single foreign molecule or a defined mixture of molecules which aggregate to form complexes in the droplet interior.

The big surprise came in 1995 when high resolution infra-red spectroscopy revealed for embedded $SF_6$ molecules a spectrum with fully resolved rotational lines of the P, Q and R branches.[1] From the relative intensities of the individual rotational lines the temperatures were measured to be 0.37 K in remarkable agreement with earlier theoretical predictions. Subsequent IR-spectroscopy on linear and symmetric top molecules confirmed that the same rotational Hamiltonian as in the gas phase could explain all the spectral features in droplets. Greatly increased moments of inertia were observed for slow rotor molecules ($B < 1\,\mathrm{cm}^{-1}$) while fast rotors were not affected. Evidence that the free rotations are related to superfluidity came from an experiment in which the spectrum of OCS in non-superfluid $^3$He droplets showed a broad unresolved band structure as expected for a classical fluid.

Since then a large body of experimental studies of more than 50 different molecules and more than 80 different clusters have been reported. Extensive theoretical and computational work has contributed substantially to the present, in many respects detailed, microscopic understanding of the solvation structure around the embedded molecules and to a lesser extent the coupling with the superfluid environment and $^4$He collective excitations. At present, it is generally accepted that the phenomena of free rotations is a new manifestation of superfluidity on a microscopic level. These experiments and the large number of other spectroscopic studies in the infra-red, visible and near UV regions have been extensively surveyed.[2–5a]

The present review is devoted to a relatively recent development, the use of helium droplets as an ultra-cold gentle nanoreactor for studying elementary chemical processes. These experiments take full advantage of the

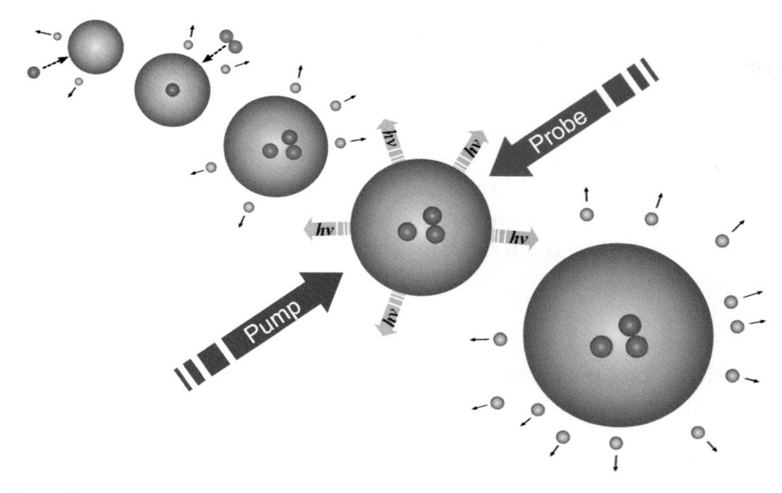


Fig. 7.1. Schematic diagram illustrating a typical experiment in which a helium droplet serves as an ultra-cold gentle nanoreactor. In some experiments (not shown) the products ejected from the droplet may also be detected. In many experiments the resonant absorption of the pump and/or probe lasers is monitored by observing the depletion of the signal of the droplets measured with a detector at the far end of the apparatus.

self-organization of the added species as well as the rapid relaxation of highly vibrationally excited species Fig. 7.1 illustrates schematically how the experiments in this novel reactor are carried out. First the droplets are doped with the reactant species, which as soon as they enter the droplet are instantaneously cooled to 0.37 K. If other species are already present they will form a van der Waals complex which may react on contact. Alternatively, a reaction can be initiated by laser excitation of one of the constituents of the complex to a specific vibrational and/or electronic state. The nature and quantum states of the newly formed products can then be probed either via laser induced fluorescence or by laser induced depletion of the droplet signal measured further downstream from the reaction region by a sensitive detector.

In the next section some important experimental aspects needed for understanding the new experiments are surveyed. A decisive result coming out of this new experimental direction is the beneficial influence of the helium environment on the microscopic dynamics. In many of the experiments the vibrations, especially of large molecules which are initially highly excited are very rapidly quenched. As a result the dynamics and the laser interrogation are greatly simplified. The downside is that translations of molecules moving very rapidly inside the droplets are appreciably slowed by collisions with the individual atoms of the droplet. To provide perspective to this new insight Sec. 7.3 briefly surveys what was previously known about the coupling of the environment to the translational, rotational and vibrational degrees of freedom on the basis of the foregoing spectroscopic studies. Then in Sec. 7.4 the most recent photodissociation and photoelectron spectroscopy experiments as well as a very recent study of photo-induced tautomerization are surveyed. In Sec. 7.5 chemical reactions including ion-molecule reactions, isomerization, and bimolecular reactions are discussed. In the final Sec. 7.6 the main new insights gained and the outlook for further studies are summarized.

## 7.2. Helium Droplet Formation and Properties

The apparatus used in a typical helium droplet experiment is in many respects very similar to that currently in use for spectroscopic studies of molecules and clusters in seeded beam free jet expansions (see the chapter by Davis, Dong and Nesbitt). The source nozzle and skimmer arrangement are essentially identical. The only major difference is that for producing helium droplets the source is typically cooled to temperatures of less than about 30 K in order that the helium gas will condense in the expansion cooling to form sufficiently large droplets. Much in the same way as in the spectroscopic studies of free molecules, the beams are detected either with a mass spectrometer equipped with an electron impact ionizer or with a bolometer. Thus the resonant photon absorption in droplets is monitored by detecting the depletion of the droplet signal resulting from the heat released and subsequent evaporation of He atoms from the droplets. In other experiments laser induced fluorescence or chemiluminescence spectroscopies provide information on intramolecular energy transfer or reactions, respectively. In one experiment femtosecond lasers are used to ionize ejected neutral products and their angular and velocity distributions are detected using ion imaging techniques. Photoelectron spectroscopy, ZEKE (Zero Electron Kinetic Energy) spectroscopy and TOF (Time Of Flight) mass spectrometry have been applied as well. In this section several of the special physical attributes of the droplets such as their sizes, cooling rates and the growth and self organisation of embedded species inside the droplets will be discussed. For more details the reader is referred to several recent reviews,[2–5a] where references to the original publications can be found.

### 7.2.1. *Droplet Formation and Droplet Sizes*

For creating droplets liquid helium cooled cryostats or closed cycle systems are used to cool the source. Because of the high gas flux at low temperatures a small nozzle with a diameter of 5$\mu$m and relatively large pumping rates ($\sim$3000 l/s) are required to maintain a sufficient vacuum in the source chamber. The sizes of the helium droplets have been extensively investigated by Toennies and co-workers in Göttingen.[6,7] For measuring the sizes of droplets with up to about $10^4$ atoms a secondary beam of a heavy rare gas with a sharp velocity distribution was directed at the droplet beam and the resulting distribution of small angular deflections ($\approx 10^{-3}$ radians) of the droplets was measured with a sensitive mass spectrometer. The heavy rare gas atoms transfer their entire momentum to the droplets and consequently the deflection distribution is directly related to the droplet size distribution.

Fig. 7.2 summarizes the measured dependence of the mean number sizes and diameters on the source temperature for four different source pressures of 20, 40, 60 and 80 bar.[6,7] In many of the experiments to be discussed in this review the droplets are produced in expansions of the gas and have $N \leq 2 \cdot 10^4$ atoms with an overall diameter of about 100Å (Fig. 7.2). The much larger droplets produced at temperatures below about 11–15 K, as seen in Fig. 7.2, is explained by the increasingly large liquid fraction in the source. The insert in Fig. 7.2 shows some typical size distributions of droplets

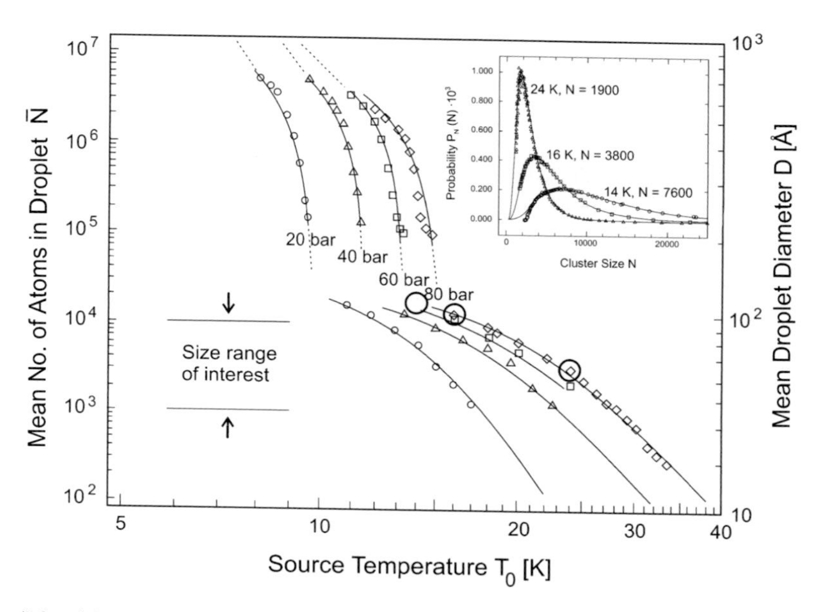


Fig. 7.2. Measured mean number droplet sizes and diameters as a function of the source temperature for four different source pressures. The insert shows some typical size distributions and log-normal best fits[6,7] for droplets produced in expansions of the gas indicated by the circles.

produced in gas expansions. The sizes are well fitted with a log-normal distribution $P(N) \propto \left(-\frac{\ln N - \mu}{2\sigma^2}\right)$ which is characterised by a half-width which is approximately equal to the mean number size $\bar{N}$. Droplets with sizes between $10^3$ and $10^4$ atoms have velocities of about 300 m/s and very sharp velocity distributions of $\Delta v/v \leq 3\%$.

Fig. 7.3 shows the calculated dependence of the average number sizes and temperatures of droplets made up of $^3$He or $^4$He atoms as a function of time after they have been formed.[8] After their formation the droplets evaporatively cool down very rapidly to temperatures of the order of about 1 K within less than $10^{-8}$ s. As the internal temperature decreases further

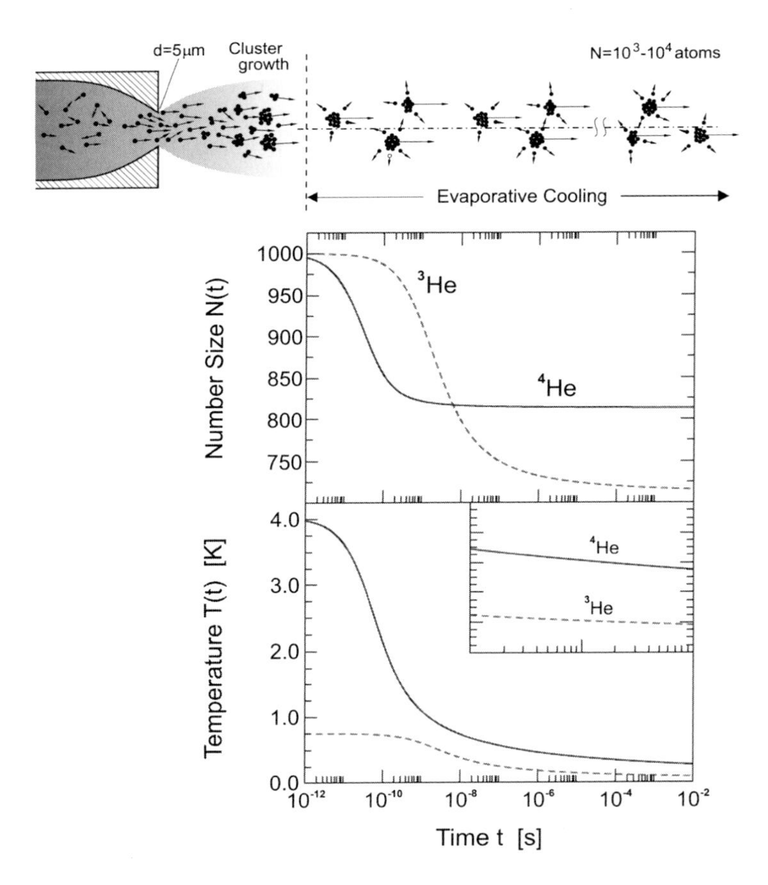


Fig. 7.3. Calculated time evolution of the mean number sizes and temperatures of $^3$He and $^4$He droplets consisting initially of 1000 atoms at an initial temperature of 4.0 K ($^4$He) and 0.8 K ($^3$He). Adapted from Ref. 8.

the rate of evaporation decreases accordingly and the rate of cooling slows down. After about $10^{-3}$ s, a typical flight time in a droplet experiment, they are predicted to have translational temperatures close to the measured rotational temperatures of 0.37 K.[1,9] This suggests that the dopant molecules are fully equilibrated after the relatively long flight times, as one might expect.

### 7.2.2. *Doping Droplets*

Since, as mentioned in the Introduction, helium droplets are definitely liquid they possess the remarkable ability to pick up all molecules which they encounter as they move through the apparatus. Because of their large size the pick up cross sections are of the order of the geometrically predicted area of $10^3$–$10^4$ Å$^2$. This, of course, means that the residual vacuum has to be very good ($<10^{-8}$ mbar) since otherwise the droplets and thus the molecules of interest will be contaminated by residual gas species. For intentional pick up a pressure of about $10^{-5}$ mbar inside a scattering chamber of several centimeters length is sufficient to embed a single molecule of interest. Thus hard-to-vaporize metals or other refractory materials, as well as fragile biomolecules, can be embedded since vapour pressures of this order-of-magnitude can be achieved even at modest temperatures.

There is now extensive evidence that all closed shell molecules, including those to be discussed here, reside in the droplet's interior and hence are classified as heliophilic. Only alkali and alkaline earth atoms and their small clusters appear to be heliophobic and remain attached at the surface of the droplets. Thus, as additional heliophilic molecules are added they will encounter and recombine with those already present in the droplet interior. This was first observed spectroscopically for $SF_6$ inside droplets consisting of $4 \cdot 10^3$ He atoms. Fig. 7.4 shows the depletion signal measured with the laser tuned to the different infra-red bands of each of the small oligomers.[10] The resulting distribution can be invariably fitted by a Poisson distribution

$$P_k(z) = \frac{z^k}{k!} \exp(-z), \tag{7.2.1}$$

where $z$, the average number of collisions of the droplet, is directly proportional to the pick-up cross section. The smooth curves in Fig. 7.4 were all fitted by Eq. (7.2.1) with a pick-up cross section of 3900 Å$^2$ which agrees nearly exactly with the cross section calculated assuming a liquid droplet with a constant bulk density.

The total number of molecules which can be picked up depends mainly on the size of the droplet. Typically the energy released on pick-up of a

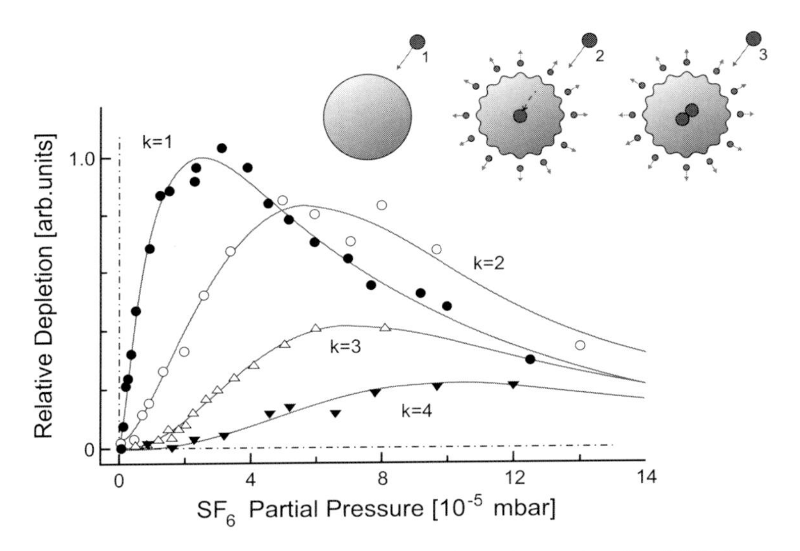


Fig. 7.4. Dependence of the depletion signals for infra-red bands of $(SF_6)_k$ clusters embedded inside helium droplets with $\bar{N} = 4.1 \cdot 10^3$atoms as a function of the $SF_6$ pressure in a pickup cell.[10] Solid curves are best fits based on a Poisson distribution (Eq. (7.2.1)) for a pick up cross section of 3900 Å$^2$. The insert shows a schematic representation of the pick-up, coagulation and resulting heating induced evaporation.[10]

single molecule leads to the evaporation of between 100 and 2000 He atoms depending on the temperature and heat capacity of the molecule. Thus large droplets with $10^7$ atoms (see Fig. 7.2) can easily pick-up $10^3$–$10^4$ atoms which aggregate to an interior cluster, as has been observed for Mg atoms. Of course, more than one species can be added either by using a gas mixture in a single scattering cell or by using two or more scattering cells in series. This has been demonstrated in past spectroscopic studies for a wide range of different combinations, including, for example, RG-$SF_6$, RG-HF, tetracene-$Ar_N$, phthalocyanine-$Ar_N$, $Mg_{1\text{-}3}$-HCN and OCS-$(H_2)_{1\text{-}17}$, where RG = Ne, Ar and Kr. Very recently several examples of unusual complexes involving embedded free radicals could also be identified.[11]

### 7.2.3. *Self-Organisation of Embedded Clusters*

One of the surprising observations made in connection with the clusters formed inside helium droplets is that they frequently adopt unique structures. This was first observed in the infra-red spectra of clusters of polar molecules such as cyanoacetylene (HCCCN).[5] Once inside the cold droplet each additional molecule is rapidly cooled translationally and subsequently aligned by the electric field of its predecessors and as a result they combine to form long chains which were not seen in seeded beam expansions. Another example is OCS with added $H_2$ molecules. The infra-red spectra indicate that as many as 16 $H_2$ molecules self-assemble to form up to 4 concentric rings around the axis of the OCS molecule. The structures of the complexes can even be influenced by changing the order in which different molecules or atoms are added. Thus the long range forces between the constituents which become active at the extremely low temperatures lead to novel structures which could not be synthesized previously. These observations open the door to a vast new field of structural and dynamical studies.

## 7.3. Translational, Rotational and Vibrational Coupling of Molecules to Their Local Helium Environment

As mentioned in the Introduction most of the early droplet experiments have been devoted to high resolution studies of ro-vibrational and vibronic transitions. These experiments have provided considerable insight into the couplings of the various molecular degrees of freedom with the helium environment albeit under far less extreme conditions compared to those which will be encountered in the next sections.

### 7.3.1. *Translational Degrees of Freedom*

Already in 1947 Landau in his seminal publications pointed out that superfluid helium would, unlike ordinary liquids, have a phonon dispersion curve which is as sharp as that found in solids. After passing through a maximum, which in helium is called the maxon, the dispersion curve does not go down to zero energy as in a solid, but instead exhibits the so-called roton minimum at an energy gap of 8.5 K (Fig. 7.5). One of the most important consequences of the sharp dispersion curve, and roton gap is that inside the superfluid an object can move without friction if its velocity is less than 58 m/s. For simplicity Landau assumed an object of infinite mass since he was mostly interested in explaining the interaction of the flowing superfluid with the walls of an enclosing channel. This fascinating idea has lead to considerable early speculation that helium droplets might be "transparent" for impinging molecules. This idea overlooks the fact that heliophilic foreign molecules are strongly bound to the droplet interior by a collective potential which is called the chemical potential. Fig. 7.6(a) shows schematically the effective chemical potentials by which three typical particles, Ar, Xe and $SF_6$ are trapped inside a helium droplet consisting of $10^3$ atoms.[12] The enlarged portion in (b) shows how this energy relates to the Boltzmann distribution of the translational energies of the embedded molecules at 0.37 K

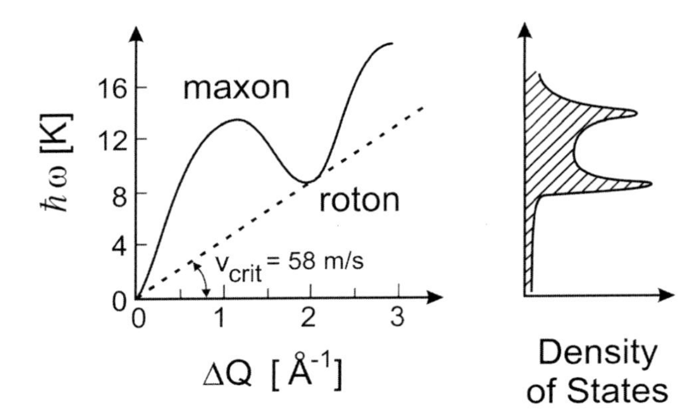


Fig. 7.5. Schematic diagram showing the sharp dispersion curve for elementary excitations in bulk superfluid $^4$He. Since a particle of infinite mass moving with a velocity of less than 58 m/s will not interact with the elementary excitations it can translate without friction.

and the energies of quantized surface modes of the droplet, the so called ripplons. The roton energy of 8.5 K (5.91 cm$^{-1}$) is also included in Fig. 7.6 since it provides an order of magnitude estimate of the limiting translational energy for frictionless motions. Obviously, particles coming from the outside will have much greater kinetic energies with respect to the bottom of the well and will not initially undergo frictionless motion. Once trapped and fully accommodated, however, their energies will be sufficiently low to allow them to move within the droplet nearly without friction.[13–15] Since the molecules are localized near the center of the droplets and the surface ripplons are largely localized at the droplet surface this coupling is expected to be very weak. There has been an extensive search for evidence of this frictionless motion in the high resolution spectra of embedded molecules.[14,15] So far there is no conclusive evidence for frictionless motion nor for the coupling of any such motion with the rotations of the chromophore or with the surface modes.

### 7.3.2. *Rotational Degrees of Freedom*

Most of the present information on the coupling of the rotational degrees of freedom to the helium environment come from the line widths of infra-red ro-vibrational transitions of small mostly diatomic and triatomic molecules. In view of the small changes in vibrational amplitudes accompanying vibrational fundamentals of only about 0.05 − 0.10 Å, the effect on the line width is considered to be negligible and the rotational coupling is the dominant mechanism determining the linewidth. This conclusion has been confirmed

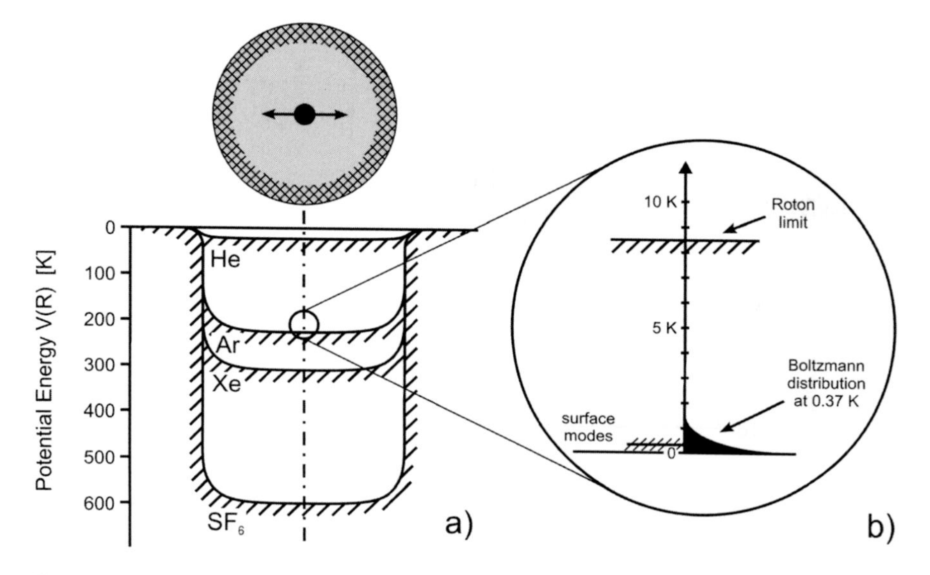


Fig. 7.6. Schematic diagram showing (a) the chemical potential holding several foreign species inside a helium droplet consisting of $10^3$ atoms. The expanded energy scale in (b) compares the Boltzmann distribution at 0.37 K with the energy of the lowest surface ripplon excitation mode and the energy of the roton. Since the foreign species are localized near the droplet center they are not expected to couple significantly to the ripplons at the surface. It should be noted that photodissociation fragments will have to surmount the chemical potential in order to be able to leave a droplet.

both by a number of observations as well as theoretical estimates. Although the inhomogeneous broadening found in solid matrices as a result of the different possible sites within the rigid matrix is not expected some inhomogeneous broadening arises from the distribution in droplet sizes and possibly from other as yet not fully identified effects. Nevertheless for many chromophores the rotational lines exhibit homogeneous broadening which can be directly attributed to rotational relaxation. Fig. 7.7 adapted from a recent review by the late Roger Miller and colleagues,[5] shows a remarkable correlation between the wide range of homogenous line widths and the rotational energies of the excited state. The comparison with the density of states of He droplet elementary excitations (for large droplets they can be approximated by the bulk density of states) strongly implies that the sharp increase in line widths is due to a coupling with the roton excitations of the superfluid (Fig. 7.5). Thus in many respects the perturbation of the local helium environment by rotations of a molecule is quite similar to that predicted for the translational motion. Fig. 7.7 also explains why the heavy molecules $SF_6$ and OCS with rotational constants of only 0.09 $cm^{-1}$ and

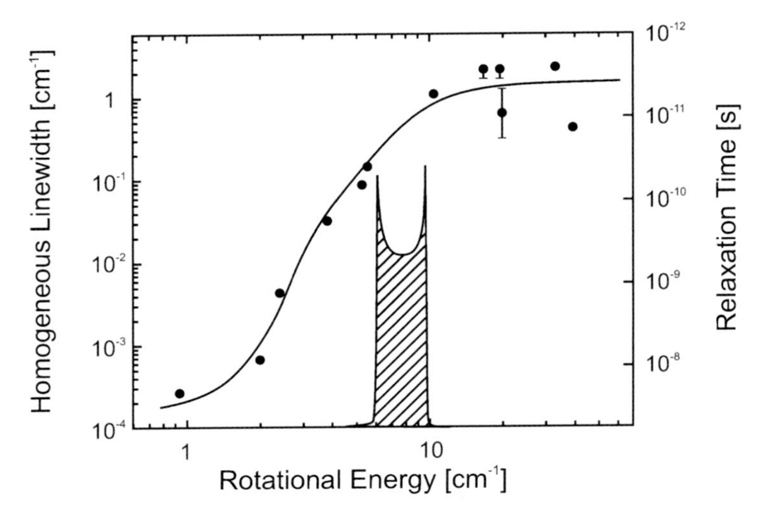


Fig. 7.7. The homogeneous linewidths for the molecules $H_2O$, HDO, HF, $C_2H_2$, $CH_4$, $C_2H_4$, CO, $NH_3$, DCN, HCN, and HCCCN are plotted as a function of the excited state rotational energy. The dramatic increase in the linewidths with increasing rotational energy correlates with the roton energy gap (Fig. 7.5), characteristic of the superfluid. Adapted from Ref. 5.

0.2 $cm^{-1}$ respectively, which lie far below the roton energy, were observed to have remarkably sharp rotational lines.

The relaxation times of between $10^{-8}$ and $10^{-12}$ s encountered in helium droplets (Fig. 7.7) are much longer than the corresponding times in classical liquids. In general a classical liquid does not allow for free rotational motion of a solvated molecule. Thus, the relaxation time is usually much shorter than the respective molecular rotational period.

### 7.3.3. *Vibronic Degrees of Freedom*

Parallel to the extensive spectroscopic investigations in the infra-red there have also been a large number of studies of vibronic transitions in the visible and near UV regions. As in other matrices the spectra consist of a zero phonon line (ZPL) involving a pure electronic transition with only a weak coupling to the matrix and a phonon wing (PW) resulting from the interaction with the internal degrees of freedom of the matrix. In helium droplets vibronic transitions have widths of the order of 0.2–1.0 $cm^{-1}$ and are usually much narrower, by up to three orders of magnitude, compared to other matrices. As first observed with glyoxal the phonon wing is separated from the zero phonon line by a gap followed by a distinct peak.[16]

A theoretical simulation revealed that the gap and subsequent peak were nearly proportional to the density of states which reflects the roton gap of the superfluid (see Fig. 7.5). This experiment provided the first convincing evidence that the elementary excitations in the droplets are similar to those in the bulk superfluid and that the droplets are indeed superfluid. They also indicate that the changes in the external electronic charge distribution accompanying an electronic transition can couple to the helium environment.

The two experiments on the planar molecules tetracene and phthalocyanine to be discussed next were the first to reveal the efficient dissipation of vibrational energy as the dominant process following electronic excitation of the dopant molecule. These experiments have also provided the first evidence for the dynamical behavior of the non-superfluid shell of He atoms surrounding the chromophore, the existence of which had been previously deduced from the anomalously large moments of inertia observed in the rotational spectra of heavy molecules (Sec. 7.3.2).

In the near UV spectrum tetracene in helium droplets shows, in addition to a broad phonon wing (PW) (Fig. 7.8b), a splitting of the zero phonon

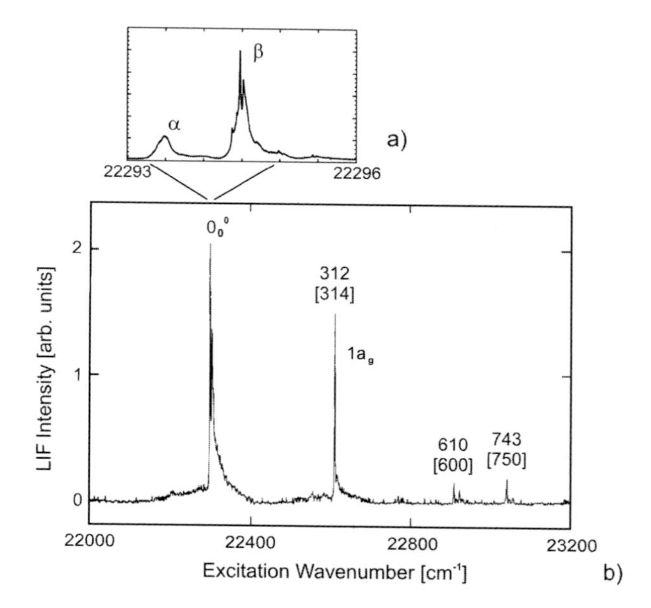


Fig. 7.8. Laser induced fluorescence (LIF) spectrum of single tetracene molecules embedded in large helium droplets ($N_4 = 1.5 \cdot 10^4$ atoms). The values in brackets are the frequencies in $cm^{-1}$ of the free molecule. Part (a) shows a high resolution scan of the $O_0^0$ zero phonon line with the $\alpha$–$\beta$ splitting. The tails in the major peaks in (b) are due to the phonon wings which are better resolved in Fig. 7.9(a) Adapted from Ref. 17.

line (ZPL) into two components, called $\alpha$ and $\beta$ which are separated by only $1.1\,\mathrm{cm}^{-1}$[17] (Fig. 7.8a) and not found in the gas phase spectrum. Pump-probe spectra and the fluorescence decay times revealed for each of the two ZPLs[18] independent sets of initial and final energy levels, which were attributed to different configurations of the non-superfluid solvation layer of He atoms surrounding the tetracene molecule. The same model served to explain the spectral shape of the PW which differed significantly from the spectrum expected from the level density of pure superfluid helium.[19] If however tetracene was complexed with a single Ar atom or in the case of naphthalene and pentacene, the ZPL consisted of only a single transition. These results suggest that the splitting of the ZPL is a special feature related to the particular shape of the tetracene — helium interaction potential. The quantitative understanding of this splitting is still under debate.[20]

In the course of this investigation the efficient cooling capacity of the helium environment could be demonstrated in a pump-probe experiment. Monitoring the depletion upon excitation at the maximum of the PW the probe-laser was scanned over a range of $120\,\mathrm{cm}^{-1}$ centered around the ZPL (Fig. 7.9a). The depletion of the pump laser was either enhanced or

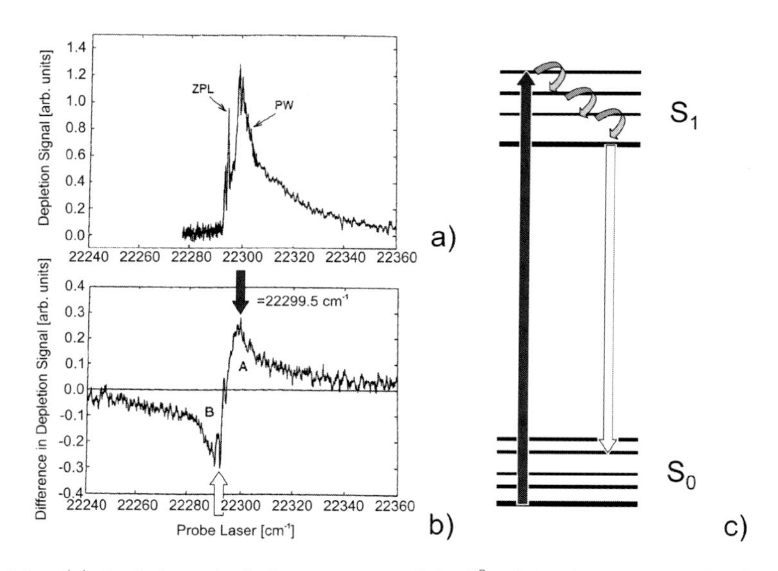


Fig. 7.9. (a) single laser depletion spectrum of the $0^0_0$ origin of tetracene in He droplets. (b) Pump-probe spectrum with the pump laser at the maximum of the phonon wing (22299.5 $\mathrm{cm}^{-1}$). The time delay between pump and probe was 18 ns. (c) Schematic diagram showing the two electronic transitions of Fig. 7.9b and some of the phonon energy levels in the $S_0$ ground and $S_1$ excited states. [Adapted from Ref. 18.]

reduced by the probe laser. This was attributed to additional excitation or a stimulated emission, respectively. The positive (A) and the negative (B) signal contributions shown in the pump-probe spectrum of Fig. 7.9(b) were almost mirror images of each other. The increased excitation (A) occurred at about the same frequency as the maximum of the PW. The similarity of the spectral shape of the reduced depletion (B) was explained by stimulated emission into the $S_0$ phonon wing preceded by relaxation of the initially excited phonon (cf. Fig. 7.9c). Based on the 18 ns time delay between the pump and probe pulses, the relaxation of the bath phonons accompanying the excitation of the $S_1$ state phonons must have been completed within this time, which was confirmed by the red shifted stimulated emission upon excitation at the maximum of the PW.

Even more direct evidence for the highly efficient dissipation of energy into the helium droplet environment comes from the dispersed emission spectra of laser excited phthalocyanine (Pc).[21] Surprisingly, in helium droplets the dispersed spectra reveal emission exclusively from the ground vibrational level of $S_1$ independent of the wavelength of excitation, which was not observed in the corresponding gas phase experiments. The same behavior was also found for excess excitation into higher electronic levels. However, in this case it is not clear whether the decay of the electronic excitation proceeds according to Kasha's rule, which implies internal conversion (IC) of electronic and vibrational energy, or whether it is mediated by the helium environment. In any case the rate constant for energy dissipation dominates over the rate constant for radiative decay. In the case of Pc, the rate constants could be extracted directly from the Lorentzian line widths of vibronic transitions in the fluorescence excitation spectrum.[22] They were found to be of the order of $10^{12}$ $s^{-1}$, almost independent of the excess excitation energy. Similar observations were also made for various organic molecules in helium droplets such as tetracene, pentacene, perylene, Mg-phthalocyanine, porphycene, flavine and van der Waals complexes of Pc with Ar, $H_2O$ and $N_2$.[23,24] For solid matrices the energy dissipation into the host system occurs at similar rates, but the much larger inhomogeneous line widths obscures most of the underlying dynamics.

These emission spectra in helium droplets revealed the dissipation of vibrational excitation energy prior to radiative decay. However, the particular mechanism for the dissipation, whether it proceeds directly into the helium environment or whether intramolecular vibrational redistribution (IVR) is preceding the dissipation, is not revealed by the data. In the case of phthalocyanine the gas phase emission spectra of excited vibronic levels indicate a radiative rate constant which exceeds that of IVR. Under the assumption that also in helium droplets IVR has a smaller rate constant than the radiative decay, the decay rate of vibronic levels as measured in helium droplets represents an upper limit for the process of energy dissipation into the helium droplet environment.

In the case of phthalocyanine no splitting of the ZPL was observed in the fluorescence excitation spectrum. However, the dispersed emission showed a line doubling. Compared to the respective gas phase spectra the emission appeared to consist of two overlapping spectra which were identical in the vibrational energies and in the Franck-Condon factors but were shifted by 10.2 $\text{cm}^{-1}$ with respect to each other (Fig. 7.10).[21] With increasing excess excitation energy $\nu_E$ the intensity of the spectrum with the larger solvent red shift increases as that of the other spectrum decreases. This observation was explained by the relaxation of a non-superfluid helium solvation layer from a metastable into a global minimum configuration within the electronically excited $S_1$ state. In the ground $S_0$ state the back relaxation of the solvation layer could also be identified in

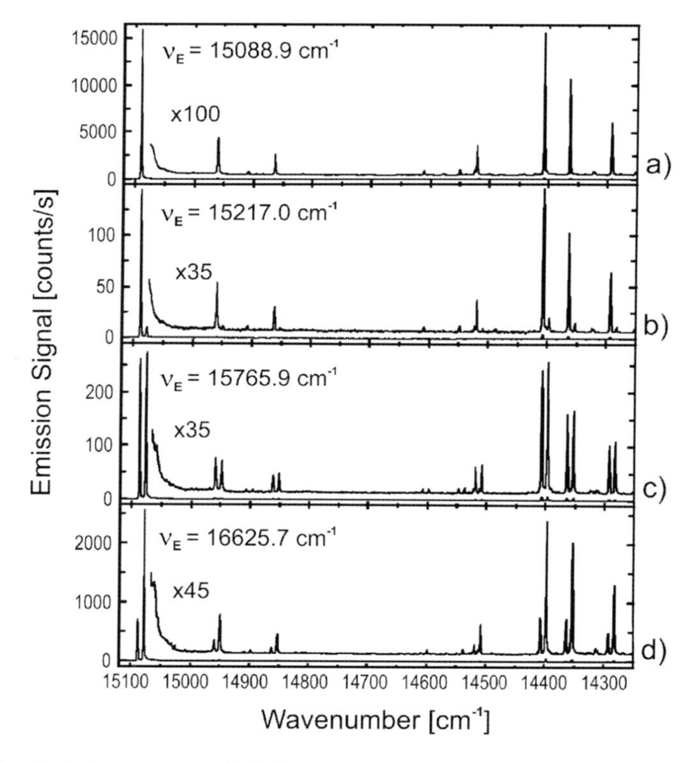


Fig. 7.10. Emission spectra of phthalocyanine in helium droplets ($N = 4 \cdot 10^4$), for four different laser excitation energies $\nu_E$. The upper traces in each frame are vertically shifted and enlarged by a factor as indicated.[21] The spectra consist of two sets of identical structures with the only difference that the line intensities of the red shifted set increases at the expense of the blue shifted set with increasing excitation energy.

a pump-probe experiment. By changing the distance between the pump and probe laser interaction regions the relaxation rate was determined to be $1.9 \cdot 10^5\,s^{-1}$.[25,26] The empirical interpretation of the observed splitting was quantitatively confirmed by PIMC simulations of the solvation layer which revealed two different arrangements — one commensurate and the other incommensurate — of 12 He atoms attached to each side of this planar molecule.[27] Similar experimental observations were made for Mg-phthalocyanine and for Pc-Ar clusters.[23]

The above experiments reveal that two major relaxation processes must be considered for molecules inside superfluid helium droplets. These are rapid relaxation of intramolecular vibrational energy by dissipation into the helium droplet, and slow relaxation of the structural configuration of a non-superfluid helium solvation layer. At the present time the extent to which the solvation layer is involved in the energy dissipation mechanisms is not yet clear. The highly efficient coupling of the dopant to the otherwise very gentle helium environment is an effect which certainly needs to be considered in photochemical experiments performed in superfluid helium droplets.

In concluding this section it is important to point out that here we have only been able to discuss selected examples of what is known about the interactions of optically excited molecules in helium droplets. Finally, we would like to mention additional important results where exceptionally narrow widths of only 15 MHz (in the tunnelling transition of $NH_3$),[28] or very broad lines ($\approx 10\,cm^{-1}$) as in butterfly vibrations of pentacene in the $S_1$ excited state,[29] or no relaxation as for the vibrationally excited HF molecule,[30] have been observed. The wealth of new data on electronic spectroscopy in helium droplets is now beginning to yield insight into the mechanisms of the solvation of molecules in superfluid helium droplets. The quantitative understanding, however, still awaits ongoing and future research.

## 7.4. Photochemistry in Helium Droplets

Since the sun's radiation is our primary source of energy, photochemistry is one of the most fundamental and important processes for understanding life on earth. From the highest altitudes down to the earth's surface gas-phase photochemistry controls the dynamic equilibrium of the atmosphere. Moreover, elementary biological processes in plants and animals such as photosynthesis and vision are photoinitiated intra- or inter-molecular processes. For understanding biological systems it is important to study not only the isolated molecules but also the influence of an environment. In this respect superfluid helium droplets provide unique conditions as an extremely cold yet gentle host, which nevertheless may influence the photochemical dynamics. Some of the first experiments on photochemistry in helium droplets which address the extent to which the superfluid environment affects the molecular dynamics are discussed below. Already these experiments have uncovered a highly efficient energy dissipation from the excited dopant molecules into the helium droplet which provides for special advantages in their laser interrogation.

### 7.4.1. *Photodissociation*

In a gas-phase photodissociation experiment the opening and closing of different break-up channels can be explored by varying the photon excitation energy. The results provide important details on the multidimensional potential surface. The interpretation is more complicated in most of the common liquid or solid matrices since the photochemical processes are usually strongly coupled to the local environment. The gentleness and highly effective cryogenic cooling in helium droplets serve as an especially simple system for studying the effect of an environment. Up to the present time the photodissociation of $H_2O$,[31] $NO_2$[32] and of $CF_3I$ and several other alkyl and methyl halides[33–36] have been studied in helium droplets. All three types of systems are well known from gas phase experiments so that the effect of the helium environment on the dissociation dynamics can be identified.

#### 7.4.1.1. *Photodissociation of $H_2O$*

The photodissociation of $H_2O$ in the gas phase below 140 nm involves three different channels each of which yield fluorescing fragments with different spectral signatures:

$$H_2O + h\nu \rightarrow OH^* + H \tag{1}$$

$$H_2O + h\nu \rightarrow OH + H^* \tag{2}$$

$$H_2O + h\nu \rightarrow H_2O^{+*} + e^- \tag{3}$$

The same system has been investigated by Möller and coworkers in helium droplets. At wavelengths above 100 nm (12.41 eV), which is *below* the ionization energy threshold for $H_2O$ at 98.4 nm (12.61 eV), the fluorescence excitation spectrum is remarkably similar to that observed for the free molecules (See Fig. 7.11).[31] In this experiment the well known Rydberg transitions of $H_2O$ were excited and the emission of $OH^*$ (channel (1)) at 315 nm was recorded. No significant emission was observed either from resonant VUV emission or of the continuous visible radiation expected if the

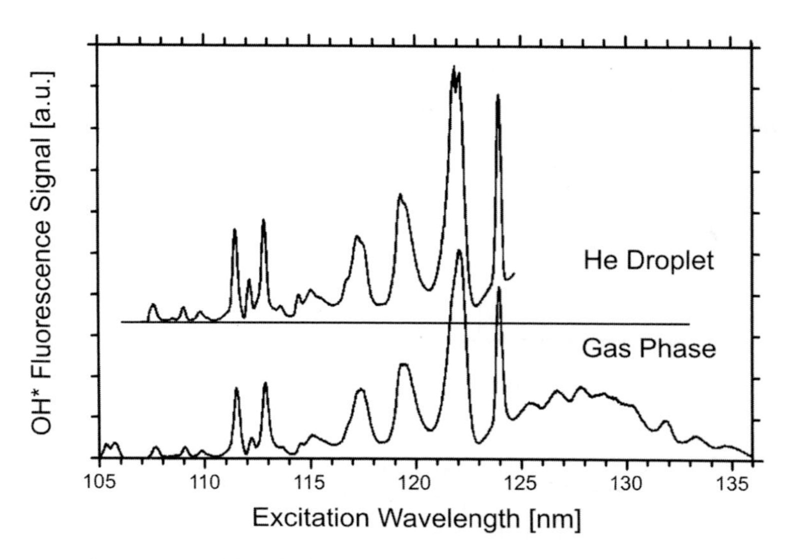


Fig. 7.11. Comparison of the UV fluorescence spectra of free water molecules with water molecules inside large He droplets ($\bar{N} = 10^4$ atoms).[31]

photofragments were to recombine. The major effect of the helium droplet environment was cryo-cooling of the embedded species which became apparent in the reduced line width of the low temperature rotational envelope as observed in a high resolution scan. At these photon energies the surrounding helium environment was of negligible influence on the dissociation dynamics of $H_2O$ and in particular the lack of the typical radiation generated by fragment recombination indicated that caging did not take place.

At energies *above* the ionization threshold of the $H_2O$ molecules, a number of remarkable phenomena were observed as illustrated in the emission spectra shown in Fig. 7.12. Under gas phase conditions and excitation at 67.1 nm (18.5 eV) channels (2) and (3) are accessible and contribute the $\beta, \gamma, \delta$ and $\varepsilon$ Balmer lines of $H^*$ and a broad structureless emission from $H_2O^{+*}$ (cf. Fig. 7.12a), respectively. In the gas phase at 57.5 nm (21.6 eV) only channel (3) is excited as identified from the broad structureless emission from $H_2O^{+*}$ (cf. Fig. 7.12c). As a great surprise, for both wavelengths the emission spectra in helium droplets showed exclusively the Balmer lines of $H^*$ (cf. Figs. 7.12b and 7.12d) which was interpreted as a suppression of the ionization channel (3) in favor of the predissociation channel (2). In addition the 57.5 nm excitation was in resonance with the pure helium droplets resulting in a rich emission in the

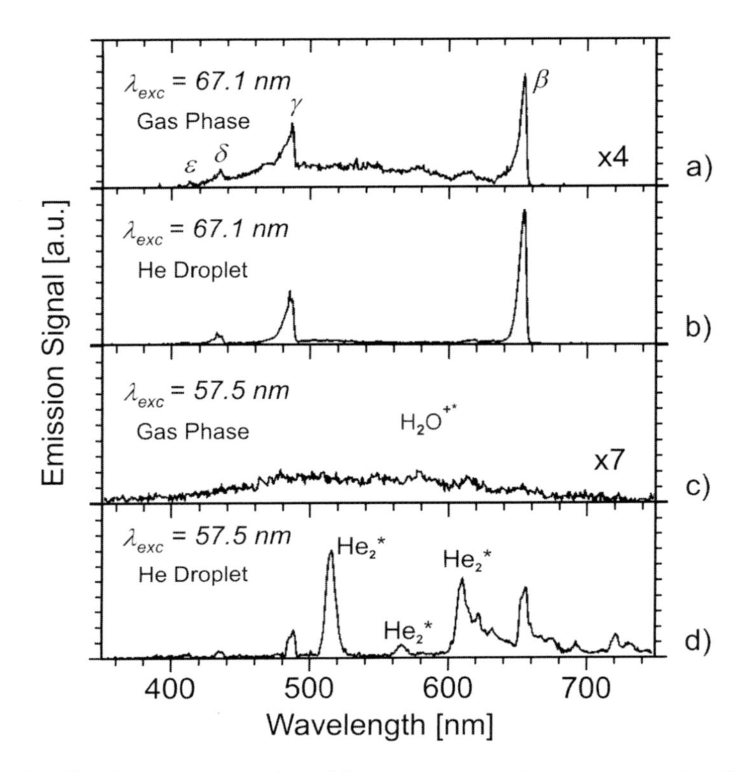


Fig. 7.12. The fluorescence spectra of free water molecules are compared with water molecules inside large He droplets at $\lambda = 67.1\,\text{nm}$ ((a) and (b) respectively), and at $\lambda = 57.5\,\text{nm}$ ((c) and (d) respectively). The free water molecule shows emission from both $H^*$ and from $H_2O^{+*}$(a) or only from $H_2O^{+*}$(c). In helium droplets only the $\delta$, $\gamma$, $\beta$ Balmer lines of $H^*$ are observed ((b) and (d)). In addition at $\lambda = 57.5\,\text{nm}$ emission features from excited helium dimers are seen (d). Adapted from Ref. 31.

visible attributed to excited helium dimers. This experiment on photodissociation of $H_2O$ in helium droplets revealed an unexpected steering of the reaction which leads to a different branching than observed in the gas phase.

#### 7.4.1.2. *Photodissociation of $NO_2$*

The gas phase photodissociation of free $NO_2$ molecules is a benchmark system which has been investigated in great detail in particular with the velocity map ion imaging technique,[37]

$$NO_2 + hv \rightarrow NO + O. \tag{4}$$

The entire dynamics including the energy partitioning among the available degrees of freedom for the free molecules are now well understood. One of the characteristic features of the $NO_2$ potential hypersurface is the barrierless pathway along the dissociation coordinate which explains the observed rapid increase of the decay rate k(E) with increasing excitation energy above the reaction threshold. This system was recently investigated in helium droplets by Wittig and coworkers for excitation energies starting from the gas phase dissociation threshold at $D_0 = 25{,}128.57\,\text{cm}^{-1}$ (3.12 eV) up to an excess energy of $4{,}300\,\text{cm}^{-1}$ (0.53 eV).[32] Upon variation of the excitation frequency no indication of dissociation was observed, based on the mass spectrometer depletion signals of the parent dopant molecule $NO_2$, or an increased signal of the NO fragment. This was explained by the dissipation of the fragment kinetic energy before reaching the droplet surface allowing the fragments to be trapped and subsequently recombine. The authors concluded from their experiments that the helium droplet acts as an effective barrier preventing the escape path of the photofragments.

#### 7.4.1.3. *Photodissociation of $CF_3I$ and $CH_3I$*

Upon excitation to the A band the free molecules are known to dissociate to an electronic ground state $CF_3$/$CH_3$ fragment and an iodine atom either in the ground $^2P_{3/2}$ or in the excited $^2P_{1/2}$ spin-orbit state. In the case of $CF_3I$ the latter branch dominates with roughly 90% over the ground state channel with about 10%. In addition the fragmentation proceeds very fast within 100 fs. Consequently, the angular distribution of the fragments is highly anisotropic, exhibiting an anisotropy parameter of about $\beta = 1.8$ as extracted from the angular intensity distribution $I(\vartheta) = 1 + \beta P_2(\cos\vartheta)$.

The escape of a photofragment from a helium cluster consisting of 200 atoms was first simulated for Cl atoms produced in the dissociation of $Cl_2$ by Takayanagi and Shiga.[38] They concluded that the kinetic energy loss is reduced when the helium is treated quantum mechanically. The first experiments in which the products of photodissociation could be observed to leave the helium droplets were recently carried out for the alkyl halides by the group of Drabbels in Lausanne.[33–36]

In order to carry out these experiments Drabbels and coworkers have constructed a sophisticated laser pump-probe helium droplet apparatus shown in Fig. 7.13 which is capable of doing ZEKE, photoelectron spectroscopy, velocity map ion imaging and TOF mass spectrometry. After doping the droplets from a secondary beam (to reduce the influx of dopant molecules into the laser excitation region) the dopant molecules are photolyzed with a 266 nm pulse of 5 ns duration. After 18 ns, enough time for the products to leave the droplet, they are non-resonantly ionized by a strong ($10^{14}$–$10^{15}\,\text{W/cm}^2$) 150 fs pulse. The newly created ions were projected on to a multichannel plate for velocity mapping recorded by a CCD

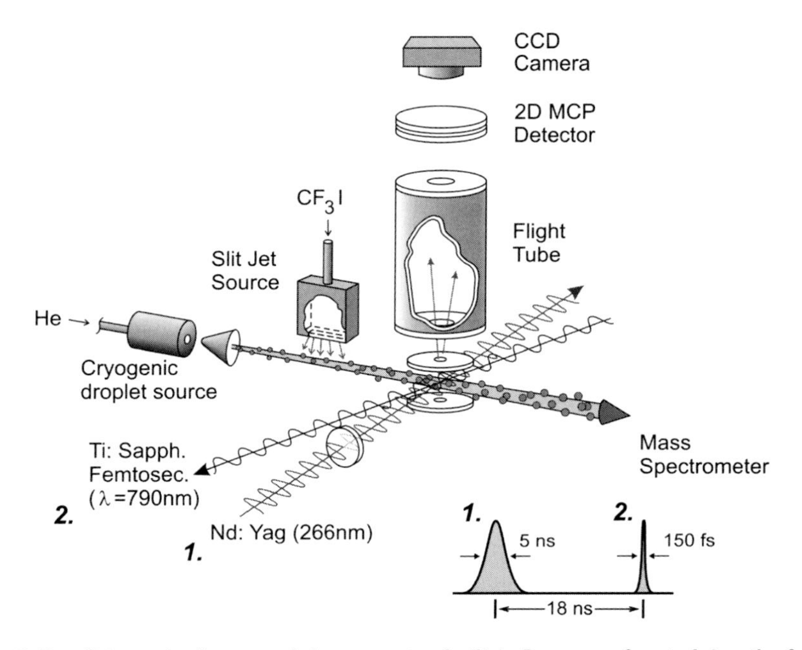


Fig. 7.13. Schematic diagram of the apparatus built in Lausanne for studying the fragments ejected from helium droplets following photodissociation of alkyl- and methyl-halide molecules embedded in helium droplets. The molecules are photolyzed by a 266 nm pulse and the fragments ionized by a powerful 150 fs pulse at 780 nm.

camera from which the full 3D velocity distribution of the ionized fragments can be extracted.

Fig. 7.14(a) shows some typical velocity map images of escaping $CF_3$ fragments after photodissociation inside droplets of three different average radii produced by varying the droplet source conditions (see Fig. 7.2). The larger velocities for the smaller droplets are explained by the smaller distances traveled before escaping from the droplet. As displayed in Fig. 7.14(b) the mean kinetic energies correspond to only a small fraction of 10–20% of the initial kinetic energy as observed under gas phase conditions. The fastest fragments had angular distributions with the same anisotropy as for the gas phase experiment while the slowest fragments showed an almost isotropic angular distribution.

In a related study of the photodissociation of $CH_3I$ into $CH_3$ and I the same group observed a signal which indicated that a fraction of the $CH_3I$ molecules appeared to recombine inside the droplets. Additionally, the iodine from both photoexcited reactions and the $CH_3$ fragment were found to leave the droplets complexed with up to a maximum of 17 helium

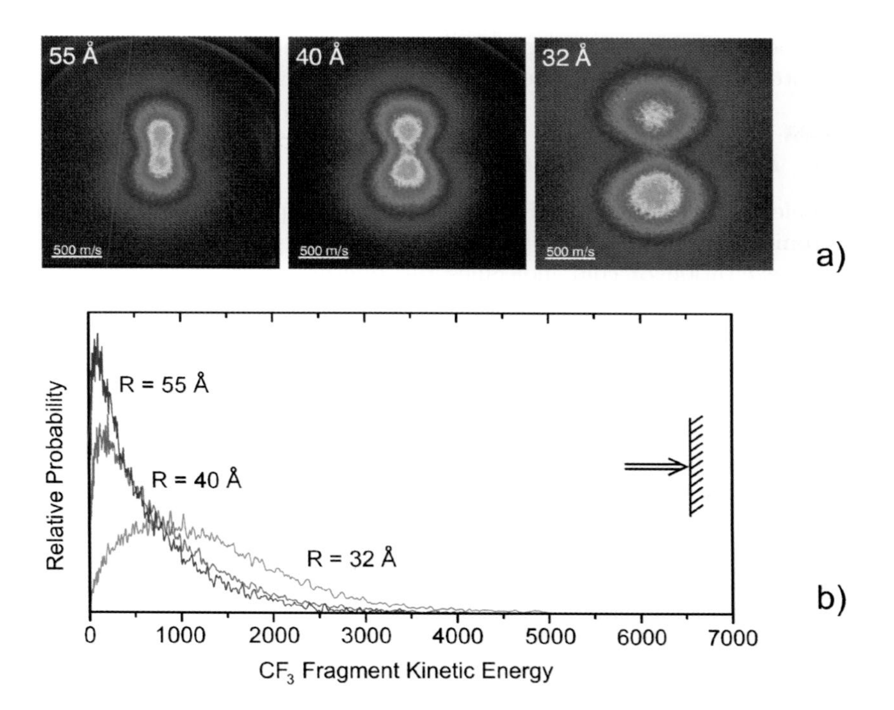


Fig. 7.14. (a) Typical velocity map ion images of $CF_3$ fragments escaping from helium droplets after photodissociation of $CF_3I$ inside helium droplets with different average radii (in color). (b) The kinetic energy distributions extracted from (a). The limiting mean initial kinetic energy averaged over the iodine $^2P_{1/2}$ and $^2P_{3/2}$ spin-orbit states, which is 6510 $cm^{-1}$, is indicated by the arrow on the right. Adapted from Ref. 33.

atoms. The actual number was found to be correlated with the velocity and angular distributions. The differences observed in the photodissociation of $CF_3I$ and $CH_3I$ could be explained by the difference in the mass ratio of the photofragments in each of the two systems.

The substantial loss of kinetic energy of the $CF_3$ fragments was reproduced quantitatively by classical Monte Carlo trajectory calculations in which the motion of the fragment through the helium droplet was described by a series of independent binary hard sphere collisions. This model was justified by the large zero-point energy and large interatomic distances in the quantum liquid. Also the observed photofragment-$He_n$ complexes could be simulated by a purely classical model for the dynamical growth of the complexes during the 5 ps needed for them to travel to the droplet surface. The angular distributions could however not be fully explained by the classical simulations. Frictionless superfluid flow as discussed in Sec. 7.3.1 is not to be expected at the large translational energies of the fast fragments encountered in these experiments.

### 7.4.2. *Photoelectron Spectroscopy*

Photoelectron spectroscopy is a powerful technique to study ionic and electronically excited levels of atoms and molecules. In the case of single photon excitation of cold molecules the photoelectron spectrum reflects the internal energy levels of the ionic system. Many experiments are performed via two photon ionization enhanced by a one-photon resonance (R2PE spectroscopy) in which transitions to intermediate electronic levels are accessed which strongly enhance the ion yield. Photoelectron spectroscopy of molecules inside superfluid helium droplets is of particular interest since the interaction of free electrons with liquid helium is known to be highly repulsive, so much so that the electrons form bubbles of about 34 Å diameter. In this section, three recent photoelectron spectra will be discussed: those of bare helium droplets,[39,40] of $Ag_8$ clusters[41] and of single aniline molecules in helium droplets.[42]

#### 7.4.2.1. *Photoionization in pure helium droplets*

Recently two experiments on photoelectron spectroscopy of bare helium droplets have been reported, one using photons in the energy range from 22.5 to 24.5 eV[39] just below the ionization threshold of the isolated helium atom at 24.6 eV and a second in the range from 24.6 eV up to 28.0 eV.[40] Surprisingly, in the first experiment photoelectrons were observed at excitation energies below the ionization threshold of helium atoms. The total electron yield started at about 23.0 eV, reaching a maximum at 23.85 eV, and then merged into the intense ionization signal of the free He atoms also present in the droplet beam. The energy distribution of the photoelectrons (Fig. 7.15) was dominated by surprisingly slow electrons with kinetic energies of less than 3 meV (32500 m/s) with an average kinetic energy of about 0.6 meV (14500 m/s) which is even smaller than the photon bandwidth of 15 meV. Moreover, the photoelectron spectrum was independent of the photon energy and showed an isotropic angular distribution. This photoelectron signal disappeared for small helium droplets confirming that it is from large droplets.

At excitation energies of 25.02 eV, just above the ionization limit of helium atoms, the photoelectron energy distribution also contained the same low energy photoelectrons as a minor component (about 2%) and a major component at higher energies consisting of a sharp peak at about 0.43 eV attributed to He atoms together with a broad tail extending to

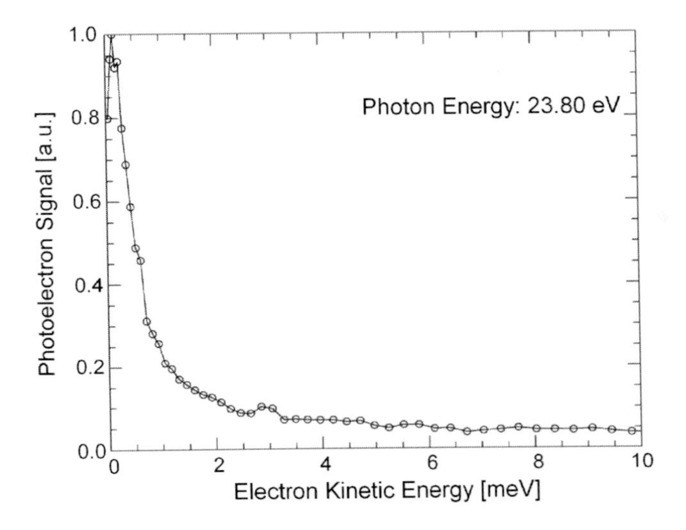


Fig. 7.15. The photoelectron kinetic energy distribution following photoexcitation at 23.8 eV, below the ionization threshold at 24.6 eV, of large pure helium droplets $N = 10^4$ atoms. Most of the electrons leave the droplet with zero kinetic energy and have an isotropic angular distribution.[39]

almost 0.5 eV above the atomic peak.[40] Both signal contributions increase with increasing droplet size as does the ratio of the low energy contribution to the high energy contribution. For a pure atomic helium beam only the sharp peak at 0.43 eV was observed but the experiment cannot, in fact, distinguish the ionization of free atoms from ionization of atoms inside the droplets. In contrast to the low energy signal the entire high energy component in the photoelectron spectrum scaled with the photon energy. The anisotropy parameter of the photoelectrons varied over the spectrum peaking with $\beta = 1$ (half the value for atomic helium) at the energy of the atomic signal dropping rapidly for higher and lower energy photoelectrons to about $\beta = 0.2$. At "zero" kinetic energy the signal was completely isotropic ($\beta = 0$).

The differences in the two signal contributions revealed different ionization mechanisms. The high energy contribution and in particular the fraction at energies above the atomic signal reflected the Franck-Condon distribution for direct ionization to form the $He_2^+$ molecule without loss of energy on the escape path of the electron through the droplet. A minor part to the low energy side of the helium atom photoelectrons and the reduced anisotropy of the atomic signal indicate that inelastic collisions occurred prior to escape. The low energy part in the spectrum was explained by an indirect ionization mechanism such as electron bubble formation or temporary trapping of the electron in a surface-bond Rydberg state along the escape trajectory.

### 7.4.2.2. *Photoionization of $Ag_8$ Clusters in Helium Droplets*

Since the observation of giant resonances in the photofragmentation of free $Ag_9^+$ clusters[43] indicating collective electronic excitations small silver clusters have attracted considerable experimental and theoretical attention. Additional interest in these small metal clusters was aroused by the unexpected observation of a very sharp absorption line of $Ag_8$ clusters inside helium droplets.[44] Because of the large capture cross sections of helium droplets (Sec. 2.2) sufficient numbers of metal atoms such as silver can be picked up by passing the droplets through an oven heated to 800°C with a vapor pressure of only $10^{-4}$ mbar.

In recent photoelectron experiments the Rostock group have been able to provide an explanation for the unsually sharp spectra. In their experiments the $Ag_8$ complex was selected out of the $Ag_n$ cluster Poisson-distribution by two photon ionization via an intermediate resonance (R2PI).[41] The remaining spectrally overlapping photoelectron signals from $Ag_2$ clusters were suppressed entirely by adjusting the pick-up conditions to favor large $Ag_8$ clusters (c.f. Fig. 7.16a).

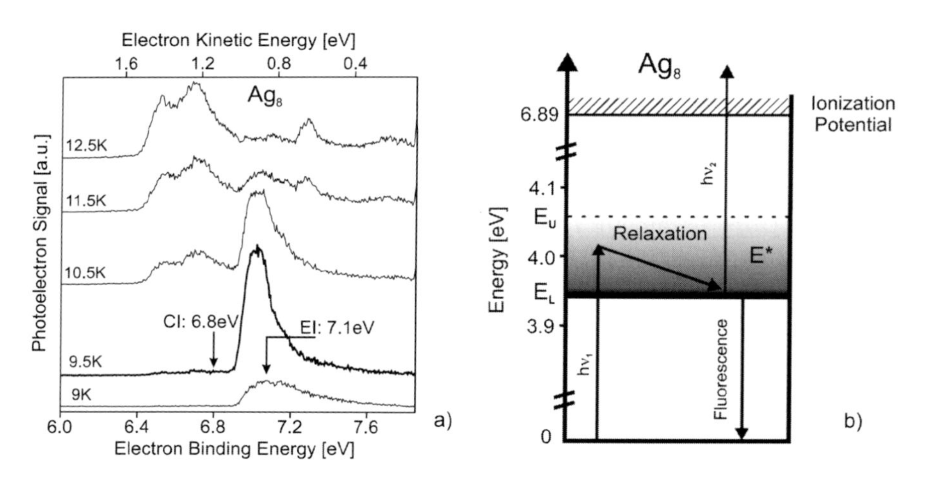


Fig. 7.16. (a) The photoelection spectra of neutral silver clusters embedded in helium droplets produced at various source temperatures. Since the probability to form $Ag_8$ clusters is greatest at low source temperatures (larger droplets) the signal between 6.9 and 7.2 eV binding energy is assigned to the $Ag_8$ clusters. (b) Schematic diagram of the two photon ionization mechanism. After excitation to the unoccupied band $E^*$ the cluster quickly relaxes to $E_L$ from which a second photon ionizes the cluster.[41]

The influence of the helium environment on the photoionization of the $Ag_8$ cluster was rather subtle. At photon energies below 3.94 eV the slope of a plot of the electron energies versus photon energy revealed a two photon process. Above 3.94 eV a linear fit was found indicating a single photon resonance[41] This was explained by noting that 3.94 eV is the threshold for fluorescence of $Ag_8$ in argon. In the helium droplet the absorption of the first photon above this threshold was followed by an extremely fast ($\approx$100 fs.) relaxation of the excess energy prior to absorption of the second photon. Thus, the final ionization step occurs effectively from the ground level $E_L$ of the intermediate electronic state of the $Ag_8$ cluster as shown in Fig. 7.16(b). This explanation was confirmed by an extrapolation of the electron kinetic energy as a function of photon energy down to zero kinetic energy which revealed the threshold for ionization from the electronically excited $Ag_8$ cluster at about 2.95 eV. The sum $3.94 + 2.95 = 6.89$ eV was assigned to the ionization potential of the $Ag_8$ cluster in good agreement with quantum chemical calculations and with electron impact ionization measurements of free $Ag_8$ clusters. It was not clear if the asymmetric line shape in the photoelectron spectrum (Fig. 7.16a) reflects the inhomogeneous droplet size distribution or the spectrum of the $Ar_8^+$ ion.

The efficient vibrational relaxation in the helium droplet environment found in these experiments is similar to spectroscopic observations for vibronically excited organic molecules in helium droplets discussed in Sec. 7.3.3. Except for the efficient relaxation of the intermediate electronic level the influence of the helium environment in this case was found to be almost insignificant.

#### 7.4.2.3. *Photoionization of Aniline in Helium Droplets*

The first measurements of the photoelectron spectrum of a molecule in helium droplets were recently reported for aniline.[42] Compared to the free molecule the R2PE spectrum in helium droplets revealed a slightly reduced ionization threshold and a significantly increased line width of asymmetric shape (Fig. 7.17). Both of these features were droplet size dependent. As for the $Ag_8$ clusters it was found that the system relaxes to the ground level of $S_1$ before the second photon is absorbed. By fitting the experimental data with the "polarizable continuum model", which treats the helium droplet as a polarizable sphere of finite radius, the ionization threshold of aniline in bulk helium was determined by extrapolation. The value obtained, 61,470 $cm^{-1}$ (7.62 eV), corresponds to a shift to the red from the gas phase value by only 834 $cm^{-1}$ (0.10 eV). This bulk shift in the ionization threshold of about 1.3%, which is even smaller in helium droplets, is of the same order of magnitude as the relative solvent shifts of the electronic origin of

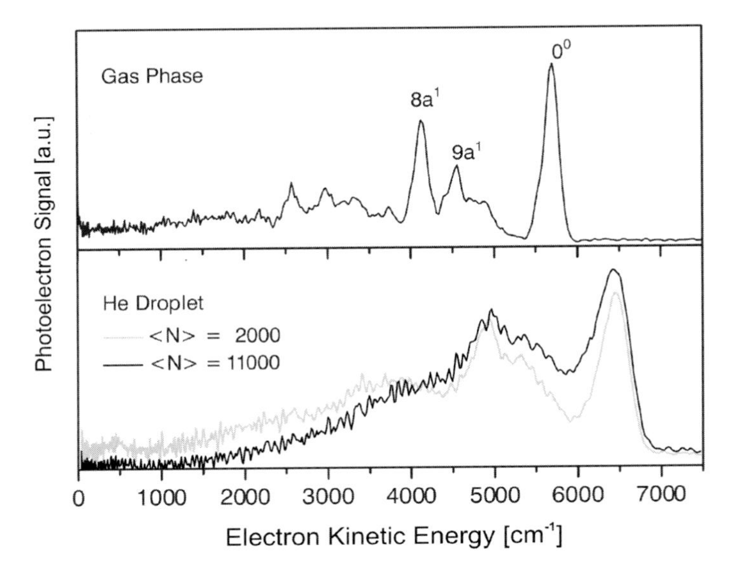


Fig. 7.17. Photoelection spectrum of aniline in the gas phase compared with the spectrum measured in helium droplets of two different sizes.[42]

molecules in helium droplets. The asymmetric line shape was attributed to the relaxation of the photoelectrons by impulsive collisions with helium atoms along the escape trajectory. An interesting droplet size effect was observed in the low energy part of the photoelectron spectrum. For all droplets with radii smaller than 38 Å, which corresponds to $\langle \mathrm{N} \rangle = 2000$, (see Fig. 7.17) photoelectrons down to zero kinetic energy were observed while for larger droplets ($\langle \mathrm{N} \rangle = 11000$ in Fig. 7.17) the minimum kinetic energy was $1{,}230(50)\,\mathrm{cm}^{-1}$ ($0.153\,\mathrm{eV}$). This was thought to be due to a localization of the slower electrons which facilitates their recombination with the nearby ion.

### 7.4.3. *Photo-Induced Tautomerization in 3-Hydroxyflavone*

Proton transfer is one of the simplest of all elementary chemical processes, the kinetics of which play an important role in many biological processes. Many examples of tautomerism (the equilibrium between two different isomers) involve proton transfer. Of the many systems studied the photon-stimulated Excited State Intramolecular Proton Transfer (ESIPT) in 3-hydroxyflavone (3-HF) ($C_{15}$ $H_{10}$ $O_3$), which is an important plant compound, has many desirable features, making it an ideal model system.

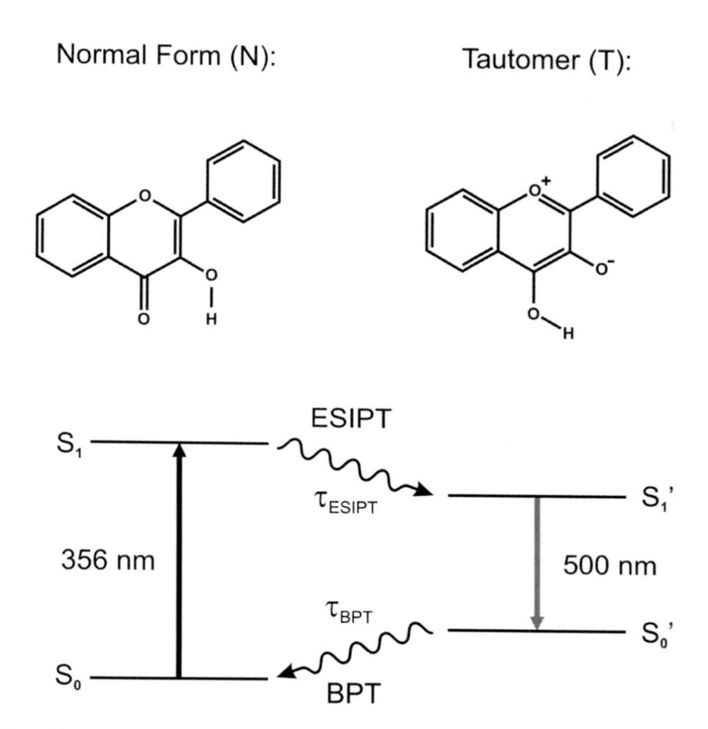


Fig. 7.18. Electronic states and optical transitions in the photochemistry of 3-hydroxyflavone. The excited $S_1$ state of the normal molecule decays spontaneously within about less than 30 fs to the $S_1'$ state of the tautomer by Excited State Intramolecular Proton Transfer (ESIPT). After radiation to $S_0'$ the tautomer decays back to the $S_0$ state of the normal molecule by Back Proton Transfer (BPT).

As illustrated in Fig. 7.18 on excitation at about 356 nm (28,081 $cm^{-1}$) it undergoes ESIPT as indicated by an emission starting at about 500 nm (20,000 $cm^{-1}$) and extending towards the red. The ground state of the tautomer then rapidly decays back to the normal state. Most of the more than 50 reported studies have been carried out either in various solvents, such as liquid decane, or in solid Ar or Schpol'ski matrices at 10 K. Hydrogen bonding protic and polar solvents inhibit ESIPT as revealed by an enhanced emission in the UV from the $S_1$ state of the normal molecule. The interpretation of the previous matrix experiments is of course complicated by the additional effects of the environment, which is also generally true for other photoactive biomolecules as in the environment of proteins. Excitation spectra of the free molecule in seeded beams have only been reported by two groups[45,46] of which one also studied the emission spectrum and 3-HF complexed with water.[46]

Recently Slenczka and coworkers[47] have embarked on an extensive experimental and theoretical study of ESIPT of 3-HF inside large He droplets ($N \approx 1.5 \cdot 10^4$ atoms). Upon excitation of 3-HF at 351 nm the emission occurred exclusively at and above 500 nm which indicates that in helium droplets proton transfer is unhindered, as in the gas phase. One of the interesting results emerging from the ongoing experiments is the observation of a large number of lines in emission (Fig. 7.19), which were not resolved in the seeded beam experiments.[46] The vibrational lines could be assigned to the vibrational modes of the ground electronic $S_0'$ state of the tautomer (Fig. 7.19) via a normal mode analysis based on quantum chemical calculations. This indicates that in the droplets the tautomer excited $S_1'$ state was completely relaxed whereas in the seeded beam experiments the spectra were smeared out by significant internal excitation of the tautomeric $S_1'$ state. The resolved vibrational fine structure reveals that the torsion vibrations of the phenyl ring appear to be completely inactive in the helium droplets. Previously there had been considerable discussion concerning the role of the phenyl ring torsions on ESIPT.

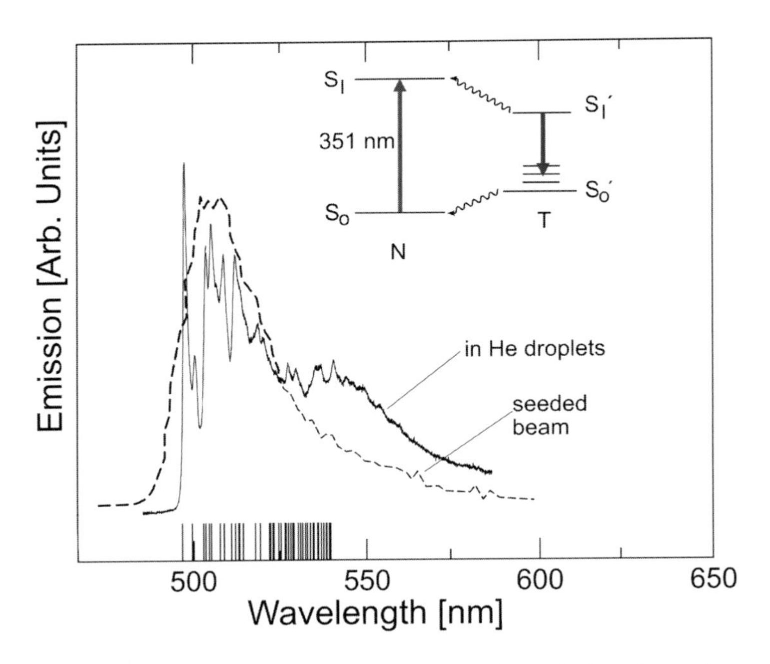


Fig. 7.19. Comparison of the emission spectrum of free 3-hydroxyflavon in a seeded beam[46] with the spectrum in helium droplets after excitation at 351 nm.[47] The vertical bars indicate the calculated vibrational levels of the $S_0'$ tautomer ground state.

Another interesting result coming from these experiments sheds new light on the Back Proton Transfer process. A careful line shape analysis of the vibronic $S_1'-S_0'$ transitions made it possible to extract the homogeneous contribution which revealed a rate constant for BPT from the $S_0'$ state of $4 \cdot 10^{12}\,\mathrm{s}^{-1}$.

An additional aspect emerging from these helium droplet experiments is the influence of a polar solvent such as $H_2O$ on the ESIPT in 3-HF. A single $H_2O$ molecule was added to the embedded 3-HF via a second pick-up cell filled with gas phase $H_2O$. The emission spectrum of the 3-HF-$H_2O$ complex revealed emission only of the tautomeric complex indicating unhindered ESIPT which was the same as for the bare 3-HF. This result is in total contradiction with the corresponding observations in the gas phase experiment.[46] Most probably the low temperature conditions in helium droplets lead to a different complex configuration from that produced in the three-body collisions occurring in a seeded beam.

## 7.5. Chemical Reactions in Helium Droplets

### 7.5.1. *Ion-Molecule Reactions*

The early mass-spectrometer studies of helium droplets[48] with electron beam ionization exposed a rich ion-molecule reaction chemistry. Usually after electron-impact ionization of free molecules there is considerable fragmentation making the identification of the neutral molecule difficult. Electron impact ionization of doped helium droplets ($\bar{N} < 10^5$) leads in addition to the usual $He_n^+$ fragments,[49,50] at least in some cases, to unfragmented ions of the embedded neutral species. This was first studied in detail for $SF_6$.[48] In He droplets the ionization mechanism is initiated by the creation of a positive $He^+$ hole inside the droplet by the impinging electron. This hole migrates facilely by a nearly random resonant charge hopping process until in a pure droplet it is finally localized by producing a strongly bound $He_2^+$ ion ($D_0 = 2.2\,\mathrm{eV}$). The release of this large amount of energy leads to a nearly complete fragmentation of the droplet. If the droplet is doped the migrant hole will more likely ionize the embedded impurity molecule to which it is attracted by induction forces.[48,51] In view of the large ionization potential of He atoms (I.P. $= 24.6\,\mathrm{eV}$) the energy released in the charge-transfer (CT) ionization of the impurity is usually very large and often sufficient to completely destroy the droplet. Despite this large energy release an atomic impurity may leave the droplet as an ion with up to about 30 attached He atoms. Rather surprisingly, polyatomic molecules or cluster ion fragments usually appear as bare ions (e.g. tetracene), or possibly as one predominate ion fragment (e.g. tryptophan) or, in other cases as protonated ion fragments (e.g. clusters of methanol or acetonitrile).[52] Presumably in this case the cage of surrounding He atoms, which might otherwise have remained attached, is evaporated by the internal energy initially deposited in the fragments.

Recently a dedicated study of ion-molecule reactions involving two different molecules inside He droplets was reported by Fárnik and Toennies.[53] In an initial series of experiments $D_2$, $N_2$ or $CH_4$ molecules were each embedded in the droplets and the ion fragment signals resulting from charge transfer (CT) with the $He^+$ hole were studied inside large droplets ($\approx 10^4$ atoms) as a function of the pick-up pressure. For added $D_2$ the following exothermic reactions could be identified

$$\mathrm{He^+ + D_2 \rightarrow He + D + D^+} \quad (\Delta E = -6.4\,\mathrm{eV})\ \text{Dissociative} - \text{CT}, \tag{1a}$$

$$\downarrow$$

$$\mathrm{D^+ + nHe \rightarrow \underline{D^+He_n}} \quad \text{Clustering}, \tag{1b}$$

and

$$\mathrm{He^+ + D_2 \rightarrow He + D_2^+} \quad (\Delta E = -9.1\,\mathrm{eV})\text{CT}, \tag{2a}$$

$$\downarrow$$

$$\mathrm{D_2^+ + D_2 \rightarrow \underline{D_3^+} + D} \quad (\Delta E = -1.9\,\mathrm{eV}), \text{Reaction} \tag{2b}$$

where the detected products are underlined. Other possible competing reactions could be ruled out on the basis of gas phase results at low collision energies.

For added $N_2$ the following sequences were found

$$\mathrm{He^+ + N_2 \rightarrow He + (N_2^+)^*}\ \text{(nearly resonant)} \quad \text{Resonant CT}, \tag{3a}$$

$$\downarrow$$

$$\mathrm{(N_2^+)^* + He \rightarrow N_2^+ + He} \quad (\Delta E = -9.00\,\mathrm{eV})\ \text{Deexcitation}, \tag{3b}$$

$$\downarrow$$

$$\mathrm{N_2^+ + nHe \rightarrow \underline{N_2^+He_n}} \quad \text{Clustering}, \tag{3c}$$

$$\downarrow$$

$$\mathrm{N_2^+ + N_2 \rightarrow \underline{N_4^+}} \quad \text{Reaction}. \tag{4}$$

A similarly simple assignment was not possible for added $CH_4$ which is consistent with the fact that five competing highly exothermic reactions between $He^+$ and $CH_4$ are known from gas phase studies. The droplet mass spectra were dominated by the $CH_2^+$, $CH_3^+$, $CH_4^+$ and $CH_5^+$ ion signals without any attached He atoms and only relatively small signals of these molecular ions with one or two attached He atoms were found. The preference for the bare ions was attributed to the large internal energy of the products which was sufficient to evaporate all the atoms of the droplets.

To isolate and investigate the ion-molecule reaction between two selected embedded molecules, such as $N_2$ and $D_2$, two pick-up chambers, one filled with $N_2$ and the other with $D_2$ were used. The observed $N_2D^+$ signal could then be attributed to the reaction.

$$N_2^+ + D_2 \rightarrow \underline{N_2D^+} + D, \quad (\Delta E = -2.6\,\text{eV}). \tag{5}$$

A competition from the other possible reaction between $D_2^+$ and $N_2$ could be ruled out by the strong dependence of the $N_2D^+$ product signal on the $D_2$ pick-up pressure. Also the gas phase charge transfer reaction of $He^+$ with $D_2$ is several orders of magnitude smaller than for $N_2$. Fig. 7.20 shows the experimental results and a comparison with a simple Monte Carlo simulation, which, among other factors, accounted for the log-normal size distribution of the droplets. The good agreement between the observed saturation of the $N_2D^+$ signal and the depletion of the reactant $N_2^+$ ion signal with increasing pressure of the reactants in the pick-up cell provided additional confidence in assigning the $N_2D^+$ product to reaction (5).

The secondary reactions of the charge transfer $CH_4$ ion fragments with $D_2$ were also investigated using two pick-up chambers. Despite the large number of reaction channels the pressure dependent results made it possible to isolate the following two ion-molecule reactions.

$$CH_4^+ + D_2 \rightarrow \underline{CH_4D^+} + D, \quad (\Delta E = -0.2\,\text{eV}) \tag{6}$$

and

$$CH_3^+ + D_2 \rightarrow \underline{CH_3D_2^+}. \quad (\Delta E = -0.7\,\text{eV}) \tag{7}$$

These experiments demonstrate that exothermic ion-molecule reactions can be initiated and studied at the very low (0.37 K) temperatures inside He droplets. The corresponding low energies are otherwise only accessible in much more elaborate and complicated merged ion-beam scattering experiments. Droplets are especially suited for investigating ion-molecule reactions since they generally have negligible activation energies. In each of the examples studied the embedded neutral molecules initially form complexes

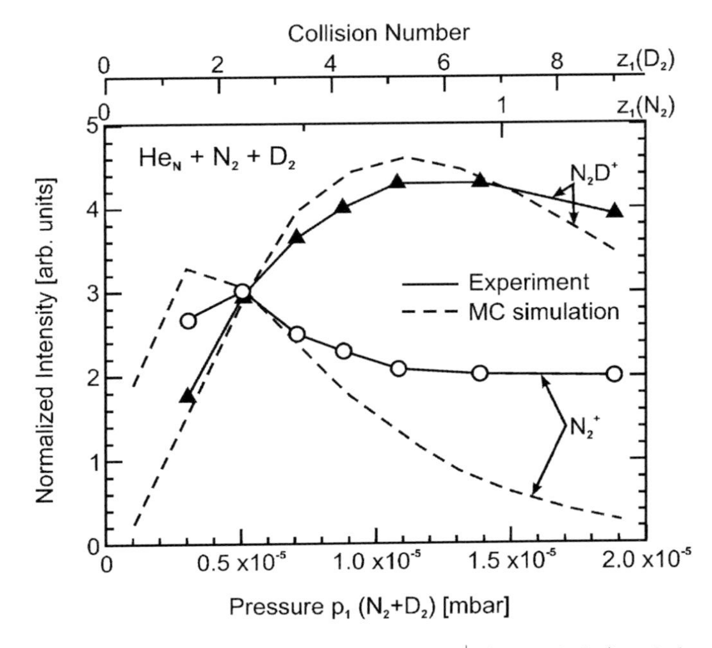


Fig. 7.20. Normalized intensities of the reactants $N_2^+$ (open circles) and the products $N_2D^+$ (solid triangles) as a function of the total pressure $p_1$ in the scattering chamber with a 75% partial $D_2$ pressure. The top scales correspond to the mean collision numbers of the droplet with the $N_2$ and $D_2$ molecules. The dashed lines are Monte Carlo simulations.[53]

in the droplet interior. There the charge transfer from a nearby $He^+$ ion ionizes one of the molecules which then proceeds to react with one of its neighbors. In some cases other molecules in the complex and the many surrounding He atoms serve as a heat sink thereby facilitating recombination reactions such as reaction (4) above.

### 7.5.2. *Spectroscopy of Entrance and Exit Channel Complexes*

The aggregation and self-organization of complexes as a result of the very low temperatures open up unique possibilities to study complexes which exhibit specific steric configurations as a result of being trapped in either their global van der Waals minimum or in other geometries favored by the directed approach of the partners inside the ultracold superfluid. The group of the late Roger Miller at the University of North Carolina has been successful in doping He droplets with a number of free radicals including the halogen atoms and several small $CH_n$ species. This has opened up a new area of chemical dynamical studies many facets of which are covered in a very recent review by Küpper and Merritt.[11] Complexes of radicals either as reactants or as reaction products can now be fabricated inside the droplets and their structures studied with high resolution infra-red spectroscopy. One of the initial experiments of this type was devoted to the oriented potentially reactive complexes $X \cdots HF$ consisting of a halogen atom and an hydrogen fluoride HF molecule:

$$X + HF \rightarrow HX + F, \tag{8}$$

where $X$ is either a Cl, Br or I atom.[54] By combining the information gained from rotationally resolved spectra as well as from Stark and pendular state spectra they were able to confirm that the complexes have $^2\Pi_{3/2}$ ground electronic states and the linear structures predicted by theory.[54]

More recently the same group extended these studies to the exit channel complexes $HCN \cdots CH_3$ and $DCN \cdots CD_3$ of the reactions $CN + CH_4\,(CD_4) \rightarrow HCN(DCN) + CH_3(CD_3)$[55] as well as the exit complexes $HF \cdots CH_3$ and $DF \cdots CD_3$ of the reactions $F + CH_4(CD_4) \rightarrow HF(DF) + CH_3(CD_3)$.[56] In both cases the observed $C_{3v}$ structure, the measured rotational constants and dipole moments were in quite reasonable agreement with MP2 calculations carried out by the same authors.

### 7.5.3. *Photon-Induced Isomerization*

The first study of a photon-induced isomerization was reported by the same group in 2004.[57] In this initial study the following isomerization reaction was investigated,

$$\underset{\text{linear}}{FH \cdots NCH} + h\nu \leftrightarrow \underset{\text{bent}}{HF \cdots HCN} \qquad (\Delta E \approx 0.124\,\text{eV}), \tag{9}$$

where the isomer on the left is linear and the isomer on the right is bent. The experiment takes advantage of the different frequencies of both the C-H and H-F stretch vibrations in the two isomers for the identification via depletion. Fortunately, both excitations also impart sufficient energy to the complex to easily exceed the gas phase isomerization barrier. Thus, they can be used to not only probe but also initiate the reaction.

Fig. 7.21 shows the apparatus developed by the Miller group. As in the spectroscopic studies laser resonant excitation is monitored by the depletion signal measured at the downstream detector, in this case, a bolometer. To collapse the rotational fine structure laser excitation is carried out in

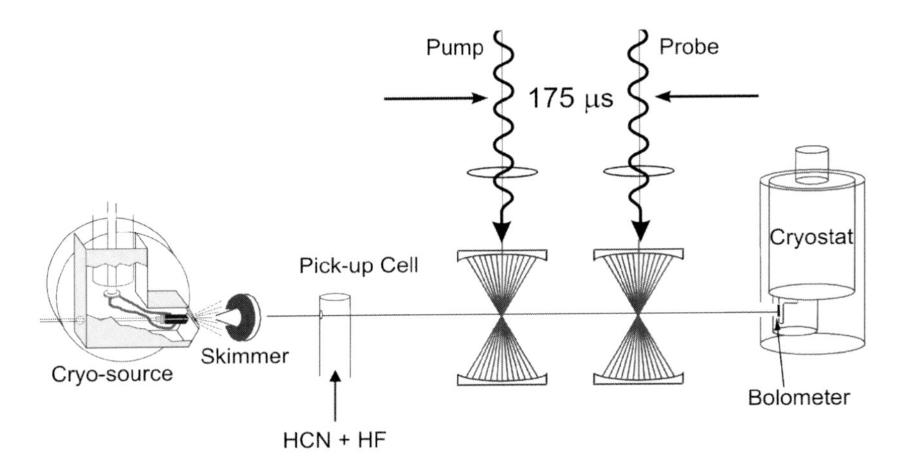


Fig. 7.21. Schematic diagram of the droplet beam apparatus used by the group of the late Roger Miller at the University of North Carolina to study isomerization of HCN···HF van der Waals complexes. The HCN and HF molecules are first picked up separately to form oriented van der Waals complexes inside the droplets. The CH or HF stretch vibrations are used to excite the complex and after a flight time of 175 $\mu$s to probe the isomers formed. The corresponding laser resonances are monitored by the increase or depletion of the droplet signal at the downstream bolometer detector. Adapted from Ref. 57.

the presence of a strong electric field (pendular state spectroscopy). For studying isomerization two multipass interaction regions separated by 8 cm which corresponded to a droplet flight time of 175 $\mu$s, were used to first excite (pump) one complex and subsequently excite (probe) either one of the possible isomers. The delay of 175 $\mu$s was chosen to provide plenty of time for isomerization to occur inside the droplet. By varying the distance between the interaction regions (as discussed in Sec. 7.3.6 for phthalocyanine) it should even be possible in future experiments to determine the time for the system to relax provided the relaxation is not too fast.

Fig. 7.22 shows a schematic diagram with the transitions on pumping the HF stretch of the linear FH···NCHcomplex and the CH and HF vibrational stretch probe signals of the linear complex, which were observed to decrease while the corresponding signals from the bent complex increased indicating isomerization to the bent complex. In the forward reaction 58% of the excited complexes were found to fall back to the initial isomer, while 29% were transferred to the bent isomer. Thus the vibrationally excited linear complex has a 2:1 probability to fall back to the same structure compared to producing the higher energy bent complex. When the bent complex was pumped all the complexes were transformed to the more tightly bound linear isomer.

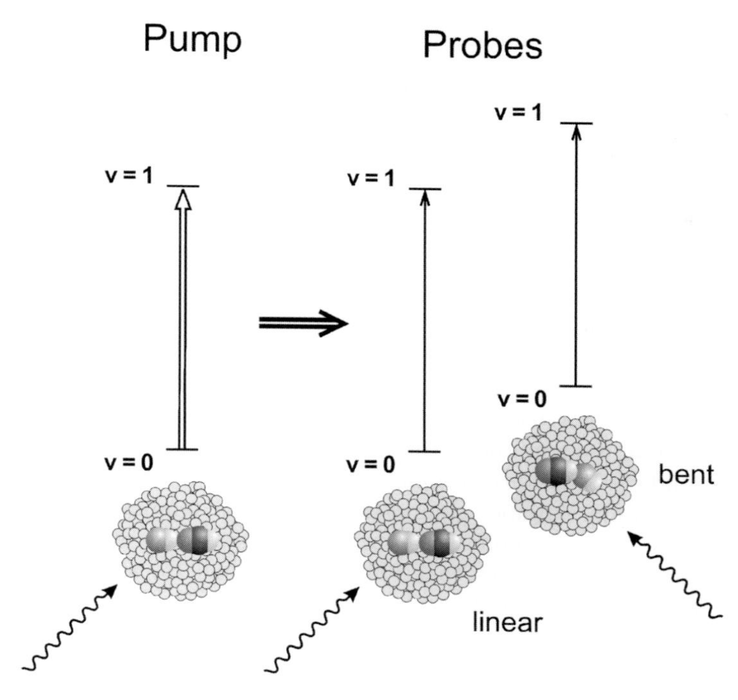


Fig. 7.22. Schematic diagram showing the isomerization reaction resulting from pumping the linear HCN-HF complex. The products are interrogated by exciting either the linear or bent complexes and monitoring the downstream depletion signal (Adapted from Ref. 57)

In addition to demonstrating new possibilities for photoisomerization of molecular complexes, these experiments were interpreted as indicating that the isomers could be dissociated before undergoing the expected recombination as a consequence of caging by the surrounding liquid helium. Of course there is also the possibility that in view of the very rapid relaxation found in many of the other systems discussed in this review that quenching was so rapid that the partners did not separate, but rearranged themselves while always remaining closely attached. Regardless of the exact mechanism these experiments are important since they point the way to more complicated and detailed pump-probe interrogation studies.

### 7.5.4. *Neutral Molecule Reactions*

So far only one bimolecular chemical reaction involving neutral species has been studied inside He droplets. For this pilot experiment the following highly exothermic chemiluminescent reaction was chosen.

$$\mathrm{Ba} + \mathrm{N_2O} \rightarrow \mathrm{BaO}^* \rightarrow \left(\mathrm{A}\ {}^1\sum_1^+, \nu'\right) + \mathrm{N_2} \quad (\Delta E = -4.1\,\mathrm{eV})$$

$$\mathrm{BaO} \rightarrow \left(\mathrm{X}\ {}^1\sum^+, \nu''\right) + h\nu \tag{8}$$

Previously, this reaction had been extensively studied in crossed-molecular beam experiments and also on the surfaces of solid Ar and Ne clusters.[58] It has the advantage that the intense chemiluminescent emission not only indicates that the reaction has occurred but also identifies the products and their internal state distributions.

The arrangement used in the He droplet experiment[59] is essentially similar to that used in the earlier studies on the surfaces of the heavy rare gas solid clusters. Large droplets ($N \cong 2 \cdot 10^5$ atoms) were first passed through a pick-up chamber equipped with a high temperature ($\approx$673 K) scattering chamber containing barium. The vapor pressure of $3.5 \cdot 10^{-5}$ mbar at this temperature assures that only at most one Ba atom was picked up. Barium being heliophobic is expected to remain on the droplet surface much in the same way, as in the earlier experiments with the solid rare gas clusters. The droplet beam then enters a chamber filled with $N_2O$ gas, which is also equipped with an optical system to collect and disperse the chemiluminiscence arising when the $N_2O$ molecules impinge on the surface and react with the Ba atoms.

Fig. 7.23 compares a series of emission spectra measured under different conditions. The spectrum in Fig. 7.23(a) which was first obtained by Ottinger and Zare in 1970 in a simple gas phase experiment[60] exhibits a broad unstructured distribution, peaked at about 500 nm. The lack of structure was attributed to a dense spectrum as expected for very hot product molecules. When the reaction occurs on the surface of solid argon clusters (Fig. 7.23b), in addition to the same broad distribution, a series of spectral lines is observed. The unresolved part is attributed to reaction products which by virtue of their large internal excitation are nearly instantly ejected and radiate their energy after they have left the surface. The superimposed spectral lines are attributed to products which remain stuck to the surface and have been accommodated internally to the Ar cluster temperature ($\approx 37 \pm 5$ K). The spectral lines correspond to Franck-Condon transitions from the ground vibrational state of the $A\ {}^1\Sigma^+$ excited state to the vibrational manifold of the ground electronic state $X\ {}^1\Sigma_1$ of the free BaO molecule except for an overall blue shift of $750\,\mathrm{cm}^{-1}$. The emission from the He droplets (Fig. 7.23c) suggests that these same processes take place here as well but with a much smaller blue shift of only $40\,\mathrm{cm}^{-1}$.

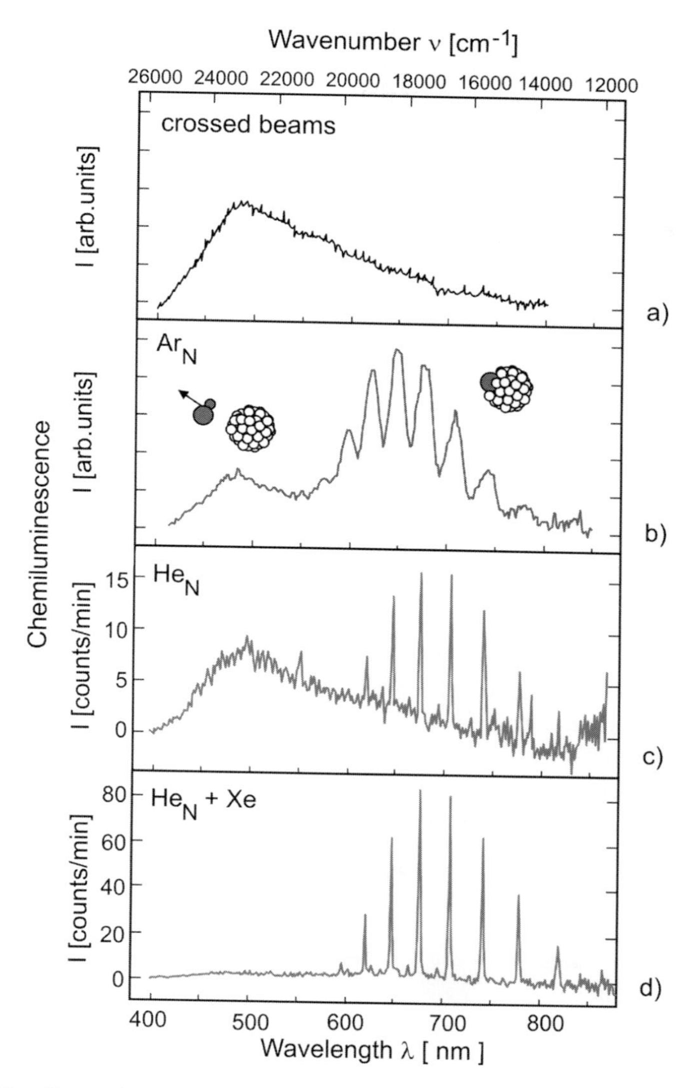


Fig. 7.23. Comparison of the chemiluminescent emission spectrum of BaO ($A\,^1\sum^+ \nu' = 0 \rightarrow X^1\sum^+ \nu''$) (a) in a crossed beam type experiment. (b) On the surface of Ar clusters ($\bar{N}_{\mathrm{Ar}} = 2 \cdot 10^3$). (c) On the surface of pure $^4$He droplets ($\bar{N}_4 = 2 \cdot 10^5$), and (d) inside pure $^4$He droplets which are first doped with about 15 Xe atoms. The cartoons in (b) illustrate the processes responsible for the major spectra features above and below $\lambda \approx 570\,\mathrm{nm}$.[59]

The only other significant difference is that the peaks in the line spectra are much narrower as expected for BaO largely accommodated to the factor $10^2$ lower droplet temperature of 0.37 K.

To take full advantage of the cold helium liquid it seemed desirable to move the site of the reaction from the surface into the interior. This was achieved by the following scheme. First about 15 Xe atoms were added to the droplets to form a cluster in the interior of the droplets. The resulting strong van der Waals attractive force acting on the Ba atoms was designed to overcome the potential barrier which prevented the surface Ba atoms from going inside. The corresponding spectrum (Fig. 7.23d) is in complete agreement with such a scenario. The broadly distributed background is no longer present as expected if the reaction occurs only inside the droplets. Fig. 7.24 shows a cartoon illustrating these various steps in the reaction when it is catalyzed by the internal Xe cluster.

A surprising feature of the spectra is the lack of a noticeable line shift compared to the surface reaction which would be expected if the emission occurred while the BaO molecule is within a few Å of the Xe cluster. This was explained by noting that the electronically excited BaO molecule is expected to repel the surrounding He atoms to form a bubble of about 10 Å diameter. The Xe cluster, on the other hand, will be surrounded by a dense shell of He atoms. Thus once excited the BaO molecule and its bubble will be repelled by the Xe cluster and its shell of He atoms. The spectra suggest that these repulsive forces act so fast that the molecule is sufficiently separated from the Xe cluster before the emission can occur.

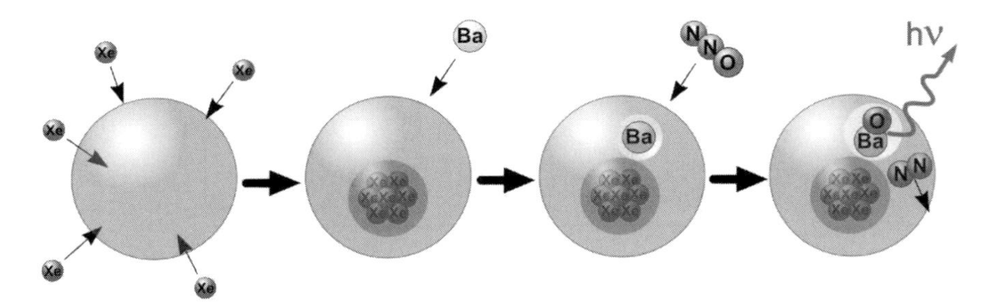


Fig. 7.24. Schematic representation of the sequence of events occurring within large ($N \approx 2 \cdot 10^5$ atoms) $^4$He droplets as they pass through the apparatus. Ba atoms arriving at the surface are drawn into the interior by the previously embedded Xe atoms. The heliophilic $N_2O$ molecules penetrate into the interior of the droplet where they can approach the Ba atom and react on contact forming a highly excited BaO($A^1\sum^+$) molecule with about 2 eV rotational-vibrational energy. This energy is rapidly dissipated by the evaporation of about 3200 He atoms from the droplet leaving the BaO molecule in the $\nu' = 0$ state. Finally, the electronically excited BaO molecules radiate to the ground state after which the surrounding bubble is expected to collapse.[59]

## 7.6. Summary and Outlook

The present review describes the beginnings of a new subfield in the ever growing saga of helium nanodroplet spectroscopy: the investigation of elementary chemical processes at subkelvin temperatures inside helium nanodroplets. Since the early 1990's when the first experiments revealed the unexpected resolution of many of the spectroscopic features found for free molecules but at much lower temperatures than previously possible, many atoms, molecules and van der Waals complexes have been studied. These early experiments also revealed that embedded complexes frequently were self-organized in either their van der Waals global or local potential minima. This combination of self-organization and high spectral resolution have made possible the exploration of chemical dynamics inside helium droplets.

The realization that helium droplets open up new opportunities for dynamical studies came initially from laser pump-probe experiments, as discussed in Sec. 7.4.1. These revealed the remarkable ability of the droplet environment to quench both the vibrational and in some cases even higher electronic excitations of relatively large organic and biological molecules to produce otherwise translationally and rotationally cold molecules in well defined electronically excited states. Although there are now many examples for this quenching the underlying coupling mechanisms are by no means fully understood. While the vibration of diatomic molecules and most of the in-plane modes of organic molecules are only weakly affected the out-of-plane vibrational modes of larger molecules are more strongly affected as evidenced by significant line broadening. The efficient coupling must also be related to the exceptionally high thermal conductivity of superfluid helium. A few experiments have been successful in quantifying the relaxation rates, but more work is still needed to better understand this phenomenon.

Related spectral studies in the near UV and visible have exposed fine structures which provide the first evidence for the internal degrees of the non-superfluid solvation layer of He atoms which surround the chromophores. The presence of solvation layers was previously inferred from the moments of inertia which in He droplets are enhanced by as much as a factor of four to five.

The rapid quenching has recently made it possible to resolve new spectral features not seen previously in the photo-stimulated tautomerization of 3-hydroxyflavone and in a study of the bimolecular reaction of $Ba+N_2O$ to produce chemiluminescent BaO, discussed in Secs. 7.4 and 7.5, respectively. Also the first successful laser pump-probe study of a simple isomerization reaction, also mentioned in Sec. 7.5, makes use of the rapid quenching. In the future this efficient quenching will certainly find wide applications in studying conformationally resolved spectroscopy of large floppy biological molecules, as recently demonstrated in helium gas seeded beams.[61]

Several examples of dynamic studies which took advantage of the new possibilities of self organization were also dealt with. Examples are the spectroscopy of entrance and exit channel complexes and the pump-probe studies of isomerization. As discussed in Sec. 7.4 the proton transfer reaction (ESIPT) in the 3-HF-$H_2O$ complex follows a different reaction path due to a uniquely different complex configuration created in the low temperature droplet environment.

Other recent sophisticated laser pump-probe experiments reviewed in Sec. 7.4.2 have been directed at determining the velocities and angular distributions of the fragments produced by photodissociation in helium droplets. In some cases molecules are ejected from the droplets after losing a significant 80% fraction of their initial kinetic energy. Atom fragments tend to pick up He atoms on their way out. Both effects could be simulated with a single binary collision model in which the liquid is treated in the same way as a dense gas. This could be another unexpected advantage resulting from the large zero point energy of the helium quantum fluid which leads to a very loose binding and large interatomic distances. In other photodissociation experiments caging and geminate recombination have also been reported. In the future more theoretical and experimental work is needed to quantitatively understand when caging, as opposed to ejection, is the dominant process.

Some example studies of photoelectron spectroscopy also discussed in Sec. 7.4 indicate that electrons may leave the droplets without appreciable energy loss. In contrast at least in one experiment a near thermalization of the ejected electrons was observed. The search for frictionless motion of atomic ions and neutral particles, as has been reported in the bulk at velocities below about 58 m/s, has not been observed inside droplets and remains a challenge for the future.

So far there have been only two studies of bimolecular reactions, one involving ion-molecule reactions and the other a chemiluminescent bimolecular reaction. Of course such experiments are limited to reactions with negligible barriers as is the case for most ion-molecule reactions. Up to the present time no firm evidence for quantum tunneling has been found, but should be expected for the long contact times of the ultra-cold complexes.

Already the few experiments reviewed here point to the great potential of helium droplets for the study of a wide range of chemical dynamic processes at ultra-low temperatures. To expand the possibilities more work is needed to understand the mechanisms for rapid, efficient quenching and how they depend on the specific dopant and its electronic structure, as well as the role of the non-superfluid layer of He atoms surrounding the chromophore.

## Acknowledgements

We thank the editor Ian Smith for a careful reading and several helpful comments and for giving us this opportunity to study and digest the most recent developments in chemical dynamics in helium droplets. We gratefully acknowledge the careful reading and useful comments by Marcel Drabbels, Rudolf Lehnig and Oleg Kornilov. We thank Feng Dong and Gary Douberly for their help in the preparation of the Figs. 7.21 and 7.22. We are most grateful to Katrin Glormann for expert help with the typing and organization of the manuscript and Sascha Warnecke for his assistance with the drawings.

## References

1. Hartmann M, Miller RE, Toennies JP, Vilesov AF. (1995) Rotationally resolved spectroscopy of $SF_6$ in liquid helium clusters: A molecular probe for cluster temperature. *Phys. Rev. Lett.* **75**: 1566–1569.
2. Special issue (2001) Helium nanodroplets: A novel medium for chemistry and physics. *J. Chem. Phys.* **115**: 22
3. Toennies JP, Vilesov AF. (2004) Superfluid helium droplets: A uniquely cold nanomatrix for molecules and molecular complexes. *Angw. Chem. Int. Ed.* **43**: 2622–2648.

3a. Macharov GN. (2004) Spectroscopy of single molecules and clusters inside helium nanodroplets. Microscopic manifestation of $^4$He superfulidity. *Physics-Usbekhi* **47**: 217–247.

4. Stienkemeier F, Lehmann KK. (2006) Spectroscopy and dynamics in helium nanodroplets. *J. Phys. B: At. Mol. Opt. Phys.* **39**: R127–R166.
5. Choi MY, Douberly GE, Falconer TM, Lewis WK, Lindsay CM, Merritt JM, Stiles PL, Miller RE. (2006) Infrared spectroscopy of helim nanodroplets: Novel methods for physics and chemistry. *Int. Rev. Phys. Chem.* **25**: 15–75.

5a. Szalewicz K. (2008) Interplay between theory and experiment in investigations of molecules embedded in superfluid helium nanodroplets. *Int. Rev. Phys. Chem.* **27**: 273–316.

6. Lewerenz M, Schilling B, Teonnies JP. (1993) A new scattering deflection method for determining and selecting the sizes of large liquid clusters of $^4$He. *Chem. Phys. Lett.* **206**: 381–387.
7. Harms J, Toennies JP, Dalfovo F. (1998) Density of superfluid helium droplets. *Phys. Rev.* **B58**: 3341–3350.
8. Barranco M, Guardiola R, Hernández S, Mayol R, Navarrro J, Pi M. (2006) Helium nanodroplets: An overview. *J. Low. Temp. Phys.* **142**: 1- 81.
9. Grebenev S, Hartmann M, Havenith M, Sartakov B, Toennies JP, Vilesov AF. (2000) The rotational spectrum of single OCS molecules in liquid $^4$He droplets. *J. Chem. Phys.* **112**: 4485–4495.

10. Hartmann M, Miller R, Toennies JP, Vilesov AF. (1996) High resolution molecular spectroscopy of van der Waals clusters in liquid helium droplets. *Science* **272**: 1631–1634.
11. Küpper J, Merritt JM. (2007) Spectroscopy of free radicals and radical containing entrance-channel complexes in superfluid helium nanodroplets. *Int. Rev. Phys. Chem.* **26**: 249–287.
12. Dalfovo F. (1994) Atomic and molecular impurities in $^4$He clusters. *Z. Phys.* **D29**: 61–66.
13. Toennies JP, Vilesov AF. (1995) Novel low-energy vibrational states of foreign particles in fluid $^4$He clusters. *Chem. Phys. Lett.* **235**: 596–603.
14. Lehmann KK. (1999) Potential of a neutral impurity in a large $^4$He cluster. *Mol. Phys.* **97**: 645–666.
15. Lehmann KK. (2000) Buoyancy correlations for the potential of an impurity in a $^4$He nanodroplet. *Mol. Phys.* **98**: 1991–1993.
16. Hartmann K, Mielke F, Toennies JP, Vilesov AF, Benedek G. (1996) Direct spectroscopic observation of elementary excitations in superfluid He droplets. *Phys. Rev. Lett.* **76**: 4560–4563.
17. Hartmann M, Lindinger A, Toennies JP, Vilesov AF. (1998) Laser-induced fluorescence spectroscopy of van der Waals complexes of tetracene-$Ar_N$ ($N \leq 5$) and pentacene-Ar within ultracold liquid He droplets. *Chem. Phys.* **239**: 139–149.
18. Lindinger A, Toennies JP, Vilesov AF. (2001) Pump-probe study of the reconstruction of helium surounding a tetracene molecule inside a helium droplet. *Z. Phys. Chem.* **3**: 2581–2587.
19. Hartmann M, Lindinger A, Toennies PJ, Vilesov AF. (2002) The phonon wings in the ($S_1 \leftarrow S_0$) spectra of tetracene, pentacene, porphin and phthalocyanine in liquid helium droplets. *Phys. Chem. Chem. Phys.* **4**: 4839–4844.
20. Whaley B, Whitley H. (2007) private communication.
21. Lehnig R, Slenczka A. (2003) Emission spectra of free base phthalocyanine in superfluid helium droplets. *J. Chem. Phys.* **118**: 8256–8260.
22. Lehnig R, Slenzka A, unpublished results.
23. Lehnig R, Slenczka A. (2004) Emission spectra of Mg-phthalocyanine and the phthalocyanine-$Ar_1$ cluster in superfluid helium droplets. *Chem. Phys. Chem.* **5**: 1014–1019.
24. Lehnig R, Slenczka A. (2005) Spectroscopic investigation of the solvation of organic molecules in superfluid helium droplets. *J. Chem. Phys.* **122**: 244317/1–244317/9.
25. Pötzl GM, Slenczka A, unpublished results.
26. Lehnig R, Slenczka A. (2004) Quantum solvation of phthalocyanine in superfluid helium droplets. *J. Chem. Phys.* **120**: 5064–5066.
27. Whitley HD, Huang P, Kwon Y, Whaley KB. (2005) Multiple solvation configurations around phthalocyanine in helium droplets. *J. Chem. Phys.* **123**: 054807/1–054807/8.
28. Lehnig R, Blinov NV, Jaeger W. (2007) Evidence for an energy level substructure of molecular states in helium droplets. *J. Chem. Phys.* **127**: 241101/1–241101/4.

29. Hartmann M, Lindinger A, Toennies JP, Vilesov AF. (2001) Hole-burning studies of the splitting in the ground and excited vibronic state of tetracene in helium droplets. *J. Phys. Chem.* A **105**: 6369–6377.
30. Nauta K, Miller RE. (2000) Metastable vibrationally excited HF($v = 1$) in helium droplets, *J. Chem. Phys.* **113**: 9466–9469.
31. Kanaev AV, Museur L, Laarmann T, Monticone S, Csatex MC, von Haeften K, Möller T. (2001) Dissociation and suppressed ionization of $H_2O$ molecules embedded in He clusters: The role of the cluster as cage. *J. Chem. Phys.* **115**: 10248–10253.
32. Stolyarov D, Polyakova E, Wittig C. (2004) Photoexcitation of $NO_2$ in $He_n$ droplets above the gas-phase dissociation threshold. *J. Phys. Chem.* A**108**: 9841–9846.
33. Braun A, Drabbels M. (2004) Imaging the translational dynamics of $CF_3$ in liquid helium droplets. *Phys. Rev. Lett.* **93**: 253401/1–253401/4.
34. Braun A, Drabbels M. (2007) Photodissociation of alkyl halides in helium nanodroplets. I, Kinetic energy transfer. *J. Chem. Phys.* **127**: 114303-1–114303-14.
35. Braun A, Drabbels M. (2007) Photodissociation of alkyl halides in helium nanodroplets. II, Solvation dynamics. *J. Chem. Phys.* 127: 114304-1–114304-9.
36. Braun A, Drabbels M. (2007) Photodissociation of alkyl halides in helium nanodroplets. III. Recombination. *J. Chem. Phys.* 127: 114305-1–114305-6.
37. Coroiu AM, Parker DH, Groenenboom GC, Barr J, Novalbos IT, Whitaker BJ. (2006) Photolysis of $NO_2$ at multiple wavelengths in the spectral region 200–205 nm. *Eur. Phys. J.* **D38**: 151–162.
38. Takayanagi T, Shiga M. (2003) Photodissociation of $Cl_2$ in helium clusters: an application of hybrid method of quantum wavepacket dynamics and path integral centroid molecular dynamics. *Chem. Phys. Lett.* **372**: 90–96.
39. Peterka DS, Lindinger A, Poisson L, Ahmed M, Neumark DM. (2003) Photoelectron imaging of helium droplets. *Phys. Rev. Lett.* **91**: 043401/1–043401/4.
40. Peterka DS, Kim JH, Wang CC, Poisson L, Neumark DM. (2007) Photoionization dynamics in pure helium droplets. *J. Phys. Chem. A* **111**: 7449–7459.
41. Radcliffe P, Przystawik A, Diederich T, Döppner T, Tiggesbäumker J, Meiwes-Broer K-H. (2004) Excited-State Relaxation of $Ag_8$ clusters embedded in helium droplets. *Phys. Rev. Lett.* **92:** 173403/1–173403/4.
42. Loginov E, Rossi D, Drabbels M. (2005) Photoelctron spectroscopy of doped helium nanodroplets. *Phys. Rev. Lett.* **95**: 163401/1–163401/4.
43. Tiggesbäumker J, Koller L, Lutz HO, Meiwes-Broer K-H. (1992) Giant resonances in silver-cluster photofragmentation. *Chem. Phys. Lett.* **190**: 42–47.
44. Federmann F, Hoffmann K, Quass N, Toennies JP. (1999) Spectroscopy of extremely cold silver clusters in helium droplets. *Eur. Phys.* **D9**: 11–14.
45. Ernesting NP, Dick B. (1989) Fluorescence excitation of isolated, jet-cooled 3-hydroxyflavone–the rate of excited-state intramolecular proton-transfer from homogenous linewidths. *Chem. Phys.* **136** (2): 181–186. Mühlpfordt A, Bultmann T, Ernesting NP. (1994) Excited-state intramolecular proton

transfer in jet-cooled 3-hydroxyflavone. Deuteration studies, vibronic double-resonance experiments, and semiempirical (AM1) calculations of potential-energy surfaces. *Chem. Phys.* **181**: 447–460.
46. Ito A, Fujiwara Y, Itoh M. (1992) Intramolecular excited-state proton transfer in jet-cooled 2-substituted 3-hydroxychromones and their water clusters. *J. Chem. Phys.* **96**: 7474–7482.
47. Lehnig R, Vdovin A, Slenczka A. unpublished results.
48. Scheidemann A, Schilling B, Toennies JP. (1993) Anomalies in the reactions of He+ with $SF_6$ embedded in large He-4 clusters. *J. Phys. Chem.* **97**: 2128–2138.
49. Buchenau H, Toennies JP, Northby JA. (1991) Excitation and ionization of He-4 clusters by electrons. *J. Chem. Phys.* **95**: 8134–8148.
50. Ruchti T, Callicoatt BE, Janda KC. (2000) Charge transfer and fragmentation of liquid helium droplets doped with xenon. *Phys. Chem. Chem. Phys.* **2**: 4057–4080.
51. Lewis WK, Lindsay CM, Bemish RJ, Miller RE. (2005) Probing charge-transfer processes in helium nanodroplets by optically selected mass spectrometry (OSMS): Charge steering by long-range interactions. *J. Am. Chem. Soc.* **127**: 7235–7242.
52. Behrens M, Fröchtenicht R, Hartmann M, Siebers JG, Buck U, Hagemeister FC. (1999) Vibrational spectroscopy of methanol and acetonitrile clusters in cold helium droplets. *J. Chem. Phys.* **111**: 2436–2443.
53. Farnik M, Toennies JP. (2005) Ion molecule reactions in $^4$He droplets: Flying nano-cryo-reactors. *J. Chem. Phys* **122**: 014307/1–014307/11.
54. Merritt JM, Küpper J, Miller RE. (2005) Entrance channel X-HF (X=Cl, Br and I) complexes studied by high-resolution infrared laser spectroscopy in helium nanodroplets. *Chem. Phys. Chem.* **7**: 67–78.
55. Rudić S, Merritt JM, Miller RE. (2006) Infrared laser spectroscopy of the $CH_3$-HCN radical complex stabilized in helium nanodroplets. *J. Chem. Phys.* **124**: 104305/1–104305/18.
56. Merritt JM, Rudic S, Miller RE. (2006) Infrared laser spectroscopy of $CH_3 \cdots HFCH_3$ in helium nanodroplets: the exit-channel complex of the F+$CH_4$ reaction. *J. Chem. Phys.* **124**: 084301/1–084301/12.
57. Douberly GE, Merritt JM, Miller RE. (2005) IR-IR double resonance spectroscopy in helium nanodroplets: Photo-induced isomerization. *Phys. Chem. Chem. Phys.* **7**: 463–468.
58. Mestdagh JM, Gaveau MA, Gée C, Sublemontier O, Visticot JP. (1997) Cluster isolated chemical reactions. *Int. Rev. Phys. Chem.* **16** (2): 215–247.
59. Lugovoj E, Toennies JP, Vielsov A. (2000) Manipulating and enhancing chemical reactions in helium droplets. *J. Chem. Phys.* **112**: 8217–8220.
60. Ottinger Ch, Zare RN. (1970) Crossed beam chemiluminescence. *Chem. Phys. Lett.* **5**: 243–248.
61. Dian BC, Longarte A, Zwier TS. (2002) Conformational dynamics in a dipeptide after single-mode vibrational excitation. *Science* **296**: 2369–2373.

# Chapter 8

# Kinematic Cooling of Molecules

Kevin E. Strecker and David W. Chandler
*Sandia National Laboratories, Combustion Research Facility*
*Livermore, Ca, 94055*

## Contents



## 8.1. Introduction

The cooling of atoms to ultracold temperatures has resulted in spectacular discoveries. The realization and study of new states of matter like Bose Einstein Condensates,[1–3] degenerate Fermi gases[4,5] and BCS[6–9] (Bardeen, Cooper Schrieffer) fluids is now "routine". In addition to these new states of matter that have been formed, cooling atoms to micro-Kelvin temperatures, and below, has enabled both ultra-high resolution spectroscopy studies[10–13] and the extraction of information about the collisional dynamics of atoms and their interactions (for example shape resonances,[14,15] Feschbach resonances[16–19] and Efimov states[20–22]). All of these areas of research have molecular analogs. The added complexity found in molecules including, permanent dipole and quadrupole moments, complex rotational and vibrational structure and chemistry, offer the possibility of rich areas of investigation. All of these areas have generally remained unexplored due to the inability to routinely make ultracold samples of molecules.

In this chapter we will focus on the production and use of cold and ultracold molecules for studies in the field of chemical dynamics of gas phase molecular species. Chemical dynamics is the detailed study of the motion of molecules and atoms on potential energy surfaces in order to learn about the details of the surface as well as the dynamics of their interactions. We want to explore new information, techniques, and insight that can be gained from the use of cold and ultracold molecules. The first step to achieve this requires us to define cold and ultracold in the context of chemical dynamics. We will then discuss the kinematic cooling technique in detail and conclude with several applications of this cooling technique and its potential for guiding and confining kinematically cooled molecules.

### 8.1.1. *Cold and Ultracold Molecules for Chemical Dynamics*

Cold and ultracold are two distinct regimes in the field of cold atoms physics; they depend on the particle momentum, $\rho$, in terms of the thermal de Broglie wavelength $\lambda_{db} = h/\rho$ in relationship to the particle density, $n$. Although the two regimes of interest are called cold and ultracold, the temperature is not the figure of merit but it is a convenient label for the energy of the system in any one dimension, given by Eq. (8.1),

$$k_B T = \frac{1}{2}mv^2 = E, \tag{8.1}$$

where $k_B$ is the Boltzmann constant. The thermal de Broglie wavelength is conventionally written in terms of the temperature as;

$$\lambda_{db} = \left(\frac{2\pi\hbar^2}{mk_B T}\right)^{\frac{1}{2}}. \tag{8.2}$$

The thermal de Broglie wavelength gives the "size" of the particle. If the de Broglie wavelength is smaller then the classical size of the particle, the particle behaves classically, and if the de Broglie wave is larger then the classical size, the particle can exhibit its quantum nature.

There is no rigorous definition for the "cold" regime. "Cold" is typically taken to be around the temperature at which the thermal de Broglie wavelength is of the same order as the size of the classical size of the particle.

However, the cold regime is also used to refer to any range of low temperatures at which new physical phenomena, not accessible by standard techniques, can be explored. For instance, if we want to measure a several millisecond excited state lifetime in a typical 500 m/s molecular beam, we would need an apparatus several meters long, such that the molecules could be excited, continue to travel and evolve for several milliseconds as we observe the decay. This is a challenging experiment. If, however, the molecules can be slowed such that they are only moving at 5 m/s, the molecules will only move a few millimeters before decaying, making the lifetime much easier to measure. Thus we would consider the molecules moving at 5 m/s to be cold regardless of the de Broglie wavelength of the molecule.

One useful definition for the cold regime is to have particles at sufficiently low temperature/energy that they can readily be manipulated, confined, or stored using external fields. For many chemical dynamics experiments the cold regime is all that is needed. The ability to trap, store and manipulate along with the extended observation times allows for ultrahigh resolution spectroscopy which allows direct counting of the density of states of the molecules and precision measurements of fundamental physical constants. Confining and orientation of the molecules prior to photo-dissociation reduces the averaging associated with photon absorption by a randomly oriented molecule yielding new details in the observed angular distributions of the photo-fragments which can reveal the underlying dynamics of the photo-dissociation process. Further, the wave nature of the particles can be exploited through the molecular analog of atom optics. The increased thermal de Broglie wavelength makes it possible to perform unique interferometric studies with molecules, such as the separation of mixtures of clusters by diffraction through a transmission grating.[23–27]

In contrast to the cold regime, the ultracold regime is more clearly defined. The term ultracold refers to a sample in which the thermal de Broglie wavelength is of the same order as the average inter-particle separation, such that

$$\lambda_{db} \geq n^{-\frac{1}{3}}, \tag{8.3}$$

where $n$ is the particle density. The physical meaning of Eq. (8.3) is that the quantum size of the particles are beginning to spatially overlap. The particle interactions are no longer classical hard sphere interactions but are now dominantly wavelike, and are governed by their respective quantum statistics. In the field of cold atom physics, this behavior leads to quantum degenerate Bose and Fermi gases,[1–5] purely quantum behavior like matter-wave lasers[28] and matter-wave solitons,[29,30] and the ability to study complex super-conducting behavior in nearly ideal dilute quantum gases.[9,31–33] As noted earlier, all these systems have molecular analogs and the specific behavior, scattering lengths, location of shape and Feshbach resonances depend strongly on the shape and structure of the potential energy surface on which the particles are interacting.

The typical first step in defining the ultracold regime is to introduce the phase space density.[34] The phase space density, which is discussed in detail in Section 8.2.1, is a measure of the probability of finding a particle at a given point in space at a given time. Mathematically the phase space density is defined as

$$p = n\lambda_{db}^{-3}. \tag{8.4}$$

As the phase space density approaches one, or $\lambda_{db} \simeq n^{\frac{1}{3}}$, there is a near unity probability to find any given particle in the ensemble at a given point in space at a given time. It follows then that if the the particles have $\frac{1}{2}$ integer total spin, they behave like fermions and repel each other such that no two identical particles occupy the same position at any given time, yielding a Fermi degenerate gas. On the other hand, if the particles have integer total spin, they behave as bosons and accumulate at the same point in space at a given time leading to Bose-Einstein condensation. With this understanding of the phase space density, the ultracold regime is defined as the energy where the phase space density of the sample $p \sim 1$.

For a gas in a harmonic confining potential, the phase space density can be directly related to the density and temperature,[34] such that

$$p = n\left(\frac{2\pi\hbar^2}{mk_BT}\right)^{-3/2}. \tag{8.5}$$

From Eq. (8.5) we note that the phase space density can be increased by manipulating the temperature, $T$, and/or the density $n$. For typical experiments in the field of cold atoms and molecules, the most effective method for increasing the phase space density is to reduce the temperature several orders of magnitude while simultaneously increasing the density, as opposed to increasing either the density or reducing the temperature. As a result, the conditions for macroscopic quantum behavior, Eq. (8.3), are satisfied at temperatures below 1mK, and hence this is referred to as the ultracold regime. This discussion is to further clarify the definition of the ultracold regime, since Eq. (8.5) states that the conditions from Eq. (8.3) can be satisfied by a dense enough gas that is not ultracold.

An example of macroscopic quantum behavior at high temperatures and exceedingly high densities can be found in nature; white dwarf stars. The white dwarf is a star that is nearing the end of its hydrogen fuel cycle and begins to implode. If the star has sufficient mass, as it implodes the density increases until the average atom-atom separation is less then $10^{-10}$ cm at a temperature of over $1 \times 10^6$ Kelvin.[35] Under these conditions the electrons delocalize from the atoms and begin to repel each other. This repulsion is a purely quantum pressure, Fermi pressure, that stabilizes the star from collapse.[35] These stars have a phase space density near unity and display macroscopic (13,000 km in diameter) quantum behavior. However, at over a million degrees these star are clearly not in the ultracold regime. The densities available in the core of a collapsing star are typically not available in the laboratory environment, but the same physics can be observed in the ultracold regime. In the experiment, a dilute sample of fermions, like lithium-6 ($^6$Li), is laser cooled and confined in a magnetic trap. The temperature of the confined gas is reduced while simultaneously increasing the density until the phase space density approaches unity. At these high phase space densities, the Fermi pressure between the identical $\frac{1}{2}$ integer spin particles stabilizes the size of the sample to the Fermi radius of $\sim$200 $\mu$m at a temperature less then 1 $\mu$K.[5]

The ultracold regime offers vast new areas of research for molecules, such as BECs with strong dipole moments, isolating and studying the effects of quantum statistical dynamics, interference, and quantum effects in chemical dynamics and reactions. However, there are technological and arguably fundamental obstacles to reaching this regime. Molecules lack a general method to cool them to the densities and temperatures needed to observe this behavior. The current experimental push is to develop methods for producing cold molecules, and confine the molecules such that they can be further cooled into the ultracold regime.

Several techniques have been developed for the production of cold molecules. Generally, each technique works on a specific class of molecules; for example Stark deceleration[36] works for molecules with large first order Stark shifts (see chapter 9 by van der Meerakker *et al.*) buffer gas cooling[37] works on molecules that have favorable elastic to inelastic collision ratios with He, and photoassociation[38] and Feshbach association[39,40] produce diatomics from atoms that can be laser cooled (see chapter 10 by Weiner). In this chapter we focus on the production of cold molecules through a kinematic method of "single collision" or "kinematic" cooling. The basis for kinematic cooling is the scattering of an atom or molecule via a single collision with an atom of appropriate mass which causes the atom or molecule to stop in the laboratory frame.

## 8.2. Kinematic Cooling Molecules

The first realization of kinematic cooling was during the performance of inelastic collision studies in a Crossed Atomic and Molecular Beam (CAMB) experiment. CAMB experiments combined with Ion Imaging have been extensively used to study the details of inelastic energy transfer for many years.[41–44] In these experiments a supersonic atomic beam is crossed at 90° with a supersonic molecular beam. The product of the scattering is detected using Resonance Enhanced Multiphoton Ionization (REMPI) spectroscopy to selectively ionize individual quantum states. The ions are subsequently projected onto a time and position sensitive detector at the end of a time-of-flight mass spectrometer. The ions are focused using electrostatic lenses onto a micro-channel plate coupled to a phosphor screen, the image is then recorded by a charge coupled device (CCD) camera. This imaging technique is known as "Velocity Mapped" Ion Imaging,[45–47] and is explained in Section 8.3.2. In performing these inelastic scattering experiments, it was noticed that under certain conditions molecules populating particular quantum states were scattered in a manner such that some of them were left with very low laboratory-frame velocities. These slow molecules are easy to identify as their signals are anomalously high in the images. The anomalously high signal is due to the low velocity of the cooled molecules. Since the cold molecules are moving slowly out of the interaction region, their density increases with respect to the uncooled molecules. The increased density of the cold molecules enhances their probability for ionization and detection, resulting in an increased signal at the detector.

Kinematic cooling occurs when an atom and a molecule scatter in a manner that the recoil velocity of the molecule is equal and opposite to the center-of-mass velocity of the collision pair. When this criterion is met the molecule will be "stationary" in the laboratory frame of reference and therefore cold. In the CAMB experimental design, kinematic cooling is achieved by scattering a molecular(or atomic) beam by an atomic beam, intersecting at 90 degrees, resulting in a subset of the scattered molecules (or atoms) in a particular quantum state, moving slowly in the laboratory frame. Conservation of momentum and energy dictate the quantum state that satisfies this condition. Conservation of energy and momentum for such a collision system is embodied in a "Newton" diagram for the system. Figure 8.1 is a generic Newton diagram for the CAMB experimental arrangement. Note that the slow molecules are located at the intersection of the atomic and molecular beams, the lab origin.

In our first experiments on the kinematic cooling process, we chose to scatter a molecular beam of NO from an atomic beam of argon. The energetics were selected to provide a velocity vector cancellation that results in the the post collision $NO_{7.5}$ (NO in the j = 7.5 rotational state) being stationary in the laboratory frame. We will show below why this quantum state of NO is essentially stationary in the laboratory reference frame,

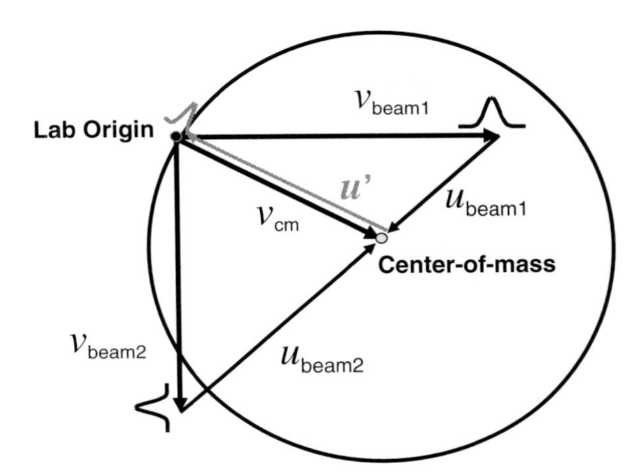


Fig. 8.1. Newton Diagram of the molecular beam scattering of Beam1 off of Beam2 where the two beams cross at the Lab Origin. Molecules scattered from the center of mass in a manner that their velocity, $u'$, is equal and opposite to the velocity of the center of mass, $v_{cm}$, will be at the Lab origin and stationary. $v$ represents laboratory frame velocities and $u$ represents center of mass frame velocities.

$v'_{\text{NO}(7.5)} = 0$ after collision with Ar. The conditions necessary for this cancelation may be derived from classical mechanics using vector algebra. The velocity cancelation occurs when the velocity of the system's center-of-mass is equal in magnitude and opposite in direction to the recoil velocity of the scattered $\text{NO}_{7.5}$ molecule in the center-of-mass frame of reference. That is, when $u'_{\text{NO}} = -v_{cm}$, where $u'_{\text{NO}}$ is the center-of-mass frame recoil velocity of the scattered $\text{NO}_{7.5}$ and $v_{cm}$ is the velocity of the NO + Ar center-of-mass in the laboratory-frame coordinate system. We have used the convention of primed variables to denote post-collision conditions and unprimed variables for the pre-collision conditions. The velocity cancelation condition follows from the relation between the laboratory-frame velocity of the scattered NO, $v'_{\text{NO}}$, and the center-of-mass frame velocity of the scattered $\text{NO}_{7.5}$: $u'_{\text{NO}} = v'_{\text{NO}} - v_{cm}$. In order to simplify the algebra, we separate the vector cancelation condition into two scalar conditions for $u'_{NO}$, one for its magnitude and one for its direction. The former condition is most simply written as a constraint on velocity,

$$v_{cm} = -u'_{\text{NO}} \tag{8.6}$$

The recoil velocity of the NO molecule is equal in magnitude to the velocity of the system's center of mass. The post collision energy of the NO molecule in the center-of-mass frame, $E'_{rel}$, is given by

$$E'_{rel} = \frac{1}{2}\mu v'^2_{rel} = \frac{1}{2}\mu \left(\frac{M}{m_{\rm Ar}} u'_{\rm NO}\right)^2 \tag{8.7}$$

The reduced mass is given by $\mu = m_{\rm NO} m_{\rm Ar}/M$, where $m_{\rm NO}$ and $m_{\rm Ar}$ are the masses of NO and Ar, and the total mass is $M = m_{\rm NO} + m_{\rm Ar}$. The zero-velocity constraint requires that $|v_{cm}| = |u'_{\rm NO}|$, so we have

$$E'_{rel} = \frac{1}{2}\mu \left(\frac{M}{m_{\rm Ar}} v_{cm}\right)^2 = E_{cm}\frac{m_{\rm NO}}{m_{\rm Ar}} \tag{8.8}$$

The energy of the center-of-mass is the sum of the lab-frame energies of the NO and Ar minus their relative translational energy. For the scattered NO to be stationary in the lab then some energy must be "hidden" in the NO molecule since the mass of the NO and Ar are not equal. This excess energy can be "hidden" internally in the molecule as rotational, vibrational or electronic energy. Under our experimental conditions, the excess energy of the NO goes into rotational energy. When the amount of rotational energy, $E'_{rot}$, of the scattered NO molecules equals the difference between the initial and final relative energies the NO molecules will be stationary in the laboratory frame. Thus, from Eq. (8.8) it follows that,

$$\begin{aligned} E'_{rot} = & \left[\frac{\mu}{m_{\rm Ar}} E_{\rm Ar} + \frac{\mu}{m_{\rm NO}} E_{\rm NO}\right] \\ & - \left[\frac{m_{\rm NO}}{m_{\rm Ar}} \left(E_{\rm NO} + E_{\rm Ar} - \frac{\mu}{m_{\rm NO}} E_{\rm NO} - \frac{\mu}{m_{\rm Ar}} E_{\rm Ar}\right)\right] \end{aligned} \tag{8.9}$$

Substituting the energy terms with their definitions in terms of mass and velocity as given above, and solving for NO internal energy yields,

$$E'_{\rm NO_{int}} = E'_{rot}({\rm NO}_{v'=0,j'}) = \left(1 - \frac{m_{\rm NO}}{m_{\rm Ar}}\right) \cdot E_{trans}({\rm NO}_{v'=0,j=0.5}), \tag{8.10}$$

here $E_{trans}({\rm NO}_{v'=0,j=0.5})$ is the lab-frame translational energy of the NO in the molecular beam, $E_{\rm NO}$ in Eq. (8.9). We expand the labels in Eq. (8.10) to emphasize the rotational energy transfer. $E'_{rot}({\rm NO}_{v'=0,j'})$ is the amount of energy that must be deposited into rotation of the NO molecule in order to produce molecules that are stationary in the laboratory frame of reference. On inspection, it may seem surprising that only the energy of the NO and not that of Ar appears in this constraint. This is a direct algebraic consequence of the separation of the vector constraint into scalar constraints but this observation has a significant impact on the efficiency of kinematic cooling. It means that to first order the velocity spread of the atomic beam does not impact the final velocity spread of the cold molecules. Collisions of a molecule in the molecular beam with a stationary set of atoms will have the same kinematic effect on the molecule as a collision with a supersonic atomic beam and the same final quantum state will be the one that stops in both situations. Below we outline a more detailed analysis of the velocity spread expected for the cooled molecules.

In order to predict the distribution in the laboratory-frame velocity of the translationally cold NO from the spread in the velocities of the NO and Ar beams we perform a simple analysis of the behavior of $v'_{\rm NO} = u'_{\rm NO} + {\rm v}_{cm}$ around $u'_{\rm NO} = -v_{cm}$ that characterizes the cold molecules. The center-of-mass energy is the sum of the mass-weighted energies of argon and nitric oxide

$$E_{cm} = \frac{m_{\rm Ar}}{M}E_{\rm Ar} + \frac{m_{\rm NO}}{M}E_{\rm NO} = \frac{1}{M}(m_{\rm Ar}E_{\rm Ar} + m_{\rm NO}E_{\rm NO}). \tag{8.11}$$

Solving for the square of the velocity of the center of mass, we have

$$v_{cm}^2 = \frac{2}{M}\left(\frac{m_{\rm Ar}}{M}E_{\rm Ar} + \frac{m_{\rm NO}}{M}E_{\rm NO}\right) = \frac{m_{\rm Ar}^2}{M^2}v_{\rm Ar}^2 + \frac{m_{\rm NO}^2}{M^2}v_{\rm NO}^2. \tag{8.12}$$

The post-collision translational energy of NO follows from Eq. (8.6) and is given by

$$E'_{\rm NO} = \frac{m_{\rm Ar}}{M}E'_{rel} = \frac{m_{\rm Ar}}{M}(E_{rel} - E'_{int}). \tag{8.13}$$

Solving for the square of the final center-of-mass velocity of the recoiling NO, we have

$$u'^2_{\rm NO} = \frac{2}{m_{\rm NO}}\left(\frac{m_{\rm Ar}}{M}E'_{rel}\right) = \frac{m_{\rm Ar}^2}{M^2}v_{\rm Ar}^2 + \frac{m_{\rm NO}^2}{M^2}v_{\rm NO}^2 - \frac{2m_{\rm Ar}}{m_{\rm NO}M}E'_{int}. \tag{8.14}$$

The resonance condition is $u'_{\rm NO} = v_{cm}$, or $v'_{\rm NO} = u'_{\rm NO} - v_{cm}$. In the vicinity of the resonance condition, it can be shown that

$$\frac{\partial(v'_{\rm NO})}{\partial v_{\rm Ar}} = \frac{\partial(u'_{\rm NO} - v_{cm})}{\partial v_{\rm Ar}} = \frac{\partial(u'_{\rm NO})}{\partial v_{\rm Ar}} - \frac{\partial(v_{cm})}{\partial v_{\rm Ar}} = 0 \tag{8.15}$$

and

$$\frac{\partial(v'_{\rm NO})}{\partial v_{\rm NO}} = \frac{\partial(u'_{\rm NO} - v_{cm})}{\partial v_{\rm NO}} = \frac{\partial(u'_{\rm NO})}{\partial v_{\rm NO}} - \frac{\partial(v_{cm})}{\partial v_{\rm NO}} = \frac{v_{\rm NO}}{v_{cm}}\left(\frac{m_{\rm Ar}^2 - m_{\rm NO}^2}{M^2}\right). \tag{8.16}$$

For finite differences in the scattering velocity of NO, we have a relationship between the post-collision velocity spread, $\Delta v'_{\rm NO}$, and the velocity spread in the NO molecules before collision:

$$\Delta v'_{\rm NO} = \frac{v_{\rm NO}}{v_{cm}} \cdot \frac{m_{\rm Ar}^2 - m_{\rm NO}^2}{(m_{\rm Ar} + m_{\rm NO})^2} \cdot \Delta v_{\rm NO} \tag{8.17}$$

To first order, the spread in $v'_{\rm NO}$ has only a weak dependence on the velocity spread of the Ar atom beam, $v_{\rm Ar}$, through the center-of-mass velocity. Interestingly the final spread of velocities of the NO is inversely proportional to the center-of-mass velocity so the faster the atomic beam moves the narrower the final velocity distribution.[48] As Eq. (8.10) indicates, the velocity spread is primarily dependent upon the velocity spread of the NO molecular beam, $\Delta v_{\rm NO}$. The dependence on the spread in $v_{\rm NO}$ is interesting as it shows a kinematic compression of the velocity spread. For collision of a room temperature expansion creating a molecular beam of 95% Ar and 5% NO in collision with a room temperature expansion creating an atomic beam of neat Ar crossing at 90° Eq. (8.17) predicts that $\Delta v'_{\rm NO} = 0.2\Delta v_{\rm NO}$. We predict an 80% compression of the velocity spread of the initial molecular beam for the cold molecules. A molecular beam spread of 40 m/s then leads to an 8 m/s spread of the cold molecules. Assuming that the velocity of the NO beam is set such that the distribution is centered at zero in the lab then the average speed predicted is approximately 4 m/s. For NO this average velocity corresponds to a temperature of approximately 30 mK.

How many molecules scatter to the intersection of the crossed atomic and molecular beams is a function of the differential cross section of the scattering process. We can predict the angle at which the molecules must scatter in order to find themselves at the scattering center by use of the Law of Cosines applied to the velocities,

$$v_{\rm NO}^2 = v_{cm}^2 + u_{\rm NO}^2 - 2u_{\rm NO}v_{cm}\cos(\theta_{cm}), \tag{8.18}$$

where $\theta_{cm}$ is the angle between the velocity of the center of mass and the pre-collision velocity of the NO molecular beam in the center-of-mass frame of reference. To derive an expression for the angular constraint, we begin by writing the velocity of the center of mass and the initial NO molecular beam velocity in the center-of-mass frame of reference, in terms of the lab-frame velocity of the NO molecules,

$$v_{cm}^2 = \frac{2E_{cm}}{M} = \frac{2}{M}\cdot\left(\frac{m_{\rm NO}v_{\rm NO}^2}{2} + \frac{m_{\rm Ar}v_{\rm Ar}^2}{2} - \frac{\mu v_{\rm NO}^2}{2} - \frac{\mu v_{\rm Ar}^2}{2}\right), \tag{8.19}$$

and

$$u_{\rm NO}^2 = \left[\frac{m_{\rm Ar}}{M}\cdot\sqrt{v_{\rm NO}^2 + v_{\rm Ar}^2}\right]^2 = \frac{m_{\rm Ar}v_{\rm NO^2}^2}{M^2} + \frac{m_{\rm Ar}v_{\rm Ar^2}^2}{M^2}. \tag{8.20}$$

Solving Eq. (8.18) for $\cos(\theta_{cm})$ and substituting for the velocities defined in Eqs. (8.19) and (8.20), it is seen that the condition for the scattering angle which must be met in order for the NO to have zero laboratory-frame velocity is:

$$\cos(\theta_{cm}) = \frac{E_{\rm Ar} - E_{\rm NO}}{\sqrt{\left(E_{\rm NO} + \frac{m_{\rm NO}}{m_{\rm Ar}}E_{\rm Ar}\right)\left(E_{\rm NO} + \frac{m_{\rm Ar}}{m_{\rm NO}}E_{\rm Ar}\right)}}, \tag{8.21}$$

where $E_{NO}$ and $E_{Ar}$ are the kinetic energies of NO and Ar, respectively. Since the angular distribution of the inelastically scattered $NO_{7.5}$ from Ar is rather broad,[48] this condition is easy to achieve, and is indeed obtained for our experimental conditions. More generally, the translational energy of the NO beam can be selected in order to ensure that the required $|u'_{NO}|$ corresponds to the energy of an NO rotational quantum state, and the translational energy of the Ar beam can be selected to ensure that sufficient scattering events satisfy the angular condition.

Fundamentally the cold molecules scatter to the position of the scattering event in the laboratory and consequently the cold molecules are found where the scattering event occurs. This highlights the largest problem to overcome in order to make usable samples of these cold molecules. They are stopped at a location where the collision with the hot beams is maximized. We have spent considerable effort to overcome this problem which is discussed in Sec. 8.3.3.

Equation (8.10) has a remarkable consequence for the scattering of partners with identical masses ($ND_3$ colliding with Ne, for instance). As the mass of the two collisions partners becomes equal, the collision can remove all kinetic energy leaving the scattered molecules in the ground rotational state. Further, for molecules colliding with scattering partners of equal mass, the molecules that satisfy the condition of $v_{cm} = -u'_{molecule}$ are those that scatter elastically. Under this condition the velocity spread, for those molecules that scatter back to their point of collision, is exactly zero, since all elastic scattering products must lie on a Newton sphere that intersects the origin regardless of collision energy. Those scattered in the appropriate direction, $v_{cm} = -u'_{molecule}$, will be completely stationary in the laboratory, and a collision energy can be chosen to optimize scattering amplitude in the direction of the laboratory origin, $v' = 0$.

### 8.2.1. *Cooling in Kinematic Cooling*

In this section we take a closer look at phase space density and what it means to cool. In Sec. 8.1.1 we discussed the phase space density in terms of the temperature and density of the sample of atoms or molecules being studied. The present discussion will focus on the inter particle energy, which is directly related to the temperature of the ensemble. The phase space density discussion is in terms of the energy in order to clarify the distinction made between cooling and slowing. Cooling is a process that decreases the laboratory frame velocity while simultaneously removing inter-particle energy, increasing the ensemble phase space density. Slowing is a process that decreases the laboratory frame velocity of the ensemble but does not increase the phase space density.

In the field of cold atoms and molecules the velocity and temperature are often used interchangeably. However, the laboratory frame velocity is a per particle characteristic while the temperature and density are properties of the ensemble. As the velocity of a single particle is reduced, the thermal de Broglie wavelength increases. If we look at a collection of particles, and reduce their velocity, the thermal de Broglie wavelength of each particle increases. If we reduce the velocity of the particles and the inter-particle energy, the temperature is also reduced and the average density is increased. This process increases the phase space density of the sample.

The per particle phase space density is defined as the probability of having a particle with a given momentum $\rho$ at position $r$ at time $t$. If we average the per particle phase space density over the entire ensemble we obtain the phase space density. The phase space density is an important quantity as it is a measure of the relative energy between particles in a sample, which dictates the behavior of the system. A system with high relative particle energies behaves like a classical gas, while a system with low relative particle energy behave like a quantum gas.

A good thought experiment to clarify this is to think about a Bose-Einstein condensate, a sample of atoms that behaves like a macroscopic quantum wave packet, and put it on a rocket. Even though it is moving at over 300 m/s, the atoms relative energies have not changed, and they still behave as a macroscopic quantum wave packet. On the other hand, if we have a classical gas pulse, from a pulsed molecular beam valve for instance, moving at over 300 m/s, and now we, the observers, get on the rocket and move with the gas pulse, even though the average position of the gas pulse is stationary, which is quite cold in our reference frame, the particles still behave classically because of their relative energy. It is the relative particle energies in the center of mass of the system under study that determines if the system behaves classically or quantum mechanically. The physics of the system must be independent of the reference frame of the observer.

The importance of the phase space density is to determine if our so called "cooling technique", kinematic cooling, actually increases the phase space density or if it is a slowing technique that preserves phase space density. The statistical mechanics approach to this question is to look at the forces acting on the system and to see if that system obeys "Liouville's theorem" or not.

Liouville's theorem states: that under conservative interactions, phase space acts as an incompressible fluid. This means that, if the forces acting on the system are conservative, then phase space density is conserved and cannot be increased. However, it can be "squeezed" like an incompressible fluid. That is we can remove momentum, $\vec{\rho}$, "squeezing" the system, which responds as an incompressible fluid and modifies the spatial distribution,

$\vec{r}$ to keep the phase space density, $(p(\vec{\rho}, \vec{r}, t = 0))$, constant (slowing). If the forces acting on the system are nonconservative, Liouville's theorem does not apply and the forces can increase the phase space density of the sample (cooling). Mathematically, Liouville's theorem is given by Eq. (8.22), Haung[49] gives a good mathematical derivation of Liouville's theorem.

$$\nabla_q \cdot \dot{\vec{q}} = \frac{\partial}{\partial \vec{r}} \left( \frac{\partial \vec{r}}{\partial t} \right) + \frac{\partial}{\partial \vec{\rho}} \left( \frac{\partial \vec{\rho}}{\partial t} \right) = 0 \tag{8.22}$$

The approach is to determine the force(s) acting on our system, write down the equations of motion, and solve Eq. (8.22). Specifically, we look at the time evolution of the phase space density, $(p(\vec{\rho}, \vec{r}, t))$, and determine if it is increasing, decreasing, or staying the same?

Before we look at the evolution of the phase space density for kinematic cooling, we are going to look at a system where it is accepted that cooling occurs, 1D atomic laser cooling.

In 1D laser cooling the force on the atoms is $F_{Laser} = -\Gamma v$, where the velocity dependence arises from the Doppler shift of the atom and the laser detuning, such that the faster the atom moves the more resonant the cooling light, the more photons the atom scatters, the larger the force is on the atom. Metcalf[50] shows that in this system the phase space density evolves according to; $\partial p(\vec{\rho}, \vec{r}, t)/\partial t = \Gamma p(\vec{\rho}, \vec{r}, t)$. This result shows that the phase space density of this system increases in time, thereby cooling the system.

In kinematic cooling, the force that an atom or molecule feels is an impulse from the collision with an atom. The change in velocity $\Delta v$ is given by $v_F - v_0$, where $v_F$ is determined from Eq. (8.11); $v_F = \frac{2}{m}[E_0(1 - \frac{m_1}{m_2}) + E_{Rot}]^{\frac{1}{2}}$. The force is then; $F = m \cdot \Delta v/dt$. For simplicity we will neglect the rotational energy and focus on the translational energy, simplifying $\Delta v = -v_0 \cdot (\frac{m_1}{m_2})$. If we let the collision time be $\delta T$, the force becomes; $F = -v_0 \frac{m_1^2}{m_2 \cdot \delta T}$. Setting $\frac{m_1^2}{m_2 \cdot \delta T} = \Gamma$ yields $F = -\Gamma v$, a similar velocity dependent force to that in 1-D laser cooling. Physically, this means that the amount of velocity removed via one collision increases with the initial velocity of the molecule. This leads to the bunching of the low velocity molecules in the collision region which is evident in the images from the narrowing of the velocity distribution near the laboratory origin, see Fig. 8.2.

It should be noted, that while there is nothing non-conservative in the atom-molecule collision, the collisions that produce the cold molecules also produce the hottest atoms. Just as in the case of laser cooling, the absorption and spontaneous emission cycle conserves energy, and while the emitted photons are in random directions, conservation laws determine the energy and direction of the emitted photon on any one absorption and

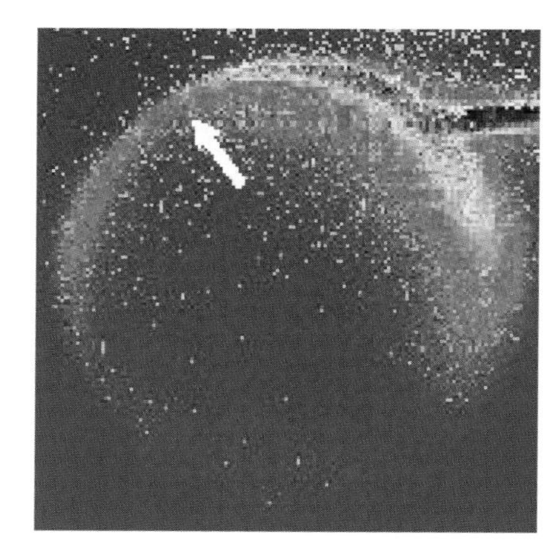

Fig. 8.2. Velocity-mapped ion image of $NO_{7.5}$ scattered off of Ar at extremely high beam densities in order to suppress the build up of cold NO. The arrow indicates the laboratory origin where the cold molecules are formed, from the image one can see the velocity compression associated with the scattering from the narrowing of the scattering ring near the laboratory origin.

emission cycle. Since the photon is absorbed from a laser and randomly emitted into the continuum the entropy of the photon increases. In a similar fashion, the atom used to cool the molecule receives a velocity dependent increase in energy, increasing the entropy of the atom. The atom/molecule system still obeys energy conservation, but just as the emitted photon is neglected in laser cooling, the atom that was used to cool is neglected.

There are two key side effects of the velocity dependent force. First, kinematic cooling results in real cooling, not just a rotation of position-momentum phase space, yielding an increased phase space for the cold molecules. Second, since there is dissipation, if the collisions occur in a region containing a trap for the molecules, the trap can be continuously loaded without the worry of how to load pre-cooled molecules into a conservative potential well.

## 8.3. Kinematic Cooling Techniques

### 8.3.1. *Crossed Atomic and Molecular Beams*

Crossed atomic and molecular beam (CAMB) experiments have been used for several decades to study collisional dynamics in gas phase chemistry. Combining CAMB experiments with the Velocity Mapped Ion Imaging technique, forms a powerful new tool that allows experimentalists to directly

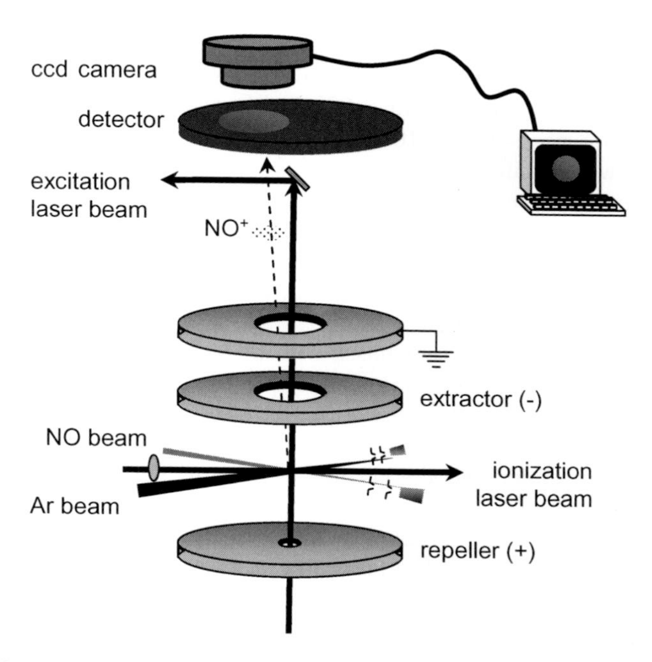


Fig. 8.3. Schematic representation of crossed molecular beam scattering apparatus with laser ionization detection of the scattering product. Ion Imaging is used to measure the velocity of the laser-produced ions.

image the velocity/momentum distribution at all angles from a scattering event. Figure 8.3 is a simplified schematic of our CAMB apparatus which we use to produce cold molecules. The source for the molecules and atoms are two doubly-skimmed, pulsed beams. One beam is typically a pure atomic beam and the other beam is typically 1% $\rightarrow$ 5% of a molecular species mixed with an atomic gas. This mixing is done for several reasons. The most important reason is that when the pulsed beam valve is opened the particles undergo multiple collisions as they are injected into the vacuum region and these collisions determine the final velocity of the molecule. When the molecules are injected into the vacuum they undergo a supersonic expansion, and the final velocity of the particles is directly related to the mass of the dominant species. That is to say, if we have 5% NO mixed with He, after the expansion the NO will be moving close to the same velocity as the He, since it dominantly is colliding with the He, moving roughly 1500 m/s. If we have the same mixture but swap Xe for the He, the NO will have a velocity around 350 m/s. So, by mixing the molecule of interest with an atomic species, one can control the velocity of the molecule which, for the production of cold molecules, dictates the necessary mass of the collision partner and the final rotational energy of the cooled molecules.

The two beams intersect at ~90° inside the interaction region with typical vacuum pressures of $10^{-9}$ Torr. The beams cross between plates that form a weak uniform electric field of ~200 V/cm. The crossing of the molecular and atomic beams is further intersected by an ionization laser. The ionization laser(s) ionizes the molecules using resonance enhanced multi-photon ionization, REMPI. Once ionized the molecules are extracted from the interaction region by the uniform electric field into the Velocity Mapped Ion-Imaging system.

Our typical NO-Ar system has 2 atm of Ar backing both valves with one doped with 5% NO. Two laser beams are used to ionize NO, one laser beam is resonant with a transition in the (0,0) band of the A-X system near 226 nm, thereby selecting a particular rotational state which is then ionized near threshold with a ~300 nm laser beam. This technique is known as $(1 + 1')$REMPI. Images similar to Fig. 8.2 are taken for each product NO rotational state. By measuring the velocity profile of the molecules scattered in the direction of the laboratory origin, the crossing point of the two beams, we can measure the post collision velocity distributions for the side scattered molecules as a function of rotational state, see Fig. 8.4.

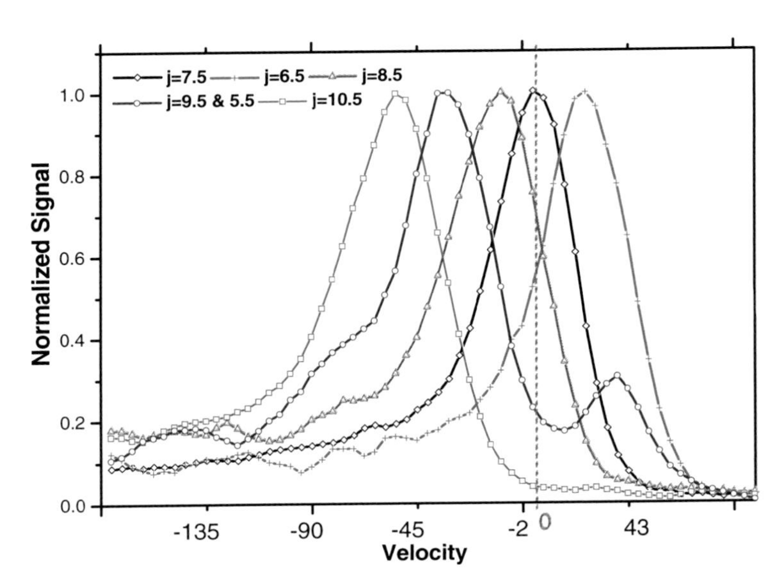


Fig. 8.4. Latent velocity of cooled NO molecules for several different rotational levels. j = 7.5 is at rest in the laboratory. j = 6.5 and j = 5.5 are moving forward in the laboratory frame while 8.5 and higher are moving backwards.

### 8.3.2. *Velocity Mapped Ion Imaging*

Velocity Mapped Ion Imaging is obtained by having a single or multiple electrostatic lens system which take ions, extracted from a uniform electric field, and focuses them onto a position sensitive ion detector. Figure 8.5 shows a typical Velocity Mapped Ion Image of the j = 7.5 state of NO with the relevant velocity axis overlaid. As discussed in Sec. 8.3.1 neutral molecules are ionized using REMPI, forming ions in a uniform electric field which accelerates the ions out of the interaction region. In the simplest, one lens configuration, the ions pass through an electrostatic lens which focuses the ions onto a micro-channel plate (MCP)/phosphor screen detector. The flight paths of the ions can be readily calculated using their classical trajectories and electro-static interactions. Typically programs like SIMION[51] are used to model the entire system, the extraction field, the lens, and the flight path to the ion detector. The computer models help determine the gaps and necessary ratios of electric fields needed to obtain the velocity mapped condition. The programs also help determine the effectiveness of more exotic lenses and configurations. However, a clear picture of how ion imaging works can be obtained using ray optics.

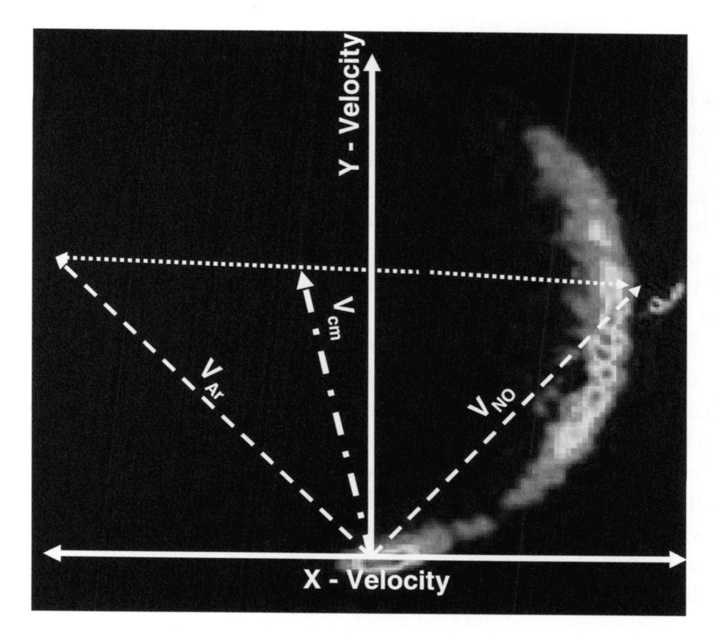


Fig. 8.5. Velocity mapped ion image of NO colliding with Ar. The image is of NO in j = 7.5 rotational state with the velocity information coarsely indicated.

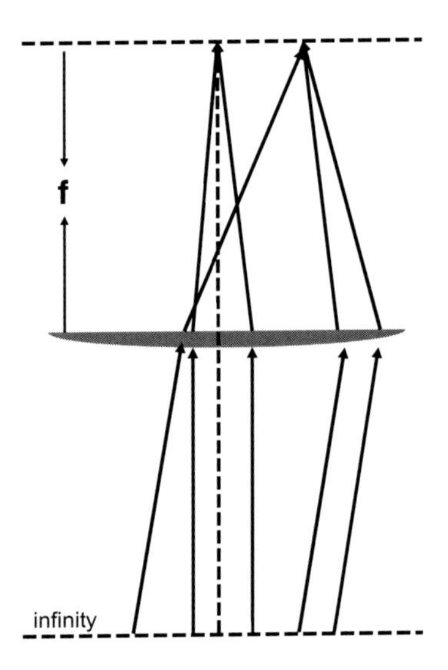


Fig. 8.6. Ray diagram for both parallel and skew rays. The parallel rays are focused on the optical axis in the focal plane, while the skew rays are focused in the focal plane, but off the optical axis. The skew in the ray is a direct measure of the ion velocity perpendicular to the optical axis.

Ions flying through electric fields behave classically and their trajectories can be well approximated by straight lines. In terms of ray optics, they fly as ideal rays (no dispersion). The electro-static lenses apply a force to the ion while the ion is inside the lens and nothing when the ion is outside the lens. The force from the electrostatic lens bends the flight path of the ion, just as optical lenses bend rays. Figure 8.6 shows a ray diagram for ions formed in a uniform electric field with one focusing optic. The ions in the uniform electric field enter the first electro-static lens like collimated rays coming from infinity. The lens bends the trajectory of the ions, "focusing" ions at the focal point of the lens. If the ions have a velocity component perpendicular to the lens, the entering ray looks like a skew ray passing through the lens. The ray is still focused at f, but the focal point is shifted off the optical axis. In this fashion, the electro-static lens transforms velocities perpendicular to the lens into spatial position in the focal plane of the lens. Figure 8.5 depicts a velocity mapped image of $NO_{7.5}$ after a collision with Ar, the relevant scattering velocities are labelled.

While velocity mapped imaging gives a direct picture of momentum space, there are some serious issues using this technique with cold molecules. The two most important are the electron recoil during the ionization step and charge repulsion from producing multiple ions per laser shot. While both of these issues are experimentally controllable, current commercially available laser systems, and ion detectors coupled to phosphor screens limit the typical velocity resolution to ~1 m/s. Our current simple 1 lens design has a measured resolution between 2.5 m/s and 4 m/s for NO.

### 8.3.3. *Production of Cold Nitric Oxide*

In Sec. 8.2 we discussed the production of cold molecules from a single collision with an atom. It was noted that the cold molecules are necessarily formed at the crossing of the atomic and molecular beams, where the scattering occurs. To date we have put considerable effort to understanding the practical and experimental limits of the crossed molecular beam apparatus for producing cold molecules and what modifications we need and can make in order to produce and confine useful amounts of cold molecules generated from this kinematic cooling technique.

The most vital insight was gained comparing NO-Ar scattering in our doubly skimmed CAMB apparatus with the singly skimmed CAMB apparatus of Dr. A. Suits at Wayne State University. The main difference between the two apparatuses is the density of the molecular beams and the overall vacuum pressure in the interaction region. All other differences in the apparatus are in the ion imaging technique. This difference in the atomic and molecular beam densities allows Dr. Suits to observe much weaker scattering phenomena than is accessible with our apparatus, such as reactive scattering.[52] We performed the same experiment, NO/Ar scattering, in both apparatuses. Figure 8.2 shows the results from the singly skimmed apparatus, while Fig. 8.5 show the results of the doubly skimmed apparatus. At the high beam densities in the single skimmed apparatus, we observe that the bright spot from the cold molecules is missing, or greatly reduced.

The steady state density of cold molecules is the result of a balance between the rate of production, $R_{pro}$, and some destruction rate, $R_{des}$, of the molecules. If $R_{pro} > R_{des}$, we would expect a build up of the cold molecules and a corresponding high intensity of molecules near the laboratory origin. While if $R_{pro} < R_{des}$, then there is no build up or enhanced signature of the cold molecules. In order to maximize the number of cold molecules favorable molecular beam conditions need to be met that minimize $R_{des}$ and maximize $R_{pro}$.

The problem and main difficulty to this cooling technique is that $R_{pro}$ derives from single collisions between $NO_{0.5}$ and Ar, with a probability factor from the total differential cross section. $R_{des}$ comes from any other collisions of the nascent $NO_{cold} + X$, where X can be NO in any rotational state, Ar, or any background gas. In order to separate the cold NO from the molecular beam, we must minimize $R_{des}$ most importantly when the molecular beams are turning off as $R_{pro} \rightarrow 0$.

A typical supersonic beam of Ar with about 5% of NO doped into it has a translational energy of $439.69 \pm 17.55\,\mathrm{cm}^{-1}$. For this translational energy Eq. (8.10) dictates that an internal energy in the NO of $110.34 \pm 4.17\,\mathrm{cm}^{-1}$ is required for an NO molecule to become stationary upon colliding with an Ar atom. The rotational energy of NO in the $j' = 7.5$ quantum state is $108.417\,\mathrm{cm}^{-1}$. Since the energy spread in the NO beam is larger than the energy difference, $NO_{7.5}$ molecules that are scattered in the correct direction, given by Eq. (8.14), will be nearly stationary in the laboratory frame. For scattering between NO and Ar in the neat Ar beam, the initial relative collision energy is $472.29 \pm 30.1\,\mathrm{cm}^{-1}$; if $108.417\,\mathrm{cm}^{-1}$ is taken up by rotation of the $NO_{7.5}$ molecule then $363.62\,\mathrm{cm}^{-1}$ remains for translational energy of the recoiling partners in the center-of-mass frame. We can check this by measuring the radius of the scattering ring in Fig. 8.5. We measure a translational energy of the recoiling $NO_{7.5}$ to be $205.63 \pm 10.9\,\mathrm{cm}^{-1}$. This corresponds to an NO + Ar recoil energy of $360.278 \pm 10.9\,\mathrm{cm}^{-1}$, in agreement with the expected value.

We have experimentally determined that the effects of $R_{des}$ can be minimized by reducing the density of the atomic and molecular beams. Ideally, the same steady state population is achieved if the beam pulses are lengthened until the steady state condition, $R_{pro} = R_{des}$, is reached. After the steady state condition is reached we want the molecular beams to turn off as quick as possible to minimize the impact of $R_{des}$. We obtain the best experimental results by reducing the densities of the atomic and molecular beams. This increases the time to reach the steady state condition while minimizing $R_{des}$. With these conditions we can observe the cold molecule signal as a function of time. Figure 8.7 shows successive ion images as a function of time after the atomic and molecular beams have been closed. We observe a spot due to the cold molecules and some effusive beams coming from the remnant gas in the source chamber of the apparatus. Figure 8.8 shows the differing lifetimes for the cold NO in $j = 7.5$ and $j = 10.5$. The decays in Fig. 8.8 are fit with dual exponentials. There are initial rapid decays attributed to secondary collisions with the atomic and molecular beams as they turn off. The secondary slow decays are attributed to fly-out of the molecules from the interaction region. The measured 1/e decay times are $124\,\mu\mathrm{s}$ for the $j = 7.5$ state and $9\,\mu\mathrm{s}$ for the $j = 10.5$ state. The interaction region is measured to be $540\,\mu\mathrm{m}$. The fly-out is modeled using a Monte-Carlo simulation to account for the interaction volume and the detection overlap. The models yields a final velocity of $\sim 5\,\mathrm{m/s}$ for the $j = 7.5$ state and $\sim 60\,\mathrm{m/s}$ for the 10.5 state. Equating these velocities to 1-D temperatures, one finds a temperature of $\sim 45\,\mathrm{mK}$ for the $j = 7.5$ state and

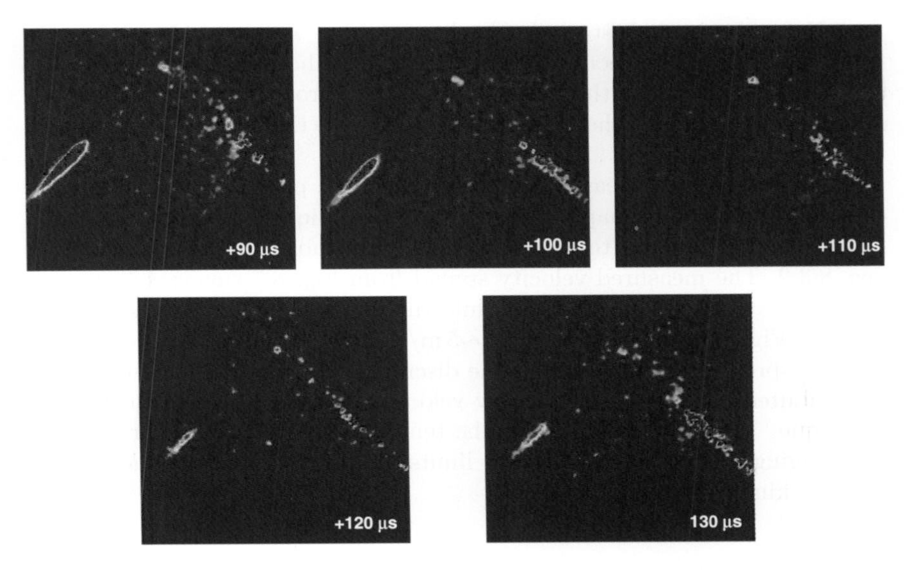


Fig. 8.7. Time sequence of cold NO from 90 to 130 $\mu$s after the peak of the scattering off of Ar.

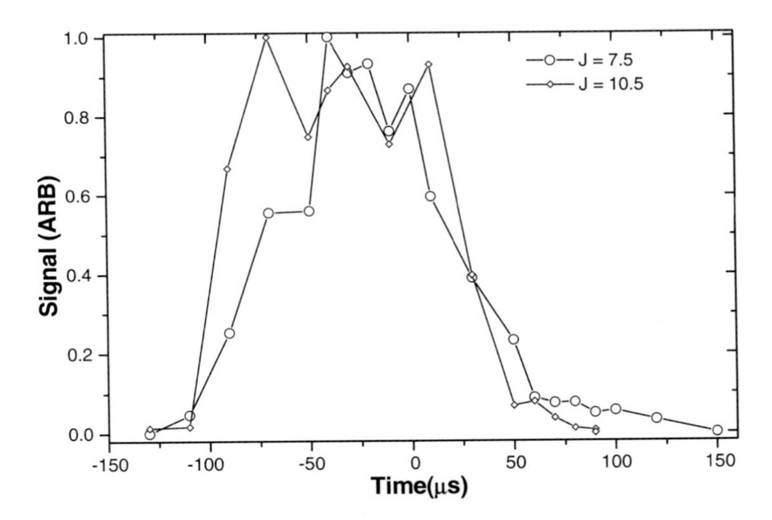


Fig. 8.8. Production and decay of cold molecules in the j = 7.5 and j = 10.5 rotational states from collisions of Ar.

~3.6 K for the j = 10.5 state. From these results, we are able to achieve a condition with the molecular beams where even though the cold molecules are being produced at the intersection of two crossed supersonic beams, approximately 10% of the cold molecules survive the turn off of the beam pulses.

It is difficult to measure the low velocities produced via kinematic cooling. The velocity-mapped ion imaging technique has limitations that make it very difficult to measure such slow molecules as discussed in Sec. 8.3.2. The measured velocity spread from Fig. 8.7 suggests a velocity of ~14 m/s for the NO j = 7.5, while the fit to the decay of the j = 7.5 signal in Fig. 8.8 yields a velocity of ~5 m/s, in agreement with the expected velocity spread from Eq. (8.17). The discrepancy in the two measurements is attributed to artifacts in the low velocity resolution of the ion imaging technique. Therefore, we consider the temperatures measured through the ion imaging technique to be upper limits for the temperatures achievable through kinematic cooling.

### 8.3.4. *Production of Cold $NH_3$ and HBr*

Kinematic cooling relies only on the mass of the particle and not on the specific internal structure of the particle, such as dipole moments, ground state or electronic structure. As discussed in Sec. 8.2, kinematic cooling relies on the mass ratio between the particle you want to stop and the colliding particle. The rate of the production of cold molecules mirrors the differential cross-section for the collision. While this seems an obvious statement, the implication is that for two mass-degenerate particles with equal velocities, for example ammonia ($ND_3$) scattering off neon (Ne), the cold molecules are those elastically scattered at 90°, which has a near zero probability. To circumvent this problem, the Ne beam can be sped up and/or the $ND_3$ be slowed down by seeding them in the appropriate carrier gases, as discussed in Sec. 8.3.1. By slowing down the $ND_3$ and speeding up the Ne, the molecules scattered against the center-of-mass motion are shifted toward the forward scattered portion of the differential cross-section, which typically results in a higher production rate.

Motivated by Dr. Meijer's success of trapping Stark decelerated $ND_3$ (see chapter 9),[36] we have looked at the possibility of kinematically cooling $ND_3$ and $NH_3$. $ND_3$ is an excellent candidate for kinematic cooling since it is nearly mass-degenerate with Ne, and a single $NH_3 \leftrightarrow$ Ne collision removes 85% of the energy. The first test experiment was to produce cold $NH_3$ in the first rotational state. $NH_3$, a symmetric top, has three rotational constant; $A = 9.44\,\text{cm}^{-1}$, $B = 9.44\,\text{cm}^{-1}$, $C = 6.19\,\text{cm}^{-1}$. If we follow through the calculation of Sec. 8.2, we find that for a beam of $NH_3$ with an initial velocity of 420 m/s, the side-scattered molecules have ~18.8 cm$^{-1}$ of internal energy

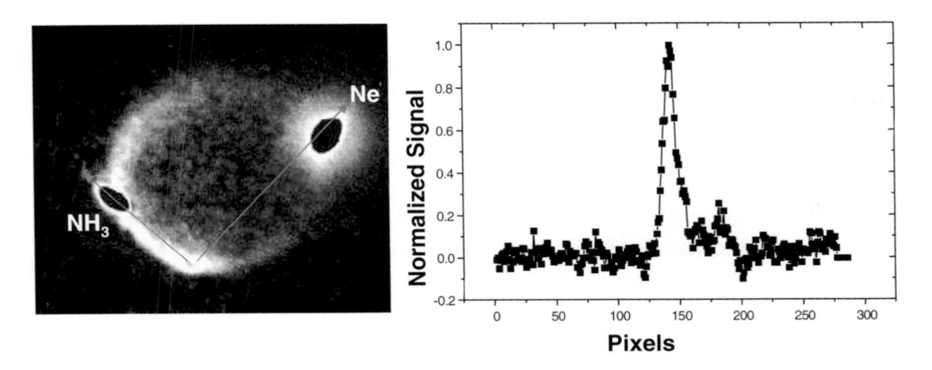


Fig. 8.9. Left side of the figure shows the differential cross-section for $NH_3$ j = 1 from $NH_3$ j = 0 doped in Kr scattering off of Ne doped in He. The right side is a cross-section of the stationary molecules. The molecules fit to a 15m/s velocity spread (200 mK) which is dictated by our imaging resolution and electron recoil during detection

which is degenerate with the $j = 0 \rightarrow 1$ rotor spacing. Coincidentally, supersonically expanded Krypton (Kr) has a velocity around 420 m/s. So, by doping about 1% $NH_3$ in Kr we can kinematically cool $NH_3$, $j = 1$. Figure 8.9 shows the results of colliding $NH_3$ doped in Kr with Ne doped in Helium (He), where the Ne is doped in He to speed up the Ne and therefore shift the differential cross-section for the collision. With this technique we have produced $NH_3$ in the $j = 1$ state with a measured velocity of 15 m/s. According to the calculations of Sec. 8.2 the velocity should be much less, but due to the resolution of our detection and electron recoil during the (2 + 1) REMPI ionization near 334 nm we are limited in our ability to accurately measure these low velocities.

Another molecule of interest is HBr (and DBr). HBr has a large dipole moment, but due to its ground state configuration HBr has a very small first order Stark shift, which makes it hard to Stark decelerate and electrostatically trap. However, HBr is near mass degenerate with Kr and as is discussed in Sec. 8.4, DBr is nearly mass degenerate with Rubidium Rb.

There are a few points of interest when dealing with heavy molecules and atoms. A typical gas bottle of Kr has 4 isotopes ($^{82}Kr = {}^{83}Kr = 11\%, {}^{84}Kr = 57\%, {}^{86}Kr = 17\%$) and HBr has 2 isotopes ($H^{79}Br$ and $H^{81}Br$). Even with a mass-mismatch larger than that of $NH_3$ and Ne, a single $H^{79}Br \leftrightarrow {}^{84}Kr$ collision removes nearly 96% of the $H^{79}Br$ energy. For HBr doped into Xenon (Xe), with a typical supersonic expansion velocity of 350 m/s, single HBr $\leftrightarrow$ Kr collisions will slow ground state $j = 0$ HBr down to 16 m/s. But there are better isotope match ups; $H^{81}Br$ colliding with $^{84}Kr$ results in a 8 m/s final velocity. The main point is that with heavy molecules the kinematic cooling technique can work over a large mass mismatch, since the energy removed scales as the ratio of the two (large) masses, making this an ideal technique for the initial cooling of heavy molecules, like YbF, for experiments searching for the electron dipole moment.

In order to kinematically cool HBr via collisions with Kr we need the post collision HBr to be in the j = 1 state. The rotational constant for HBr is $B = 8.46\,cm^{-1}$. In order to stop the j = 1 molecules, we are going to look at the collision of $H^{79}Br$ with $^{86}Kr$. Even with a 6 amu mass difference we still need to seed the $H^{79}Br$ in He to achieve an initial velocity of ~1000 m/s. Speeding up the molecule shifts the differential cross section away from forward scattered, typically reducing the scattering probability. To counter this we also seed the Kr in He. Figure 8.10 shows a cut through the origin of the scattered $H^{79}Br$. We observe a width of ±10 m/s or nearly 500 mK. The final temperature is nearly a factor of ten greater than the predicted temperature. However, due to the presence of the other Kr isotopes there are three other potential collision channels which could be blurring the image. We have successfully demonstrated that the kinematic cooling technique works for HBr by producing a sample of cold HBr with an upper limit on the temperature of ~500 mK.

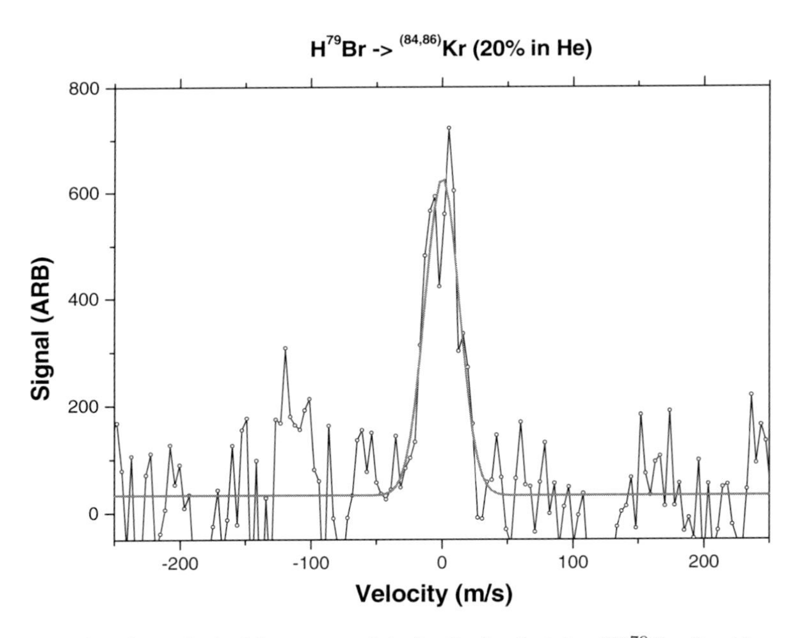


Fig. 8.10. Cut through the laboratory origin for the j = 1 state of $H^{79}Br$ after kinematic cooling from colliding with 5% Kr doped in He. The image fits to 10 m/s spread, which gives a 1D temperature of ~500 mK.

Kinematic cooling works for cooling a wide range of molecules and atoms irrespective of the unique properties of the individual molecules and atoms. While this technique is general, it does rely on a favorable mass ratio between the colliding particles, favorable rotational spacings, and favorable differential collision cross-sections. However, as we have demonstrated, even with fixed mass ratios and rotational spacings, the initial velocities and differential cross-sections can be manipulated in order to produce cold molecules.

### 8.3.5. *Techniques for Trapping Kinematically Cooled Molecules*

Producing cold or slow moving molecules gives experimentalists the ability to study energy selective collisions[53] and high resolution spectroscopy, and to orient molecules and manipulate the electronic states of molecules with external fields. The ability to manipulate molecules with external fields leads to trapping and/or confining molecules. Molecules confined in traps have been held for several seconds.[36] This opens the doors to a new arena of physics and chemistry that can be studied; collisional processes with small cross-sections, ultra-high resolution spectroscopy ($10^{-10}\,\text{cm}^{-1}$), measuring reaction barriers, and the density of states near the continuum to name a few. However, all this relies on effective methods for not only producing cold molecules and developing some type of trap for the molecules, but also in loading the molecules into the trap.

In the field of ultracold atoms, atoms are typically trapped at initial velocities of around 10 m/s or so, depending on the apparatus, in a magneto optical trap or MOT. The MOT is a hybrid of a laser molasses[50] and magnetic quadrupole trap. The magnetic trap defines the spatial location of the trap and the molasses quickly ($\sim$KHz) reduces the temperature of the atoms into the $100\,\mu$K regime where the atoms can be magnetically confined. They are then further cooled by forced evaporation to a few $\mu$K or less. At these low temperatures atoms can easily be confined by a number of techniques including electric, magnetic, and/or optical fields. At these low temperatures, atoms can even be confined using a focused far off resonance red detuned laser beam which act on the polarizability of the atoms to confine them at the laser focus.[54]

Ideally one would like the ability to manipulate molecules in the same fashion as atoms. However, without the ability to laser cool, the requirements of the initial trap become much more involved. Instead of the initial trap being at $\sim$1 mK, we now need the initial trap to be in the 10 to 100 mK range. Most closed shell ground state molecules have small Zeeman shifts which makes magnetic trapping, even with super-conducting magnets difficult. Molecules with permanent dipole moments can in principle be trapped by electrostatic fields, which has been accomplished for molecules with large first order Stark shifts.[36] A near resonance microwave trap has been proposed[55] which couples to the $j' = 0$ to $j = 1$ rotational transition and is capable of producing trap depths of a few Kelvin. This, in principle, is a general trap but the required microwave frequency is $\nu = 2B$, where B is the rotational constant. This is a drawback as only a few diatomic molecules have rotational constants small enough such that the required frequency lies in the a microwave band where the required power is available. As the field of cold molecules stands, there does not appear to be a general trapping technique. Each system of interest must be analyzed to find the best technique for trapping.

### 8.3.5.1. *Magnetic Trapping*

For nitric oxide (NO) the best trap is a magnetic trap. NO has a $^2\Pi_{\frac{1}{2}}$ ground state with no first order Zeeman shift. The spin-orbit excited state is $^2\Pi_{\frac{3}{2}}$ with a $\sim 1.6\,\mu B$ Zeeman shift, where $\mu B$ is the Bohr magneton ($\mu B = 1.4\,\text{MHz/G}$). Further, as the rotational levels of the ground state increase, the ground states and spin-orbit exited states mix, increasing the effective Zeeman shift for the $^2\Pi_{\frac{1}{2}}$ states. Following the calculations done by Takazawa,[56] we calculate the Zeeman shift for different rotational levels of NO. Figure 8.11 shows the Zeeman structure for $^2\Pi_{\frac{1}{2}}$ $j = 7.5$ and $^2\Pi_{\frac{3}{2}}$ $j = 1.5$. The energy shift can be converted into temperature using the relationship; 20.84 MHz/mK. At magnetic fields near 3000 G the $^2\Pi_{\frac{1}{2}}$ $j = 7.5$ level of NO has nearly a 50 mK energy shift, while the $^2\Pi_{\frac{3}{2}}$ $j = 1.5$ state has nearly a 200 mK energy shift. Recall from Sec. 8.3.3

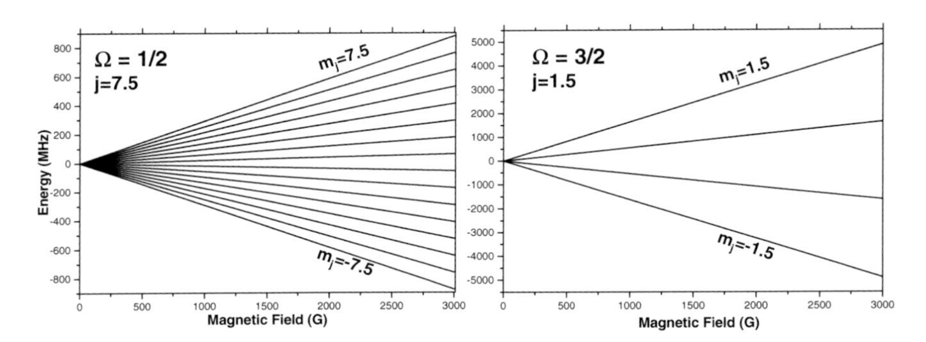


Fig. 8.11. Calculated Zeeman shifts for; Left frame in NO $^2\Pi_{\frac{1}{2}}$ $j = 7.5$ and right frame is the spin orbit exited state of NO, the $^2\Pi_{\frac{3}{2}}$ $j = 1.5$.

that kinematic cooling produces NO in j = 7.5 with a temperature of ~35 mK, so we would need around 2000 G to produce a trap capable of confining the NO.

One drawback to attempting to trap the higher rotational states is that in a magnetic field the rotational states split into $m_j = \pm j$ components. For j = 7.5 there are 16 $m_j$ projections, half are anti-confined and only $m_j = 7.5$ has the maximum positive energy shift. The $^2\Pi_{\frac{3}{2}}$ j = 1.5 state has a large energy shift and only 4 $m_j$ projections. Further, the energy that rotationally excites the $NO_{7.5}$ is 105 cm$^{-1}$, while the spin-orbit constant is 123 cm$^{-1}$. This implies that by increasing the initial velocity of the NO $^2\Pi_{\frac{3}{2}}$ j = 1.5, it can be kinematically cooled to near zero velocity in the laboratory frame where it can be trapped. The drawback to cooling the spin-orbit excited state is that the collisional cross-section to scatter from the ground state j = 0.5 to the spin-orbit excited j = 1.5 state is a factor of 10 less.[57]

To conventionally produce the 3000 G field needed to trap NO would require very high current densities and massive heat dissipation (like water or liquid nitrogen cooling), in the high vacuum region where the cold molecules are formed. Another solution is to use permanent magnets. Readily available rare-Earth magnets can provide magnetic fields nearing 5000 G (0.5T), which can be used to form a trap. Our first trap for NO is designed using two Neodymium-Iron-Boron (NdFeB) magnets with their poles facing, to form a quadrupole filed. Figure 8.12 shows the measured magnetic field and a vector plot simulation of the magnetic fields. Using just two rare-Earth magnets in this configuration we can construct quadrupole

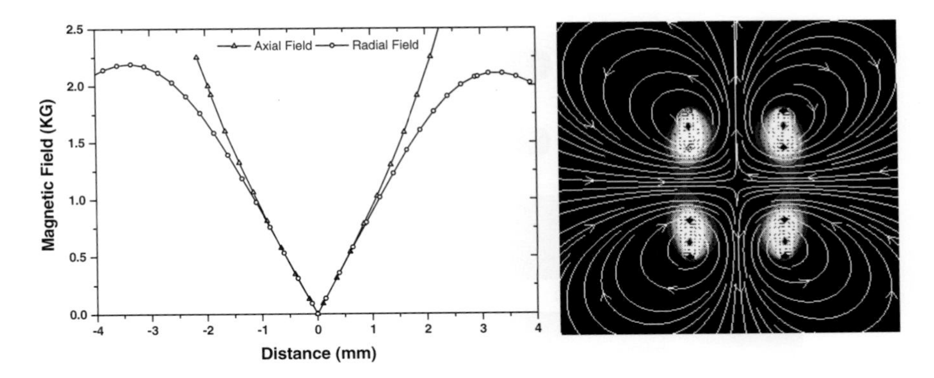


Fig. 8.12. Left figure is the measured magnetic field, in Gauss, from a pair of commercially available permanent magnets with their poles facing each other. The right frame is a calculation simulating this magnet configuration. From both frames we find that we get near quadrupole field configuration.

traps with just over 2000 G of radial confinement and near 3000 G of axial confinement, which should be sufficient for magnetic trapping of NO.

Although there are issues with trapping NO magnetically, compared to the other trapping techniques an electrostatic trap at 15 KV/cm only produces a 3 mK trap and a microwave trap would need to operate at 150 GHz. Following the Zeeman calculations,[56] it should be possible to confine kinematically cooled NO using conventional magnetic fields.

#### 8.3.5.2. *Electrostatic Trapping*

The best trap for trapping polar molecules with strong, first-order, Stark shifts is the electrostatic trap. Meijer's group showed that a quadrupole electric trap can be used to confine $ND_3$,[36] and recently they have trapped OH,OD, and NH in similar traps.[58–60] We have collaborated with the Meijer group to design an electrostatic trap for confining kinematically cooled $ND_3$. The major advantage of kinematic cooling for loading traps follows from Sec. 8.2.1; the collisions provide dissipation, and if the crossing of the beams is in the trapping region, the trap can be continuously loaded.

The electrostatic trap is depicted in Fig. 8.13. The left frame is a SIMION model of the trap and ion optics, while the right frame shows the constructed trap. The trap is constructed to allow the $ND_3$ and Ne beams to intersect in the trapping region with each other and an ionization laser. Before the molecules are ionized the trap must be turned off so that only a

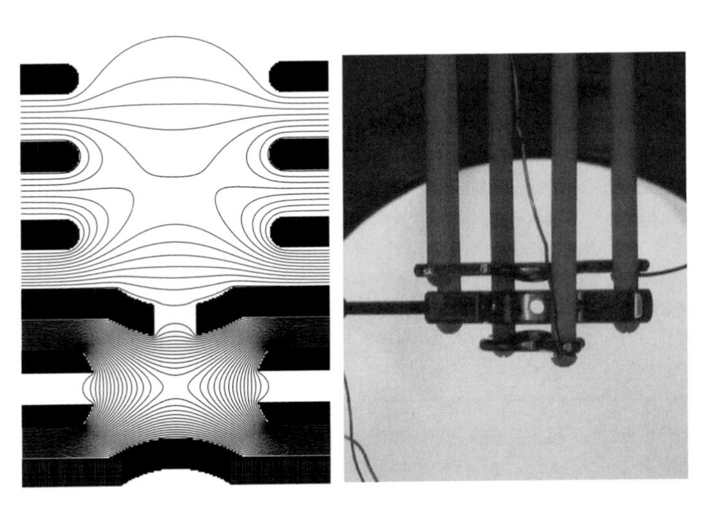

Fig. 8.13. SIMION contour plot of a quadrupole electrostatic trap with ion optics attached. The contour lines are representative of the electric potential inside the system. The scale on the ion optics has been increased by a factor of 100 to be visible alongside the trapping potential.

weak extraction field exists. The external ion optics then velocity map the extracted ions on to the detector; see Sec. 8.3.2.

Figure 8.13 shows a scaled representation of the electric fields inside the electro-static trap and ion optics. For clarity the trapping fields have been reduced by a factor of 100 so that the extraction and trapping field can be shown simultaneously. Under operational conditions the two end caps are held near ground while 10 KV is placed on the ring electrode. The electro-static trap generates a 200 mK deep trap for the $ND_3$ j = 0 state. Unlike the magnetic trapping of NO, as the $ND_3$ rotates, the rotation averages out the effect of the electric dipole moment, decreasing the trapping potential. As a result, the trap is most efficient for the ground rotational state, avoiding the complication of spreading the population out over several $m_j$ levels.

### 8.3.6. *Slow Molecule Source*

In kinematic cooling of molecules there are two ways to produce stationary lab frame molecules; first is to have an exact mass match between the molecule and the scattering partner, second is without an exact mass match one must tune the initial energy of the molecule, such that the excess post collision energy (Eq. (8.10)) is degenerate with an accessible internal state. If these conditions are not met, either the wrong rotational state, a poor choice of collision partner, or the wrong initial energy, then the kinematically cooled molecules will still have a velocity compression but the finial velocity distribution will not be centered around zero, laboratory frame, velocity. The final velocity can be found by subtracting the post collision rotational energy from the total post collision energy given by Eq. (8.10) such that

$$v_f^2 = \frac{2}{m_{\mathrm{NO}}}\left[E_{trans(\mathrm{NO}_{v'=0,j=0.5})}\left(1-\frac{m_{\mathrm{NO}}}{M_{\mathrm{Ar}}}\right) - j\cdot(j+1)\cdot B\right], \quad (8.23)$$

where $B$ is the rotational constant. For the NO–Ar system, the j = 7.5 rotational energy, $j \cdot (j+1) \cdot B$, is degenerate with the post collision energy, $E_{trans(\mathrm{NO}_{v'=0,j=0.5})}(1 - \frac{m_{\mathrm{NO}}}{M_{\mathrm{Ar}}})$, of the NO and the post collision average velocity of the $NO_{7.5}$, $v_f$, is zero. The finite post collision velocities are shown in Fig. 8.4. For states with j $\leq$ 6.5, not all of the post collision energy goes into rotation and the molecules have a finite velocity in the laboratory frame. Similarly, molecules with j $\geq$ 8.5 have put too much energy into rotation and have a post collision velocity in the laboratory frame.

We have used a simplified one-dimensional description for the scattering events, but in the experiments, molecules scatter in all three dimensions. The result is that some molecules with low post collision energies are scattered straight up or down out of the plane of the collision. This can be experimentally verified with a slight modification to the experimental setup

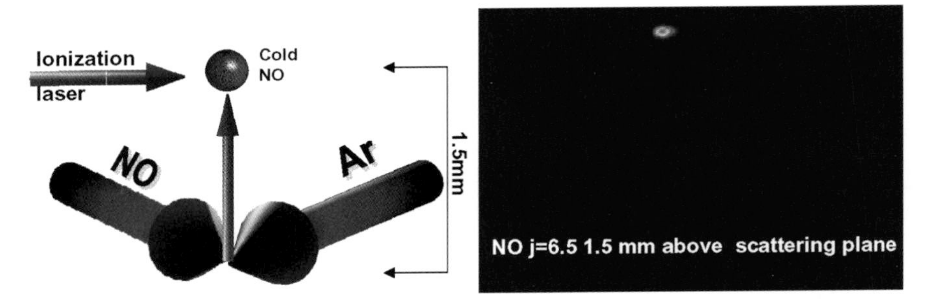


Fig. 8.14. Image of j = 6.5 rotational level of NO 30 $\mu$s after the peak of the scattering with the ionization laser 1.5 mm above the scattering plane.

as depicted in Fig. 8.3; we translate the focus of the ionization laser 1.5 mm above the scattering region.

Figure 8.14 shows the results of this experimental configuration designed for detection of $NO_{6.5}$. We observe only the kinematically cooled j = 6.5 molecules scattered directly up out of the scattering region. By looking at the time evolution of the j = 6.5 signal, shown in Fig. 8.15, we observe a time delay between the arrival of the j = 6.5 molecules, 1.5 mm out of the scattering plane relative to the arrival of the molecular beam pulses. Knowing

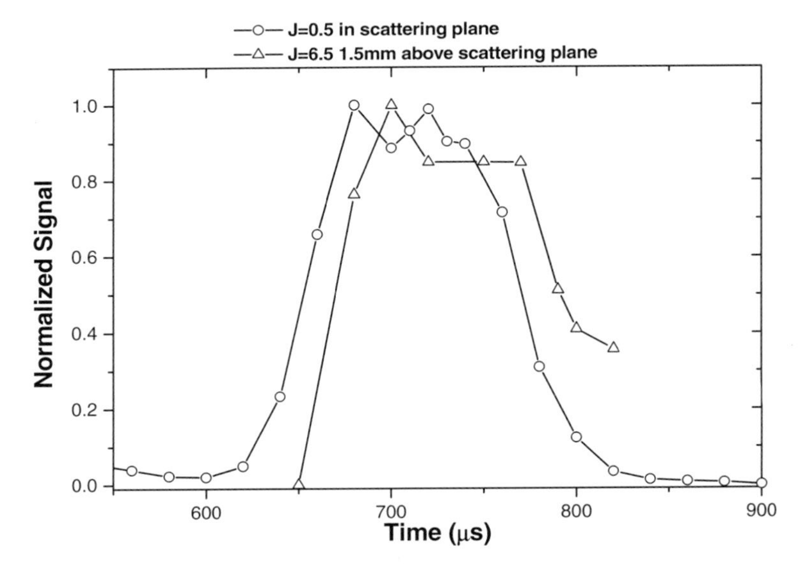


Fig. 8.15. Time evolution of j = 6.5 rotational level of NO 1.5 mm above the scattering plane, compared to the time evolution of the NO j = 0.5 in the plane of the scattering.

the delay time and the distance from the scattering center we can estimate an upper limit of 30 m/s for the j = 6.5 state. This is a very crude estimate since the scattering region is not a point source but between 700 and 900 $\mu$m wide and the laser focus is nearly 100 $\mu$m. Due to the large sizes of the laser and scattering region compared with our 1.5 mm separation it is very difficult to get quantitative information about the group velocity and velocity spread of the j = 6.5 molecules. However, qualitatively we observe a packet of j = 6.5 flying up out of the scattering plane into our detection laser.

The consequence of producing a packet of cold molecules with a finite center of mass velocity is that it can be used as a source of cold molecules. They can be guided, scattered, used as a molecular fountain, and manipulated in many other ways for high resolution spectroscopy and precise energy control in studying collisional and chemical dynamics.

## 8.4. Kinematic Cooling of Molecules by Collision in a MOT

The crossed molecular beam experimental arrangement is only one possible experimental realization of kinematic cooling. A second that is presently being tested in our laboratory is to have a collision of a hot molecule and a stationary, or cold, atom. If the mass defect between the molecule and atom is small, then some fraction of the collisions will result in the molecule transferring enough momentum to the atom such that the post collision molecule is cooled to a temperature where the molecule can be trapped. Let us consider the CAMB experiments discussed thus far, and now reduce the velocity of the atomic beam until it is stationary in the laboratory frame. With the atomic beam stationary, the necessary 90° scattering angle between the atomic and molecular beams becomes undefined, and all approach angles are equally valid for kinematic cooling. As a result, the molecules can come in from any, or every, angle and scatter off the stationary atomic "beam" and have some probability of undergoing a kinematic cooling collision.

The technique for producing stationary samples of alkali atoms is a magneto-optical trap, or MOT.[50] MOTs can be produced with the alkali atoms, metastable halides, some lanthanides, and several other atoms.[50] The technique for producing a MOT relies on the atoms having closed or near closed electronic transitions at experimentally accessible laser wavelengths. With the appropriate laser and magnetic field configuration atoms can be cooled and confined with typical temperatures below 1 mK and peak densities of $\sim 10^{10}$ atoms/cm$^3$.

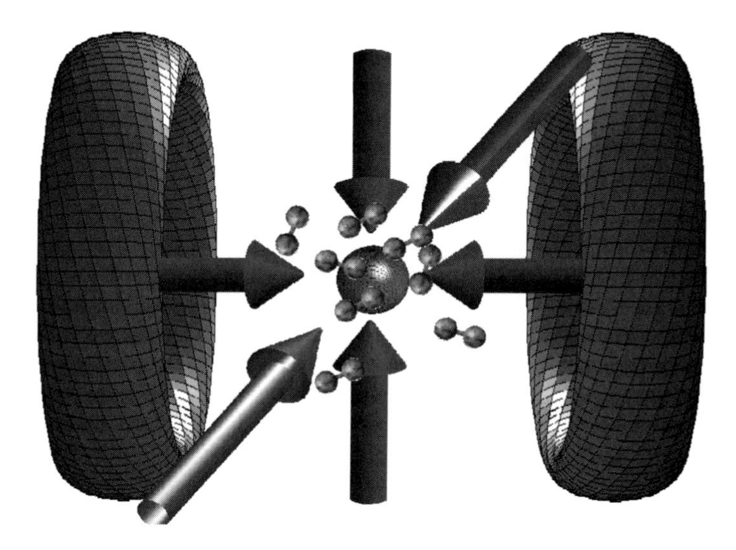

Fig. 8.16. Schematic of a magneto-optical trap (MOT) used for kinematic cooling of molecules. A MOT is formed from 6 counter propagating lasers and a set of anti-Helmhotz coils. The MOT keeps an atomic sample near 1mK while molecules impinge on the trapped atoms, some fraction of the molecules are kinematically cooled.

The kinematic cooling technique we propose is to collide hot molecules with cold atoms in a magneto-optical trap. Since no specific impact angle is required, the molecules can impact the MOT from any angle, thus the molecules can either impinge on the MOT in the form of a molecular beam or as background gas. This technique is illustrated in Fig. 8.16.

The dynamics of colliding a hot molecule with a cold atom are similar to the dynamics in a CAMB experiment. The relevant energy for the collision, as was discussed in Sec. 8.2.1, is the relative energy between the hot molecules and cold atoms, which is not a cold collision. It should be noted that this is distinct from buffer gas cooling[37] since we want the cold atom and the hot molecule to be nearly mass-degenerate such that a single collision between the atom and molecule removes enough energy to trap the molecules. Neglecting reactive and inelastic processes, if the cooled molecules undergo subsequent collisions with the cold atoms, they will be further cooled. This later process is more akin to buffer gas cooling and may have the same obstacles, namely inelastic collision channels which reheat the molecules.

Here we will make an order of magnitude calculation for the probability of cooling molecules via a single collision with an atom in a Magneto-Optical Trap. A simple calculation, assuming hard sphere collisions and

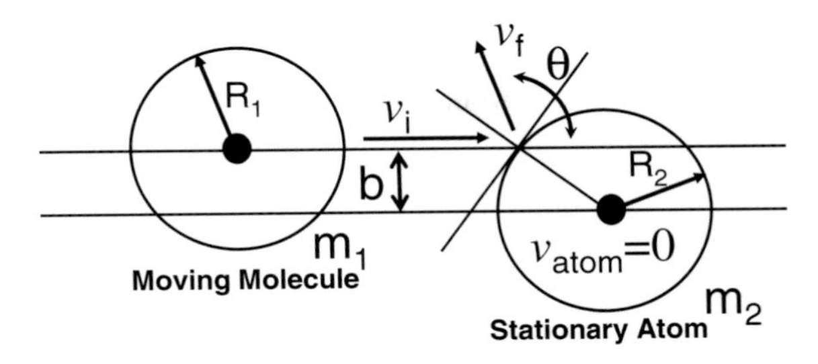


Fig. 8.17. Diagram of a collision between a molecule of mass $m_1$, radius $R_1$, moving with a velocity $v_i$ impacting on a stationary atom with mass $m_2$, and radius $R_2$. The impact parameter for the collision is given by $b$. After the collision, the atom is scattered in a direction determined by $\theta$ with a velocity given by $v_f$.

spherical particles, will give us an estimate of the density of molecules that we can expect to trap with this technique and therefore the feasibility of this experiment. This collision is shown schematically in Fig. 8.17. We assume that one particle (the molecule), (mass $\mathrm{m_1}$ and hard sphere radius $\mathrm{R_1}$) is moving and strikes a stationary particle (the atom) (mass $\mathrm{m_2}$ and hard sphere radius $\mathrm{R_2}$) with impact parameter $b$. For such a situation one can derive an equation for the amount of energy transferred upon collision as a function of the mismatch between the masses of the particles and the impact parameter of the collision. To first order, the amount of energy transferred to the initially stationary atom, $E(trans)_{SA}$, is given by

$$E(trans)_{SA} = \frac{m_1 v_i^2}{2} \cdot \left(1 - \frac{f}{2}\right)\left(1 - \frac{\Delta^2}{4}\right), \tag{8.24}$$

where $v_i$ is the velocity of the moving molecule, $f$ is the fraction of collisions with a scattering cross section from $b = 0$ to $b = b_{max}$ or the area associated with a collision having impact parameter $b_{max}$ compared to the area associated with a collision having the hard sphere radius $(R_1 + R_2)$, and $\Delta$ is the mass defect defined by $(m_1 - m_2)/m_1$. This simple formula tells us that if the mismatch between the mass of the particles is zero then $\Delta = 0$. Further, if the impact parameter $\mathrm{b} = 0$ (a direct hit) then all of the initial energy of the molecule, $(m_1 u_{12}^2/2)$, is completely transferred to the initially stationary atom and the molecule is left at rest in the laboratory frame. Note that at this point $f$ can never truly equal zero as there is no collision that truly has $\mathrm{b} = 0$. If the mismatch in the mass of the molecule and the atom is 10% then Eq. (8.24) tells us that the maximum amount of energy that one can transfer with a perfect collision ($b = 0$, $f = 0$) collision is still 99.75% ($2/4 = (0.1)2/4 = 2.5 \cdot 10^{-3}$). If we need to remove 99% of the energy from the moving molecule, $1 - \mathrm{f}/2 = 0.99$, and the mass of the atom and molecule are equal, then f = 0.02 or 2% of the collisions will remove 99% of all the energy from the molecule. So to first order, if the mass of the molecule is within 10% of the mass of the atom and we need to remove 99% of the energy from the molecule in order for the molecule to be trapped, then approximately 2% of our collisions will be effective at slowing the molecule, compared to 98% of the collisions that will eject a molecule and atom from the trap and not stop. This implies that at equilibrium the density of trapped molecules will be about 1% of the density of cold atoms, if we only need to remove 99% of the translational energy from the molecule in order to trap the molecule. This is a somewhat surprisingly large fraction of collisions that are effective at transferring 99% of the energy and must be taken as an upper limit as we have assumed only hard sphere collisions.

The technique of kinematic cooling relies specifically on the ratio of the masses of the two colliding partners, and while we have been concentrating on this technique for cooling molecules, the technique also works for cooling atoms. We are therefore currently building an apparatus to demonstrate the kinematic cooling technique with a MOT by forming a MOT with one isotope of rubidium and simultaneously kinematically cooling and trapping another isotope of rubidium. The apparatus consists of a $^{87}$Rb MOT that is actively loaded via a Zeeman slower [50] from a thermal oven. Our atomic candidate to kinematically cool is $^{85}$Rb. This experiment will prove the principle of kinematic cooling without several of the complications associated with the trapping and detection of most molecules. The atomic oven has the natural abundance of each rubidium isotope (72% $^{85}$Rb and 28% $^{87}$Rb). As the MOT fills with Zeeman slowed $^{87}$Rb collisions will occur with the more abundant unslowed $^{85}$Rb in the thermal atomic beam. A modulated laser beam resonant on the cycling transition ($^5S_{\frac{1}{2}}$ $F_2$–$^5P_{\frac{3}{2}}$ $F_3$) for $^{85}$Rb will be used to monitor the amount of $^{85}$Rb that is trapped. This experiment will demonstrate the ability to collect particles of similar mass in a MOT via kinematic cooling.

This technique can then be generalized to kinematically cool molecules via collisions with atoms in a MOT. The first molecular candidates we propose to scatter from a Rb MOT are DBr (mass 83), $^{85}$RbD (mass 87), $MnO_2$ (mass 87), and HBr (mass 82). While there are several candidates for scattering off rubidium, the NIST chemical handbook lists over 100 molecules that have masses near 85 or 87 amu.

There are several concerns with scattering from alkalis. The least studied to date is that of a reactive collision between the alkali and the molecule. The same weakly bound electron that gives the simple electronic structure can also lead to large reactive scattering cross-sections. For initial kinematic cooling, which only requires a single collision, reactions are not a large concern. However, if the goal is to have the molecules thermalize with the MOT, then multiple collisions are needed and a reactive collision can occur. As an example, consider the reaction of Rb atoms with the HBr molecule. The bond energy of the HBr is 28087 $cm^{-1}$. The bond energy of RbBr is 31765 $cm^{-1}$. The reaction channel is exothermic with over 3670 $cm^{-1}$ of energy released. Energetically, the initial collision can lead to a reaction if the cross-section is sufficiently large. If the HBr-Rb can survive several collisions, such that they thermalize, any subsequent reactive collision must occur at milli-Kelvin (6.95 $10^{-4}$ $cm^{-1}$) or lower temperatures. Then any reaction barrier larger then 6.95 $10^{-4}$ $cm^{-1}$ will further suppress any reaction.

Another concern is matching spatial overlap of the impinging molecules and the MOT. Typical MOTs are about 1 cm in diameter, while a typical skimmed supersonic molecular beam is 1 mm or less. To better overlap the MOT and the molecules, the molecules can be bled in from a background, however, the molecules will then have a thermal rotational state distribution. This is a disadvantage for kinematically cooling and trapping a specific quantum state of the molecule. However, monitoring the rotational state distribution inside the MOT compared to the impinging thermal distribution can reveal details about the collisional dynamics.

Kinematic cooling of molecules via collisions with Magneto-Optically trapped atoms provides a straightforward, yet undemonstrated, method for producing cold molecules in environments where they can undergo thermalizing collisions with cold atoms and potentially be further sympathetically cooled[5,61] into the ultracold regime.

## 8.5. Conclusion

Kinematic cooling is a cooling technique that slows supersonic molecules via a single collision in which the kinetic energy of the molecule is transferred nearly completely to the atom. This technique was first realized in the scattering of nitric oxide (NO) off argon (Ar) in a crossed atomic and molecular beam apparatus (CAMB). The molecules have been shown to have a final velocity spread of ~30 m/s centered around the origin in the laboratory frame. The cold molecules are demonstrated to persist for over 150 $\mu$s after the atom and molecular beams shut off, separating the cold molecules from the hot beams. Recently this technique has been demonstrated on hydrogen bromide (HBr) scattering off krypton (Kr) and with ammonia ($NH_3$) scattering off neon (Ne). Further, this technique is currently being extended to collisions between hot molecules and pre-cooled atoms in a Magneto-Optical Trap (MOT), where secondary collisions between the atoms and cooled molecules may further buffer gas cool the molecules into the micro-Kelvin regime.

The kinematic cooling technique is not only useful for cooling molecules, but it is easily extended to cooling any atom, since it works on the basis of collisions between particles with equal or near equal masses. It can be applied to cooling atoms with complicated electronic structure which hinder direct laser cooling.

Kinematic cooling is also a useful technique for producing packets of slow moving molecules with narrow velocity distributions. These packets have the potential to be used in spectroscopy as well as in scattering and collisional dynamics experiments, where the narrow velocity distribution and the energy tunablity can allow for the exploration of threshold energy transfer processes,[60] or simply decreasing the energy uncertainty in a collision process allowing for a more accurate description of the potential energy surfaces.

There are now several techniques for producing cold molecules, which are discussed in this book. Each technique brings its advantages and disadvantages for producing cold molecules. No technique to date works for every molecule, but each technique addresses either differing classes or subsets of molecules that can be cooled, and when the community is viewed as a whole, the experientialist is now able to produce a wide range of diatomic and polyatomic molecules in the milli-Kelvin regime. The goal for the near future is to continue to push the existing technology and explore new technology, in order push the achievable temperatures into the micro-Kelvin regime and increase the density of the cold molecules to where their properties and collisional dynamics can be manipulated and studied with the precision and detail only achievable in atoms today.

## References


1. Anderson MH, Ensher JR, Matthews MR, Wieman CE, Cornell AE. (1995) Observation of Bose-Einstein condensation in a dilute atomic vapor. *Science* **269**: 198.
2. Andrews MR, Mewes M-O, van Druten NJ, Durfee DS, Kurn DM, Ketterle W. (1996) Direct, nondestructive observation of a Bose condensate. *Science* **273**: 84–87.
3. Bradley CC, Sackett CA, Tollett JJ, Hulet RG. (1995) Evidence of Bose-Einstein condensation in an atomic gas with attractive interactions. *Phys. Rev. Lett.* **75**: 1687.
4. DeMarco B, Jin DS. (1999) Onset of Fermi degeneracy in a trapped atomic gas. *Science* **285**: 1703.
5. Truscott AG, Strecker KE, McAlexander WI, Partridge GB, Hulet RG. (2001) Observation of Fermi pressure in a gas of trapped atoms. *Science* **291**: 2570.

6. Bartenstein M, Altmeyer A, Riedl S, Jochim S, Chin C, Denschlag J, Grimm R. (2004) Collective excitations of a degenerate gas at the BEC-BCS crossover. *Phys. Rev. Lett.* **92**: 203201.
7. Greiner M, Regal C, Jin D. (2005) Probing the excitation spectrum of a Fermi gas in the BEC-BCS crossover regime. *Phys. Rev. Lett.* **94**: 070403.
8. Bourdel T, Khaykovich L, Cubizolles J, Zhang J, Chevy F, Teichmann M, Tarruell L, Kokkelmans S, Salomon C. (2004) Experimental study of the BEC-BCS crossover region in lithium 6. *Phys. Rev. Lett.* **93**: 050401.
9. Partridge GB, Strecker KE, Kamar RI, Jack MW, Hulet GR. (2005) Molecular probe of pairing in the BEC-BCS crossover. *Phys. Rev. Lett.* **95**: 020404.
10. Hudson J, Sauer B, Tarbut M, Hinds E. (2002) Measurement of the electron electric dipole moment using YbF molecules. *Phys. Rev. Lett.* **89**: 023003.
11. Veldhoven JV, Kupper J, Bethlem H, Sartakov B, Roij AV, Meijer G. (2004) Decelerated molecular beams for high-resolution spectroscopy - the hyperfine structure of $ND_3$. *Europ. Phys. J. D* **31**: 337.
12. Bollinger J, Itano W, Wineland D, Heinzen D. (1996) Optimal frequency measurements with maximally correlated states. *Phys. Rev. A* **54**: R4649.
13. Weiner J, Bagnato V, Zilio S, Julienne PS. (1999) Experiments and theory in cold and ultracold collisions. *Rev. Mod. Phys.* **71**: 1.
14. Williams C, Tiesinga E, Julienne P. (1996) Hyperfine structure of the $Na_2$ $O_g^-$ long range molecular state. *Phys. Rev. A* **53**, R1939.
15. Stock R, Deutsch I, Bolda E. (2003) Quantum state control via trap-induced shape resonance in ultracold atomic collisions. *Phys. Rev. Lett.* **91**: 183201.
16. Volz T, Durr S, Syassen N, Rempe G, Kempen EV, Kokkelmans S. (2005) Feshbach spectroscopy of a shape resonance. *Phys. Rev. A* **72**: 010704.
17. Kerman A, Chin C, Vuletic V, Chu S, Leo P, Williams C, Julienne P. (2001) Determination of Cs-Cs interaction parameters using Feshbach spectroscopy. *Comptes Rendus de l'Academie des Sciences, Serie IV* (*Physique, Astrophysique*) **2**: 633.
18. Houbiers M, Stoof HTC, McAlexander WI, Hulet RG. (1998) Elastic and inelastic collisions of $^6$Li atoms in magnetic and optical traps. *Phys. Rev. A* **57**: R1497.
19. Inouye S, Andrews MR, Stenger J, Miesner H-J, Stamper-Kurn DM, and Ketterle W. (1998) Observation of Feshbach resonances in a Bose-Einstein condensate. *Nature* **392**: 151.
20. Baccarelli I, Delgado-Barrio G, Gianturco F, Gonzalez-Lezana T, Miret-Artes S, Villarreal P. (2000) Searching for Efimov states in triatomic systems: The case of $LiHe_2$. *Europ. Phys. Lett.* **50**: 567.
21. Braaten E, Hammer H, Kusunoki M. (2003) Efimov states in a bose-einstein condensate near a Feshbach resonance. *Phys. Rev. Lett.* **90**: 170402.
22. Stoll M, Kohler T. (2005) Production of three-body Efimov molecules in an optical lattice. *Phys. Rev. A* **72**: 22714.
23. Bruhl R, Kalinin A, Kornilov O, Toennies J, Hegerfeldt G, Stoll M. (2005) Matter wave diffraction from an inclined transmission grating: Searching for the elusive He-4 trimer Efimov state. *Phys. Rev. Lett.* **95**: 063002.

24. Kalinin A, Kornilov O, Schollkopf W, Toennies J. (2005) Observation of mixed Fermionic-Bosonic helium clusters by transmission grating diffraction. *Phys. Rev. Lett.* **95**: 113402.
25. Vale J, Hall B, Lau D, Jones M, Retter J, Hinds E. (2002) Atom chips. *Europhys. News* **33**.
26. Lapshin D, Balykin V, Letokhov V. (1997) Diffraction of atoms by an evanescent wave grating formed by a periodic surface structure. *Laser Physics* **7**: 361.
27. Roach T. (2004) Atom wave diffraction in an accelerating potential. *J. Phys. B* **37**: 3551.
28. Mewes M-O, Andrews MR, Kurn DM, Durfee DS, Townsend CG, Ketterle W. (1997) Output coupler for Bose-Einstein condensed atoms. *Phys. Rev. Lett.* **78**(4): 582.
29. Strecker KE, Partridge GB, Truscott AG, Hulet RG. (2002) Formation and propagation of matter-wave soliton trains. *Nature* **417**: 150–153.
30. Khaykovich L, Schreck F, Ferrari G, Bourdel T, Cubizolles J, Carr L, Castin Y, Salomon C. (2002) Formation of a matter-wave bright soliton. *Science* **296**: 1290–1293.
31. Zwierlein M, Abo-Shaeer J, Schirotzek A, Schunck C, Ketterle W. (2005) Vortices and superfluidity in a strongly interacting Fermi gas. *Nature* **435**: 1047–1051.
32. Chin C, Bartenstein M, Altmeyer A, Riedl S, Jochim S, Denschlag JH, Grimm R. (2004) Observation of the pairing gap in a strongly interacting Fermi gas. *Science* **305**: 1128.
33. Partridge GB, Li W, Kamar RI, Liao Y, Hulet RG. (2006) Pairing and phase separation in a polarized Fermi gas. *Science* **311**: 503.
34. Pethick C, Smith H. (2002) *Bose Einstein Condensation in Dilute Gases* (Cambridge University Press, UK), 1st edition.
35. Shapiro S, Teukolsky S. (1983) *Black Holes, White Dwarfs, Neutron Stars; The Physics of Compact Objects* (Wiley, New York)
36. Bethlem H, Berden G, Crompvoets F, Jongma R, van Roij A, Meijer G. (2000) Electrostatic trapping of ammonia molecules. *Nature* **406**, 491–494.
37. deCarvalho R, Doyle J, Friedrich B, Guillet T, Kim J, Patterson D, Weinstein J. (1999) Buffer-gas loaded magnetic traps for atoms and molecules: A primer. *Europ. Phys. J. D* **7**: 289–309.
38. Wynar R, Freeland RS, Han DJ, Ryu C, Heinzen DJ. (2000) Molecules in a Bose-Einstein condensate. *Science* **287**: 1016–1019.
39. Regal C, Ticknor C, Bohn J, Jin D. (2003) Creation of ultracold molecules from a Fermi gas of atoms. *Nature* **424**: 47–50.
40. Strecker KE, Partridge G, Hulet RG. (2003) Conversion of an atomic Fermi gas into a long-lived molecular Bose gas. *Phys. Rev. Lett.* **91**(8): 135–138.
41. Suits A, Bontuyan L, Houston P, Whitaker B. (1992) Differential cross sections for state-selected products by direct imaging: Ar+NO. *J. Chem. Phys.* **96**: 8618–20.

42. Lorenz K, Westley M, Chandler D. (2000) Rotational state-to-state differential cross sections for the HCl-Ar collision system using velocity-mapped ion imaging. *Phys. Chem. Chem. Phys.* **2**: 481–494.
43. Kohguchi H, Suzuki T, Alexander M. (2001) Fully state-resolved differential cross sections for the inelastic scattering of the open-shell NO molecule by Ar. *Science* **294**: 832–834.
44. Gijsbertsen A, Linnartz H, Rus G, Wiskerke A, Stolte S, Chandler D, Klos J. (2005) Differential cross sections for collisions of hexapole state-selected NO with He. *J. Chem. Phys.* **123**: 224305–1–14.
45. Chandler D, Parker D. (1999) *Advances in Photochemistry.* In: Neckers DC, Volman DH (eds), Vol. 25 (Wiley, New York).
46. Parker D, Eppink A. (1997) Photoelectron and photofragment velocity map imaging of state-selected molecular oxygen dissociation/ionization dynamics. *J. Chem. Phys.* **107**: 2357–2362.
47. Parker D, Eppink A. (1997) Velocity map imaging of ions and electrons using electrostatic lenses: Application in photoelectron and photofragment ion imaging of molecular oxygen. *Rev. Sci. Instr.* **68**: 3477–3484.
48. Elioff M, Valentini J, Chandler D. (2004) Formation of NO($j' = 7.5$) molecules with sub-kelvin translational energy via molecular beam collisions with argon using the technique of molecular cooling by inelastic collisional energy-transfer. *Europ. Phys. J. D* **31**: 385–393.
49. Huang K. (1987) *Statistical Mechanics* (John Wiley & Sons, New York), 2nd edition.
50. Metcalf HJ, van der Straten P. (1999) *Laser Cooling and Trapping* (Springer-Verlag, New York).
51. Dahl DA, Delmore JE, Appelhans AD. (1990) SIMION PC/PS2 electrostatic lens design program. *Rev. Sci. Instr.* **61**: 607–609.
52. Wen L, Cunshun H, Patel M, Wilson D, Suits A. (2006) State-resolved reactive scattering by slice imaging: A new view of the $Cl+C_2H_6$ reaction. *J. Chem. Phys.* **124**: 11102–1–4.
53. Gilijamse J, Hoekstra S, van de Meerakker S, Groenenboom G, Meijer G. (2006) Near-threshold inelastic collisions using molecular beams with a tunable velocity. *Science* **313**: 1617–1620.
54. Chu S, Bjorkholm JE, Ashkin A, Cable A. (1986) Experimental observation of optically trapped atoms. *Phys. Rev. Lett.* **57**: 314–317.
55. DeMille D, Glenn D, Petricka J. (2004) Microwave traps for cold polar molecules. *Europ. Phys. J. D* **31**: 375–384.
56. Takazawa K, Abe H. (1999) Electronic spectra of gaseous nitric oxide in magnetic fields up to 10 T. *J. Chem. Phys.* **110**: 9492–9499.
57. Elioff MS, Chandler DW. (2002) State-to-state differential cross sections for spin-multiplet-changing collisions of NO($X^2\pi_{1/2}$) with argon. *J. Chem. Phys.* **117**: 6455–6462.
58. van de Meerakker S, Jongma R, Bethlem H, Meijer G. (2001) Accumulating NH radicals in a magnetic trap. *Phys. Rev. A* **64**: 41401–1–4.

59. van de Meerakker S, Smeets P, Vanhaecke N, Jongma R, Meijer G. (2005) Deceleration and electrostatic trapping of OH radicals. *Phys. Rev. Lett.* **94**: 023004/1–4.
60. Hoekstra S, Gilijamse J, Sartakov B, Vanhaecke N, Scharfenberg L, van de Meerakker S, Meijer G. (2007) Optical pumping of trapped neutral molecules by blackbody radiation. *Phys. Rev. Lett.* **98**: 133001/1–4.
61. Myatt C, Burt R, Ghrist, Cornell E, Wieman C. (1997) Production of two overlapping Bose-Einstein condensates by sympathetic cooling. *Phys. Rev. Lett.* **78** (4): 586–589,

# Chapter 9

# Manipulation of Molecules with Electric Fields

Sebastiaan Y.T. van de Meerakker

*Fritz-Haber-Institut der Max-Planck-Gesellschaft, Faradayweg 4-6 D-14195 Berlin, Germany*

Hendrick L. Bethlem

*Laser Centre Vrije Universiteit, De Boelelaan 1081, NL-1081 HV Amsterdam, The Netherlands*

Gerard Meijer

*Fritz-Haber-Institut der Max-Planck-Gesellschaft, Faradayweg 4-6 D-14195 Berlin, Germany*

## Contents

## 9.1. Introduction

Atomic and molecular beams have played central roles in many experiments in physics and chemistry — from seminal tests of fundamental aspects of quantum mechanics to molecular reaction dynamics — and have found a wide range of applications.[1] Nowadays, sophisticated laser-based methods exist to perform sensitive and quantum state selective detection of the atoms and molecules in the beams. In the early days, such detection methods were lacking and the particles in the beam were detected, for instance, by a "hot wire" (Langmuir-Taylor) detector, by electron-impact ionization or by deposition and *ex-situ* investigation of the particles on a substrate at the end of the beam-machine.[2] To achieve quantum state selectivity in the overall detection process, these methods were combined with inhomogeneous magnetic and/or electric field sections to influence the trajectories of the particles on their way to the detector. This was the approach used by Otto Stern and Walther Gerlach in 1922,[3] and the key concept of their experiment, i.e. the sorting of quantum states via space quantization, has been extensively used ever since. The original experimental geometries were devised to create strong magnetic or electric field gradients on the beam axis to efficiently deflect particles. In 1939, Rabi introduced the molecular beam magnetic resonance method by using two magnets in succession to produce inhomogeneous magnetic fields of oppositely directed gradients. In this set-up the deflection of particles caused by the first magnet is compensated by the second magnet, such that the particles are directed on a sigmoid path to the detector. A transition to "other states of space quantization" induced in between the magnet sections can be detected via the resulting reduction of the detector signal.[4] Later, both magnetic[5,6] and electric[7] field geometries were designed to focus particles in selected quantum states onto the detector. An electrostatic quadrupole focuser, i.e. an arrangement of four cylindrical electrodes with alternating positive and negative voltages, was used to couple a beam of ammonia molecules into a microwave cavity. Such an electrostatic quadrupole lens focuses ammonia molecules that are in the upper level of the inversion doublet while it simultaneously defocuses those that are in the lower level. The inverted population distribution of the ammonia molecules in the microwave cavity that is thus produced led to the invention of the maser by Gordon, Zeiger and

Townes in 1954–1955.[8,9] Apart from the spectacular observation of the amplification of the microwaves by stimulated emission, these focusing elements more generally enabled the recording, with high resolution and good sensitivity, of microwave spectra in a molecular beam. By using several multipole focusers in succession, with interaction regions with electro-magnetic radiation in between, versatile set-ups to unravel the quantum structure of atoms and molecules were developed. In scattering experiments, multipole focusers were exploited to study steric effects, i.e. to study how the orientation of an attacking molecule affects its reactivity.[10] Variants of the molecular beam resonance methods as well as scattering machines that employed state-selectors were implemented in many laboratories, and have yielded a wealth of detailed information on stable molecules, radicals and molecular complexes, thereby contributing enormously to our present understanding of intra- and inter-molecular forces.

The manipulation of beams of atoms and molecules with electric and magnetic fields is thus about as old as the field of atomic and molecular beams itself, and it actually has been crucial for the success of the latter field. In his autobiography, Norman Ramsey recalls that Rabi was rather discouraged about the future of molecular beam research when he arrived in Rabi's lab in 1937, and that this discouragement only vanished when Rabi invented the molecular beam magnetic resonance method.[11] However, even though the manipulation of beams of molecules with external fields has been used extensively and with great success in the past, this manipulation exclusively involved the transverse motion of the molecules. It was only in 1999, that it was experimentally demonstrated that appropriately designed arrays of electric fields in a so-called "Stark decelerator" can also be used to influence and control the longitudinal (forward) velocity of the molecules in a beam, *e.g.* to decelerate a beam of neutral polar molecules.[12] Since then, the ability to produce focused packets of state-selected accelerated or decelerated molecules has made a whole variety of new experiments possible. How to achieve full control over the 3D motion of neutral polar molecules with electric fields, and how this can be used to advantage in a variety of novel experiments, is the subject of this Chapter.

For the ease of discussion, we restrict ourselves from now on to the interaction of molecules with a permanent electric dipole moment with electric fields, but the same arguments and principles hold of course for the interaction of particles with a magnetic dipole moment with magnetic fields. In a quadrupole or hexapole focuser, the magnitude of the electric field is zero on the symmetry axis, and this axis is normally made to coincide with the molecular beam axis. Close to this axis, the electric field strength is — to a good approximation — cylindrically symmetric, and it increases with

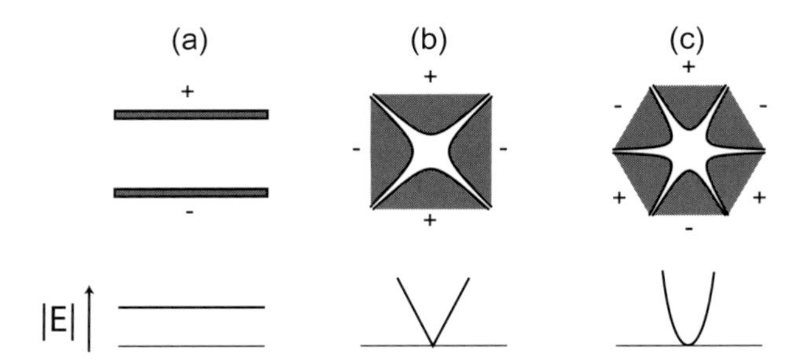


Fig. 9.1. The electrode geometries for generating a dipole (a), quadrupole (b) and hexapole (c) field. The magnitude of the electric field is shown along a line through the center of each geometry in the lower part. Whereas the dipole geometry generates a constant electric field, the quadrupole (hexapole) geometry generates an electric field that increases linearly (quadratically) away from the center.

distance $r$ from the axis proportional to $r$ or $r^2$, respectively, as shown in Fig. 9.1.

For polar molecules in a so-called low-field seeking quantum state, i.e. with their space-fixed dipole moment anti-parallel to the external electric field, their Stark energy increases with increasing electric field strength. The force that a molecule experiences in an electric field is given by the negative gradient of the Stark energy, which implies that in the multipole field there always is a restoring force towards the molecular beam axis for a molecule in a low-field seeking state. A multipole focuser acts as a perfect lens when this restoring force is linearly proportional to $r$. For a molecule whose Stark energy scales linearly with the strength of the electric field, a hexapole focuser acts as a perfect lens, whereas a quadrupole focuser is needed to perfectly focus a molecule whose Stark energy scales quadratically with the electric field strength. The depth of the confining potential depends on the properties of the molecule as well as on the magnitude of the electric fields that can experimentally be realized. In Fig. 9.2, the Stark shifts are shown for the relevant low rotational levels of OH, $CO(a^3\Pi_1)$, $ND_3$, and $H_2CO$, molecules that have been used in the studies presented in this Chapter.

It can be seen from these graphs that with electric field strengths of up to 100 kV/cm, field strengths that can be routinely realized in the laboratory, potential wells with a depth of typically $1\,\text{cm}^{-1}$ ($\equiv$1.44 K) can be created. This is enough to transversally confine these molecules around the beam axis as, in a typical molecular beam experiment, the transverse velocity distribution is centered around zero m/s with a full-width-at-half-maximum (FWHM) spread of several tens of m/s, corresponding to a sub-$\text{cm}^{-1}$ transverse kinetic energy spread. Normally, the multipole focusers are made just

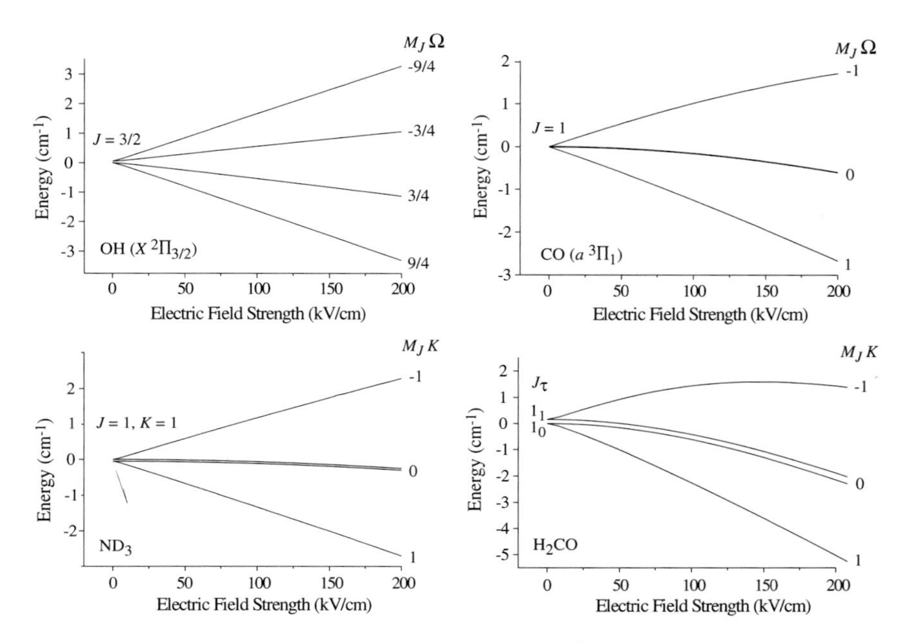


Fig. 9.2. Stark energy curves of OH($X\,^2\Pi_{3/2}$), CO($a\,^3\Pi_1$), $ND_3$, and $H_2CO$ in low rotational states. In zero field, there are two closely spaced levels with opposite parity. This zero field energy splitting is caused by Λ-doubling (OH, CO), inversion doubling ($ND_3$) or $K$-doubling ($H_2CO$). In an electric field, the opposite parity levels are coupled, leading to a large linear Stark splitting. Levels that experience a positive (negative) Stark shift in an increasing electric field are referred to as "low-field seeking" ("high-field seeking"). If the Stark shift becomes comparable to the rotational energy splitting, the interaction with higher rotational states, which pushes the lowest rotational state down, needs to be taken into account. Ultimately, all states become "high-field seeking" in sufficiently high electric fields.

long enough to focus a beam of molecules into the interaction or detection point somewhere downstream, and inside the multipole the molecules move on a short part of a sinusoidal path. This is schematically shown in Fig. 9.3.

The position of the focus can be varied by varying the voltage on the multipole electrodes, or, while using a fixed voltage, by changing the duration that these voltages are switched on; the latter approach has the additional advantage that chromatic aberration can be avoided. Also shown in this Figure is the situation when the focusing lens is not perfect, due to a non-linear behavior of the force. If the molecules were to spend longer inside the focusing lens or if the trapping potential were to be made steeper, they would undergo multiple transverse oscillations. The latter is the case, for instance, in a guide[13] or when a long multipole focuser is bent into

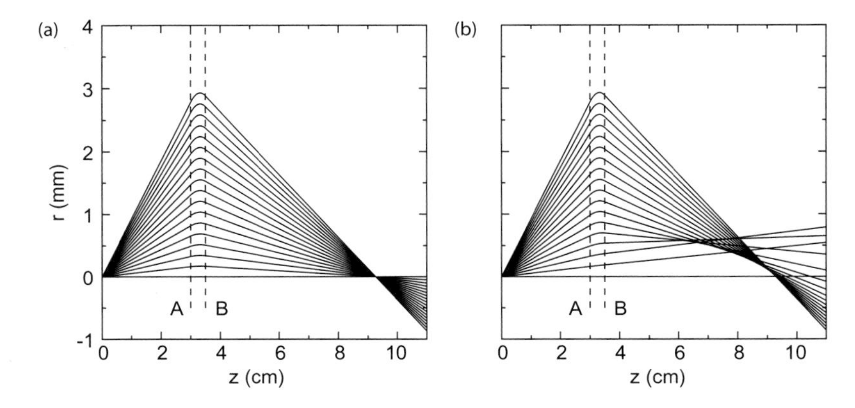


Fig. 9.3. Trajectories in the $(r, z)$-plane of molecules in a low-field seeking state, flying with a fixed velocity through a hexapole. The hexapole is positioned along the $z$-axis, and switched on (off) when the molecules are at position A (B). The molecules are assumed to originate from a single point. In (a), the situation is shown when the force in the hexapole is perfectly linear. In this case, the molecular beam is focused into a single point again. In (b), the situation is shown when a zero field energy splitting is included. In this case, the force is non-linear at low electric field strengths, resulting in a blurring of the focus.

a torus, confining the particles in a circular orbit, as in an electrostatic storage ring.[14]

In multipole focusers, but also in the deflection elements that have been used in atomic and molecular beam experiments in the past, the field gradients are perpendicular to the beam axis, and the velocity component of the particles along the beam axis is not changed. In a typical molecular beam experiment the forward velocity distribution is centered at a high velocity (300–2000 m/s) with a FWHM spread of about 10% of this central velocity, although the latter can also be considerably smaller. Even at the low end of this velocity range, the kinetic energy of the molecules whose characteristics are depicted in Fig. 9.2 is on the order of $100\,\mathrm{cm}^{-1}$. This is much larger than the depth of any single potential well that can ever be realized for these molecules, and direct longitudinal confinement of the molecules in a static potential well is therefore impossible. The FWHM longitudinal velocity spread in the beam, on the other hand, corresponds to a kinetic energy spread on the order of only $1\,\mathrm{cm}^{-1}$, comparable to the FWHM of the transverse kinetic energy spread. As potentials with a depth of $1\,\mathrm{cm}^{-1}$ can be produced, such a potential — with a gradient along the molecular beam axis — can longitudinally confine the molecules, *provided that this potential moves along with the molecular beam.* A multipole focuser could be used to produce a strong gradient of the electric field along the molecular beam when it would be mounted with its symmetry axis perpendicular to the molecular beam axis. If this multipole focuser could be moved along the molecular beam axis with the central velocity of the molecules in the beam, it would longitudinally confine the molecules; the molecules on the beam axis would oscillate, both in position and velocity, around the center of the multipole. Although the central velocity of the molecules in the beam would not be changed in this way, a packet of molecules would stay together in the forward direction, and such a device would therefore have the same advantages for the longitudinal motion of the molecules in a beam as a multipole focuser normally has for the transverse motion. Moreover, if the velocity of this moving potential well could be gradually changed, (a fraction of) the molecules in the beam could be brought to any desired final velocity. To decelerate the molecules in the beam, for instance, the potential well would gradually move more slowly, such that the molecules in the beam would spend more time on the leading slope of the potential well, thereby feeling a force opposing their motion; to accelerate the molecules, the potential well would gradually speed up, pushing the molecules ahead on the trailing slope of the potential well. The hypothetical situation depicted here is actually almost exactly what happens in the Stark decelerator.[12] However, rather than mechanically moving an electrode geometry that generates the confinement potential along the molecular beam, a static array of electrode pairs that create field gradients along the molecular beam axis is used. By switching the fields on adjacent pairs of electrodes on and off at the appropriate times, effectively a travelling potential well is created.[15] In this Stark decelerator the molecules in the beam can either be transported along the beam axis at a constant velocity or gradually brought to any desired final velocity.

In the first experimental demonstration of the Stark deceleration of a beam of neutral polar molecules, a beam of metastable $CO(a^3\Pi_1, J = 1)$ molecules was slowed down from 225 m/s to 98 m/s.[12] Experiments of this kind had been considered and tried before. Electric field deceleration of neutral molecules was first attempted by John King at MIT in 1958. He intended to produce a slow ammonia beam to obtain a maser with an ultra-narrow linewidth. However, in the physical chemistry community the experimental efforts of Lennard Wharton, to demonstrate electric field *acceleration* of a molecular beam, are much better known. In the sixties, at the University of Chicago, he constructed an eleven meter long molecular beam machine for the acceleration of LiF molecules in high-field seeking states from 0.2 to 2.0 eV, aiming to use these high energy beams for reactive scattering studies.[16] Both of these experiments were unsuccessful, and were not continued after the PhD students were finished.[17,18] Whereas interest in slow molecules as a maser medium declined owing to the invention of the laser, the molecular beam accelerator was made obsolete by gas dynamic acceleration of heavy species in seeded supersonic He and $H_2$ beams.[19]

The state-selected molecular beams that exit the Stark decelerator, with their tunable velocity and tunable velocity spread, are ideally suited for many applications. These decelerated beams can be used, for instance, for high-resolution spectroscopic studies,[20,21] taking advantage of the increased interaction times. We also anticipate that these beams are advantageous for future molecular interferometry and molecular optics experiments. These beams enable the study of (in)elastic collisions and reactive scattering as a function of collision energy, down to zero collision energy.[22] Last but not least, a Stark decelerator enables 3D trapping of neutral polar molecules.

"If one extends the rules of two-dimensional focusing to three dimensions, one possesses all ingredients for particle trapping." This is literally how Wolfgang Paul stated it in his Nobel lecture,[23] and as far as the underlying physics principles of particle traps are concerned, it is indeed as simple as that. To experimentally realize the trapping of neutral particles, however, the main challenge is to produce sufficiently slow particles that they can be trapped in the relatively shallow traps that can be made. When the particles are confined along a line, rather than around a point, the requirements on the kinetic energy of the particles are more relaxed, and storage of neutrons in a one meter diameter magnetic hexapole torus could thus be demonstrated first.[24] Trapping of atoms in a 3D trap only became feasible when Na atoms were laser cooled to sufficiently low temperatures that they could be confined in a quadrupole magnetic trap.[25] The Stark decelerator enabled the first demonstration of 3D trapping of neutral ammonia molecules in a quadrupole electrostatic trap[26] even before it was used in the demonstration of an electrostatic storage ring for neutral molecules.[14] These traps permit the observation of packets of molecules, isolated from their environment, for times up to several seconds. This, for instance, enables the direct measurement of lifetimes of metastable states.[27,67] These traps also hold great promise for the study of cold collisions, and, more generally, for the further development of the field of cold molecules, with the production and study of quantum degenerate gases of polar molecules as an important goal.

In the remainder of this Chapter, Stark deceleration of a molecular beam is presented in more detail, followed by a description of the process of trapping neutral polar molecules. An overview of the applications of the slow beams and trapped samples of molecules that can be produced is given, before the Chapter is concluded. The experiments described in this Chapter to exemplify the operational characteristics of the various components, have been performed in different molecular beam machines and using different molecules. The deceleration and 3D trapping of molecules in low-field seeking states is explained using the OH radical as a model

Fig. 9.4. Photograph of the Stark decelerator. The molecular beam passes through the $4 \times 4$ mm$^2$ opening between the electrodes, shown enlarged in the inset. The forward velocity of the molecules is influenced by switching, at the appropriate times, a voltage difference of 40 kV between opposing electrodes. (Reproduced from S.Y.T. van de Meerakker *et al.*[28] with permission. ©2006 by Annual Reviews www.annualreviews.org.)

system; a photograph of the Stark decelerator that is used in these experiments is shown in Fig. 9.4. The buncher, used for longitudinal focusing of a molecular beam, and the storage ring and synchrotron are demonstrated using the $ND_3$ molecule. The $ND_3$ molecule, decelerated to a near standstill while in the low-field seeking state and then transferred by microwaves to a high-field seeking state, is also used in the demonstration of an AC trap for neutral polar molecules. In order to explain the operation of the decelerator for molecules in high-field seeking states, the so-called alternating gradient (AG) decelerator, the experiments on metastable CO in the appropriate quantum state are highlighted. Throughout this Chapter we restrict ourselves to the discussion and demonstration of the basic concepts of the various elements; for more details the reader is referred to the original literature.

## 9.2. Stark Deceleration of Neutral Polar Molecules

### 9.2.1. *The Stark Decelerator*

The Stark decelerator for neutral polar molecules is the equivalent of a linear accelerator for charged particles. The quantum-state specific force that a polar molecule experiences in an electric field is exploited in a Stark decelerator. This force is rather weak, typically some eight to ten orders of magnitude weaker than the force that the parent ion would experience in an equivalent electric field. Nevertheless, this force suffices to achieve complete control over the motion of polar molecules using techniques akin to those used to control charged particles.

In a Stark decelerator, the longitudinal velocity of a beam of polar molecules is manipulated using longitudinally inhomogeneous electric fields. Let's consider an electric field stage composed of two opposing electrodes that are connected to power supplies of opposite polarity, as is shown in Fig. 9.5.

A polar molecule in a low-field seeking state experiences the increasing electric field when it approaches the plane of the electrodes as a potential hill, and will lose kinetic energy on the upward slope of the potential hill. When the molecule leaves the region of high field again, however, it will regain the same amount of kinetic energy. The acceleration on the downward slope of the hill can be avoided if time-varying inhomogeneous electric fields are used; when the electric field is abruptly switched off before the molecule has left the region of high electric field, the velocity of the molecule will not return to its original value. As discussed already in the previous section, however, the effect of a single electric field stage on the forward velocity of the molecules in the beam is rather small. To obtain a significant change in the velocity, the process needs to be repeated many times. In a Stark decelerator, there is an array of such electric field stages, separated by a distance $L$, as shown in Fig. 9.6.

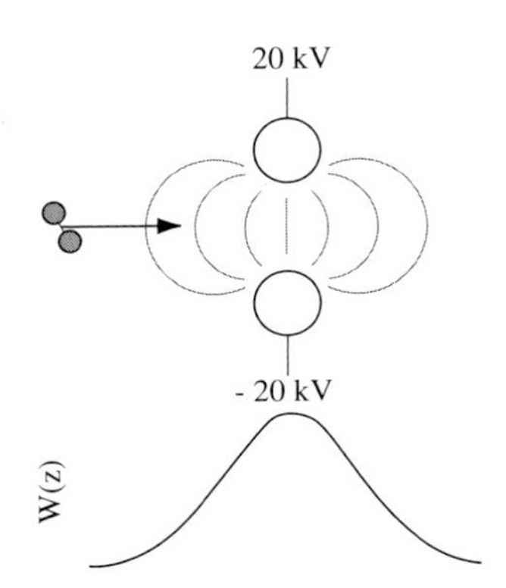


Fig. 9.5. The potential energy $W(z)$ of a polar molecule as a function of position $z$ along the molecular beam axis. The molecule is in a low-field seeking state and the electric field is created by two high voltage electrodes at opposite polarity.

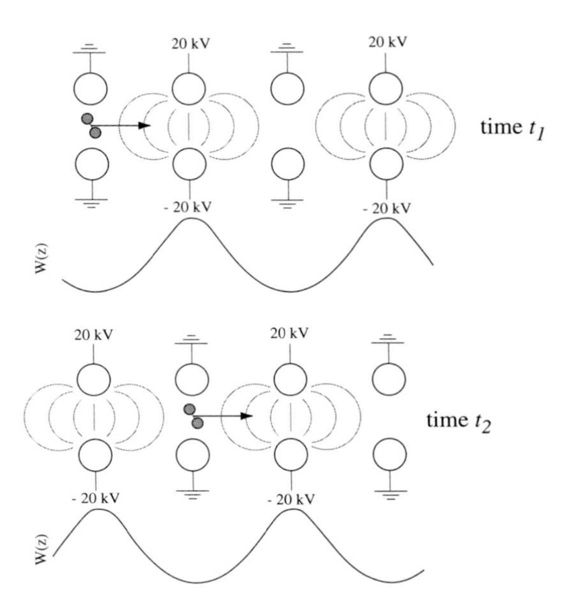


Fig. 9.6. The potential energy $W(z)$ of a polar molecule as a function of position $z$ along the axis of an array of electrode pairs with voltages applied as shown. Repeated switching between the upper and the lower configuration is performed after a time interval $t_2 - t_1$.

Each stage consists of two parallel cylindrical metal rods with radius $r$, centered a distance $2r + d$ apart. One of the rods is connected to a positive, and the other to a negative switchable power supply. Alternating rods are connected to each other. At a given time, the even stages are switched to high voltage and the odd stages are grounded. The potential energy $W(z)$ of a molecule as a function of the position $z$ along the beam axis is shown in Fig. 9.6 as well. The deceleration procedure is now straightforward: when the molecule has reached a position that is close to the top of the first potential hill, the stages that were at high voltage are switched to ground, and vice versa. After switching, the molecule will find itself again in front of a potential hill and will again lose kinetic energy when climbing this hill. When the molecule has reached the high electric field region, the voltages are switched back to the original configuration. By repeating this process many times, the velocity of the molecule can be reduced in a stepwise fashion.

The amount of kinetic energy that is lost per stage, and thus the velocity with which the molecule exits the decelerator, depends on the exact position of the molecule at the time that the fields are being switched. An important property of the Stark decelerator is that the deceleration process does not only work for one molecule in the beam, but that it works for all molecules that at the entrance of the decelerator have a position and velocity that are within the so-called acceptance of the decelerator. Together, these properties allow one to decelerate (or to accelerate) part of the beam to any desired velocity, while keeping the selected part of the beam together as a compact packet.

Note that it is essential that molecules remain in the same quantum state throughout the deceleration process. In order to achieve this, the orientation of a molecule needs to adiabatically follow the field. This requires that the electric field varies slow enough that the radiation generated during the transient time of switching cannot induce a transition in the molecule.

### 9.2.2. *Phase Stability in the Stark Decelerator*

Central to the understanding of the operation principles of a Stark decelerator are the concepts of the synchronous molecule and phase stability. Referring back to Fig. 9.6, the position $z$ (with periodicity $2L$) of a molecule at the time that the fields are switched is called the "phase angle" $\phi = z\pi/L$ (with periodicity $2\pi$). We define the position with $\phi = 0°$ as the position in between two adjacent pairs of electrodes such that the electrodes at $\phi = 90°$ are grounded just after the fields are switched. By definition, a molecule with velocity $v_0$ is called *synchronous* if its phase $\phi_0$ on the potential is always the same at the time that the fields are switched, i.e. $\phi_0$ remains constant and the molecule will lose a constant amount of kinetic energy $\Delta K(\phi_0)$ per stage. The synchronous molecule achieves this by travelling exactly a distance $L$ in the time interval between two successive switch times. This means that the synchronous molecule is always "in phase" with the switching of the fields in the decelerator. A molecule that has a slightly different phase $\phi$ and/or velocity $v$ than the synchronous molecule will experience an automatic correction towards the equilibrium values $\phi_0$ and $v_0$. For instance, a molecule that has a phase that is slightly higher than $\phi_0$ at a certain switch time, will loose more kinetic energy than the synchronous molecule, and the molecule will slow down with respect to the synchronous molecule. Its phase will get smaller, until it lags behind, after which the process reverses. Molecules within a certain region in phase-space, bound by the so-called separatrix, will undergo stable phase-space oscillations around the synchronous molecule. This mechanism is referred to as phase stability, and ensures that a packet of molecules is kept together in the Stark decelerator throughout the deceleration process.

To better understand phase stability, it is helpful to consider the trajectories of the molecules along the molecular beam axis and to derive the longitudinal equation of motion for these. The most important elements of the derivation are reproduced here; more details can be found elsewhere.[29,30] A mathematically more rigorous derivation, based on a spatial and temporal Fourier representation of the potential, has also been given.[31]

The Stark energy of a molecule $W(z\pi/L)$ is symmetric around the position of a pair of electrodes and can be conveniently written as a Fourier series:

$$\begin{aligned} W(z\pi/L) &= \frac{a_0}{2} + \sum_{n=1}^{\infty} a_n \cos(n(z\pi/L + \pi/2)) \\ &= \frac{a_0}{2} - a_1 \sin\frac{z\pi}{L} - a_2 \cos 2\frac{z\pi}{L} + a_3 \sin 3\frac{z\pi}{L} + \cdots \end{aligned} \tag{9.1}$$

By definition, the synchronous molecule travels a distance $L$ in the time interval between two successive switch times. The change in kinetic energy per stage $\Delta K(\phi_0) = -\Delta W(\phi_0)$ for a synchronous molecule with phase $\phi_0$ and velocity $v_0$ at a certain switch time is then given by the difference in potential energy at the positions $\phi_0$ and $\phi_0 + \pi$:

$$\Delta W(\phi_0) = W(\phi_0 + \pi) - W(\phi_0) = 2a_1 \sin\phi_0. \tag{9.2}$$

The average force $\bar{F}$ that acts on the synchronous molecule is then given by

$$\bar{F}(\phi_0) = \frac{-\Delta W(\phi_0)}{L} = -\frac{2a_1}{L}\sin\phi_0 \tag{9.3}$$

if we only take the leading terms up to $n = 2$ in Eqn. (9.1) into account. The average force acting on a non-synchronous molecule with phase $\phi = \phi_0 + \Delta\phi$, but with velocity $v_0$, is given by $-\frac{2a_1}{L}\sin(\phi_0 + \Delta\phi)$, and to a good approximation the equation of motion with respect to the synchronous molecule is

$$\frac{mL}{\pi}\frac{d^2\Delta\phi}{dt^2} + \frac{2a_1}{L}(\sin(\phi_0 + \Delta\phi) - \sin(\phi_0)) = 0 \tag{9.4}$$

where $m$ is the mass of the molecule.

In the phase stability diagrams of Fig. 9.7, lines of constant energy are shown that result from a numerical integration of Eqn. (9.4) for OH radicals in the $J = 3/2, M\Omega = -9/4$ state, using parameters for a decelerator that is operational in our laboratory. The equilibrium phase angles $\phi_0 = 0°$ and $\phi_0 = 70°$ for the synchronous molecule are used. The solid lines indicate the trajectories in phase space that molecules will follow. The positions of the electrodes of the decelerator (at the moment the fields are being switched) are indicated by the dashed lines. Closed curves in the phase space diagram correspond to bound orbits; molecules within the "bucket" bound by the thick contour will oscillate in phase space around the phase and velocity of the synchronous molecule. Note that operation of the decelerator at $\phi_0 = 0°$

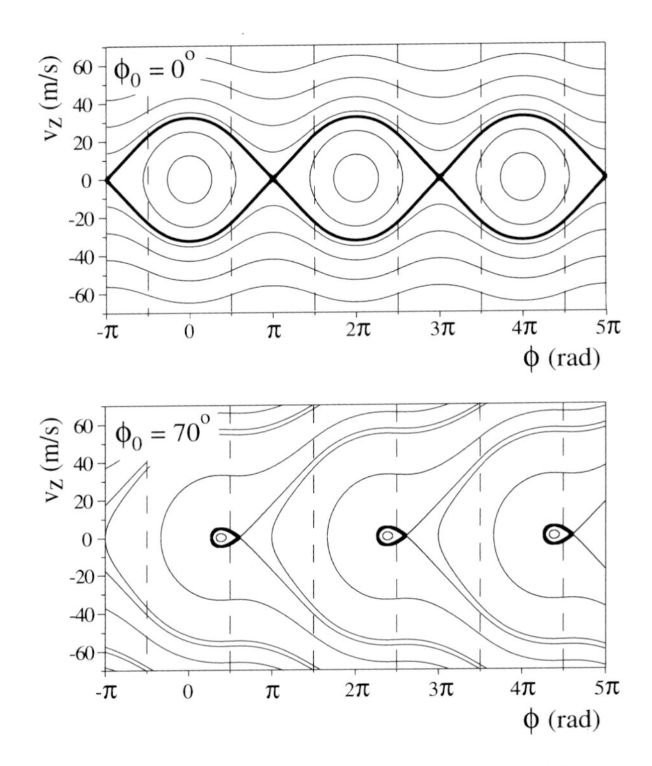


Fig. 9.7. Phase stability diagram for OH($J = 3/2, M\Omega = -9/4$) radicals when the decelerator is operated at a phase angle of $\phi_0 = 0°$ or $\phi_0 = 70°$. The positions of the electrodes of the decelerator are indicated by the dashed lines. (Reproduced from S.Y.T. van de Meerakker *et al.*[28] with permission. ©2006 by Annual Reviews www.annualreviews.org.)

corresponds to transporting (part of) the beam through the decelerator without deceleration, while the acceleration or deceleration of the beam occurs for $-90° < \phi_0 < 0$ and $0 < \phi_0 < 90°$, respectively. The separatrix defines the longitudinal acceptance of the Stark decelerator, and is shown for different phase angles $\phi_0$ in Fig. 9.8. It is seen that the acceptance is larger for smaller values of $\phi_0$, while the deceleration per stage increases for higher values of $\phi_0$. Since for deceleration both are desirable, there is a trade-off between the two.

In a more extensive description of phase stability in a Stark decelerator, it has been shown that additional phase-stable regions exist, which have indeed been experimentally observed.[30] These higher-order phase-stable regions can be understood to result from interferences of partial waves in the Fourier expansion.[31]

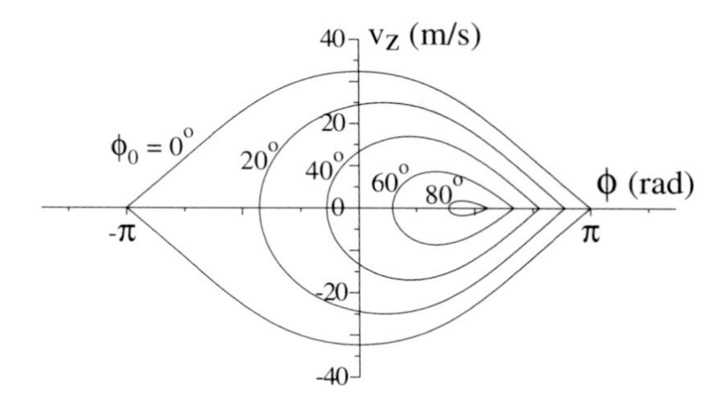


Fig. 9.8. Longitudinal acceptance of the decelerator for different values of the phase angle $\phi_0$. The velocity $v_z = 0$ corresponds to the velocity of the synchronous molecule.

### 9.2.3. *Transverse Focusing in the Stark Decelerator*

Phase stability only ensures that the packet of molecules is kept together in longitudinal phase space. However, we also require that the molecules keep together in the transverse direction. Ideally, the transverse focusing would be completely decoupled from the longitudinal motion. This would require spatially separated focusing and deceleration stages, which would make the decelerator impractically long. We have instead opted for a very compact design of the decelerator, using the electric field stages simultaneously for deceleration and transverse focusing. In the electrode geometry as shown in Fig. 9.6, molecules in low-field seeking states stay transversely confined to the molecular beam axis as the electric field is always lower on this axis than on the electrodes. In order to focus the molecules in both transverse directions, the electrode pairs that make up a deceleration stage are alternately positioned horizontally and vertically, as can be seen in the inset of Fig. 9.4.

The 3D trajectories of molecules traveling through the Stark decelerator are rather complex. In the longitudinal direction, a molecule oscillates in position and velocity around the synchronous molecule, while in the transverse direction, a molecule oscillates around the molecular beam axis. The oscillation frequencies that are involved strongly depend on the phase angle $\phi$, but are in general similar in magnitude. The influence of the transverse motion of the molecules on longitudinal phase stability has been studied. It has been found that, for high values of the phase angle $\phi_0$, the transverse motion actually enhances the region in phase-space for which phase stable deceleration occurs. For low values of $\phi_0$, however, the transverse motion reduces the acceptance of a Stark decelerator and unstable regions in phase-space appear. These effects can be quantitatively explained in terms of a coupling between the longitudinal and transverse motion, and coupled equations of motion have been derived that reproduce the observations. This coupling does not significantly deteriorate the overall performance of the Stark decelerator, provided that the number of deceleration stages is limited.[32] At low longitudinal velocities the assumptions used in the analytical model of the longitudinal and transverse motion are no longer valid. To avoid losses, therefore, care must be taken in the design of the last few electrode pairs of the decelerator.

### 9.2.4. *Stark Deceleration of OH Radicals*

A scheme of the Stark deceleration and trapping machine that has been used to decelerate and trap OH radicals in our laboratory is shown in Fig. 9.9.

A pulsed beam of OH radicals is produced by photodissociation of $HNO_3$ molecules that are co-expanded with a rare gas through a room-temperature pulsed solenoid valve. In most experiments, either Kr or Xe is used as a carrier gas, producing a beam with a mean velocity of 450 m/s or 360 m/s, respectively. In the supersonic expansion, the beam is rotationally and vibrationally cooled, and after the expansion most of the OH radicals reside in the lowest rotational ($J = 3/2$) and vibrational level of the electronic

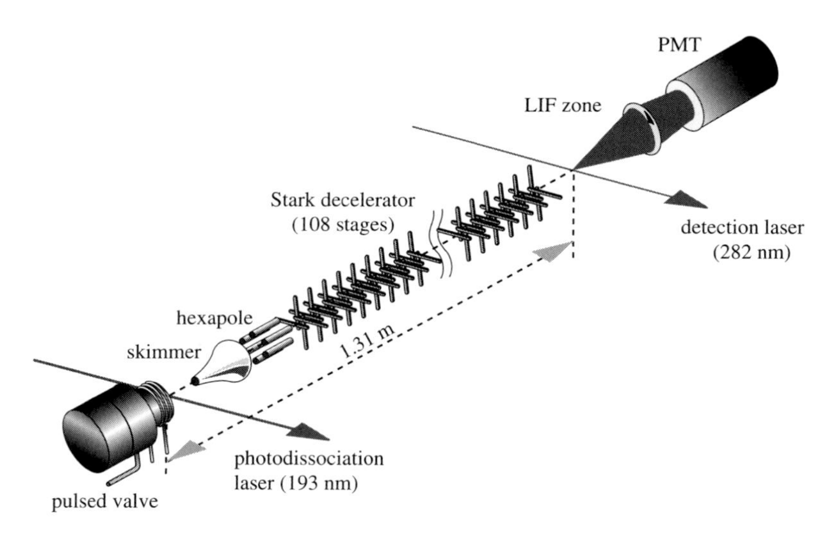


Fig. 9.9. Scheme of the experimental set-up. A pulsed beam of OH radicals is produced via ArF-laser photodissociation of $HNO_3$ seeded in a heavy carrier gas. The molecular beam passes through a skimmer, hexapole and Stark decelerator into the detection region. State-selective LIF detection is used to measure the arrival time distribution of the OH($J = 3/2$) radicals in the detection zone. (Reproduced from S.Y.T. van de Meerakker *et al.*[28] with permission. ©2006 by Annual Reviews www.annualreviews.org.)

ground state $X^2\Pi_{3/2}$. This level has a $\Lambda$-doublet splitting of $0.055\,\text{cm}^{-1}$ and the upper $\Lambda$-doublet component is split (neglecting hyperfine structure) into a $M_J\Omega = -3/4$ and a $M_J\Omega = -9/4$ component when an electric field is applied, as is shown in the Stark energy diagram in Fig. 9.2. The $M_J\Omega = -9/4$ component offers a Stark shift three times larger than the $M_J\Omega = -3/4$ component. Only the molecules that happen to be in the low-field seeking $M_J\Omega = -3/4$ or $M_J\Omega = -9/4$ components of the upper $\Lambda$-doublet level participate further in the electric field manipulation process. The molecular beam passes through a skimmer, and enters the second vacuum chamber. In the decelerator chamber, the beam of OH radicals enters a short hexapole that focusses the beam into the Stark decelerator. The 1188 mm-long Stark decelerator consists of an array of 108 equidistant electric field stages, with a center to center distance $L$ of adjacent stages of 11 mm. Each stage consists of 2 parallel 6 mm diameter polished hardened steel rods that are centered 10 mm apart, and that are located symmetrically around the beam axis. Alternating stages are rotated by 90° with respect to each other, providing a $4 \times 4\,\text{mm}^2$ spatial transverse acceptance area. A photograph of the decelerator, with a close-up of the $4 \times 4\,\text{mm}^2$ opening between the electrodes, is shown in Fig. 9.4.

The decelerator is operated using a voltage of $\pm$ 20 kV on opposing electrodes in a field stage, creating a maximum electric field strength near the electrodes of 115 kV/cm. The high voltage pulses are applied to the electrodes using fast semiconductor high voltage switches. The OH radicals are state selectively detected 21 mm downstream from the last electric field stage (1.307 m from the nozzle) using an off-resonant Laser Induced Fluorescence (LIF) detection scheme.

The performance of the Stark decelerator can be studied by recording the time-of-flight (TOF) profile of the OH radicals exiting the decelerator, i.e. by scanning the timing of the detection laser relative to the dissociation laser. Alternatively, the phase space distribution of the molecules can be studied inside the decelerator by propagating a laser beam along the molecular beam axis, and by performing spatially resolved LIF detection from there. The latter strategy has been implemented by Jun Ye and coworkers at JILA, also using the OH radical.[33] Typical TOF profiles of OH radicals at the exit of the decelerator as obtained in our laboratory are shown in Fig. 9.10.

These TOF profiles are obtained using Kr as a carrier gas, producing a molecular beam with a mean velocity of 460 m/s. The decelerator is operated at a phase angle of $\phi_0 = 70°$ and an initial velocity of the synchronous molecule of 470 m/s (curve *a*), 450 m/s (*b*), 430 m/s (*c*), and 417 m/s (*d*). With these settings, the decelerated bunch of molecules exits the decelerator with a final velocity of 237 m/s, 194 m/s, 142 m/s and

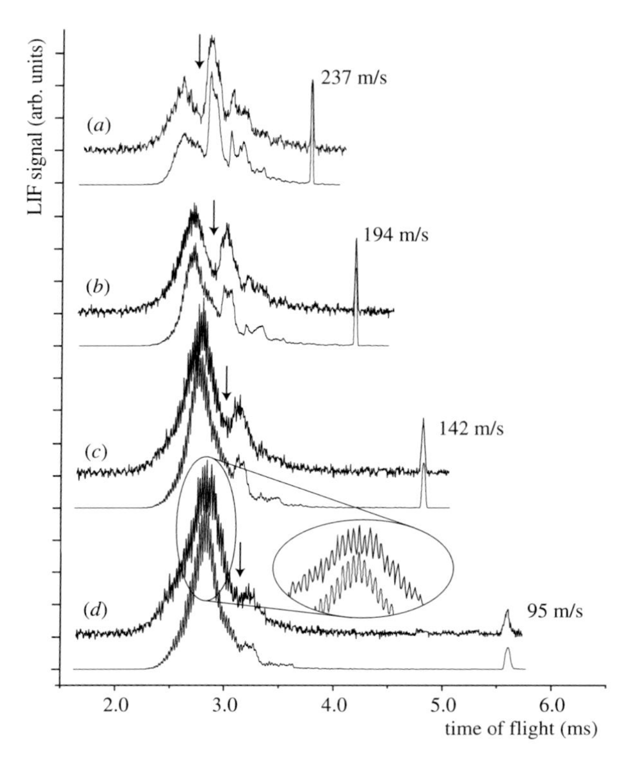


Fig. 9.10. Observed and simulated TOF profiles of a molecular beam of OH radicals exiting the Stark decelerator when the decelerator is operated at a phase angle of 70° for a synchronous molecule with an initial velocity of 470 m/s (*a*), 450 m/s (*b*), 430 m/s (*c*) and 417 m/s (*d*). The molecules that are accepted by the decelerator are split off from the molecular beam and arrive at later times, and with the final velocities as indicated, in the detection region. (Reproduced from S.Y.T. van de Meerakker *et al.*[28] with permission. ©2006 by Annual Reviews www.annualreviews.org.)

95 m/s, respectively. The holes in the profiles of the non-decelerated beams that result from the removal of the bunch of OH radicals that is decelerated, are indicated by the vertical arrows. The width of the arrival time distribution of the decelerated packet becomes larger for lower values of the final velocity. This is due to the spreading out of the beam while flying from the exit of the decelerator to the LIF zone, and due to the spatial extent of the detection laser beam. Both effects are incorporated in the simulated TOF profiles, shown underneath the observed ones. In curves (*c*) and (*d*), rich oscillatory structure on the TOF profile of the fast beam is observed, as shown enlarged in the inset. This structure results from the

modulation of the phase space distribution of the non-decelerated beam in the decelerator.[30]

### 9.2.5. *Longitudinal Focusing of a Stark Decelerated Molecular Beam*

In the Stark decelerator, the force experienced by a polar molecule is conservative, as, at any given time, this force depends only on the position of the molecule. Consequently, the product of the velocity spread and the position spread stays constant throughout the deceleration process, in accordance with Liouville's theorem. Cooling of the packet of decelerated molecules while maintaining the particle density in the packet, is therefore fundamentally impossible in a Stark decelerator. It is possible, however, to reduce the velocity spread at the expense of the position spread, or vice versa, as long as the product of the two stays constant. Either of these can be done using a so-called buncher, a second array of deceleration fields mounted some distance behind the decelerator. In the buncher a beam of polar molecules is exposed to a harmonic potential in the forward direction, i.e. along the molecular beam axis. This results in a uniform rotation of the longitudinal phase-space distribution of the ensemble of molecules. By switching the buncher on and off at the appropriate times, it can be used either to produce a narrow spatial distribution at a certain position downstream from the buncher or to produce a narrow velocity distribution. Both applications have been experimentally demonstrated, and in particular, a beam of Stark decelerated $ND_3$ molecules with a longitudinal temperature of 250 $\mu$K has been produced.[34–36]

Figure 9.11 shows schematically the bunching principle. When the Stark decelerator is operated at a phase angle of 70°, the longitudinal position spread of the decelerated ammonia molecules is about 1 mm and the longitudinal velocity spread is about 6.5 m/s. The phase-space distribution of the decelerated molecules is shown in Fig. 9.11 underneath the last stage of the decelerator. In flying from the exit of the decelerator to the buncher the packet of ammonia molecules spreads out along the molecular beam axis. This results in the elongated and tilted distribution in longitudinal phase-space as shown in Fig. 9.11. The geometry of the buncher is identical to that of the decelerator, except for an overall scaling factor. The fields are switched such that the synchronous molecule spends an equal amount of time on the downward slope and on the upward slope of the potential well, and the synchronous molecule will therefore keep its original velocity. Molecules that are originally slightly ahead of the synchronous molecule, i.e. molecules that are originally faster, will spend more time on the upward slope than on the downward slope of the potential well, and are therefore

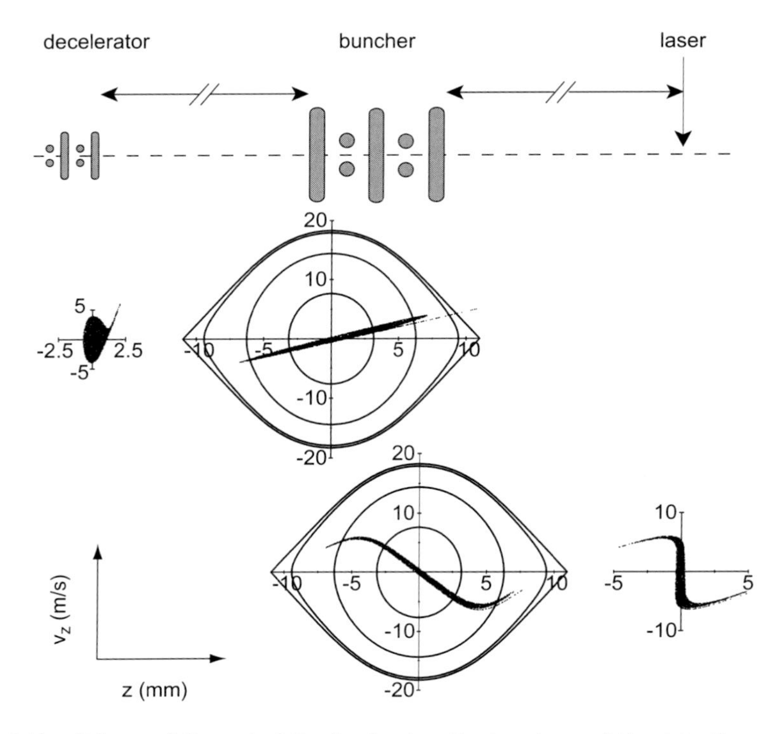


Fig. 9.11. Scheme of the end of the decelerator, the buncher and the detection region. The calculated longitudinal phase-space distribution of the ammonia molecules is given at the exit of the Stark decelerator, at the entrance and exit of the buncher and in the detection region. The position and velocity are plotted with respect to the synchronous molecule. The solid curves show lines of equal energy in the buncher potential (compare to Fig. 9.7). (Reproduced from F.M.H. Crompvoets *et al.*[34] with permission. Copyright the American Physical Society.)

decelerated relative to the synchronous molecule. Molecules that are originally slightly behind the synchronous molecule, i.e., molecules that are originally slower, will spend less time on the upward slope than on the downward slope of the potential well, and are therefore accelerated relative to the synchronous molecule. The calculated distribution at the time that the buncher is switched off is also shown in Fig. 9.11. In this figure, the contours of equal energy for the ammonia molecules in the potential well are shown, relative to the position in phase-space of the synchronous molecule. As the potential well is approximately harmonic, the phase space distribution of the molecules rotates uniformly. When the buncher is switched off, the slow molecules are ahead while fast molecules are lagging behind, leading to a longitudinal spatial focus at some position downstream. The angle over which the packet is rotated, and thus the exact position where the packet will come to a spatial focus can be changed by varying the voltage on the buncher electrodes or by varying the duration that the buncher fields are on.[35] In Stark deceleration beamlines, hexapoles and bunchers are used to map the phase space distribution of the molecules at the exit of one element onto the entrance of the next one in the transverse and longitudinal direction, respectively.

### 9.2.6. *Decelerating Molecules in High-Field Seeking States*

Summarizing the previous sections, polar molecules in low-field seeking states are decelerated by letting them fly from a region of low electric field into a region of high electric field. Transverse stability is achieved by using an electrode geometry that produces a minimum of the electric field on the molecular beam axis, thereby continuously focusing the beam. It might appear to be straightforward to apply the above method to molecules in high-field seeking states by simply letting the molecules fly out of, instead of into, the region of a high electric field. For the motion of the molecules in the forward direction, this is indeed true. However, Maxwell's equations do not allow for a maximum of the electric field in free space,[37] e.g. on the molecular beam axis. Therefore, transverse stability cannot be maintained easily; molecules in high-field seeking states have a tendency to crash into the electrodes where the electric fields are the highest. This fundamental problem can be overcome by using alternating gradient (AG) focusers.[38]

Figure 9.12(a) shows the general form of an AG decelerator. The AG lenses are formed from a pair of cylindrical electrodes to which a voltage difference is applied. Molecules in high-field seeking states will be defocused in the plane containing the electrode center lines while being focused in the orthogonal plane. As the molecules move down the beamline the orientation of the lenses, and thus the focusing and defocusing directions, alternate. In any transverse direction the defocusing lenses have less effect than the focusing lenses, not because they are weaker (they are not), but because the molecules are closer to the axis inside a defocusing lens than inside a focusing lens.

Molecules in high-field seeking states are accelerated while entering the field of an AG lens and are decelerated while leaving the field. By simply switching the lenses on and off at the appropriate times, AG focusing and deceleration of polar molecules can be achieved simultaneously. Fig. 9.12(b) shows the potential energy along the $z$-axis of a single lens for a metastable $CO(a\,^3\Pi_1, v = 0, J = 1)$ molecule in the $M\Omega = +1$ high-field seeking state. The molecules enter each lens with the electric fields turned off so that their speed is not influenced. The fields are then suddenly turned on, and

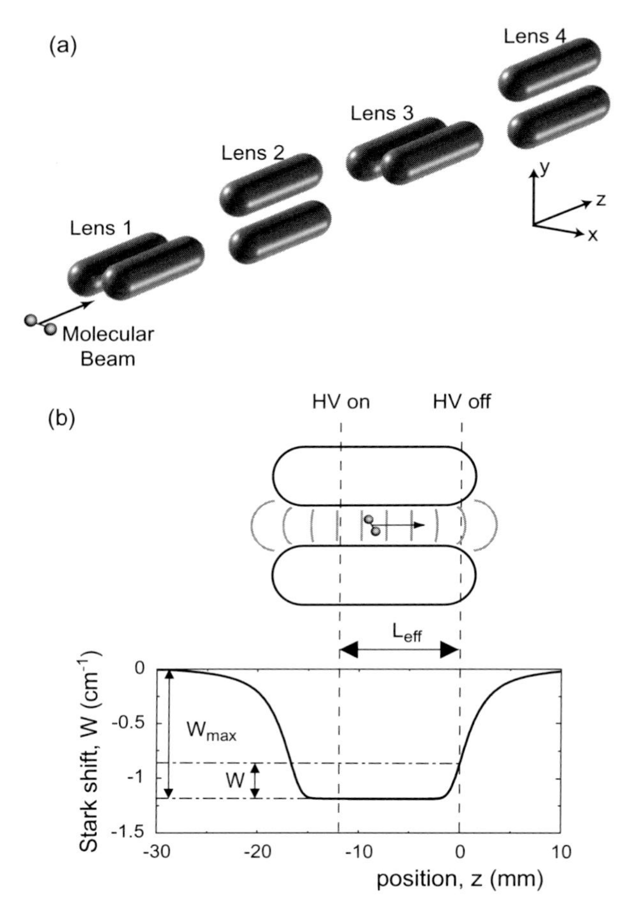


Fig. 9.12. (a) Layout of an alternating gradient decelerator for polar molecules showing the first four deceleration stages. Each electrode pair acts both to focus and to decelerate the molecules. (b) Cross-section of a single lens formed from two 20 mm long rods with hemispherical ends, 6 mm in diameter and spaced 2 mm apart. The potential energy is shown along the $z$-axis for metastable CO molecules in the $a^3\Pi_1$, $J = 1$, $M\Omega = +1$ level, when the potential difference between the electrodes is 20 kV. The high-voltage switching procedure is indicated: the voltages are turned on when the bunch of molecules reaches the "HV on" position, and are turned off once they reach the "HV off" position. (Reproduced from H.L. Bethlem *et al.*[39] with permission. Copyright IOP Publishing Ltd.)

the high-field seeking molecules are decelerated as they leave the lens and move from a region of high field to one of low field. This process is repeated until the molecules reach the desired speed.

A prototype machine of this type consisting of 12 AG lenses has been used to decelerate high-field seeking metastable CO molecules[40] and, in the group of Hinds at the Imperial College London, ground state YbF molecules.[41] The transverse acceptance of this AG decelerator was determined to be about a factor of 100 smaller than for a decelerator for molecules in low-field seeking states having the same aperture.[39] The acceptance can be increased by using a more sophisticated lens design consisting of four electrodes rather than two. At Imperial College London a new AG decelerator was constructed to bring YbF molecules to a standstill, whereas at the Fritz-Haber-Institut AG focusing and deceleration are being applied to molecules like benzonitrile and tryptophan.

### 9.2.7. *Other Types of Decelerators*

Inspired in part by the manipulation of polar molecules in a Stark decelerator, studies on the manipulation of atoms and molecules in high Rydberg states with electric fields are now being actively pursued. Compared to the polar molecules discussed thus far, atoms or molecules in a Rydberg state offer a much larger electric dipole moment. Hence, these particles can be manipulated using only modest electric field strengths in a single, or a few electric field stages. Energy level crossings in the dense Rydberg manifold limit the magnitude of the electric field strengths that can actually be applied. These methods have been pioneered using $H_2$ molecules[42] and Ar atoms.[43] Using a Rydberg decelerator, $H$ atoms could be stopped and reflected from an electrostatic mirror.[44] A disadvantage of these experiments is that the lifetime of the Rydberg states limits the time that is available to bring the molecules to rest and to store and study them in a trap.

Magnetic and optical analogues of the Stark decelerator have been developed as well. In a proof-of-principle experiment, ground-state hydrogen atoms have been decelerated using six pulsed magnetic field stages.[45] This experiment demonstrated the possibility of switching magnetic fields fast enough to allow phase-stable deceleration. In principle, this technique can be used to decelerate and trap all species with a nonzero electron spin. The interaction of polarizable molecules with a high-intensity pulsed optical lattice, produced by two counter-propagating laser beams, can also be used to slow down molecules. By chirping one of the beams, the lattice velocity can be reduced from the mean velocity of a molecular beam to any desired final velocity. In a proof-of-principle experiment, a single stage optical Stark decelerator has been used to reduce the velocity of a beam of benzene molecules from 320 m/s to 295 m/s.[46] The technique has also been used to decelerate and accelerate NO molecules.[47,48]

## 9.3. Trapping Neutral Polar Molecules

### 9.3.1. *DC Trapping of Molecules in Low-Field Seeking States*

The first electrostatic trap for polar molecules was demonstrated in 2000 using decelerated $ND_3$ molecules.[26] This trap, consisting of a ring electrode and two hyperbolic end-caps in a quadrupole geometry, was originally proposed by Wing for Rydberg atoms.[49] Such a quadrupole trap can be mounted directly behind the decelerator, and, through holes in the end-caps, a slow packet of molecules can be loaded into the trap using the procedure that is schematically shown in Fig. 9.13. The specific parameters in this Figure apply to the OH trapping experiments.[50] In this case, the Stark decelerator is programmed to produce a packet of molecules with a velocity of approximately 20 m/s. The slow beam of OH radicals is loaded into the electrostatic trap with voltages of 7 kV, 15 kV and −15 kV on the first end-cap, the ring electrode and the second end-cap, respectively. This "loading geometry" of the trap is shown on the left in Fig. 9.13. In the loading geometry, a potential hill in the trap is created that is higher than the remaining kinetic energy of the molecules. The OH radicals therefore come to a standstill near the center of the trap. Then, the trap is switched into the "trapping geometry," shown on the right side in Fig. 9.13; the first end-cap is switched from 7 kV to −15 kV to create a (nearly) symmetric 500 mK deep potential well.

A typical TOF profile that is obtained in an OH deceleration and trapping experiment is shown in Fig. 9.14. For this, the fluorescence signal of the OH radicals in the trap is detected through the hole in the end-cap. This TOF profile is recorded using Kr as carrier gas, and the profile is therefore complementary to the series of TOF profiles shown in Fig. 9.10. The hole in the TOF profile of the fast beam due to the removal of OH radicals that are decelerated is indicated by a vertical arrow. These OH radicals come to a standstill around 7.4 ms after their production. At that time, indicated by another arrow, the trap electrodes are switched into the trapping configuration. After some initial oscillations, a steady LIF signal is observed from the OH radicals in the trap, indicating that the molecules are confined. In the inset, the signal of the trapped OH radicals is shown on a 10 second timescale, from which a $1/e$ trap lifetime of 1.6 s is deduced.

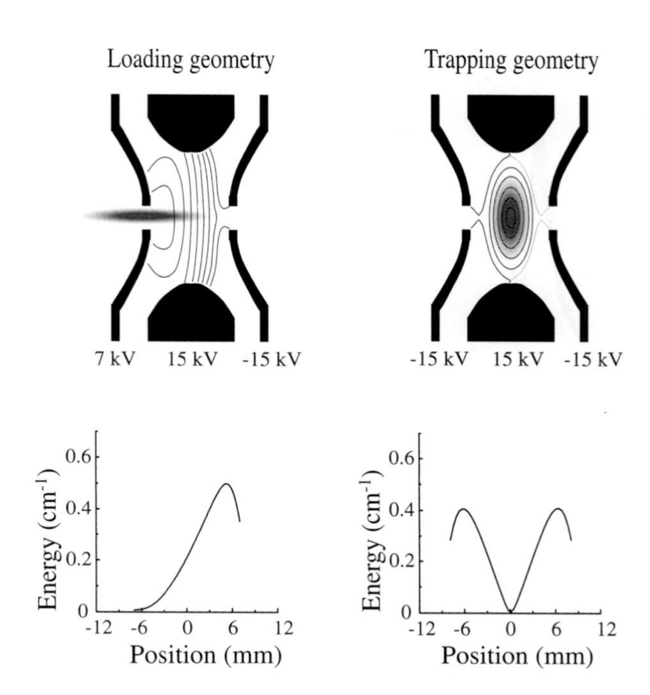


Fig. 9.13. Scheme of the loading procedure of the electrostatic quadrupole trap. In the "loading geometry," the voltages on the trap electrodes are set such that a potential hill is created in the trap that is higher than the remaining kinetic energy of the incoming molecules. At the time that the molecules come to a standstill in the center of the trap, the trap is switched into the "trapping geometry." In this geometry a (nearly) symmetric 500 mK deep potential well is created in which the molecules are confined. (Reproduced from S.Y.T. van de Meerakker *et al.*[28] with permission. ©2006 by Annual Reviews www.annualreviews.org.)

It is noted that when Xe is used as a carrier gas, the most intense part of the molecular beam pulse can be selected and trapped.[28] The efficiency of the trap loading process could be increased by 40% by feedback control optimization using evolutionary strategies.[51]

Electrostatic traps with other electrode geometries have been developed and tested as well. A four electrode trap geometry that combines a dipole, quadrupole and hexapole field has been tested using decelerated $ND_3$ molecules. By applying different voltages to the electrodes, a double-well or a donut trapping potential can be created. Rapid switching between the various trap potentials offers good prospects for studying collisions as a function of collision energy at low temperatures.[52]

Confinement of Stark decelerated molecules in combined magnetic and electric fields has recently been demonstrated in the JILA group for the OH

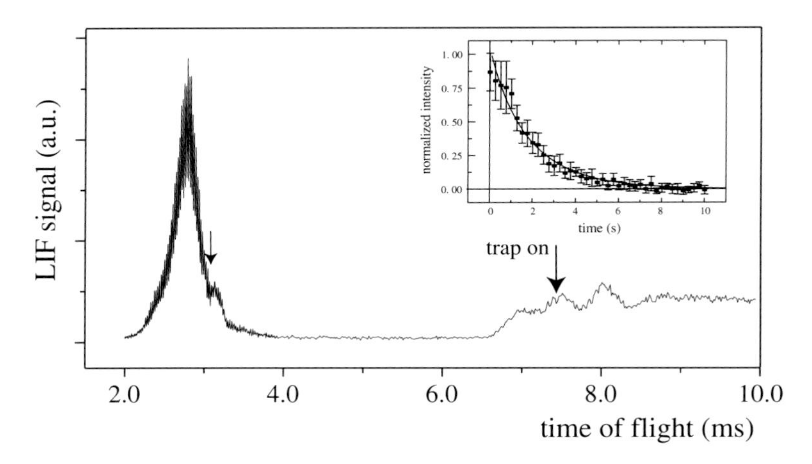


Fig. 9.14. Measured TOF profile of OH ($J = 3/2$) radicals for a typical trapping experiment. The time at which the trap is switched on is indicated. In the inset, the signal of the trapped OH radicals is shown on a 10 second timescale. (Figure partially reproduced from S.Y.T. van de Meerakker *et al.*[50] with permission. Copyright the American Physical Society.)

radical. An electric field was superimposed on a magnetic field to create an overall magnetoelectric trapping potential.[53] The adjustable electric field might be advantageous in the study of low energy dipolar interactions.

Rather then actively bringing a state-selected packet of molecules from a high velocity to a standstill to load a trap, a mere selection of slow molecules from an effusive source using a bent quadrupole guide has also been used to produce translationally cold molecules.[13,54]

### 9.3.2. *The Storage Ring and the Molecular Synchrotron*

In its simplest form a storage ring is a trap in which the particles — rather than having a minimum potential energy at a single location in space — have a minimum potential energy on a circle. The advantage of a storage ring over a trap is that packets of particles with a non-zero mean velocity can be confined. While circling the ring, these particles can be made to interact repeatedly, at well defined times and at distinct positions, with electromagnetic fields and/or other particles.

In the top part of Fig. 9.15 the experimental setup that has been used in the demonstration of an electrostatic storage ring for neutral molecules is shown.[55] A beam of ammonia molecules is decelerated to a velocity of 92 m/s, longitudinally cooled to a temperature of 300 $\mu$K and focused into a 25 cm diameter hexapole ring. The density of ammonia molecules inside

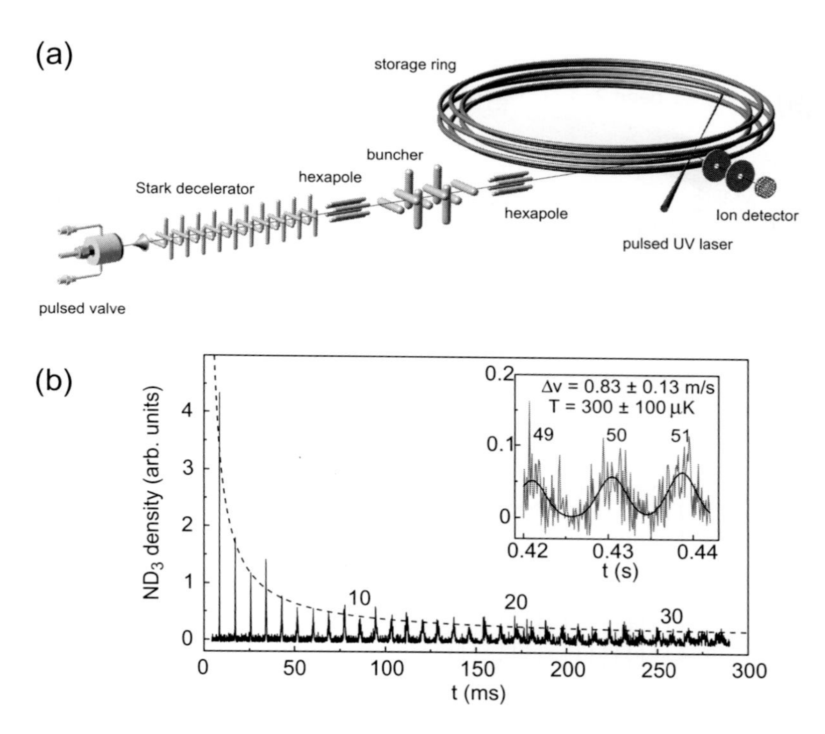


Fig. 9.15. (a) Scheme of the experimental set-up. A pulsed beam of $ND_3$ molecules is decelerated to 92 m/s using a Stark decelerator, cooled to 300 $\mu$K using a buncher and focused into a hexapole torus storage ring. (b) Density of ammonia molecules at the detection zone inside the ring as a function of storage time up to the 33rd round trip. Due to the spreading of the packet, the peak density decreases as $1/t$ (shown as the dashed line). In the inset, a measurement of the ammonia density after 49–51 round trips is shown, together with a multi-peak Gaussian fit. (Reproduced from F.M.H. Crompvoets *et al.*[55] with permission. Copyright the American Physical Society.)

the storage ring is probed via a laser-ionization detection scheme. In the bottom part of Fig. 9.15 the ion signal is shown as a function of the storage time in the ring; the origin of the time axis is at the time when the high voltages on the ring are switched on. Peaks are observed when the packet of molecules passes the detection zone. Upon making successive round trips the packet of molecules gradually spreads out as a result of the residual velocity spread until it fills the entire ring.

In order to counteract the spreading of the packet in the ring we have constructed a storage ring consisting of two half-rings separated by a 2 mm gap, schematically shown in the top part of Fig. 9.16. By appropriately switching the voltages applied to the electrodes as the molecules pass

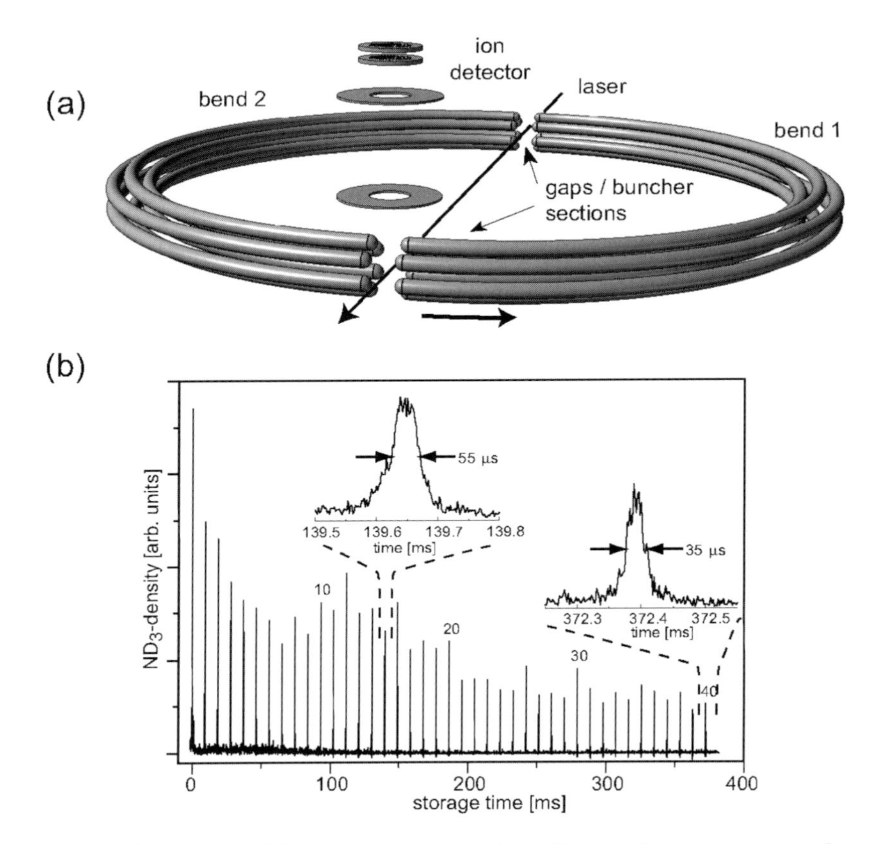


Fig. 9.16. (a) Scheme of the molecular synchrotron. The synchrotron consists of two hexapole half-rings with a 12.5 cm radius separated by a 2 mm gap. (b) Density of ammonia molecules at the detection zone inside the synchrotron as a function of storage time up to the 40th round trip. Expanded views of two TOF profiles are shown as insets, more clearly illustrating the absolute widths of these peaks. (Reproduced from C.E. Heiner *et al.*[56] with permission. Copyright McMillan Publishers Ltd.)

through the gaps between the two half-rings, molecules can be accelerated, decelerated and bunched. This structure is the neutral analog of a synchrotron for charged particles. The lower part of Fig. 9.16 again shows the density of molecules stored in the ring as a function of storage time. It is seen that, after an initial decrease during the first 25 round trips, the width of the stored packet now stays constant. Bunching ensures a high density of stored molecules, and in addition makes it possible to inject multiple — either co-linear or counter propagating — packets into the ring without affecting the packet(s) already stored.[56]

### 9.3.3. *AC Trapping of Molecules in High-Field Seeking States*

Trapping molecules in high-field seeking states is of particular interest for two reasons: (i) The ground state of a system is always lowered by an external perturbation. Therefore, the ground state of any molecule is high-field seeking. In the ground state, trap loss due to inelastic collisions is absent, making it possible to cool these molecules further using evaporative or sympathetic cooling. This is particularly relevant as the dipole-dipole interaction is predicted to lead to large cross sections for inelastic collisions for polar molecules in excited ro-vibrational states.[57] (ii) Molecules composed of heavy atoms or many light atoms, such as polycyclic hydrocarbons, have small rotational constants. Consequently all states of these molecules become high-field seeking in relatively small magnetic or electric fields.

The problem of trapping molecules in high-field seeking states is essentially the same as that of decelerating molecules in high-field seeking states discussed in Section 9.2.6. Ideally, one would like to have an electrode geometry that creates a maximum electric field strength at some position away from the electrodes, however this is prohibited by the Earnshaw Theorem.[37] Although it is not possible to generate a maximum of the electric field in free space, it is possible to generate an electric field that has a saddle point by superimposing an inhomogeneous electric field and a homogeneous electric field. In such a field, molecules are focused along one direction, while being defocused along the other. By reversing the direction of the inhomogeneous electric field, the focusing and defocusing directions can be exchanged. If this is done periodically, molecules will be further away from the saddle point along the focusing direction and closer to the saddle point along the defocusing direction, leading to a net time-averaged focusing force in all directions. Such a trap works both for molecules in high-field seeking states and for molecules in low-field seeking states. There are three possible electrode geometries that can be used to create the desired electric 3D trapping field.[58] By now all these geometries have been demonstrated for atoms and/or molecules.

Figure 9.17(a) schematically shows the electrode geometry of our cylindrical AC trap that has been used to trap both $ND_3$ molecules[58,59] and Rb atoms.[60] The trap has hexapole symmetry. Consequently, when positive and negative voltages are applied alternately to the four electrodes of the trap a perfect hexapole field is obtained. In order to create a saddle point, the voltage applied to one of the end-caps is increased, while the voltage on the other end-cap is decreased. This adds a dipole term to the electric field. If the direction of the hexapole term is reversed, the focusing and defocusing directions are exchanged. Figures 9.17(b) and (c) show the strength

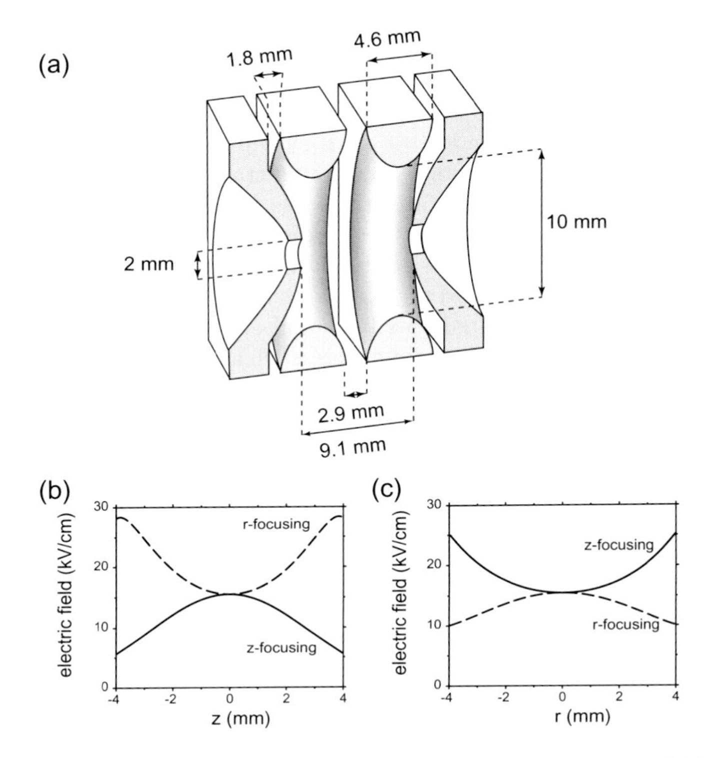


Fig. 9.17. (a) Scheme of the cylindrically symmetric AC trap. The electric field strength along the symmetry axis ($z$-axis) is shown in (b), whereas in (c) this is shown along an axis perpendicular to the symmetry axis, through the trap center. When voltages of 5, 7.5, $-7.5$ and $-5$ kV are applied to the electrodes, molecules in high-field seeking states are focused towards the center along the $z$-axis and defocused in the radial direction. When voltages of 11, 1.6, $-1.6$, and $-11$ kV are applied to the electrodes, molecules in high-field seeking states are radially focused and defocused along the $z$-axis. (Reproduced from H.L. Bethlem *et al.*[58] with permission. Copyright the American Physical Society.)

of the electric field along the symmetry axis ($z$-axis) and along an axis perpendicular to this, respectively, when the field is either focusing along $z$ or focusing radially.

$ND_3$ molecules are loaded into the AC trap by decelerating them to a standstill while in their low-field seeking state, and by then applying a microwave pulse that pumps a fraction (about 20%) of the molecules into the high-field seeking state. Figure 9.18 shows the density of $ND_3$ molecules at the center of the AC trap as a function of the switching frequency for molecules in either the high-field seeking or the low-field seeking component of the ground state of para-ammonia.[58]

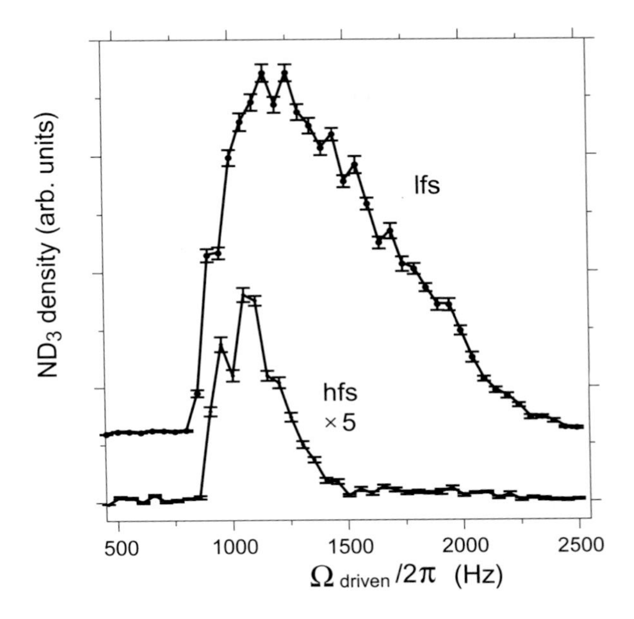


Fig. 9.18. Density of $^{15}ND_3$ molecules in low-field seeking (lfs) and high-field seeking levels (hfs) of the $|J, K\rangle = |1, 1\rangle$ state at the center of the trap as a function of the switching frequency. The measurements are performed 80 ms after the molecules are loaded into the trap. The signal of the high-field seekers is scaled up by a factor of 5 to compensate for the fact that the initial density of high-field seekers is only about 20% of that of the low-field seekers. (Reproduced from H.L. Bethlem *et al.*[58] with permission. Copyright the American Physical Society.)

For clarity, the signal for molecules in low-field seeking states has been vertically offset. The signal for the high-field seekers is scaled up by a factor of five, to correct for the smaller initial density of high-field seekers. At low frequencies of the applied voltages, the trajectories of the molecules in the AC trap are unstable and no signal is observed. Above a frequency of about 900 Hz the trap becomes abruptly stable. Maximum signal is observed at 1100 Hz. When the frequency is increased further, the molecules have less time to move between switching times, and the net time-averaged force on the molecules decreases. As a result the depth of the trap decreases. Higher order terms in the trapping potential give rise to a (frequency independent) potential that reduces the trap depth for molecules in high-field seeking states and increases the trap depth for molecules in low-field seeking states. Therefore the signal of the high-field seekers drops faster with increasing frequency than the signal of the low-field seekers.

It is useful to compare the AC trap with the DC trap that was discussed in Section 9.3.1. The electrostatic traps are on the order of 1 K deep (depending on the molecular species and details of the trap design), and have a trap volume of typically $1\,cm^3$. The AC electric trap has a trap depth of about 1–10 mK and a trap volume of about $10^{-2}\,cm^3$.[58]

### 9.3.4. *Trap Lifetime Limitations*

Referring back to Fig. 9.14, the duration for which the OH radicals are stored in the electrostatic trap is on the order of a few seconds. Molecules can leave the trap via several distinct mechanisms. As there is a zero electric field at the center of the DC quadrupole or hexapole traps, the molecules can undergo transitions to degenerate quantum states that cannot be trapped. In practice, however, these Majorana transitions are not (yet) a limitation on the lifetime, both because the volume in which these transitions could occur is very small and because, when hyperfine structure is included, molecules are often in states that are exclusively low-field seeking, i.e. there is no degeneracy.[61] Molecules can collide with particles in the residual gas in the vacuum chamber, leading to a kinetic energy transfer that results in a direct loss of trapped molecules from the rather shallow trap. Molecules in the trap can also collide with each other. In this case, inelastic collision processes can transfer the molecules from the trapped state to a different quantum state that is either anti-trapped (high-field seeking) or that experiences a lower trapping potential. Finally, the molecules can absorb blackbody radiation from the (room temperature) surroundings, leading to a change of the internal quantum state of the molecule.

For future studies of collisions between trapped molecules, a quantitative understanding of all trap loss mechanisms is essential. The trap losses due to optical pumping by blackbody radiation and due to collisions with the background gas have been studied by monitoring the population decay of OH and OD radicals in a room-temperature electrostatic trap.[62] By comparing two isotopes of the same molecular species under otherwise identical conditions, both trap loss mechanisms, which are coupled as they both depend on the temperature, could be disentangled and quantified. The optical pumping rate by room-temperature blackbody radiation was determined as $0.49\ s^{-1}$ for the OH radical and $0.16\ s^{-1}$ for the OD radical. Trap loss due to blackbody radiation is thus a major limitation for the room-temperature trapping of OH radicals. The trapped molecules would have to be shielded from thermal radiation if longer trapping times are required. Most polar molecules exhibit strong electric dipole allowed rovibrational transitions within the room temperature blackbody spectral region. In Table 9.1 the calculated blackbody pumping rates out of the

Table 9.1. Pumping rates due to blackbody radiation at two different temperatures, out of the specified initial state, for a number of polar molecules (Reproduced from S. Hoekstra *et al.*[62])

| System | Initial state | Pumping rate ($s^{-1}$) | |
|---|---|---|---|
| | | 295 K | 77 K |
| OH/OD | $X^2\Pi_{3/2}, J = \frac{3}{2}$ | 0.49/0.16 | 0.058/0.027 |
| NH/ND | $a^1\Delta, J = 2$ | 0.36/0.12 | 0.042/0.021 |
| NH/ND | $X^3\Sigma^-, N = 0, J = 1$ | 0.12/0.036 | 0.025/0.083 |
| $NH_3/ND_3$ | $\tilde{X}^1A_1', J = 1, \lvert K\rvert = 1$ | 0.23/0.14 | 0.019/0.0063 |
| SO | $X^3\Sigma^-, N = 0, J = 1$ | 0.01 | $< 10^{-3}$ |
| $^6LiH/^6LiD$ | $X^1\Sigma^+, J = 1$ | 1.64/0.81 | 0.31/0.11 |
| CaH/CaD | $X^2\Sigma^+, N = 0, J = \frac{1}{2}$ | 0.048/0.063 | $0.0032/< 10^{-3}$ |
| RbCs | $X^1\Sigma^+, J = 0$ | $< 10^{-3}$ | $< 10^{-3}$ |
| KRb | $X^1\Sigma^+, J = 0$ | $< 10^{-3}$ | $< 10^{-3}$ |
| CO | $a^3\Pi_{1,2}, J = 1, 2$ | 0.014/0.014 | $< 10^{-3}/ < 10^{-3}$ |

specified initial quantum state is given for a number of polar molecules for which trapping is being pursued using the currently available techniques.

## 9.4. Applications of Decelerated Beams and Trapped Molecules

As already mentioned in the Introduction, the 3D focused packets of decelerated molecules with their tunable velocity and their narrow velocity spread can, for instance, be used for high-resolution spectroscopy and metrology, taking advantage of the increased interaction times. These beams also enable novel molecular beam collision studies, in particular the study of (in)elastic collision and scattering processes as a function of collision energy. Trapped samples of neutral molecules can be used to measure lifetimes of metastable states, which is of particular relevance for free radicals, for which basically no other reliable methods exist to obtain this information. In the following sections, examples of these three applications are given and future prospects are discussed.

### 9.4.1. *High Resolution Spectroscopy and Metrology*

Ultimately, the precision of any spectroscopic measurement is limited by the interaction time of the particle to be investigated with the radiation field. The interaction time that can be obtained in a normal molecular beam setup is, at best, in the millisecond range. By decreasing the velocity of the molecular beam, the interaction time and thus the obtainable resolution, can be increased by orders of magnitude. We demonstrate this here, using the proof-of-principle microwave experiments on a slow beam of ammonia molecules.[20] Similar measurements have also been performed on decelerated OH radicals.[21]

The inversion spectrum of the $|J, K\rangle = |1, 1\rangle$ level of $^{15}ND_3$ consists of 72 hyperfine transitions in a frequency interval of about 300 kHz. Due to this spectral congestion, the hyperfine structure could not be resolved in an earlier molecular beam experiment.[61] We have therefore repeated this experiment using a Stark decelerated molecular beam, using the set-up schematically shown in Fig. 9.19. A beam of ammonia molecules is decelerated from 280 m/s to either 100 m/s or 50 m/s, and focused into a

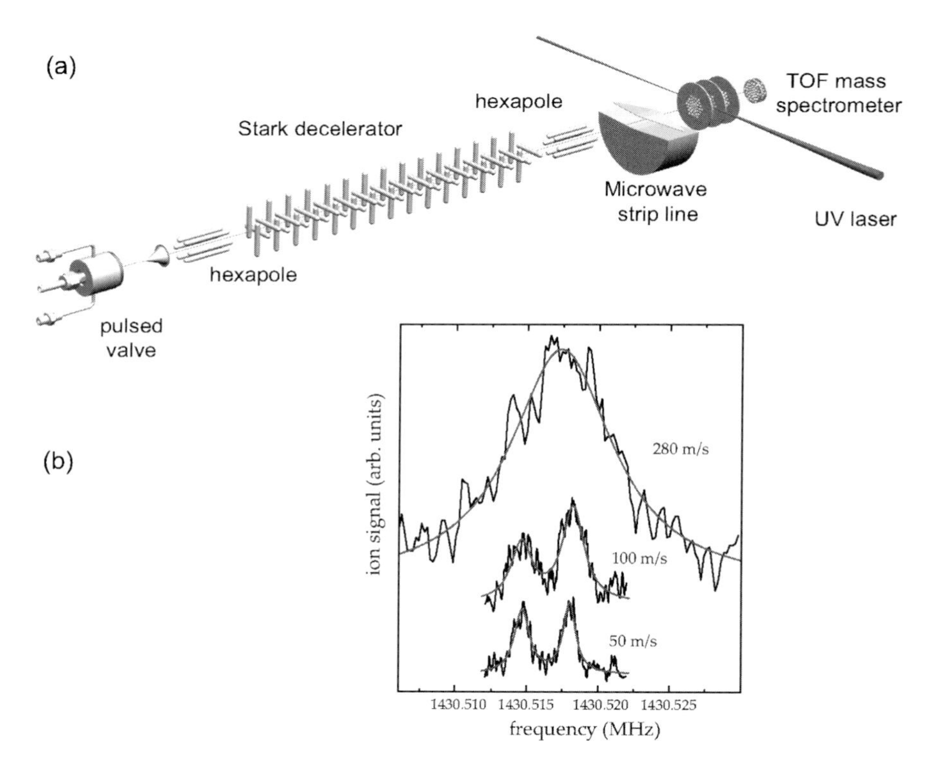


Fig. 9.19. (a) Experimental setup. A beam of $^{15}ND_3$ molecules is decelerated from 280 m/s to either 100 m/s or 50 m/s, sent through a microwave zone, and subsequently detected. (b) Upon reducing the beam velocity, the transit time limited linewidth of the microwave transitions decreases, reaching about 1 kHz at 50 m/s. (Reproduced from J. van Veldhoven *et al.*[20] with permission. Copyright EDP Sciences.)

microwave zone. The microwave zone provides a nearly rectangular field strength distribution along the molecular beam axis with a width of about 65 mm. This field drives the molecules, initially in the upper inversion level, into the lower inversion level. Shortly behind the microwave zone, molecules in the lower inversion level of the $|J, K\rangle = |1, 1\rangle$ state are detected. The lower part of Fig. 9.19 shows a small fraction of the recorded inversion spectrum using a beam of 280 m/s (upper trace), 100 m/s (middle trace) or 50 m/s (lower trace). The broad line observed in the measurements taken at 280 m/s, is fully resolved in the measurements on the decelerated beam. At 50 m/s the transit time limited linewidth is about 1 kHz, sufficient to resolve all individual hyperfine transitions. In this way we were able to determine the absolute energy of all 22 hyperfine levels in this system with an accuracy of better than 100 Hz.[20]

The accuracy can be further increased by using a longer microwave zone, slower molecules, or both. At some point, however, beam deflection due to the gravitational pull of the earth becomes an issue. A way to deal with this is to use a vertical setup, *e.g.* a molecular fountain. Such a fountain is currently being set up at the Laser Centre Vrije Universiteit in Amsterdam, in collaboration with the Fritz-Haber-Institut. In this fountain, ammonia molecules are decelerated, cooled, and subsequently launched. The molecules fly upwards some 10–50 cm before falling back under gravity, thereby passing a microwave cavity twice — as they fly up and as they fall back down. The effective interrogation time in such a Ramsey type measurement scheme includes the entire flight time between the two traversals through the driving field, which can be up to a second. This should make it possible to obtain a precision of $10^{-12}$–$10^{-14}$.

In the near future, it is unlikely that the precision of molecular frequency standards will compete with that obtained with their atomic counterparts. Nevertheless, precision measurements on molecules can be used as stringent tests of fundamental theories in physics. The sensitivity of any experiment on looking for a frequency shift due to a certain physical phenomenon depends both on the size of the shift, i.e. the inherent sensitivity of the atom or molecule, and on the ability to measure this shift. Although the precision obtained in atoms is far better than the precision obtained in molecules, in a number of cases this is compensated by the fact that the structure (and symmetry) of molecules makes them inherently more sensitive. For instance, in certain molecules like YbF and PbO, time-symmetry violating interactions leading to a permanent electric dipole moment (EDM) of the electron are three orders of magnitude stronger than in atoms.[63] The possibility of improving the resolution of the EDM experiment by the use of cold molecules, has been an important motivation for the work on AG deceleration.[41] Molecules are also used in the search for a difference in transition frequency between chiral molecules that are each other's mirror-image,[64] and also here the use of slow molecules would offer great advantages. Ammonia, in its various isotopomers, might be ideally suited to test the time-variation of the proton-to-electron mass ratio; the inversion frequency in ammonia is determined by the tunneling rate of the protons through the barrier between the two equivalent configurations of the molecule, and is exponentially dependent on the reduced mass, which is closely linked to the proton-to-electron mass ratio.

### 9.4.2. *Collision Studies at a Tunable Collision Energy*

Generally, a Stark decelerator offers new possibilities in all those experiments in which the velocity (distribution) of the beam is an important parameter. Arguably, one of the most interesting applications of Stark decelerated beams is in scattering experiments, where the velocity tunability of the beam can be exploited to accurately measure scattering resonances for instance. Studies of this kind have thus far been performed by crossing molecular beams under a variable angle.[65] With Stark decelerated beams, these experiments can be performed with an unprecedented energy resolution in a fixed experimental geometry. As the deceleration process is quantum-state specific, the bunches of slow molecules that emerge from the decelerator are extremely pure, which is of particular importance for inelastic collision studies. Moreover, the decelerated molecules are all naturally spatially oriented, allowing steric effects to be studied.

The first crossed-beam scattering experiments using Stark-decelerated beams have been performed on the OH-Xe system.[22] By changing the velocity of the OH radicals from 33 m/s to 700 m/s, while the velocity of the Xe atoms is kept fixed at 320 m/s, the total center-of-mass collision energy in this system is varied from $50\,\mathrm{cm}^{-1}$ to $400\,\mathrm{cm}^{-1}$. The collision energy is thereby varied over the energetic thresholds for inelastic scattering into the first excited rotational levels of the OH radical.

In Fig. 9.20 the measured relative inelastic cross sections are shown as a function of the center-of-mass collision energy for scattering into four different inelastic channels, indicated by arrows in the energy level scheme. The largest cross section is observed for scattering into the $(X\,^2\Pi_{3/2}, J = 3/2, e)$ state. This $\Lambda$-doublet changing collision is the only exo-energetic channel, and the relative cross section for this channel therefore approaches 100% at low collision energies. The other channels show a clear threshold behavior. These measurements provide a very sensitive probe for theoretical potential energy surfaces, from which a detailed understanding of the collision dynamics can be obtained. The solid curves that are shown in the

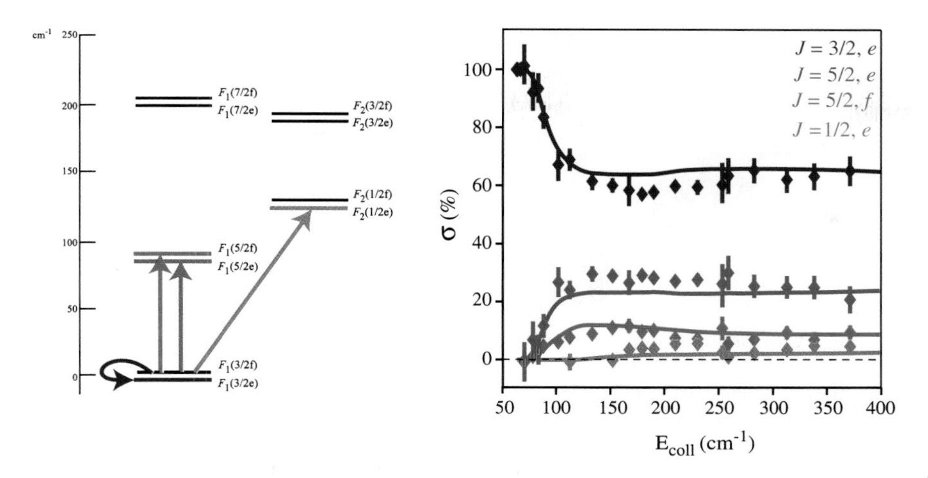


Fig. 9.20. Relative inelastic cross sections for scattering of OH ($X\,^2\Pi_{3/2}, J = 3/2, f$) radicals with Xe atoms as a function of the center-of-mass collision energy. As depicted in the energy level scheme, collisions populating the ($X\,^2\Pi_{3/2}, J = 3/2, e$), ($X\,^2\Pi_{3/2}, J = 5/2, e$), ($X\,^2\Pi_{3/2}, J = 5/2, f$), and ($X\,^2\Pi_{1/2}, J = 1/2, e$) states have been studied. The measurements are indicated by points, whereas the solid lines are the theoretically predicted curves. (Figure partially reproduced from J.J. Gilijamse *et al.*[22] with permission. Copyright AAAS.)

Figure are the result of quantum scattering calculations, which are seen to be in excellent agreement with the experiment.

In the OH-Xe experiment, the energy resolution of (only) about $13\,\mathrm{cm}^{-1}$ is almost exclusively determined by the relatively large longitudinal velocity spread in the Xe beam. A major improvement in the energy resolution is obtained when both scattering partners are fully under control. For this, a new crossed-beam scattering machine, consisting of two Stark decelerators at a 90 degree crossing angle, is currently being built at the Fritz-Haber-Institut. An artist's impression of this machine is shown in Fig. 9.21. Each decelerator consists of 300 electric field stages. With this relatively large number of stages, the efficiency of the deceleration process can be optimized, such that the particle densities required for scattering studies are obtained in the interaction region.

In this machine molecular inelastic or reactive scattering can be studied between combinations of neutral polar molecules like OH, $NH(a^1\Delta)$, $CO(a^3\Pi)$, $ND_3$, $SO_2$, and $H_2CO$. The collision energy can be continuously varied in the $1\,\mathrm{cm}^{-1}$ to $500\,\mathrm{cm}^{-1}$ range, with an overall energy resolution down to better than $1\,\mathrm{cm}^{-1}$. This should enable the accurate mapping of scattering resonances, for instance, thereby providing sensitive tests of the molecular potential energy surfaces.

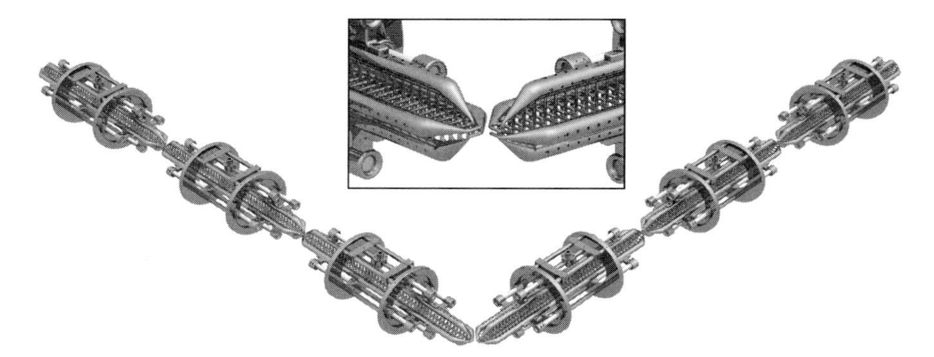

Fig. 9.21. Scheme of the crossed beam scattering machine. Each decelerator has a modular design and consists in total of 300 electric field stages. As shown in the inset, the decelerators are designed such that the exits of both decelerators are very close to the collision zone, while simultaneously providing excellent optical access for detection of the scattered products (design: Henrik Haak).

In the crossed beam geometry, the particles encounter each other only once. Alternatively, the decelerated beams can be loaded into a molecular synchrotron, positioned at the intersection point of the two beams. In a synchrotron containing 20 packets revolving in both directions, a packet having completed 100 round trips will have had 4000 encounters. We are currently constructing a 40 segment synchrotron, designed such that it can indeed be combined with the crossed beam setup, i.e. a molecular synchrotron as a future collider for neutral polar molecules.

### 9.4.3. *Direct Lifetime Measurements of Metastable States*

The long observation time afforded by a trap can be exploited to directly measure the radiative lifetime of excited rovibrational or electronic states with high precision. To demonstrate this, we have electrostatically trapped OH radicals in the first vibrationally excited $X\,^2\Pi_{3/2}, v = 1, J = 3/2$ state, and monitored the temporal decay of the trapped molecules.[27] The measured population in the vibrationally excited state as a function of the storage time in the trap is shown in the left panel of Fig. 9.22. From the observed exponential decay a radiative lifetime of this state of $59.0 \pm 2.0$ ms is deduced, in good agreement with the calculated value of 58.3 ms.[66] This experiment benchmarks the Einstein $A$ coefficients in the important Meinel system of OH.

The same experimental approach has also been applied to accurately measure the electronic lifetimes of CO molecules in the metastable $a\,^3\Pi$ state. CO molecules in this state can only decay to the $X^1\Sigma^+$ electronic

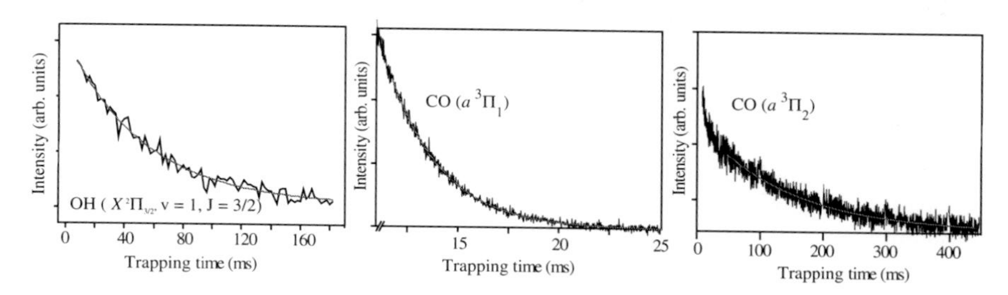


Fig. 9.22. Measured population of trapped OH ($X\,^2\Pi_{3/2}, v = 1, J = 3/2$) radicals (left panel), CO ($a\,^3\Pi_1(v = 0, J = 1)$) molecules (middle panel) and CO ($a\,^3\Pi_2(v = 0, J = 2)$) molecules (right panel) as a function of the trapping time. (Figure partially reproduced from S.Y.T. van de Meerakker *et al.*[27] and from J.J. Gilijamse *et al.*[67] with permission. Copyrights the American Physical Society and the American Institute of Physics.)

ground state (the Cameron bands), and this spin-forbidden transition is weakly allowed because of spin-orbit mixing of the $a^3\Pi$ state with $^1\Pi$ states.[68] The latter process makes the lifetime of the $a^3\Pi$ state strongly quantum state dependent. To measure these quantum state specific lifetimes, the CO molecules are laser prepared in either the $a\,^3\Pi_1(v = 0, J = 1)$ state or in the $a\,^3\Pi_2(v = 0, J = 2)$ state, both of which can be decelerated and trapped. The measured trap decay curves are shown in the middle and right panel of Fig. 9.22, yielding lifetimes of $2.63 \pm 0.02$ ms and $143 \pm 4$ ms for the $a\,^3\Pi_1(v = 0, J = 1)$ and the $a\,^3\Pi_2(v = 0, J = 2)$ states, respectively. Although the absolute values of these lifetimes were not accurately known before these measurements, the ratio of these lifetimes follows from the known energy levels of the $a^3\Pi$ state with spectroscopic precision as 1:54.7. The independently determined lifetimes of the two different quantum states are seen to agree perfectly with this ratio.[67]

## 9.5. Conclusions and Outlook

The merging of molecular beam methods with those of accelerator physics has yielded new tools to manipulate the motion of molecules. Over the last few years decelerators, lenses, bunchers, traps, storage rings and a synchrotron for neutral molecules have been demonstrated. Molecular beams with a tunable velocity and with a tunable width of the velocity distribution can now be produced. We expect this new molecular beam technology to become a valuable tool in a variety of chemical physics experiments, ranging from ultra-high resolution spectroscopy to crossed beam (reactive) scattering experiments. The decelerated beams have enabled the loading of molecules into a variety of traps. In these traps, electric fields are used to keep the molecules confined in a region of space where they can be studied in complete isolation from the environment. This enables the investigation of molecular properties in unprecedented detail.

The Stark deceleration and trapping technique can be applied to any polar molecule, provided it has a sufficiently large Stark shift in experimentally attainable electric fields. In Tables 9.2 and 9.3 selections of polar molecules with their relevant properties are listed that are suited for deceleration and trapping in low-field seeking and high-field seeking states, respectively. Both the production in a molecular beam and the detection methods for these molecules are well known. Molecules that have already been Stark decelerated and/or trapped are indicated, and the reference to the original experiments is given.

The trapping of neutral polar molecules holds great promise in the investigation of molecular interactions and quantum collective effects at ultralow temperatures. For this, the phase-space density needs to be further increased, i.e. the number density needs to be made higher and/or the temperature needs to be reduced. The most straightforward way to increase the number density of trapped molecules would be the accumulation of several packets of molecules in the trap. Simply re-loading the trap, however, requires opening up the trapping potential thereby losing or heating the molecules that are already stored. Two different schemes that work specifically for the NH radical[77] and the SO molecule,[73] respectively, have been proposed to circumvent this fundamental obstacle. The phase-space density of the trapped gas can also be increased by decreasing the temperature of the molecules. Various cooling schemes have been proposed to achieve temperatures below 1 mK. The most promising scheme is *sympathetic cooling*, in which the cold molecules are brought into contact with an ultracold atomic gas and equilibrate with it via elastic collisions. Most suited for sympathetic cooling are high-field seeking molecules that are trapped in an AC trap, eliminating possible inelastic collision channels that lead to trap loss. Experiments to spatially overlap magnetically trapped atoms with molecules in an AC trap are underway.[60]

When further cooling of the trapped molecules is achieved, a wealth of new experiments become possible. At sufficiently low temperatures, the de Broglie wavelength of the molecules becomes comparable to, or even larger than, the inter-particle separation. In this exotic regime, quantum degenerate effects dominate the dynamics of the particles, and a Bose Einstein condensate can be formed. Particularly interesting for these experiments is the presence of a permanent electric dipole moment in the molecules. The anisotropic, long-range, dipole-dipole interaction is predicted to give rise to new and rich physics in these cold dipolar gases.[78]

Table 9.2. A selection of polar molecules suited for deceleration and trapping experiments, with their relevant properties (see also Ref. 29). The Stark shift at 200 kV/cm is indicated. If the Stark energy reaches a maximum at smaller field strengths, indicated by the symbol *, the maximum Stark shift is given.

| Molecule | State | Stark dec. (√) and trapped (†) | Multi-plicity of hyperfine levels | Stark shift at 200 kV/cm ($cm^{-1}$) | # Stages required in present setup | Dipole moment (D) |
|---|---|---|---|---|---|---|
| CH | $\lvert X^2\Pi_{1/2}, J = 1/2, M\Omega = -1/4\rangle$ | | 4 | 1.54 | 90 | 1.46 |
| | $\lvert X^2\Pi_{3/2}, J = 3/2, M\Omega = -9/4\rangle$ | | 4 | 1.88 | 71 | |
| CF | $\lvert X^2\Pi_{1/2}, J = 1/2, M\Omega = -1/4\rangle$ | | 4 | 0.44 | 533 | 0.65 |
| | $\lvert X^2\Pi_{1/2}, J = 3/2, M\Omega = -3/4\rangle$ | | 4 | 0.32 | 845 | |
| $CH_2F_2$ | $\lvert J_\tau M\rangle = \lvert 2_{-2}0\rangle$ | | | 1.52* | 168 | 1.96 |
| $CH_3F$ | $\lvert JKM\rangle = \lvert 100\rangle$ | | | 0.54* | 217 | 1.86 |
| | $\lvert JKM\rangle = \lvert 1 \pm 1 \mp 1\rangle$ | | | 1.05* | 99 | |
| CO | $\lvert a^3\Pi_{\Omega=1}, J = 1, M\Omega = -1\rangle$ | √[12] †[67] | 2 | 1.71 | 89 | 1.37 |
| | $\lvert a^3\Pi_{\Omega=2}, J = 2, M\Omega = -4\rangle$ | √[67] †[67] | 2 | 2.87 | 63 | |
| $H_2CO$ | $\lvert J_\tau M\rangle = \lvert 1_1 1\rangle$ | √[69] | 6 | 1.44* | 130 | 2.34 |
| $D_2CO$ | $\lvert J_\tau M\rangle = \lvert 1_1 1\rangle$ | | 6 | 1.11* | 155 | 2.34 |
| $H_2O$ | $\lvert J_\tau M\rangle = \lvert 1_1 1\rangle$ | | 6 | 0.45 | 1081 | 1.82 |
| $D_2O$ | $\lvert J_\tau M\rangle = \lvert 1_1 1\rangle$ | | 6 | 0.72 | 667 | 1.85 |
| HDO | $\lvert J_\tau M\rangle = \lvert 1_1 1\rangle$ | | 12 | 0.73 | 558 | 1.85 |
| HCN | $\lvert (v_1, v_2^l, v_3), J, M\rangle = \lvert (0, 0^0, 0), 1, 0\rangle$ | | 6 | 0.92* | 177 | 3.01 |
| | $\lvert (v_1, v_2^l, v_3), J, Ml\rangle = \lvert (0, 1^1, 0), 1, -1\rangle$ | | 12 | 1.80* | 79 | |
| | $\lvert (v_1, v_2^l, v_3), J, Ml\rangle = \lvert (0, 1^1, 0), 2, -1\rangle$ | | 12 | 2.58 | 150 | |
| | $\lvert (v_1, v_2^l, v_3), J, Ml\rangle = \lvert (0, 1^1, 0), 2, -2\rangle$ | | 12 | 1.66* | 109 | |

(*Continued*)

Table 9.2. (Continued)

| Molecule | State | Stark dec. (√) and trapped (†) | Multiplicity of hyperfine levels | Stark shift at 200 kV/cm ($cm^{-1}$) | # Stages required in present setup | Dipole moment (D) |
|---|---|---|---|---|---|---|
| LiH | $\|X^1\Sigma^+, J=1, M=0\rangle$ | | 8 | 3.11 | 45 | 5.88 |
| LiD | $\|X^1\Sigma^+, J=1, M=0\rangle$ | | 12 | 2.67 | 34 | |
| NH | $\|a\ ^1\Delta, J=2, M=2\rangle$ | √[70] †[71] | 12 | 3.34 | 48 | 1.49 |
| $^{14}NH_3$ | $\|JKM\rangle = \|1 \pm 1 \mp 1\rangle$ | √[29] | | 2.11 | 79 | 1.47 |
| $^{15}NH_3$ | $\|JKM\rangle = \|1 \pm 1 \mp 1\rangle$ | | | 2.11 | 84 | 1.47 |
| $^{14}ND_3$ | $\|JKM\rangle = \|1 \pm 1 \mp 1\rangle$ | √[26] †[26] | 48 | 2.29 | 65 | 1.50 |
| $^{15}ND_3$ | $\|JKM\rangle = \|1 \pm 1 \mp 1\rangle$ | √[29] †[29, 52] | 32 | 2.29 | 68 | 1.50 |
| NO | $\|X^2\Pi_{1/2}, J=1/2, M\Omega=-1/4\rangle$ | | 6 | 0.17 | 1179 | 0.16 |
| $N_2O$ | $\|(v_1, v_2^l, v_3), J, Ml\rangle = \|(0,1^1,0),1,-1\rangle$ | | 18 | 0.26 | 1041 | 0.17 |
| OCS | $\|(v_1, v_2^l, v_3), J, M\rangle = \|(0,0^0,0),1,0\rangle$ | | 1 | 0.13* | 1172 | 0.72 |
| | $\|(v_1, v_2^l, v_3), J, M\rangle = \|(0,0^0,0),2,0\rangle$ | | 1 | 0.43 | 1407 | |
| | $\|(v_1, v_2^l, v_3), J, Ml\rangle = \|(0,1^1,0),1,-1\rangle$ | | 2 | 0.25* | 647 | 0.70 |
| | $\|(v_1, v_2^l, v_3), J, Ml\rangle = \|(0,1^1,0),2,-1\rangle$ | | 2 | 0.56 | 759 | |
| | $\|(v_1, v_2^l, v_3), J, Ml\rangle = \|(0,1^1,0),2,-2\rangle$ | | 2 | 0.24* | 779 | |
| OH | $\|X^2\Pi_{3/2}, J=3/2, M\Omega=-9/4\rangle$ | √[72] †[50, 53] | 4 | 3.22 | 56 | 1.67 |
| OD | $\|X^2\Pi_{3/2}, J=3/2, M\Omega=-9/4\rangle$ | √[62] †[62] | 6 | 3.22 | 58 | 1.65 |
| SH | $\|X\ ^2\Pi_{3/2}, J=3/2, M\Omega=-9/4\rangle$ | | 4 | 1.51 | 227 | 0.76 |
| SD | $\|X\ ^2\Pi_{3/2}, J=3/2, M\Omega=-9/4\rangle$ | | 6 | 1.49 | 235 | 0.76 |
| $SO_2$ | $\|J_\tau M\rangle = \|1_0 0\rangle$ | √[73 ] | 1 | 1.47 | 343 | 1.59 |

Table 9.3. A selection of molecules suited for AG deceleration and AC trapping, with their relevant properties (see also Ref. 39).

| Molecule | State | AG dec. (√) or trapped (†) | Stark shift ($cm^{-1}$) at 100 kV/cm | Effective dipole ($cm^{-1}$/kV/cm) at 100 kV/cm | Rotational constants ($cm^{-1}$) A/B/C | Mass (amu) |
|---|---|---|---|---|---|---|
| CO ($a^3\Pi_1$) | $\|J=1, M\Omega=+1\rangle$ | √[40] | −1.25 | 0.0135 | −/1.68/− | 28 |
| OH ($X^2\Pi_{3/2}$) | $\|J=3/2, M\Omega=+9/4\rangle$ | | −1.62 | 0.0191 | −/18.52/− | 17 |
| CaF | $\|J=1/2, M\Omega=+1/4\rangle$ | √[75] | −3.43 | 0.0420 | −/0.34/− | 59 |
| YbF ($X^2\Sigma^+$) | $\|J=1/2, M\Omega=+1/4\rangle$ | √[41] | −4.91 | 0.0569 | −/0.24/− | 193 |
| $^{15}ND_3$ | $\|J=1, MK=+1\rangle$ | †[59, 76] | −1.27 | 0.0134 | −/5.14/3.12 | 20 |
| pyridazine ($C_4H_5N$) | $\|J_{K_aK_c}\|M\|\rangle = \|0_{00}0\rangle$ | | −5.59 | 0.0624 | 0.21/0.20/0.10 | 80 |
| benzonitrile ($C_7H_5N$) | $\|J_{K_aK_c}\|M\|\rangle = \|0_{00}0\rangle$ | √[74] | −6.71 | 0.0711 | 0.19/0.051/0.040 | 103 |
| tryptophan ($C_{11}H_{12}N_2O_2$) I | $\|J_{K_aK_c}\|M\|\rangle = \|0_{00}0\rangle$ | | −6.25 | 0.0646 | 0.041/0.013/0.012 | 216 |
| II | | | −4.72 | 0.0494 | 0.039/0.014/0.012 | |
| III | | | −1.71 | 0.0183 | 0.033/0.017/0.013 | |
| IV | | | −11.68 | 0.120 | 0.032/0.016/0.013 | |
| V | | | −12.28 | 0.126 | 0.043/0.011/0.0096 | |
| VI | | | −11.37 | 0.116 | 0.045/0.011/0.0095 | |

The different tools that were developed in the past to transversally manipulate molecular beams, dating back to the Stern and Rabi era, have proven to be crucial for developments beyond molecular physics alone. The complete control over the full three-dimensional motion of molecules, now adds a new dimension to the long and rich history of the manipulation of molecules with electric fields.

## Acknowledgements

The experiments that are described in this Chapter are the result of almost ten years of research efforts by a large group of people. These efforts started at the University of Nijmegen, the Netherlands. In the year 2000, the group moved to the FOM Institute for Plasmaphysics "Rijnhuizen" in Nieuwegein, the Netherlands, and it moved again in 2003 to its present home at the Fritz-Haber-Institut der Max-Planck-Gesellschaft in Berlin, Germany. We are greatly indebted to the technical and scientific staff at all three institutions. In particular, we thank all the students, postdocs, senior scientists and research technicians that have been involved in this work, and without whom these experiments would not have been possible.

## References

1. Scoles G. (ed). *Atomic and molecular beam methods.* vol. 1 & 2, (Oxford University Press, New York, NY, USA, 1988 & 1992). ISBN 0195042808.
2. Friedrich B, Herschbach D. (2003) Stern and Gerlach: How a bad cigar helped reorient atomic physics. *Phys. Today.* **56**: 53–59.
3. Gerlach W, Stern O. (1922) The experimental evidence of direction quantisation in the magnetic field. *Zeitschrift Für Physik.* **9**: 349–352.
4. Rabi II, Millman S, Kusch P, Zacharias JR. (1939) The molecular beam resonance method for measuring nuclear magnetic moments. *Phys. Rev.* **55**: 526–535.
5. Friedburg H, Paul W. (1951) Optische Abbildung mit neutralen Atomen. *Die Naturwissenschaften* **38**: 159–160.
6. Bennewitz H, Paul W. (1954) Eine Methode zur Bestimmung von Kernmomenten mit fokussiertem Atomstrahl. *Z. Phys.* **139**: 489–497.
7. Bennewitz HG, Paul W, Schlier C. (1955) Fokussierung polarer Moleküle. *Z. Phys.* **141**: 6–15.
8. Gordon JP, Zeiger HJ, Townes CH. (1954) Molecular microwave oscillator and new hyperfine structure in the microwave spectrum of $NH_3$. *Phys. Rev.* **95**: 282–284.
9. Gordon JP, Zeiger HJ, Townes CH. (1955) The maser — New type of microwave amplifier, frequency standard, and spectrometer. *Phys. Rev.* **99**: 1264–1274.

10. Levine R, Bernstein R. (1987) *Molecular reaction dynamics and chemical reactivity*, Oxford University Press, New York.
11. Ramsey NF. http://nobelprize.org/nobel_prizes/physics/laureates/1989/ramsey-autobio.html.
12. Bethlem HL, Berden G, Meijer G. (1999) Decelerating neutral dipolar molecules. *Phys. Rev. Lett.* **83**: 1558–1561.
13. Rangwala SA, Junglen T, Rieger T, Pinkse PWH, Rempe G. (2003) Continuous source of translationally cold dipolar molecules. *Phys. Rev. A.* **67**: 043406.
14. Crompvoets FMH, Bethlem HL, Jongma RT, Meijer G. (2001) A prototype storage ring for neutral molecules. *Nature* **411**: 174.
15. Bethlem HL, Berden G, van Roij AJA, Crompvoets FMH, Meijer G. (2000) Trapping neutral molecules in a traveling potential well. *Phys. Rev. Lett.* **84**: 5744–5747.
16. Wolfgang R. (1968) Chemical accelerators. *Sci. Am.* **219**(4): 44.
17. Golub R. (1967) *On decelerating molecules.* PhD thesis, MIT, Cambridge, USA.
18. Bromberg EEA. (1972) *Acceleration and alternate-gradient focusing of neutral polar diatomic molecules.* PhD thesis, University of Chicago, USA.
19. Abuaf N, Andres JBARP, Fenn JB, Marsden DGH. (1967) Molecular beams with energies above one volt. *Science* **155**: 997–999.
20. van Veldhoven J, Küpper J, Bethlem HL, Sartakov B, van Roij AJA, Meijer G. (2004) Decelerated molecular beams for high-resolution spectroscopy: The hyperfine structure of $^{15}ND_3$. *Eur. Phys. J. D.* **31**(2): 337–349.
21. Hudson ER, Lewandowski HJ, Sawyer BC, Ye J. (2006) Cold molecule spectroscopy for constraining the evolution of the fine structure constant. *Phys. Rev. Lett.* **96**(14): 143004.
22. Gilijamse JJ, Hoekstra S, van de Meerakker SYT, Groeneboom GC, Meijer G. (2006) Near-threshold inelastic collisions using molecular beams with a tunable velocity. *Science* **313**: 1617–1620.
23. Paul W. (1990) Electromagnetic traps for charged and neutral particles. *Angew. Chem. Int. Ed. Engl.* **29**: 739–748.
24. Kügler K-J, Paul W, Trinks U. (1978) A magnetic storage ring for neutrons. *Phys. Lett. B.* **72**: 422–424.
25. Migdall AL, Prodan JV, Phillips WD, Bergeman TH, Metcalf HJ. (1985) First observation of magnetically trapped neutral atoms. *Phys. Rev. Lett.* **54**: 2596.
26. Bethlem HL, Berden G, Crompvoets FMH, Jongma RT, van Roij AJA, Meijer G. (2000) Electrostatic trapping of ammonia molecules. *Nature* **406**: 491–494.
27. van de Meerakker SYT, Vanhaecke N, van der Loo MPJ, Groenenboom GC, Meijer G. (2005) Direct measurement of the radiative lifetime of vibrationally excited OH radicals. *Phys. Rev. Lett.* **95**(1): 013003.
28. van de Meerakker SYT, Vanhaecke N, Meijer G. (2006) Stark deceleration and trapping of OH radicals. *Ann. Rev. Phys. Chem.* **57**: 159–190.
29. Bethlem HL, Crompvoets FMH, Jongma RT, van de Meerakker SYT, Meijer G. (2002) Deceleration and trapping of ammonia using time-varying electric fields. *Phys. Rev. A.* **65**(5): 053416.

30. van de Meerakker SYT, Vanhaecke N, Bethlem HL, Meijer G. (2005) Higher-order resonances in a Stark decelerator. *Phys. Rev. A.* **71**(5): 053409.
31. Gubbels K, Meijer G, Friedrich B. (2006) Analytic wave model of Stark deceleration dynamics. *Phys. Rev. A.* **73**: 063406.
32. van de Meerakker SYT, Vanhaecke N, Bethlem HL, Meijer G. (2006) Transverse stability in a Stark decelerator. *Phys. Rev. A.* **73**(2): 023401.
33. Bochinski JR, Hudson ER, Lewandowski HJ, Ye J. (2004) Cold free-radical molecules in the laboratory frame. *Phys. Rev. A.* **70**(4): 043410.
34. Crompvoets FMH, Jongma RT, Bethlem HL, van Roij AJA, Meijer G. (2002) Longtudinal focusing and cooling of a molecular beam. *Phys. Rev. Lett.* **89**(9): 093004.
35. Heiner CE, Bethlem HL, Meijer G. (2006) Molecular beams with a tunable velocity. *Phys. Chem. Chem. Phys.* **8**: 2666–2676.
36. Crompvoets FMH, Bethlem HL, Meijer G. (2006) A storage ring for neutral molecules. *Advances in Atomic, Molecular, and Optical Physics.* **52**: 209–287.
37. Wing WH. (1984) On neutral particle trapping in quasistatic electromagnetic fields. *Prog. Quant. Electr.* **8**: 181–199.
38. Auerbach D, Bromberg EEA, Wharton L. (1966) Alternate-gradient focusing of molecular beams. *J. Chem. Phys.* **45**: 2160.
39. Bethlem HL, Tarbutt MR, Küpper J, Carty D, Wohlfart K, Hinds EA, Meijer G. (2006) Alternating gradient focusing and deceleration of polar molecules. *J. Phys. B.* **39**(16): R263–R291.
40. Bethlem HL, van Roij AJA, Jongma RT, Meijer G. (2002) Alternate gradient focusing and deceleration of a molecular beam. *Phys. Rev. Lett.* **88**(13): 133003.
41. Tarbutt MR, Bethlem HL, Hudson JJ, Ryabov VL, Ryzhov VA, Sauer BE, Meijer G, Hinds EA. (2004) Slowing heavy, ground-state molecules using an alternating gradient decelerator. *Phys. Rev. Lett.* **92**(17): 173002.
42. Yamakita Y, Procter SR, Goodgame AL, Softley TP, Merkt F. (2004) Deflection and deceleration of hydrogen Rydberg molecules in inhomogeneous electric fields. *J. Chem. Phys.* **121**: 1419.
43. Vliegen E, Wörner HJ, Softley TP, Merkt F. (2004) Nonhydrogenic effects in the deceleration of Rydberg atoms in inhomogeneous electric fields. *Phys. Rev. Lett.* **92**(3): 033005.
44. Vliegen E, Merkt F. (2006) Normal-incidence electrostatic Rydberg atom mirror. *Phys. Rev. Lett.* **97**(3): 033002.
45. Vanhaecke N, Meier U, Andrist M, Meier BH, Merkt F. (2007) Multistage Zeeman deceleration of hydrogen atoms. *Phys. Rev. A.* **75**(3): 031402.
46. Fulton R, Bishop AI, Barker PF. (2004) Optical Stark decelerator for molecules. *Phys. Rev. Lett.* **93**: 243004.
47. Fulton R, Bishop AI, Shneider MN, Barker PF. (2006) Controlling the motion of cold molecules with deep periodic optical potentials. *Nature Physics* **2**: 465–468.
48. Fulton R, Bishop AI, Shneider MN, Barker PF. (2006) Optical Stark deceleration of nitric oxide and benzene molecules using optical lattices. *J. Phys. B: At. Mol. Opt. Phys.* **39**: S1097–S1109.

49. Wing WH. (1980) Electrostatic trapping of neutral atomic particles. *Phys. Rev. Lett.* **45**: 631–634.
50. van de Meerakker SYT, Smeets PHM, Vanhaecke N, Jongma RT, Meijer G. (2005) Deceleration and electrostatic trapping of OH radicals. *Phys. Rev. Lett.* **94**: 023004.
51. Gilijamse JJ, Küpper J, Hoekstra S, Vanhaecke N, van de Meerakker SYT, Meijer G. (2006) Optimizing the Stark-decelerator beamline for the trapping of cold molecules using evolutionary strategies. *Phys. Rev. A.* **73**(6): 063410.
52. van Veldhoven J, Bethlem HL, Schnell M, Meijer G. (2006) Versatile electrostatic trap. *Phys. Rev. A.* **73**: 063408.
53. Sawyer BC, Lev BL, Hudson ER, Stuhl BK, Lara M, Bohn JL, Ye J. (2007) Magnetoelectrostatic trapping of ground state OH molecules. *Phys. Rev. Lett.* **98**: 253002.
54. Rieger T, Junglen T, Rangwala SA, Pinkse PWH, Rempe G. (2005) Continuous loading of an electrostatic trap for polar molecules. *Phys. Rev. Lett.* **95**: 173002.
55. Crompvoets FMH, Bethlem HL, Küpper J, van Roij AJA, Meijer G. (2004) Dynamics of neutral molecules stored in a ring. *Phys. Rev. A.* **69**(6): 063406.
56. Heiner CE, Carty D, Meijer G, Bethlem HL. (2007) A molecular synchrotron. *Nature Physics* **3**: 115–118.
57. Bohn JL. (2001) Inelastic collisions of ultracold polar molecules. *Phys. Rev. A.* **63**: 052714.
58. Bethlem HL, van Veldhoven J, Schnell M, Meijer G. (2006) Trapping polar molecules in an ac trap. *Phys. Rev. A.* **74**(6): 063403.
59. van Veldhoven J, Bethlem HL, Meijer G. (2005) AC electric trap for ground-state molecules. *Phys. Rev. Lett.* **94**: 083001.
60. Schlunk S, Marian A, Geng P, Mosk AP, Meijer G, Schöllkopf W. (2007) Trapping of Rb atoms by ac electric fields. *Phys. Rev. Lett.* **98**: 223002.
61. van Veldhoven J, Jongma RT, Sartakov B, Bongers WA, Meijer G. (2002) Hyperfine structure of $ND_3$. *Phys. Rev. A.* **66**(3): 32501.
62. Hoekstra S, Gilijamse JJ, Sartakov B, Vanhaecke N, Scharfenberg L, van de Meerakker SYT, Meijer G. (2007) Optical pumping of trapped neutral molecules by blackbody radiation. *Phys. Rev. Lett.* **98**(13): 133001.
63. Hudson JJ, Sauer BE, Tarbutt MR, Hinds EA. (2002) Measurement of the electron electric dipole moment using YbF molecules. *Phys. Rev. Lett.* **89**: 023003.
64. Daussy C, Marrel T, Amy-Klein A, Nguyen CT, Bordé CJ, Chardonnet C. (1999) Limit on the parity nonconserving energy difference between the enantiomers of a chiral molecule by laser spectroscopy. *Phys. Rev. Lett.* **83**: 1554–1557.
65. Macdonald R, Liu K. (1989) State-to-state integral cross sections for the inelastic scattering of $CH(X\,^2\Pi)$ + He: Rotational rainbow and orbital alignment. *J. Chem. Phys.* **91**: 821–838.
66. van der Loo MPJ, Groenenboom GC. (2007) Theoretical transition probabilities for the OH Meinel system. *J. Chem. Phys.* **126**: 114314.

67. Gilijamse JJ, Hoekstra S, Meek SA, Metsälä M, van de Meerakker SYT, Meijer G, Groenenboom GC. (2007) The radiative lifetime of metastable CO $(a\,^3\Pi, v = 0)$. *J. Chem. Phys.* **127**: 221102.
68. James T. (1971) Transition moments, Franck-Condon factors, and lifetimes of forbidden transitions. Calculation of the intensity of the Cameron system of CO. *J. Chem. Phys.* **55**: 4118–4124.
69. Hudson ER, Ticknor C, Sawyer BC, Taatjes CA, Lewandowski HJ, Bochinski JR, Bohn JL, Ye J. (2006) Production of cold formaldehyde molecules for study and control of chemical reaction dynamics with hydroxyl radicals. *Phys. Rev. A.* **73**(6): 063404.
70. van de Meerakker SYT, Labazan I, Hoekstra S, Küpper J, Meijer G. (2006) Production and deceleration of a pulsed beam of metastable NH $(a\,^1\Delta)$ radicals. *J. Phys. B.* **39**: S1077–S1084.
71. Hoekstra S, Metsälä M, Zieger P, Scharfenberg L, Gilijamse JJ, Meijer G, van de Meerakker SYT. (2007) Electrostatic trapping of metastable NH molecules. *Phys. Rev. A.* **76**: 063408.
72. Bochinski JR, Hudson ER, Lewandowski HJ, Meijer G, Ye J. (2003) Phase space manipulation of cold free radical OH molecules. *Phys. Rev. Lett.* **91**(24): 243001.
73. Jung S, Tiemann E, Lisdat C. (2006) Cold atoms and molecules from fragmentation of decelerated $SO_2$. *Phys. Rev. A.* **74**: 040701.
74. Wohlfart K, Grätz F, Filsinger F, Haak H, Meijer G, Küpper J. (2008) Alternating-gradient focusing and deceleration of large molecules. *Phys. Rev. A.* **77**: 031404.
75. Tarbutt MR. Private communications.
76. Schnell M, Lützow P, van Veldhoven J, Bethlem HL, Küpper J, Friedrich B, Schleier-Smith M, Haak H, Meijer G. (2007) A linear AC trap for polar molecules in their ground state. *J. Phys. Chem. A.* **111**: 7411–7419.
77. van de Meerakker SYT, Jongma RT, Bethlem HL, Meijer G. (2001) Accumulating NH radicals in a magnetic trap. *Phys. Rev. A.* **64**: 041401.
78. Baranov M, Dobrek L, Góral K, Santos L, Lewenstein M. (2002) Ultracold dipolar gases — a challenge for experiments and theory. *Phys. Scr.* **T102**: 74–81.

# Chapter 10

# Cold Collisions, Quantum Degenerate Gases, Photoassociation, and Cold Molecules

John Weiner

*Université Paul Sabatier Toulouse, France*
*and*
*IFSC/CePOF Universidade de São Paulo, Avenida Trabalhador São-carlense, 400-CEP 13566-590 São Carlos SP, Brazil*

## Contents



## 10.1. Introduction

Cold collisions, cold quantum gases and cold molecules occupy a strategic position at the intersection of several powerful themes of current research in chemical physics, in atomic, molecular and optical physics, and condensed matter.[1,2] The nature of these collisions has important consequences for optical manipulation of inelastic and reactive processes, precision measurement of molecular and atomic properties, matter-wave coherences and quantum-statistical condensates of dilute, weakly interacting atoms. This crucial position explains the wide interest and explosive growth of the field since its inception in 1987. Obviously due to continuing rapid developments the very latest results cannot appear in book form, and therefore the the purpose of this chapter is to outline the principal developments in ultracold photoassociation collisions and cold molecule formation that constitute an important line of continuing research. Other aspects of cold collisions such as fine- and hyperfine-structure-changing inelastic collisions, while interesting in themselves, are less important to the quest for mesoscopic quantum matter in the form of molecular quantum degenerate gases.

## 10.2. Recent History: The Scope of the Problem

In the 1980's the first successful experiments[3] and theory,[4] demonstrating that light could be used to cool and confine atoms to submillikelvin temperatures, opened several exciting new chapters in atomic, molecular, and optical (AMO) physics. Atom interferometry,[5,6] matter-wave holography,[7] optical lattices,[8] and Bose-Einstein condensation in dilute gases[9,10] all exemplified significant new physics where collisions between atoms cooled with light play a pivotal role. The nature of these collisions has become the subject of intensive study not only because of their importance to these new areas of AMO physics but also because their investigation has lead to new insights into how cold collision spectroscopy can lead to precision measurements of atomic and molecular parameters and how radiation fields can manipulate the outcome of a collision itself. As a general orientation Fig. 10.1 shows how a typical atomic de Broglie wavelength varies with temperature and where various physical phenomena situate along the scale. With de Broglie wavelengths on the order of a few thousandths of a nanometer, conventional gas-phase chemistry can usually be interpreted as the interaction of classical nuclear point particles moving along potential surfaces defined by their associated electronic charge distribution. At one time liquid helium was thought to define a regime of cryogenic physics, but it is clear from Fig. 10.1 that optical and evaporative cooling have created "cryogenic" environments below liquid helium by many orders of magnitude. At the level of Doppler cooling and optical molasses[a] the

[a] A good introduction to the early physics of laser cooling and trapping can be found in two special issues of the Journal of the Optical Society of America B. These are: "The

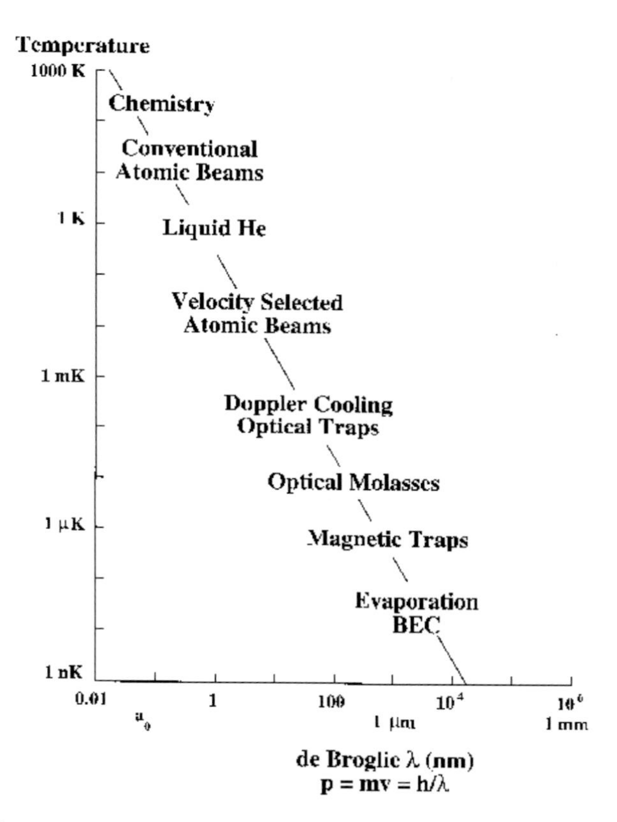


Fig. 10.1. Illustrative plot of various physical phenomena along a scale of temperature (energy divided by Boltzmann's constant $k_B$ ) plotted against de Broglie wavelength for atomic sodium.

de Broglie wavelength becomes comparable to or longer than the chemical bond, approaching the length of the cooling optical light wave. Here we can expect wave and relativistic effects such as resonances, interferences, and interaction retardation to become important. Following Suominen,[13] we will term the Doppler cooling and optical molasses temperature range, roughly between 1 mK and 1 $\mu$K, the regime of cold collisions. Most collision phenomena at this level are studied in the presence of one or more light fields used to confine the atoms and to probe their interactions. Excited quasi-molecular states often play an important role. Below about 1 $\mu$K, where

Mechanical Effects of Light," *J. Opt. Soc. Am. B* **2**(11), November 1985 and "Laser Cooling and Trapping of Atoms," *J. Opt. Soc. Am. B* **6**(11), November 1989. Two more recent reviews,[11,12] update a decade of developments since the early work recounted in the *J. Opt. Soc. Am. B* special issues.

evaporative cooling and Bose-Einstein condensation[b] (BEC) become the focus of attention, the de Broglie wavelength grows to a scale comparable to the mean distance separating atoms at the critical condensation density; quantum degenerate states of the atomic ensemble begin to appear. In this regime ground state collisions only take place through radial (not angular) motion and are characterized by a phase shift,or scattering length, of the ground-state wave function. Since the atomic translational energy now lies below the kinetic energy transferred to an atom by recoil from a scattered photon, light can play no role; and collisions occur in a temperature range from $1\,\mu\text{K} \rightarrow 0$ and in the dark.

Many early experiments were carried out in optical traps and optically slowed atomic beams at the temperature of "cold collisions". In 1986, soon after the first successful experiments reported optical cooling and trapping in alkali gasses, Vigué published a paper discussing the possible consequences of binary collisions in a cold or ultracold gaseous medium.[14] He concluded that long-range resonance dipole interaction "...cast strong doubts on the possibility of observing collective quantum effects with laser-cooled alkali-metal vapors." In retrospect we now see that he was too pessimistic. Fig. 10.2 shows schematically the general features of a cold binary collision. Two $S$ ground-state atoms interact at long range through electrostatic dispersion forces, and approach along an attractive $C_6/R^6$ potential. The first excited states, correlating to an S + P asymptote and separated from the ground state by the atomic excitation energy $\hbar\omega_0$, interact by resonance dipole-dipole forces and approach either along an attractive or repulsive $C_3/R^3$ curve. If the colliding atoms on the ground state potential encounter a *red-detuned* optical field $\hbar\omega_1$, the probability of a free-bound transition to some vibration-rotation level $v$ of the attractive excited state will maximize around the Condon point $R_C$ where $\hbar\omega_1$ matches the potential difference. The quasimolecule finds itself photoassociated and vibrating within the attractive well. A second field $\hbar\omega_2$ can then further excite or even ionize the photoassociated quasimolecule, or it can relax back to some distribution of the continuum and bound levels of the ground state with a spontaneous emission rate $\Gamma$. Photoassociation is discussed in Secs. 10.4–10.6. If the atoms colliding on the ground state potential interact with a *blue-detuned* optical field $\hbar\omega$, the probability of transition to some continuum level of the repulsive excited state will maximize around the Condon point $R_C$ which will also be quite close to the turning point of the nuclear motion on the repulsive excited state. The atoms approach no further and begin to separate along the repulsive curve leading to the S + P asymptote.

[b]For an introduction to research in alkali-atom Bose-Einstein condensation see the special issue on Bose-Einstein condensation in the Journal of Research of the National Institute of Standards and Technology **101**(4), July–August 1996.

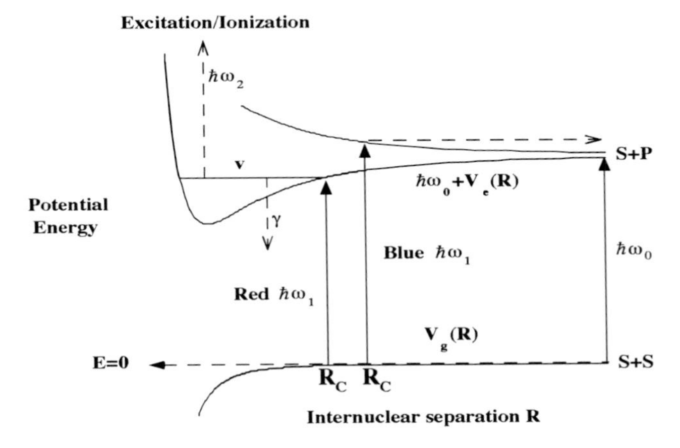


Fig. 10.2. Schematic of a cold collision. Light field $\hbar\omega_1$, red detuned with respect to the S+P asymptote, excites the quasimolecule in a free-bound transition around the Condon point $R_C$ and can lead to excitation or ionization with the absorption of a second photon $\hbar\omega_2$. Light field $\hbar\omega_1$, blue detuned with respect to the S + P asymptote, prevents atoms from approaching significantly beyond the Condon point. Blue-detuned excitation leads to optical shielding and suppression of inelastic and reactive collision rates.

The blue-detuned field "shields" the atoms from further interaction at more intimate internuclear separation and "suppresses" the rate of various inelastic and reactive processes. Cold collisions in the presence of optical fields or in the ground state reveal new physics in domains where atomic scales of length, time, and spectral line width are reversed from their conventional relations familiar to the chemical kineticist. In addition to the atomic de Broglie wave increasing to a length several hundred times that of the chemical bond, the collisional interaction time grows to several times the spontaneous emission lifetime, and the inhomogeneous Doppler line width at cold and ultracold temperatures narrows to less than the natural width of an atomic dipole transition. The narrow, near-threshold continuum state distribution means that at most a few partial waves will contribute to a scattering event. Averaging therefore does not obscure matter wave effects such as resonances, nodes and antinodes in scattering wave function amplitudes, and channel-opening threshold laws. In the cold collision regime, Doppler broadening is narrow compared to the radiative natural width; and therefore permits ultrahigh precision, free-bound molecular spectroscopy and efficient participation of the entire atom ensemble in the excitation process. Long collision duration ensures interacting partners sufficient time to exchange multiple photons with modes of an externally applied radiation field. The frequency, intensity, and polarization of the optical field can in turn modify effective interaction potentials and control the probability of inelastic, reactive, and elastic final product channels.

Three questions have motivated significant developments in cold collisions: (1) How do collisions lead to loss of atom confinement in traps? (2) How can photoassociation spectroscopy yield precision measurements of atomic properties and insight into the quantum nature of the scattering process itself? (3) How can optical fields be used to control the outcome of a collisional encounter? The first question will be largely ignored in this chapter. The answer to the second question has proved to be fruitful not only for molecular spectroscopy but has also contributed to real advances in the precision measurement of *atomic* properties.

This third question has had an enduring appeal to chemical physicists, and we review here some of the early history to put in perspective current developments and to emphasize the importance of the ultracold regime. In the 1970's, after the development of the $CO_2$ laser, researchers in atomic and chemical physics immediately thought to use it to influence or control inelastic processes and chemical reaction. However, early attempts to induce reactivity by exciting well-defined, localized molecular sites such as double bonds or functional groups failed because the initial optical excitation diffused rapidly into the rotations, vibrations, and torsional bending motions of the molecular nuclei. The unfortunate result was that the infrared light of the $CO_2$ laser essentially heated the molecules much as the familiar and venerable Bunsen burner. Enthusiasm for laser-controlled chemistry cooled when researchers realized that this rapid energy diffusion throughout the molecular skeleton blocked significant advance along the road to optical control of reactivity. In the early 1980's the development of the pulsed dye laser, tunable in the visible region of the spectrum, together with new proposals for "radiative collisions", in which the electrons of the molecule interact with the light rather than the nuclei, revived interest. A second round of experiments achieved some success in optically transforming reactants to products; but in general the high peak powers necessary to enhance reactivity significantly interfered with the desired effects by inducing nonlinear, multiphoton excitation and ionization in the molecule itself. The necessity for high peak power in turn arose from two crucial factors: (1) Doppler broadening at ambient temperature permits only a few per cent of all molecular collisions to interact with an applied optical field mode so the product "yield" is low, and (2) the optical field had to influence the strong chemical binding interaction in order to affect atomic behavior during the collisional encounter. This requirement implied the need for power densities greater than a few megawatts per square centimeter. Power densities of this order are well above the threshold for multiphoton absorption and ionization, and these processes quickly convert atoms or molecules from a neutral gas into an ionized plasma. It appeared depressingly difficult to control collisions with light without first optically destroying them.

However, atomic deceleration and optical cooling brightened this discouraging picture, by narrowing the inhomogeneous Doppler broadening to less than a natural line width of an atomic transition and transferring the optical-particle interaction from the "chemical" zone of strong wave function overlap to an outer region where weak electrostatic terms characterize the collision. In this weakly interacting outer zone only hundreds of milliwatts per square centimeter of optical power density suffice to profoundly alter the inelastic and reactive collision rate constant. Furthermore, although a conventional atomic collision lasts only a few hundred femtoseconds, very short compared to the tens of nanoseconds required before excited molecules or atoms spontaneously emit light, in the ultracold regime particles move much more slowly, taking up to hundreds of nanoseconds to complete a collisional encounter. The long collision duration leaves plenty of time for the two interacting partners to absorb energy from an external radiation field or emit energy by spontaneous or stimulated processes.

To the three earlier questions motivating studies of cold collisions may now be added a fourth relevant to the ultracold regime: What role do collisions play in the attainment and behavior of boson and fermion gases in the quantum degenerate regime? Quantum statistical effects have been observed and studied in the superfluidity of liquid helium and in the phenomena of metallic and high-temperature superconductivity. These dramatic and significant manifestations of quantum collective effects, are nevertheless difficult to study at the atomic level because the particles are condensed and strongly interacting. Observation and measurement in weakly interacting dilute gases, however, relate much more directly to the simplest, microscopic models of this behavior. The differences between model "ideal" quantum gases and "real" quantum gases begin with binary interactions between the particles, and therefore the study of ultracold collisions is a natural point of departure for investigation. Collisions determine two crucial aspects of BEC experiments: (1) the evaporative cooling rate necessary for the attainment of BEC depends on the elastic scattering cross section, proportional to the square of the $s$-wave scattering length, and (2) the sign of the scattering length indicates the stability of the condensate: positive scattering lengths lead to large stable condensates while negative scattering lengths do not. The ability to produce condensates by sympathetic cooling also depends critically on the elastic and inelastic (loss and heating) rate constants among different states of the colliding partners. Although a confined Bose atom condensate bears some analogy to an optical cavity (photons are bosons), the atoms interact through collisions; and these collisions limit the coherence length of any "atom laser" coupled out of the confining "cavity" or BEC trap. Another important point, relating back to optical control, is that the amplitude and sign of the scattering length depends sensitively on the fine details of the ground potentials. The possibility of manipulating these potentials and consequently collision rates with external means, using optical, magnetic, or radio frequency fields, holds the promise of tailoring the properties of the quantum gas to enhance stability and coherence. Finally it now appears that the limiting loss process for dilute gaseous Bose-Einstein condensates are three-body collisions which also measure the third-order coherence, thus providing a critical signature of true quantum statistical behavior. Understanding the quantum statistical collective behavior of ultracold dilute gases will drive research in ultracold collisions for years to come.

## 10.3. Cold Collision Theory

### 10.3.1. *Basic Concepts of Scattering Theory*

Let us first consider some of the basic concepts that are needed to describe the collision of two ground state atoms. We initially consider the collision of two distinguishable, structureless particles $a$ and $b$ with interaction potential $V_g(R)$ moving with relative momentum $\mathbf{k}$, where $\mathbf{R}$ is the vector connecting $a$ and $b$. We will generalize below to the cases of identical particles and particles with internal structure. The collision energy is

$$E = \frac{\hbar^2 k^2}{2\mu}$$

where $\mu$ is the reduced mass of the two particles. If there is no interaction between the particles, $V_g = 0$, the wave function describing the relative motion of the two particles in internal states $|0_a\rangle$ and $|0_b\rangle$ is:

$$\Psi_g^+(\mathbf{R}) = e^{i\mathbf{k}\cdot\mathbf{R}}|0_a 0_b\rangle. \tag{10.1}$$

If the interaction potential is nonzero, the collision between the particles results in a scattered wave, and at large $R$ beyond the range of the potential the wave function is represented as:

$$\Psi_g^+(\mathbf{R}) \sim \left\{ e^{i\mathbf{k}\cdot\mathbf{R}} + \frac{e^{ikR}}{R} f(E, \hat{\mathbf{k}}, \hat{\mathbf{k}}_\mathbf{s}) \right\} |0_a 0_b\rangle, \tag{10.2}$$

where $\hat{\mathbf{k}}$ is a unit vector indicating the direction of $\mathbf{k}$ and $\hat{\mathbf{k}}_\mathbf{s}$ is a unit vector indicating the direction of the scattered wave with amplitude $f(E, \hat{\mathbf{k}}, \hat{\mathbf{k}}_\mathbf{s})$.

The overall effect of the collision is described by a cross section $\sigma(E)$. In a gas cell, for which all directions $\hat{\mathbf{k}}$ are possible, $\sigma(E)$ is determined by integrating over all scattered directions and averaging over all values of initial $\hat{\mathbf{k}}$:

$$\sigma(E) = \int_{4\pi} \frac{d\hat{\mathbf{k}}}{4\pi} \int_{4\pi} d\hat{\mathbf{k}}_{\mathbf{s}} |f(E, \hat{\mathbf{k}}, \hat{\mathbf{k}}_{\mathbf{s}})|^2. \tag{10.3}$$

Since $f$ has units of length, the cross section has units of $(\text{length})^2 = \text{area}$. Equation (10.3) simplifies for a spherically symmetric potential, for which $f$ depends only on the angle $\theta$ between $\hat{\mathbf{k}}$ and $\hat{\mathbf{k}}_{\mathbf{s}}$:

$$\sigma(E) = 2\pi \int_0^{\pi} d\theta \sin(\theta) |f(E, \theta)|^2 \tag{10.4}$$

The object of scattering theory is to calculate the scattering amplitude and cross section, given the interaction potentials between the two atoms. The first step in reducing the problem to practical computation is to introduce the partial wave expansion of the plane wave:

$$e^{i\mathbf{k}\cdot\mathbf{R}} = 4\pi \sum_{\ell=0}^{\infty} \sum_{m=-\ell}^{\ell} i^{\ell} Y_{\ell m}^{*}(\hat{\mathbf{k}}) Y_{\ell m}(\hat{\mathbf{k}}_{\mathbf{s}}) j_{\ell}(kR), \tag{10.5}$$

where $Y_{\ell m}$ is a spherical harmonic and the function $j_\ell$ has the following form as $R \to \infty$:

$$j_\ell(kR) \sim \frac{\sin(kR - \frac{\pi}{2}\ell)}{kR}. \tag{10.6}$$

The complete wave function at all R is also expanded in a partial wave series:

$$\Psi_g^{+}(\mathbf{R}) = 4\pi \sum_{\ell=0}^{\infty} \sum_{m=-\ell}^{\ell} i^{\ell} Y_{\ell m}^{*}(\hat{\mathbf{k}}) Y_{\ell m}(\hat{\mathbf{k}}_{\mathbf{s}}) \frac{F_\ell^{+}(E, R)}{R} |0_a 0_b\rangle, \tag{10.7}$$

where $F_\ell^{+}(E, R)$ is determined from the Schrödinger equation,

$$\frac{d^2 F_\ell^{+}(E, R)}{dR^2} + \frac{2\mu}{\hbar^2}\left(E - V_g(R) + \frac{\hbar^2 \ell(\ell+1)}{2\mu R^2}\right) F_\ell^{+}(E, R) = 0. \tag{10.8}$$

By imposing the following boundary condition on $F_\ell^{+}(E, R)$ as $R \to \infty$, the asymptotic wave function has the desired form, Eqn. (10.2), representing an incident plane wave plus scattered wave:

$$\frac{F_\ell^{+}(E, R)}{R} \sim \sin\left(kR - \frac{\pi}{2}\ell + \eta_\ell\right) \frac{e^{i\eta_\ell}}{kR} \tag{10.9}$$

$$\sim j_\ell(kR) + \frac{i}{2} \frac{e^{i(kR - \frac{\pi}{2}\ell)}}{kR} T_\ell(E). \tag{10.10}$$

Here $\eta_\ell$ is the phase shift induced by the interaction potential $V_g(R)$, and $T_\ell(E) = 1 - e^{2i\eta_\ell}$ is the T-matrix element from which the amplitude of the scattered wave is determined,

$$f(E, \hat{\mathbf{k}}, \hat{\mathbf{k}}_{\mathbf{s}}) = \frac{2\pi i}{k} \sum_{\ell=0}^{\infty} \sum_{m=-\ell}^{\ell} i^\ell Y_{\ell m}^*(\hat{\mathbf{k}}) Y_{\ell m}(\hat{\mathbf{k}}_{\mathbf{s}}) T_\ell(E) \tag{10.11}$$

Using the simpler form of Eqn. (10.4), the cross section becomes

$$\sigma(E) = \frac{\pi}{k^2} \sum_{\ell=0}^{\infty} (2\ell+1) |T_\ell(E)|^2 \tag{10.12}$$

$$= \frac{4\pi}{k^2} \sum_{\ell=0}^{\infty} (2\ell+1) \sin^2 \eta_\ell. \tag{10.13}$$

The cross section has a familiar semiclassical interpretation. If the interaction potential $V_g(R)$ vanishes, the particle trajectory is a straight line with relative angular momentum $\mathbf{R} \times \mathbf{p} = bp$, where $p$ is linear momentum and $b$ is the distance of closest approach. If we take the angular momentum to be

$$bp = \hbar\sqrt{\ell(\ell+1)} \approx \hbar \left(\ell + \frac{1}{2}\right)$$

where the "classical" $\ell$ here is not quantized, then b is the classical turning point of the repulsive centrifugal potential $\hbar^2 \ell(\ell+1)/2\mu R^2$ in Eqn. (10.8). The semiclassical expression for the cross section, analogous to Eqn. (10.12), has the form of an area,

$$\sigma(E, \text{semiclassical}) = 2\pi \int_0^\infty bP(b, E) db \tag{10.14}$$

weighted by $P(b, E) = |T_\ell(E)|^2$. An important feature of cold collisions is that only a very few values of the $\ell$ can contribute to the cross section, because the classical turning points of the repulsive centrifugal potential are at large values of $R$. Only collisions with the very lowest $\ell$ values allow the atoms to get close enough to one another to experience the interatomic interaction potential. We will discuss below the specific quantum properties associated with discrete values of $\ell$. Semiclassical theory is useful for certain types of trap loss collisions in relatively warm traps where light is absorbed by atom pairs at very large $R$, but a quantum treatment always becomes necessary at sufficiently low collision energy.

In the simple introduction above, the cross section in Eqn. (10.13) represents elastic scattering, for which the internal states of the particles do not change, and their relative kinetic energy $E$ is the same before and after the collision. In general, the atoms will have nonzero internal angular momentum due to hyperfine structure in the case of alkali atoms or due to electronic structure in the case of rare gas metastable atoms. In a field-free region, these internal states are characterized by total angular momentum $F$ and projection $M$ on a space-fixed quantization axis. Often an external field, either a magnetic or an optical or radio frequency electromagnetic field, is present in cold collision experiments, and these $FM$ states are modified by the external field. Species like hydrogen and alkali atoms, which have $^2$S ground states and nonvanishing nuclear spin $I$, have two hyperfine components with $F = I + \frac{1}{2}$ and $F = I - \frac{1}{2}$. Fig. 10.3 shows the Zeeman splitting of the two $F = 1$ and $F = 2$ hyperfine components of ground state $^{87}$Rb atoms, and Fig. 10.4 shows the ground state interaction potentials for two interacting Rb atoms. If the atomic field-modified states, commonly called field-dressed states, are represented by $|\alpha_i\rangle$ for atom $i = a, b$, then the general collision process represents a transition from the state $|\alpha_a\rangle|\alpha_b\rangle$

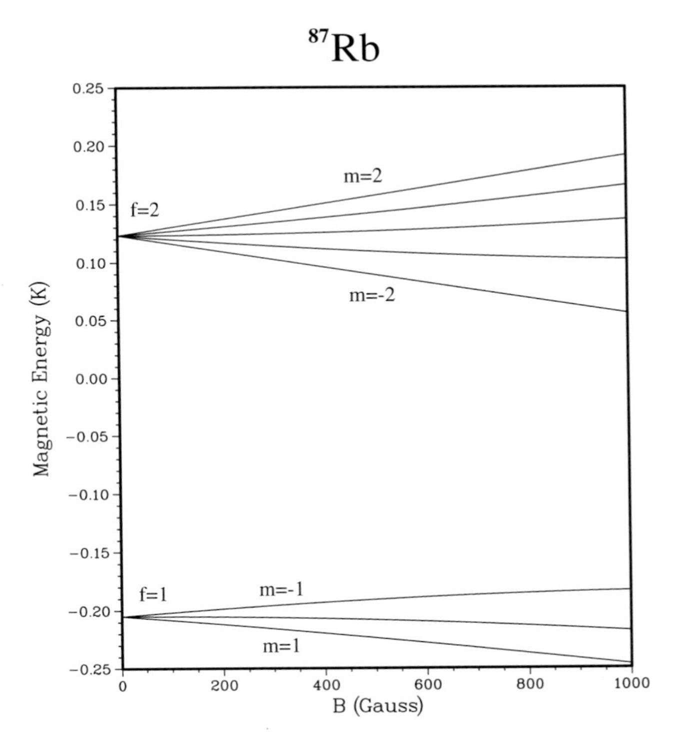


Fig. 10.3. Ground hyperfine levels of the $^{87}$Rb atom versus magnetic field strength. The Zeeman splitting manifolds is evident. The energy has been divided by the Boltzmann constant in order to express it in temperature units.

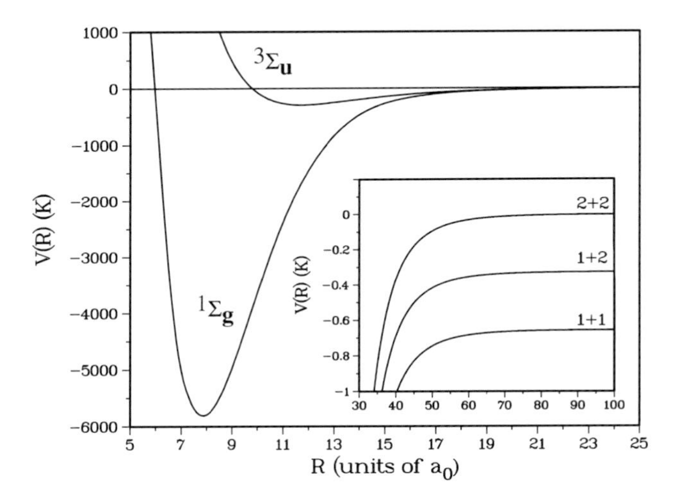


Fig. 10.4. The ground state potential energy curves of the $Rb_2$ molecule. The potentials have been divided by the Boltzmann constant in order to express them in units of temperature. The full figure shows the short range potentials on the scale of chemical bonding. The inset shows a blowup at long range, showing the separated atom hyperfine levels $F_a + F_b = 1+1$, $1+2$, and $2+2$. The upper two potentials in the inset correlate adiabatically with the $^3\Sigma_u$.

to the state $|\alpha'_a\rangle|\alpha'_b\rangle$, represented by the general transition amplitude

$$f(E, \hat{\mathbf{k}}, \hat{\mathbf{k}}_\mathbf{s}, \alpha_a\alpha_b \to \alpha'_a\alpha'_b)$$

and the T-matrix element

$$T(E, \alpha_a\alpha_b \to \alpha'_a\alpha'_b).$$

These are the most complete transition amplitudes that describe an observable transition, specified in terms of quantities that can be selected experimentally. The most complete T-matrix elements that can be calculated from the multicomponent version of Eqn. (10.8) are also specified by the initial and final partial waves:

$$T(E, \alpha_a\alpha_b\ell m \to \alpha'_a\alpha'_b\ell' m').$$

There are three distinct physical axes which define these amplitudes: a space-fixed axis which defines the space projection quantum numbers $M_a$ and $M_b$, the asymptotic direction of approach $\hat{\mathbf{k}}$, and the asymptotic direction of separation $\hat{\mathbf{k}}_\mathbf{s}$. In beam experiments all three axes can be different. Nearly all the work on cold atom collisions is carried out in a homogeneous gas, where neither $\hat{\mathbf{k}}$ nor $\hat{\mathbf{k}}_\mathbf{s}$ are selected or measured, and a cell average cross section defined as in Eqn. (10.12) for the general $\sigma(E, \alpha_a\alpha_b \rightarrow \alpha'_a\alpha'_b)$ is appropriate. Some experiments, especially in the context of Bose-Einstein condensation, have a well-defined local space-fixed axis, and select particular values of $M_a$ and $M_b$, but most of the work we will describe is for unpolarized gases involving an average over $M_a$ and $M_b$.

Instead of cross sections, it is usually preferable to give a rate coefficient for a collision process. The rate coefficient is directly related to the number of collision events occurring in a unit time in a unit volume. Consider the reaction

$$\alpha_a + \alpha_b \rightarrow \alpha'_a + \alpha'_b, \tag{10.15}$$

where the quantum numbers are all assumed to be different. If the density of species $\alpha_i$ in a cell is $n_{\alpha_i}$, the rate of change of the density of the various species due to collision events is

$$-\left(\frac{n_{\alpha_a}}{dt} + \frac{n_{\alpha_b}}{dt}\right) = \left(\frac{n_{\alpha'_a}}{dt} + \frac{n_{\alpha'_b}}{dt}\right) = K(T, \alpha_a\alpha_b \rightarrow \alpha'_a\alpha'_b)n_{\alpha_a}n_{\alpha_b}. \tag{10.16}$$

The rate coefficient $K(T, \alpha_a\alpha_b \rightarrow \alpha'_a\alpha'_b)$ is related to the cross section through

$$K(T, \alpha_a\alpha_b \rightarrow \alpha'_a\alpha'_b) = \langle\sigma(E, \alpha_a\alpha_b \rightarrow \alpha'_a\alpha'_b)v\rangle\,, \tag{10.17}$$

where the brackets imply an average over the distribution of relative collision velocities $v$.

The general theory of collisions of degenerate species is well-understood. The basic multichannel theory can be found, for example, in Mies[15,16] or Bayless *et al.*[17] There is a considerable body of work on the theory of collisions of cold spin-polarized hydrogen atoms, as a consequence of the quest to achieve Bose-Einstein condensation in such a system. The Eindhoven group led by B. Verhaar has been especially active in developing the theory. The paper by Stoof, Koelman, and Verhaar[18] gives an excellent introduction to the subject. The theory of cold collisions in external magnetic fields and optical or radio frequency electromagnetic fields has also been developed.

It is important to distinguish between two different kinds of collisions: elastic and inelastic. As mentioned above, an elastic collision is one in which the quantum states $\alpha_a, \alpha_b$ of each atom remain unchanged by the collision. These collisions exchange momentum, thereby aiding the thermalizing of the atomic sample. These are "good" collisions that do not destroy the trapped states, and they are necessary for the process of evaporative cooling we will describe later. An inelastic collision is one in which one (or more) of these two quantum numbers changes in the collision. Most cold collision studies have dealt with inelastic events instead of elastic ones, that is, the collision results in hot atoms or untrapped species or even ionic species.

As we will see in the next section, the quantum threshold properties of elastic and inelastic collisions are very different.

The basic difference between collisions of different atomic species and identical atomic species is the need to symmetrize (bosons) or antisymmetrize (fermions) the wave function with respect to exchange of identical particles. Most studies have concentrated on the symmetrization of boson states because of the interest in BEC. Other than this symmetrization requirement, the theory is the same for the two cases. Symmetrization has two effects: the introduction of factors of two at various points in the theory, and the exclusion of certain states since they violate the exchange symmetry requirement. Such symmetry restrictions are well-known in the context of diatomic molecular spectroscopy, leading, for example, to ortho- and para-species of molecular hydrogen and to every other line being missing in the absorption spectrum of molecular oxygen, due to the zero nuclear spin of the oxygen atom.[19] In the case of atomic collisions of identical species, if the two quantum numbers are also identical, $\alpha_a = \alpha_b$, only *even* partial waves $\ell$ are possible if the particles are composite bosons, and only *odd* partial waves are possible if the particles are composite fermions. If the two quantum numbers are not identical, $\alpha_a \neq \alpha_b$, both even and odd partial waves can contribute to collision rates. The effect of this symmetry is manifestly present in photoassociation spectra, where for example, half the number of lines appear in a doubly spin polarized gas (where all atoms are in the same quantum state) as contrasted to an unpolarized gas (where there is a distribution of quantum states). These spectra will be described in Sec. 10.6 below.

Stoof *et al.*,[18] give a good discussion of how to modify the theory to account for exchange symmetry of identical particles. Essentially, they set up the states describing the separated atoms, the so-called channel states of scattering theory, as fully symmetrized states with respect to particle exchange. T-matrix elements and event rate coefficients, defined as in Eqn. (10.17), are calculated conventionally for transitions between such symmetrized states. The event rate coefficients are given by

$$K(\{\gamma\delta\} \to \{\alpha\beta\}) = \left\langle \frac{\pi\hbar}{\mu k} \sum_{\ell'm'} \sum_{\ell m} |T(E, \{\gamma\delta\}\ell'm', \{\alpha\beta\}\ell m)|^2 \right\rangle \quad (10.18)$$

where the braces $\{\ldots\}$ signify symmetrized states, and the T-matrix as defined in this review is related to the unitary S-matrix by $\mathbf{T} = \mathbf{1} - \mathbf{S}$. Then collision rates are unambiguously given by:

$$\frac{dn_\alpha}{dt} = \sum_\beta \sum_{\{\gamma\delta\}} (1 + \delta_{\alpha\beta})(K(\gamma\delta \to \alpha\beta)n_\gamma n_\delta - K(\alpha\beta \to \delta\gamma)n_\alpha n_\beta). \quad (10.19)$$

This also works for the case of elastic scattering of identical particles in identical quantum states: $K(\alpha\alpha \to \alpha\alpha)$ must be multiplied by a factor of 2 to get the rate of momentum transfer ($\hat{\mathbf{k}}$ scatters to $\hat{\mathbf{k_s}} \neq \hat{\mathbf{k}}$) since two atoms scatter per collision event. Gao[20] has also described the formal theory for collisions of cold atoms taking into account identical particle symmetry.

### 10.3.2. *Quantum Collisions as Energy Approaches Zero*

Cold and ultracold collisions have special properties that make them quite different from conventional room temperature collisions. This is because of the different scales of time and distance involved. Here we will examine the consequence of the long de Broglie wavelength of the colliding atoms. The basic modification to collision cross sections when the de Broglie wavelength becomes longer than the range of the potential was described in Ref. 21 in the context of cold neutron scattering, and has been widely discussed in the nuclear physics literature.[22,23] Such quantum threshold effects only manifest themselves in neutral atom collisions at very low temperature, typically $\ll 1\,\mathrm{K}$,[24,25] but they play a very important role in the regime of laser cooling and evaporative cooling that has been achieved.

Fig. 10.1 in the Introduction gives a qualitative indication of the range of $\lambda$ as temperature is changed. Even at the high temperature end of the range of laser cooling, $\lambda$ on the order of 100 $\mathrm{a_0}$ is possible, and at the lower end of evaporative cooling, $\lambda > 1\,\mu\mathrm{m}$ (20,000 $\mathrm{a_0}$). These distances are much larger than the typical lengths associated with chemical bonds, and the delocalization of the collision wave function leads to characteristic behavior where collision properties scale as some power of the collision momentum $k = \sqrt{2\mu E} = 2\pi/\lambda$ as $k \to 0$, depending on the inverse power $n$ of the long range potential, which varies as $R^{-n}$.

In the case of elastic scattering, Mott and Massey[26] show that the phase shift $\eta_\ell$ in Eqns. ( 10.9) and (10.13) has the following property as $k \to 0$: if $2\ell < n - 3$,

$$\lim_{k\to 0} k^{2\ell+1} \cot \eta_\ell = -\frac{1}{A_\ell}, \tag{10.20}$$

where $A_\ell$ is a constant, whereas if $2\ell > n - 3$,

$$\lim_{k\to 0} k^{n-2} \cot \eta_\ell = \text{constant}. \tag{10.21}$$

For neutral ground state atoms, this ensures that the phase shift vanishes at least as fast as $k^3$ for all $\ell \geq 1$. Thus all contributions to the cross section vanish when $k$ becomes sufficiently small, except the contribution from the $s$-wave, $\ell = 0$. Since the $s$-wave phase shift varies as $-kA_0$ as $k \to 0$, we see from Eqn. (10.13) that the elastic scattering cross section for identical particles approaches

$$\sigma(E) \to 8\pi A_0^2, \tag{10.22}$$

where the factor of 8 instead of 4 occurs due to identical particle symmetry. Thus, the cross section for elastic scattering becomes constant in the low energy limit. The quantity $A_0$ is the $s$-wave scattering length, an important parameter in the context of Bose-Einstein condensation. Note that the rate coefficient for elastic scattering vanishes as $T^{1/2}$ in the limit of low temperature, since $K = \langle \sigma v \rangle$.

Fig. 10.5 illustrates the properties of a very low collision energy wave function, taken for a model potential with a van der Waals coefficient $C_6$ and mass characteristic of Na atom collisions. The upper panel illustrates the wave function at long range, on the scale of $1\,\mu$m. The lower panel shows a blowup of the short range wave function for the case of three different potentials, with three different scattering lengths, negative, zero, and positive. The figure shows the physical interpretation of the scattering length, as the effective point of origin of the long wavelength asymptotic sine wave at $R = A_0$, since the long range wave function is proportional to $\sin(k(R - A_0))$.

Inelastic collisions have very different threshold properties from elastic ones. As long as the internal energy of the separated atoms in the exit channel is lower than for the entrance channel, so that energy is released in the collision, the transition matrix element varies as,[21,22,24]

$$T(E) \propto k^{\ell+\frac{1}{2}}, \tag{10.23}$$

where $\ell$ is the entrance channel partial wave index. Using this form in Eqn. (10.12) shows that the cross section vanishes at least as fast as $k$ for all $\ell > 0$, but it varies as $1/k$ for the s-wave. This variation (sometimes called the $1/v$ law) was given by Ref. 21 and is well-known in nuclear physics. Although the cross section for an inelastic, energy-releasing, collision becomes arbitrarily large as $k \to 0$, the rate coefficient $K$ remains finite, and approaches a nonvanishing constant.

The range of $k$ where these limiting threshold laws become valid depends strongly on the particular species, and even on the specific quantum numbers of the separated atoms. Knowledge of the long range potential alone does not provide a sufficient condition to determine the range of $k$ in which they apply. This range, as well as the scattering length itself, depends on the actual phase shift induced by the whole potential, and is very sensitive to uncertainties in the short range part of the potential in the chemical bonding region.[27] On the other hand, a necessary condition for threshold law behavior can be given based solely on the long range

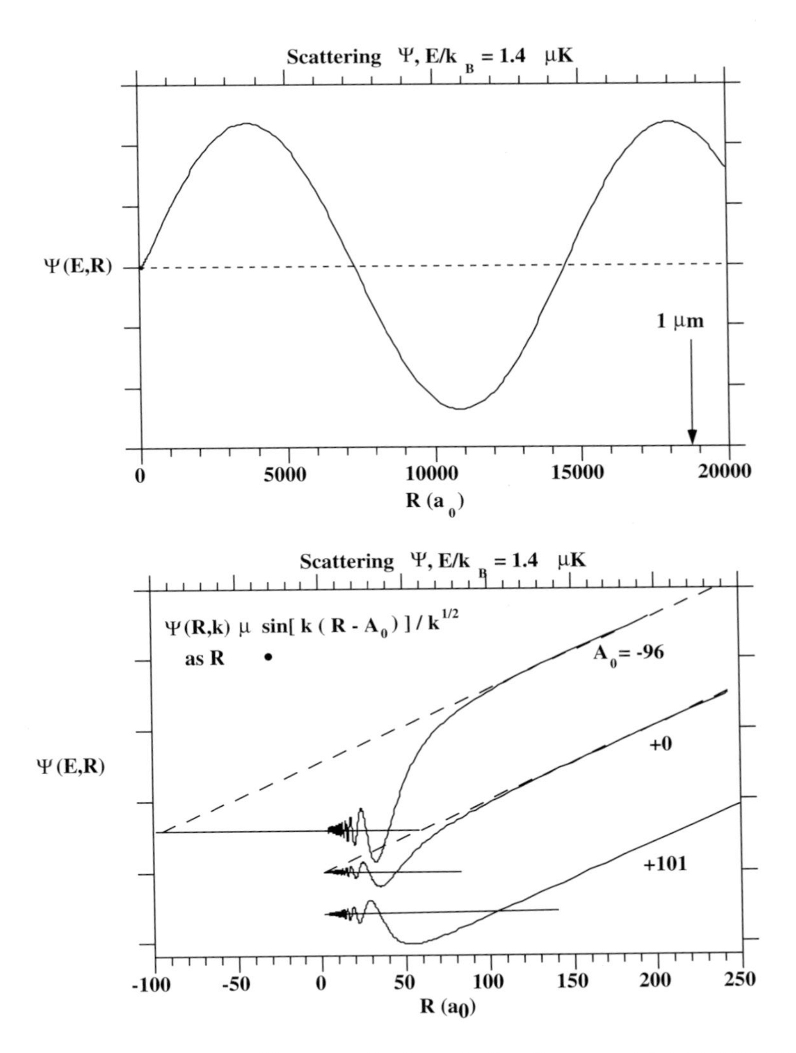


Fig. 10.5. The upper panel illustrates the long de Broglie wave at long range, on the scale of 1 $\mu$m. The lower panel shows a blowup of the short range wave function for the case of three different potentials, with three different scattering lengths, negative, zero, and positive.

potential.[24,25] This condition is based on determining where a semiclassical, WKB connection breaks down between the long range asymptotic $s$-wave and the short range wave function which experiences the acceleration of the potential. Consider the ground state potential $V_g(R)$ as a function of $R$.

The long range potential is,

$$V_g(R) = -\frac{C_n}{R^n} \tag{10.24}$$

where $n$ is assumed to be $\geq 3$. Let us first define

$$E_Q = \frac{\hbar^2}{\mu}\left[\left(2\frac{n+1}{3}\right)^{2n}\left(\frac{n-2}{6n}\right)^{n}\left(\frac{2n+2}{n-2}\cdot\frac{\hbar^2}{\mu C_n}\right)^{2}\right]^{\frac{1}{n-2}} \tag{10.25}$$

$$R_Q = \left(\frac{n-2}{2n+2}\cdot\frac{C_n}{E_Q}\right)^{\frac{1}{n}} \,. \tag{10.26}$$

The properties of the wave function $\Psi(E,R)$ depend on the values of $E$ and $R$ relative to $E_Q$ and $R_Q$. When $E \gg E_Q$, the energy is high enough that it is always possible to make a semiclassical WKB connection between the wave function in the long range zone, $R \gg R_Q$, and the wave function in the short range zone, $R \ll R_Q$. In this case, the WKB representation of the $s$-wave wave function is a good approximation at all $R$, and there are no special threshold effects. On the other hand, $E \ll E_Q$ satisfies the condition that there is some range of distance near $R \approx R_Q$ were the WKB approximation fails, so that no semiclassical connection is possible between long and short range wave functions. This is the regime where a quantum connection, with its characteristic threshold laws, is necessary. When $E \ll E_Q$ the wave function for $R \gg R_Q$ is basically the asymptotic wave with the asymptotic phase shift,[28] whereas the wave function for $R \ll R_Q$ is attenuated in amplitude, relative to the normal WKB amplitude, by a factor proportional to $k^{\ell+1/2}$.[24]

Julienne *et al.*,[25] calculated the values of $E_Q$ and $R_Q$ for a number of alkali and rare gas metastable species that can be laser cooled. These are shown in Table 10.1. The alkali values were based on known values of $C_6$ for the $n = 6$ van der Waals potentials, and estimates were made for the metastable rare gas $C_6$ coefficients, for which the weak $C_5/R^5$ quadrupole-quadrupole contribution to the potential for metastable rare gases was ignored in comparison to the $C_6$ contribution. The value of $R_Q$ is small enough that relativistic retardation corrections to the ground state potential are insignificant. The $R_Q$, $E_Q$ parameters for each rare gas are

Table 10.1. Characteristic threshold parameters $R_Q$ and $E_Q$.

| Species | $R_Q$ (Bohr) | $E_Q/k_B$ (mK) | Species | $R_Q$ (Bohr) | $E_Q/k_B$ (mK) |
|---|---|---|---|---|---|
| Li | 32 | 120 | He* | 34 | 180 |
| Na | 44 | 19 | Ne* | 40 | 26 |
| K | 64 | 5.3 | Ar* | 60 | 5.7 |
| Rb | 82 | 1.5 | Kr* | 79 | 1.6 |
| Cs | 101 | 0.6 | Xe* | 96 | 0.6 |

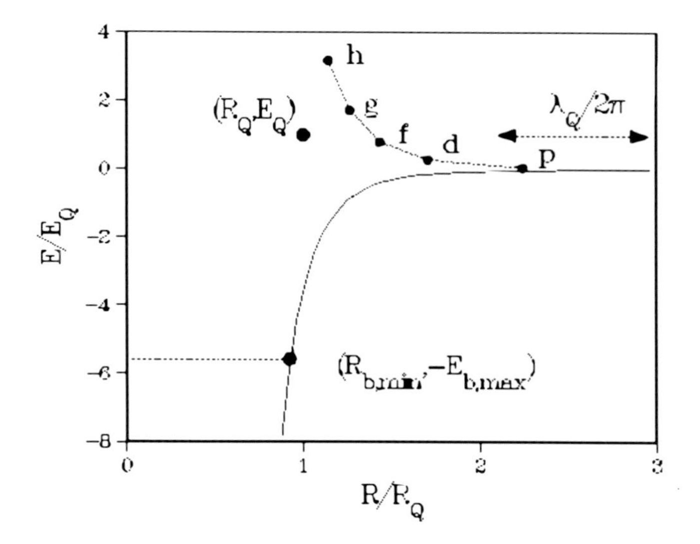


Fig. 10.6. The position of the maximum in the barrier due to the centrifugal potential for $\ell \neq 0$, as well as the maximum binding energy and minimum outer classical turning point of the last bound state in the van der Waals potential, all have identical scalings with $C_6$ and mass as $R_Q$ and $E_Q$, scaling as $(\mu C_6)^{1/4}$ and $\mu^{-3/2}C_6^{-1/2}$ respectively.

similar to those of the neighboring alkali species in the periodic table. Fig. 10.6 shows a plot of the ground state potential expressed in units of $R_Q$ and $E_Q$. The position of the barrier maximum due to the centrifugal potential for $\ell \neq 0$, as well as the maximum binding energy and minimum outer classical turning point of the last bound state in the van der Waals potential, scale with $C_6$ and mass as $R_Q$ and $E_Q$ do, scaling as $(\mu C_6)^{1/4}$ and $\mu^{-3/2}C_6^{-1/2}$ respectively. These other scalable parameters are also indicated on Fig. 10.6, which gives a universal plot for any of the species in Table 10.1. Fig. 10.6 also shows $\lambda/2\pi = 0.0857R_Q$ for $E = E_Q$. The centrifugal barriers for the lowest few partial waves are lower than $E_Q$ , but lie outside $R_Q$. The positions $R_C(\ell)$ and heights $E_C(\ell)$ of the centrifugal barriers are $R_C(\ell) = 2.67R_Q/[\ell(\ell+1)]^{1/4}$ and $E_C(\ell) = 0.0193E_Q[\ell(\ell+1)]^{3/2}$ respectively. The $p$- and $d$-wave barriers have heights of $0.055E_Q$ and $0.28E_Q$ respectively. $E_Q$ is small enough that several partial waves typically contribute to ground state collisions of the heavier species at typical magneto-optical trap (MOT) temperatures of $100\,\mu$K to 1 mK.

The distance $R_Q$, closely related to the mean scattering length $\bar{a} = 0.967R_Q$ defined by Gribakin and Flambaum,[27] is solely a property of the long range potential and atomic mass. On the other hand, the approach to the quantum threshold laws is in a range $0 < E \ll E_Q$, where the actual range depends on the effect of the whole potential, as measured by the threshold phase shift proportional to the actual $s$-wave scattering length $A_0$. Let $A_b$ be defined by $E_b = \hbar^2/2\mu A_b^2$, where $E_b$ is the position of the last bound state of the potential just below threshold ($A_b > 0$) or the position of a virtual bound state just above threshold ($A_b < 0$). When $E_b < E_Q$, the range over which the threshold laws apply can be estimated from $0 < k \leq |A_b|^{-1}$, or $0 < E \ll E_b$. Thus, the threshold laws apply only over a range comparable to the lesser of $E_Q$ and $E_b$. In the limit where the last bound state is very near threshold, Ref. 27 finds that $A_0 = \bar{a} + A_b$; this reduces to the result of Mies *et al.*,[29] $A_0 \approx A_b$, when $|A_0| \gg \bar{a}$. The special threshold properties of the last bound state in the potential has recently been described by Boisseau *et al.*[30] and Trost *et al.*[31] The usual quantization rules in a long range potential[32] do not adequately characterize the threshold limit of the last bound state.

Since the scattering length can be arbitrarily large, corresponding to the case where the last bound state is arbitrarily close to threshold, $E_b$ can be arbitrarily small. Therefore, the actual range over which the threshold law expressions in Eqns. (10.22) and (10.23) apply can in principle be less than the actual temperature range in cold or even ultracold experiments. In fact, the Cs atom in its doubly spin polarized hyperfine level seems not yet to be in the threshold limit even at a few $\mu$K.[33,34]

To show how the interatomic potential affects the scattering length, we can imagine artificially modifying the ground state potential by adding some small correction term $\lambda V_{corr}(R)$ to it which has a short range compared to $R_Q$. Fig. 10.7 shows that we can make $A_0$ vary between $-\infty$ and $+\infty$ by making relatively small changes in the potential. The two singularities correspond to the case where the potential just supports another bound state. Thus, when $\lambda V_{corr}$ changes across a singularity, the number of bound states in the potential changes by exactly one. The scattering length is negative when a new bound state is just about to appear, and positive when a new bound state has just appeared. Scattering lengths are sufficiently sensitive to potentials that, with rare exceptions, it is impossible to predict them accurately from first principles, and therefore they must be measured. How this can be done accurately will be described in the section on photoassociation spectroscopy.

### 10.3.3. *Relations between Phase Shift, Scattering Length, and Bound States*

In the ultracold regime, where only $s$-wave elastic scattering occurs, the magnitude and sign of the scattering length determines all the collisional properties. These properties are crucially important to the stability and

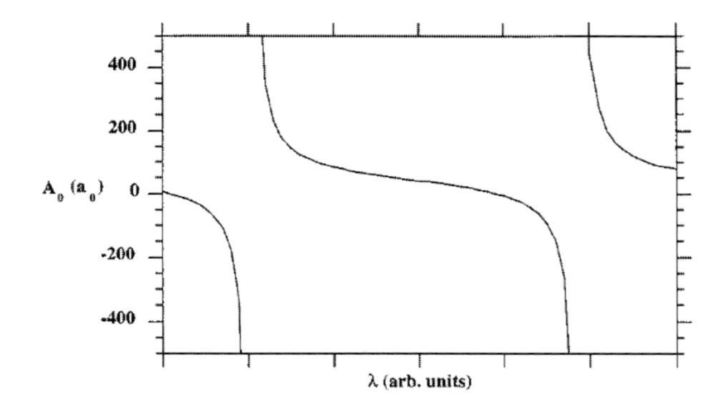


Fig. 10.7. The scattering length $A_0$ varies between $-\infty$ and $+\infty$ by making relatively small changes ($\lambda V_{corr}$) in the ground-state potential. The two singularities correspond to the case where the potential just supports another bound state.

internal dynamics of Bose-Einstein condensates. It is worthwhile therefore to review the relationship between the phase shift, the scattering length, and how they behave in the presence of bound states just at the potential threshold. We follow the discussion by Joachain,[35] adapting it to the notation already established.

For $s$-wave scattering we need only consider the radial part of the Schrödinger equation,

$$-\frac{\hbar^2}{2m}\left[\frac{1}{R^2}\frac{d}{dr}\left(R^2\frac{d}{dR}\right)-\frac{l(l+1)}{R^2}\right]\mathcal{R}_l(k,R)+V(R)\mathcal{R}_l(k,r)=E\mathcal{R}_l(k,R) \tag{10.27}$$

where $\mathcal{R}_l(k,R)$ are the radial solutions. The Schrödinger equation can be cast into a more tractable form by making two judicious substitutions. In place of $V(R)$ we use a "reduced" potential

$$U(R)=\frac{2m}{\hbar^2}V(R) \tag{10.28}$$

and introduce a new radial function

$$u_l(k,R)=R\mathcal{R}_l(k,R) \tag{10.29}$$

With these substitutions the Schrödinger radial equation takes on the simpler form

$$\left[\frac{d^2}{dR^2}+k^2-\frac{l(l+1)}{R^2}-U(R)\right]u_l(k,R)=0 \tag{10.30}$$

where we have used $E=\hbar^2k^2/2m$. Now Eqn. (10.30), with $U(R)=0$, assumes the form of the spherical Bessel differential equation that has two sets of linearly independent solutions useful for scattering problems. The first set comprises the spherical Bessel and Neumann functions, $j_l(kR)$ and $n_l(kR)$, respectively. The second set is the spherical Hankel functions, $h_l^{(1)}$, $h_l^{(2)}$. The Bessel and Neumann functions are real. Their asymptotic forms are

$$\underset{kR \to \infty}{j_l(kR)} \quad \to \frac{1}{kR} \sin\left(kR - \frac{1}{2} l\pi\right) \tag{10.31}$$

$$\underset{kR \to \infty}{n_l(kR)} \quad \to -\frac{1}{kR} \cos\left(kR - \frac{1}{2} l\pi\right) \tag{10.32}$$

The Hankel functions are complex exponential functions. Their asymptotic forms are

$$\underset{kR \to \infty}{h_l^{(1)}(kR)} \to -i \frac{e^{i(kR - 1/2 l\pi)}}{kR} \tag{10.33}$$

$$\underset{kR \to \infty}{h_l^{(2)}(kR)} \to i \frac{e^{-i(kR - 1/2 l\pi)}}{kR} \tag{10.34}$$

The real and complex set of solutions are related by linear combinations reminiscent of the relations between ordinary sin and cos functions and their exponential representations.

Near the origin the Bessel function is a *regular* solution; its limiting value is unity for $l = 0$ and it vanishes for all higher partial waves. The Neumann function, as well as the two Hankel functions, are singular with a pole of order $l + 1$. As already indicated by Eqn. (10.6) the $U(R) = 0$ solution to the radial equation must be $j_l(R)$ since it does not have unphysical singularities at the scattering center. When a scattering potential is added to Eqn. (10.30), the general solution must be some linear combination of one of the two sets of linearly independent functions, the real set, $j_l$, $n_l$, or the complex set $h_l^{(1)}$, $h_l^{(2)}$. Here we choose the real set and assume that the potential has a *finite range* such that there is a distance from the scattering center $R = a$ where $U(R)$ becomes negligible. This condition applies to atom-atom or ion-atom collisions, but is not satisfied for collisions between charged species. We can write the general solution $u_l(kR)$ in this asymptotic region as a linear combination of the two asymptotic forms of $j_l(kR)$ and $n_l(kR)$,

$$u_l(k, R) = kR\left[C_l^{(1)}(k) j_l(kR) + C_l^{(2)}(k) n_l(kR)\right] \tag{10.35}$$

Now nothing prevents us from writing the "mixing coefficients" as

$$C_l^{(1)}(k) = \cos \eta_l(k) \quad \text{and} \quad C_l^{(2)}(k) = -\sin \eta_l(k) \tag{10.36}$$

so that

$$\tan \eta_l(k) = -\frac{C_l^{(2)}}{C_l^{(1)}} \tag{10.37}$$

and

$$\underset{R \to \infty}{u_l(k,R)} \to \mathcal{C}_l(k) \sin\left[kR - \frac{1}{2}l\pi + \eta_l(k)\right] \tag{10.38}$$

Thus the effect of the scattering potential is to "mix in" some Neumann function to the general solution, and the result of this mixing is simply the phase shift $\eta_l(k)$. The factor $\mathcal{C}_l(k)$ is a normalization constant for $u_l(k,R)$, and it is clear from the linear independence of $j_l(kR)$ and $n_l(kR)$ that

$$[\mathcal{C}_l(k)]^2 = \left[C_l^{(1)}\right]^2 + \left[C_l^{(2)}\right]^2$$

It is easy to show that the radial solution can also be written as

$$\underset{R \to \infty}{u_l(k,R)} \to \tilde{\mathcal{C}}_l \left[\sin\left(kR - \frac{l}{2}\pi\right) + \tan \eta_l(k) \cos\left(kR - \frac{l}{2}\pi\right)\right] \tag{10.39}$$

and from Eqn. (10.29) the solutions to the Schrödinger equation

$$\underset{R \to \infty}{\mathcal{R}_l(k,R)} \to \tilde{\mathcal{C}}_l(k)\left[j_l(kR) - \tan \eta_l(k) n_l(kR)\right] \tag{10.40}$$

Now we make use of the *finite range* of the potential $U(R)$. At internuclear distances $R > a$ we will consider the potential to be zero so that the asymptotic form Eqn. (10.40) is the relevant solution to the Schrödinger equation. At shorter distances we assume a well-defined potential form with known solutions $\mathcal{R}_l$ for Eqn. (10.30). At the boundary $R = a$ we apply joining conditions to the interior and exterior solutions. Smooth joining requires that the solutions and their derivatives match at the boundary $R = a$. A convenient way to express this matching is the *logarithmic derivative*

$$\gamma_l = \left[\frac{d\mathcal{R}_l/dR}{\mathcal{R}_l}\right]_{R=a} \tag{10.41}$$

Using Eqn. (10.40) for the exterior solution we express the logarithmic derivative at the boundary as

$$\gamma_l = \frac{k\left[j_l'(ka) - \tan \eta_l(k) n_l'(ka)\right]}{j_l(ka) - \tan \eta_(k) n_l(ka)} \tag{10.42}$$

where the $j_l'$, $n_l'$ indicate differentiation with respect to $R$. From this joining condition we obtain an expression for the phase shift in terms of the asymptotic solutions and $\gamma_l$

$$\tan \eta_l(k) = \frac{kj_l'(ka) - \gamma_l(k)j_l(ka)}{kn_l'(ka) - \gamma_l(k)n_l(ka)} \tag{10.43}$$

It can be shown[35] that in the limit $k \to 0$ the tangent of the phase shift takes on the limiting form (for $l = 0$)

$$\underset{k \to 0}{\tan \eta_0} \to -ka \left[ \frac{a\hat{\gamma}_0}{1 + a\hat{\gamma}_0} \right] \tag{10.44}$$

where

$$\lim_{k \to 0} \gamma_l \to \hat{\gamma}_0$$

We identify the factor

$$a \left[ \frac{a\hat{\gamma}_0}{1 + a\hat{\gamma}_0} \right]$$

with the *scattering length* $A_0$ and write

$$A_0 = -\lim_{k \to 0} \frac{\tan \eta_0(k)}{k} \tag{10.45}$$

Except for isolated points of *zero-energy resonances* (discussed in Sec. 10.3.4) the phase shift goes to zero with $k$. Therefore we can write

$$A_0 = -\lim_{k \to 0} \frac{\sin \eta_0(k)}{k} \tag{10.46}$$

So from Eqn. (10.13) the total scattering cross section at very low energy approaches

$$\sigma = 4\pi A_0^2 \tag{10.47}$$

In the case of identical particles, as already pointed out in Eqn. (10.22), an extra factor of 2 must be inserted in the cross section formula.

### 10.3.4. *Scattering Length in a Square-well Potential*

To illustrate the behavior of the phase shift and scattering length as the collision energy approaches zero, we calculate them for the simple case of an attractive square-well potential. We take

$$U(R) = -U_0 \quad R < a$$
$$U(R) = 0 \qquad R > a$$

In the interior region, for $s$-wave scattering, the radial equation will have the form

$$\left[\frac{d^2}{dR^2} + \kappa^2 - U_0(R)\right] u_0(k, R) = 0 \tag{10.48}$$

where $\kappa = \sqrt{k^2 + U_0}$. The solution of the Schrödinger equation must look like the "regular" solution,

$$\mathcal{R}_0 = C_0\, j_0(\kappa R)$$

In the exterior region we should have, according to Eqn. (10.40),

$$\mathcal{R}_0(R) = \tilde{C}_0(k)\,[j_0(kR) + \tan\eta_0\, n_0(kR)]$$

This 1-D scattering problem is depicted in Fig. 10.8. Application of the joining conditions at the boundary, Eqn. (10.43), results in a expression for the phase shift

$$\eta_0 = -ka + \tan^{-1}\left[\frac{k}{\kappa}\tan\kappa a\right] \tag{10.49}$$

As $k \to 0$

$$\eta_0 \simeq ka\left[\frac{\tan\kappa a}{\kappa a} - 1\right] \tag{10.50}$$

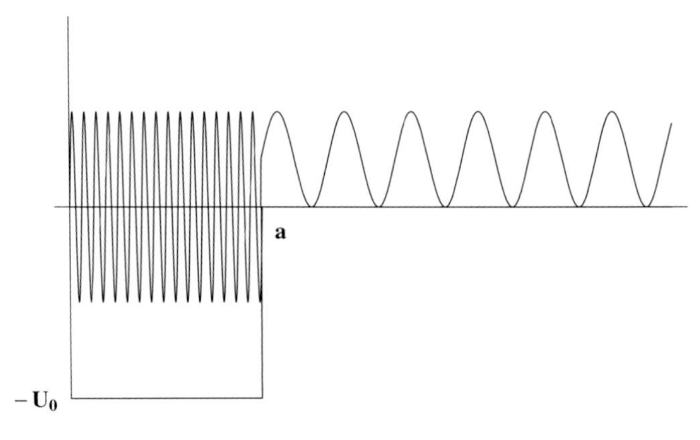


Fig. 10.8. Scattering from a 1-D "square-well" potential: the waves in the inner and outer regions must join smoothly at $R = a$. In reduced units the kinetic energy in the outer region is $k^2$ and in the inner region $\kappa^2$. The two are related by $\kappa = \sqrt{k^2 + U_0}$.

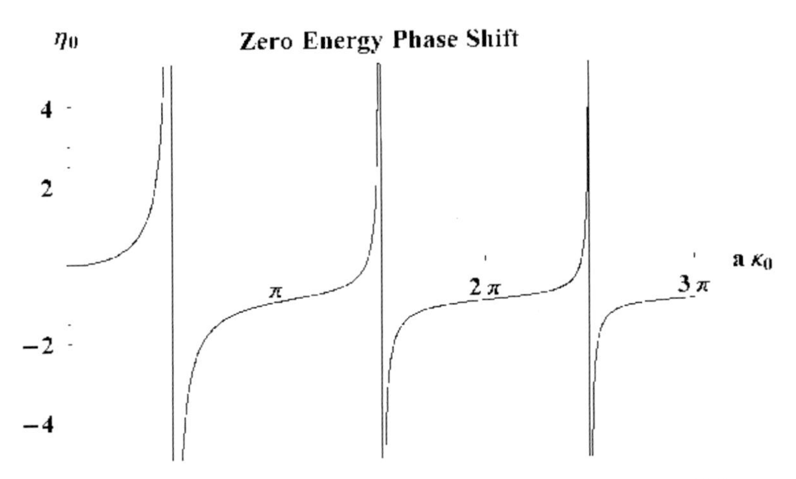


Fig. 10.9. Behavior of the phase shift as $a\kappa_0$ increases. The term $\kappa_0 = \sqrt{U_0}$ and is the limiting expression for $\kappa$ as the collision energy goes to zero. The phase shift diverges at $\pi/2$ (modulo $\pi$) as $a\kappa_0$ increases. For fixed potential width the increase of $a\kappa_0$ can be interpreted as increasing well depth.

Fig. 10.9 shows the behavior of the phase shift as the potential well depth increases. These periodic divergences in the phase shift are directly related to the appearance of bound states in the potential. The condition for the appearance of bound states is that the matching conditions, Eqn. (10.41), be satisfied for a wave solution in the outer zone ($R > a$) which falls off exponentially with distance beyond $a$ instead of propagating indefinitely. Straight-forward application of the logarithmic derivative matching shows that the condition for bound states is given by

$$\xi \cot \xi = -\zeta \tag{10.51}$$

with $\xi = \kappa a$ and $\zeta = ka$. Equation (10.51) is transcendental and does not have closed-form solutions. The position of the bound states can be found graphically, however, by plotting each side of the equation with respect to $\kappa a$ and $ka$ and determining where the functions intersect. Fig. 10.10 illustrates this procedure. The key point is that in the limit $k \to 0$ the $s$-wave bound states appear at the same positions of $\kappa a$ as the divergences in the phase shift plotted in Fig. 10.9. In other words, phase shift divergence is a signature of the appearance of bound states in the attractive scattering potential. This condition is sometimes called the *zero-energy resonance.* Since the scattering length is simply related to the phase shift, it is not surprising that the scattering length also exhibits divergence when bound states appear. Fig. 10.11 shows the behavior of the scattering length as a function of $\kappa a$ as the collision energy approaches zero. For fixed $a$ this

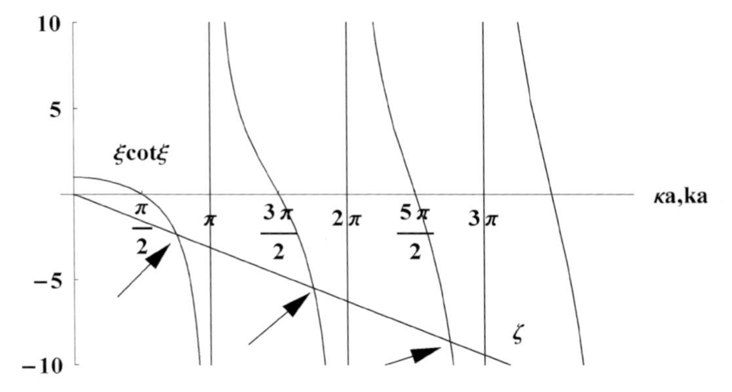


Fig. 10.10. Plot of the left side of Eqn. (10.51) vs. $\kappa a$ and the right side of Eqn. (10.51) vs. $ka$. The points of intersection determine the positions of the bound states. Note that as $k \to 0$, the straight line $\zeta$ approaches the axis, and the bound states appear at $\kappa_0 a = \pi/2$ (modulo $\pi$).

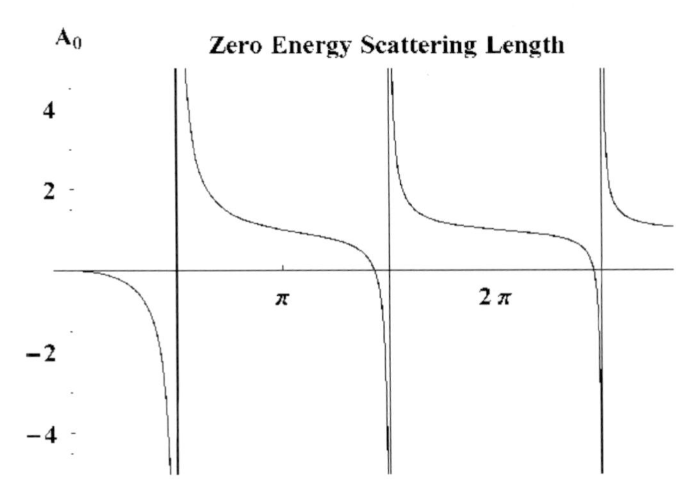


Fig. 10.11. Behavior of the scattering length as a function of $\kappa a$. Divergences are the signature for the appearance of bound states in the square well potential. The values of $\kappa a$ at which the divergences appear correspond to a collision energy approaching zero.

variation shows how the scattering length varies with increasing potential well depth. When $U_0$ is too shallow to support any bound states, the scattering length is negative. At the appearance of the first bound state at threshold, $A_0$ diverges, changes sign and reappears as a positive scattering length. Increasing the potential depth still further continues to reduce $A_0$

until it passes smoothly through zero and again diverges negatively at the threshold for the second bound state.

### 10.3.5. *Collisions in a Light Field*

Much of the work on collisions of cooled and trapped atoms has involved collisions in a light field. This is the subject of the present section and has been the subject of earlier reviews.[36,37] Therefore, it is useful to provide an overview of some of the concepts that are used in understanding cold collisions in a light field. Fig. 10.2 in the Introduction gives a schematic description of the key features of these collisions. Fig. 10.2 shows ground and excited state potentials, $V_g(R)$ and $V_e(R)$, and indicates the optical coupling between ground and excited states induced by a light field at frequency $\omega_1$, which is detuned by $\Delta = \omega_1 - \omega_0$ from the frequency of atomic resonance $\omega_0$. Detuning can be to the red or blue of resonance, corresponding to negative or positive $\Delta$ respectively. The figure illustrates that the excited state potential, having a resonant-dipole form that varies as $C_3/R^3$, which can be attractive or repulsive, is of much longer range than the ground state van der Waals potential. Fig. 10.2 also indicates the Condon point $R_C$ for the optical transition. This is the point at which the quasi-molecule comprised of the two atoms is in optical resonance with the light field, i.e.,

$$V_e(R_C) - V_g(R_C) = \hbar\omega_1. \tag{10.52}$$

The Condon point is significant because in a semiclassical picture of the collision, it is the point at which the transition from the ground state to the excited state is considered to occur. The laser frequency $\omega_1$ can be readily varied in the laboratory, over a wide range of red and blue detuning. Thus, the experimentalist has at his disposal a way of selecting the Condon point and the upper state which is excited by the light. By varying the detuning from very close to atomic resonance to very far from resonance, the Condon point can be selected to range from very large $R$ to very small $R$.

Although the semiclassical picture implied by using the concept of a Condon point is very useful, a proper theory of cold collisions should be quantum mechanical, accounting for the delocalization of the wave function. If the light is not too intense, the probability of a transition, $P_{ge}(E, \omega_1)$, is proportional to a Franck-Condon overlap matrix element between ground and excited state wave functions:

$$P_{ge}(E, \omega_1) \propto |\langle \Psi_e(R)|\Omega_{eg}(R)|\Psi_g(E, R)\rangle|^2 \tag{10.53}$$

$$\approx |\Omega_{eg}(R_C)|^2 |\langle \Psi_e(R)|\Psi_g(E, R)\rangle|^2 \tag{10.54}$$

where the optical Rabi matrix element $\Omega_{eg}(R)$ measures the strength of the optical coupling. Julienne[28] showed that a remarkably simple reflection approximation to the Franck-Condon factor, based on expanding the integrand of the Franck-Condon factor about the Condon point, applies over a wide range of ultracold collision energies and red or blue detunings. This approximation, closely related to the usual stationary phase approximation of line broadening theory, shows that the Franck-Condon factor (in the cases where there is only one Condon point for a given state) is proportional to the ground state wave function at the Condon point:

$$F_{ge}(E,\Delta) = |\langle\Psi_e(R)|\Psi_g(E,R)\rangle|^2 \tag{10.55}$$

$$= \frac{1}{D_C}|\Psi_g(E,R_C)|^2 \qquad \text{blue detuning} \tag{10.56}$$

$$= h\nu_v\frac{1}{D_C}|\Psi_g(E,R_C)|^2 \quad \text{red detuning} \tag{10.57}$$

Here $D_C = |d(V_e - V_g)/dR|_{R_C}$ is the slope of the difference potential, and the wave functions have been assumed to be energy-normalized, that is, the wave functions in Eqns. (10.7) to (10.9) are multiplied by $\sqrt{\frac{2\mu k}{\pi\hbar^2}}$ so that

$$|\langle\Psi_g(E,R)|\Psi_g(E',R)\rangle|^2 = \delta(E-E'). \tag{10.58}$$

The formulas for red and blue detuning differ because they represent free-bound and free-free transitions respectively; here $\nu_v$ is the vibrational frequency of the bound vibrational level excited in the case of red detuning. Fig. 10.12 shows the validity of this approximation over a range of Condon points and collision energies for a typical case. We thus see that it is legitimate to use the semiclassical concept of excitation at a Condon point in discussing cold collisions in a light field. The excited state potential can be written at long range as,[38]

$$V_e(R) = f\hbar\Gamma_A\left(\frac{\lambda_A}{2\pi R}\right)^3, \tag{10.59}$$

where $\lambda_A$ and $\Gamma_A$ are the respective transition wavelength and spontaneous decay width of the atomic resonance transition, and where $f$ is a dimensionless factor on the order unity. Thus, in a MOT, where $|\Delta|$ is on the order of $\Gamma_A$, the Condon point will have a magnitude on the order of $\lambda_A/2\pi$, that is, on the order of 1000 $a_0$ for a typical laser cooling transition. Optical shielding experiments are typically carried out with detunings up to several hundred $\Gamma_A$, corresponding to much smaller Condon points, on the order of a few hundred $a_0$. Photoassociation spectroscopy experiments use quite large detunings, and typically sample Condon points over a range from short range chemical bonding out to a few hundred $a_0$.

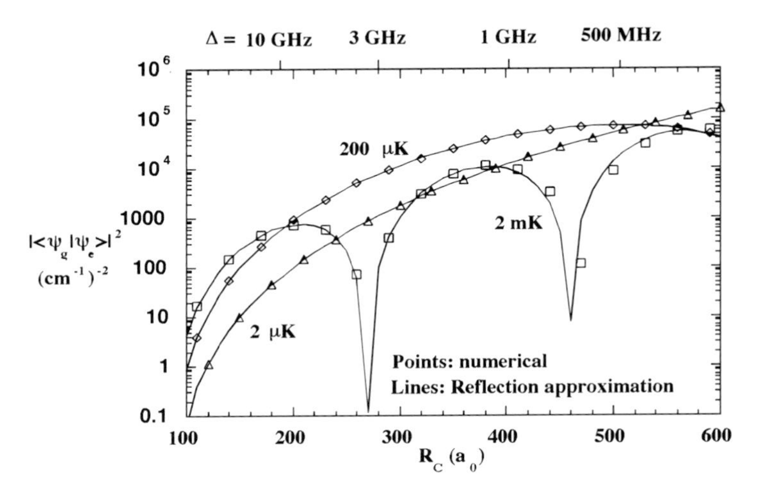


Fig. 10.12. The plots show the validity of the reflection approximation over a range of Condon points and collision energies for a typical case.

Because of the very different ranges of $\Delta$ and $R_C$ in the various kinds of cold collisions in a light field, a number of different kinds of theories and ways of thinking about the phenomena have been developed to treat the various cold collision processes. Since $R_C$ for MOT trap loss collisions is typically much larger than the centrifugal barriers illustrated in Fig. 10.6, many partial waves contribute to the collision, the quantum threshold properties are not evident at all, and semiclassical descriptions of the collision have been very fruitful in describing trap loss. The main difficulty in describing such collisions is in treating the strong excited state spontaneous decay on the long time scale of the collision. We will see that a good, simple quantum mechanical description of such collisions is given by extending the above Condon point excitation picture to include excited state decay during the collision. Quantitative theory is still made difficult by the presence of complex molecular hyperfine structure, except in certain special cases. Since the Condon points are at much smaller $R$ in the case of optical shielding, excited state decay during the collision in not a major factor in the dynamics, and the primary challenge to theory in the case of optical shielding is how to treat the effect of a strong radiation field, which does strongly modify the collision dynamics. Simple semiclassical pictures seem not to work very well in accounting for the details of shielding, and a quantum description is needed. The theory of photoassociation is the most quantitative and well developed of all. Since the Condon point is typically inside the centrifugal barriers and $R_Q$, only the lowest few partial waves contribute to the spectrum. Although the line shapes of the free-bound transitions exhibit the strong influence of the quantum threshold laws, they can be quantitatively described by Franck-Condon theory, which can include ground and excited state hyperfine structure. We will show how photoassociation spectroscopy has become a valuable tool for precision spectroscopy and determination of ground state scattering lengths.

## 10.4. Photoassociation at Ambient and Cold Temperatures

The first measurement of a free-bound photoassociative absorption appeared long before the development of optical cooling and trapping, about two decades ago, when Scheingraber and Vidal[39] reported the observation of photoassociation in collisions between magnesium atoms. In this experiment fixed ultraviolet emission lines from an argon ion laser excited free-bound transitions from the thermal continuum population of the ground $X^1\Sigma_g^+$ state to bound levels of the $A^1\Sigma_u^+$ state of $Mg_2$. Scheingraber and Vidal analyzed the subsequent fluorescence to bound and continuum states from which they inferred the photoassociative process. The first unambiguous photoassociation *excitation spectrum*, however, was measured by Inoue *et al.*[40] in collisions between Xe and Cl at 300 K. In both these early experiments the excitation was not very selective due to the broad thermal distribution of populated continuum ground states. Jones *et al.*[41] with a technically much improved experiment reported beautiful free-bound vibration progressions in KrF and XeI $X \rightarrow B$ transitions; and, from the intensity envelope modulation, were able to extract the functional dependence of the transition moment on the internuclear separation. Although individual vibrational levels of the $B$ state were clearly resolved, the underlying rotational manifolds were not. Jones *et al.*[41] simulated the photoassociation structure and line shapes by assuming a thermal distribution of rotational levels at 300 K. Photoassociation and dissociation processes prior to the cold and ultracold epoch have been reviewed by Tellinghuisen.[42]

A decade after Scheingraber and Vidal reported the first observation of photoassociation Thorsheim *et al.*[43] proposed that high-resolution free-bound molecular spectroscopy should be possible using optically cooled and confined atoms. Fig. 10.13 shows a portion of their calculated $X \rightarrow A$ absorption spectrum at 10 mK for sodium atoms. This figure illustrates how cold temperatures compress the Maxwell-Boltzmann distribution to the point where individual rotational transitions in the free-bound absorption are clearly resolvable. The marked differences in peak intensities indicate scattering resonances, and the asymmetry in the line shapes, tailing off to the red, reflect the thermal distribution of ground-state collision energies

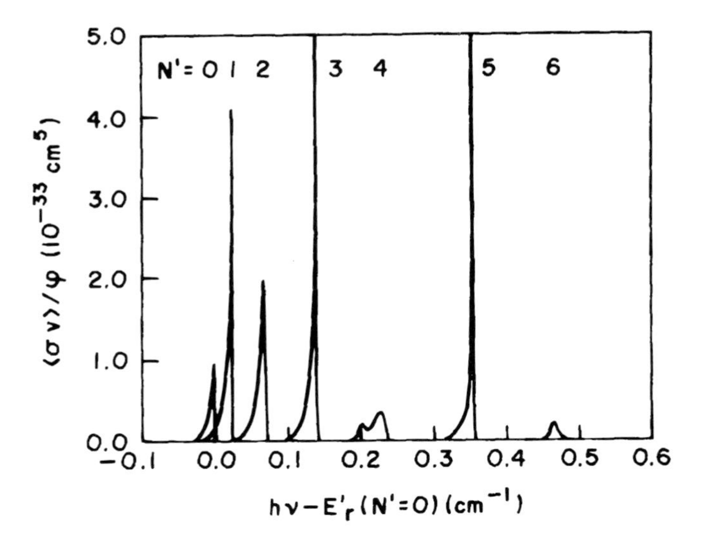


Fig. 10.13. Calculated free-bound photoassociation spectrum at 10 mK from Thorsheim *et al.*[43]

at 10 mK. Fig. 10.14 plots the photon-flux-normalized absorption rate coefficient for singlet $X^1\Sigma_g^+ \to A^1\Sigma_u^+$ and triplet $a^3\Sigma_u^+ \to 1^3\Sigma_g^+$ molecular transitions over a broad range of photon excitation, red-detuned from the Na ($^2$S$\to^2$P) atomic resonance line. The strongly modulated intensity envelopes are called Condon modulations, and they reflect the overlap between the ground state continuum wave functions and the bound excited vibrational wave functions. These can be understood from the reflection approximation described in Eqn. (10.57). We shall see later that these Condon modulations reveal detailed information about the ground state scattering wave function and potential from which accurate $s$-wave scattering lengths can be determined. Thorsheim *et al.*[43] therefore predicted three of the principal features of ultracold photoassociation spectroscopy later to be exploited in many follow-up experiments: (1) precision measurement of vibration-rotation progressions from which accurate excited-state potential parameters can be determined, (2) line profile measurements and analysis to determine collision temperature and threshold behavior, and (3) spectral intensity modulation from which the ground state potential, the scattering wave function, and the $s$-wave scattering length can be characterized with great accuracy.

An important difference distinguishes ambient temperature photoassociation in rare-gas halide systems and sub-milliKelvin temperature photoassociation in cooled and confined alkali systems. At temperatures found

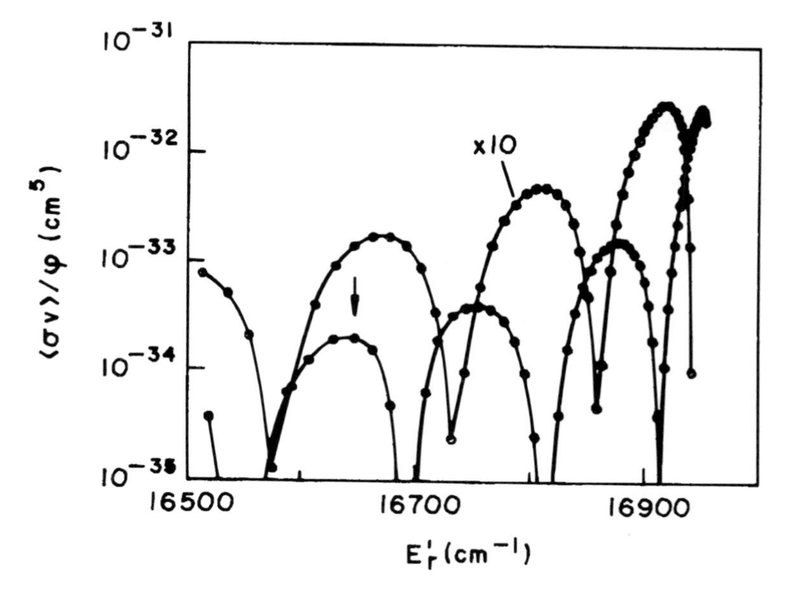


Fig. 10.14. Calculated absorption spectrum of photoassociation in Na at 10 mK, showing Condon fluctuations. The curve labeled 'x10' refers to $^3\Sigma_g \rightarrow {}^3\Sigma_u$ while the other curve refers to $^1\Sigma_g \rightarrow {}^1\Sigma_u$ transitions, from Thorsheim *et al.*[43]

in MOTs (and within selected velocity groups or collisions within atomic beams) the collision dynamics are controlled by long-range electrostatic interactions, and Condon points $R_C$ are typically at tens to hundreds of $a_0$. In the case of the rare-gas halides the Condon points are in the short-range region of chemical binding, and therefore free-bound transitions take place at much smaller internuclear distances, typically less than ten $a_0$. For the colliding A,B quasimolecule the pair density $n$ as a function of $R$ is given by

$$n = n_A n_B 4\pi R^2 e^{-\frac{V(R)}{kT}} dR \tag{10.60}$$

so the density of pairs varies as the square of the internuclear separation. Although the pair-density $R$-dependence favors long-range photoassociation, the atomic reactant densities are quite different with the $n_A n_B$ product on the order of $10^{35}\,\mathrm{cm}^{-6}$ for rare-gas halide photoassociation and only about $10^{22}\,\mathrm{cm}^{-6}$ for optically trapped atoms. Therefore the effective pair density available for rare-gas halide photoassociation greatly exceeds that for cold alkali photoassociation, permitting fluorescence detection and dispersion by high resolution (but inefficient) monochromators.

### 10.5. Associative and Photoassociative Ionization

Conventional associative ionization (AI) occurring at ambient temperature proceeds in two steps: excitation of isolated atoms followed by molecular autoionization as the two atoms approach on excited molecular potentials. In sodium for example,[44]

$$Na + \hbar\omega \rightarrow Na^* \tag{10.61}$$

$$Na^* + Na^* \rightarrow Na_2^+ + e \tag{10.62}$$

The collision event lasts a few picoseconds, fast compared to radiative relaxation of the excited atomic states (~tens of nanoseconds). Therefore the incoming atomic excited states can be treated as stationary states of the system Hamiltonian, and spontaneous radiative loss does not play a significant role. In contrast, cold and ultracold photoassociative ionization (PAI) must always start on ground states because the atoms move so slowly that radiative lifetimes become short compared to the collision duration. The partners must be close enough at the Condon point, where the initial photon absorption takes place, so that a significant fraction of the excited scattering flux survives radiative relaxation and goes on to populate the final inelastic channel. Thus PAI is also a two-step process: (1) photoexcitation of the incoming scattering flux from the molecular ground state continuum to specific vibration-rotation levels of a bound molecular state; and (2) subsequent photon excitation either to a doubly-excited molecular autoionizing state or directly to the molecular photoionization continuum. For example in the case of sodium collisions the principal route is through doubly-excited autoionization,[45]

$$Na + Na + \hbar\omega_1 \rightarrow Na_2^* + \hbar\omega_2 \rightarrow Na_2^{**} \rightarrow Na_2^+ + e \tag{10.63}$$

whereas for rubidium atoms the only available route is direct photoionization in the second step,[46]

$$Rb + Rb + \hbar\omega_1 \rightarrow Rb_2^* + \hbar\omega_2 \rightarrow Rb_2^+ + e \tag{10.64}$$

Collisional ionization can play an important role in plasmas, flames, atmospheric and interstellar physics and chemistry. Models of these phenomena depend critically on the accurate determination of absolute cross sections and rate coefficients. The rate coefficient is the quantity closest to what an experiment actually measures and can be regarded as the product of the cross section and the velocity averaged over the collision velocity distribution,

$$K = \int_0^\infty v\sigma(v) f(v) dv \tag{10.65}$$

The velocity distribution $f(v)$ depends on the conditions of the experiment. In cell and trap experiments it is usually a Maxwell-Boltzmann distribution at some well-defined temperature, but $f(v)$ in atomic beam experiments, arising from optical excitation velocity selection, deviates radically from the normal thermal distribution.[47–49] The actual signal count rate, $\frac{d(X_2^+)}{dt}$, relates to the rate coefficient through

$$\frac{1}{V\alpha}\frac{d(X_2^+)}{dt} = K[X]^2 \tag{10.66}$$

where $V$ is the interaction volume, $\alpha$ the ion detection efficiency, and $[X]$ the atom density. If rate constant or cross section measurements are carried out in crossed or single atomic beams[47,50,51]; special care is necessary to determine the interaction volume and atomic density. PAI was the first measured collisional process observed between cooled and trapped atoms.[52] The experiment was performed with atomic sodium confined in a hybrid laser trap, utilizing both the spontaneous radiation pressure and the dipole force. The trap had two counterpropagating, circularly polarized Gaussian laser beams brought to separate foci such that longitudinal confinement along the beam axis was achieved by the spontaneous force and transversal confinement by the dipole force. The trap was embedded in a large (~1 cm diameter) conventional optical molasses loaded from a slowed atomic beam. The two focused laser beams comprising the dipole trap were alternately chopped with a $3\,\mu$sec "trap cycle", to avoid standing-wave heating. This trap cycle for each beam was interspersed with a $3\,\mu$sec "molasses cycle" to keep the atoms cold. The trap beams were detuned about 700 MHz to the red of the $3s^2S_{1/2}(F=2) \rightarrow 3p^2P_{3/2}(F=3)$ transition while the molasses was detuned only about one natural line width (~10 MHz). The atoms captured from the molasses ($\sim 10^7$ cm$^{-3}$) were compressed to a much higher excited atom density ($\sim 5 \times 10^9$cm$^{-3}$) in the trap. The temperature was measured to be about $750\,\mu$K. Ions formed in the trap were accelerated and focused toward a charged-particle detector. To assure the identity of the counted ions, Gould *et al.*[52] carried out a time-of-flight measurement; the results of which, shown in Fig. 10.15, clearly establish the $Na_2^+$ ion product. The linearity of the ion rate with the square of the atomic density in the trap, supported the view that the detected $Na_2^+$ ions were produced in a binary collision. After careful measurement of ion rate, trap volume and excited atom density, the value for the rate coefficient was determined to be $K = (1.1^{+1.3}_{-0.5}) \times 10^{-11}$ cm$^3$ sec$^{-1}$. Gould *et al.*[52] following conventional wisdom, interpreted the ion production as coming from collisions between two *excited* atoms,

$$\frac{dN_I}{dt} = K\int n_e^2(\vec{r})d^3\,\vec{r} = K\,\bar{n}_e\,N_e \tag{10.67}$$

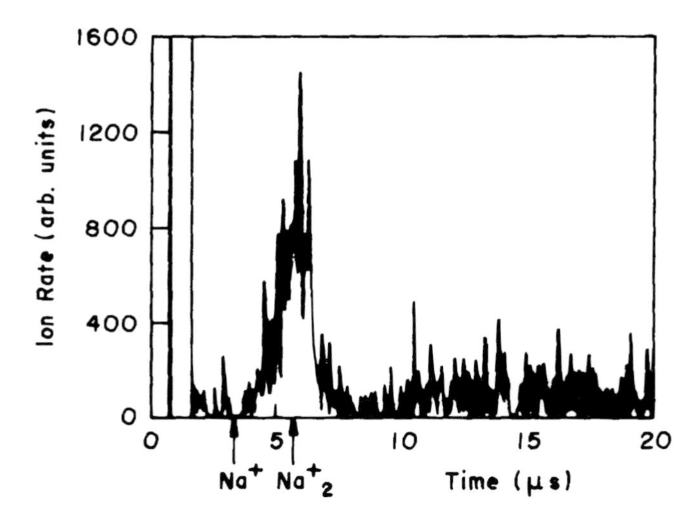


Fig. 10.15. Time-of-flight spectrum clearly showing that the ions detected are $Na_2^+$ and not the atomic ion, from Gould *et al.*[52]

where $\frac{dN_I}{dt}$ is the ion production rate, $n_e(\mathbf{r})$ the excited-state density, $N_e$ the number of excited atoms in the trap ($= \int n_e(\mathbf{r}) d^3\mathbf{r}$), and $\hat{n}_e$the "effective" excited-state trap density. The value for $K$ was then determined from these measured parameters. Assuming an average collision velocity of $130\,\text{cm}\,\text{s}^{-1}$, equivalent to a trap temperature of $750\,\mu$K, the corresponding cross section was determined to be $\sigma = (8.6^{+10.0}_{-3.8}) \times 10^{-14}\,\text{cm}^2$. In contrast, the cross section at ~575 K had been previously determined to be $\sim 1.5 \times 10^{-16}\,\text{cm}^2$.[53–55] Gould *et al.*[52] rationalized the difference in cross section by invoking the difference in de Broglie wavelengths, the number of participating partial waves, and the temperature dependence of the ionization channel probability. The quantal expression for the cross section in terms of partial wave contributions $l$ and inelastic scattering probability $|S_{12}|^2$ is

$$\sigma_{12}(\epsilon) = \left(\frac{\pi}{k^2}\right) \sum_{l=0}^{\infty} (2l+1) \left|S_{12}(\epsilon, l)\right|^2 \qquad (10.68)$$
$$\cong \pi \left(\frac{\lambda_{dB}}{2\pi}\right)^2 (l_{\max}+1)^2 P_{12}$$

where $\lambda_{dB}$ is the entrance channel de Broglie wavelength and $P_{12}$ is the probability of the ionizing collision channel averaged over all contributing partial waves of which $l_{\max}$ is the greatest. The ratio of the $(l_{\max}+1)^2$ values at 575 K and $750\,\mu$K is about 400 and the de Broglie wavelength ratio factor varies inversely with temperature. Therefore in order that the cross section ratio be consistent with low- and high-temperature experiments, $\frac{\sigma_{12}(575\,\text{K})}{\sigma_{12}(750\,\mu\text{K})}$ $\sim 1.7\times 10^{-3}$,[52] concluded that $P_{12}$ must be about three times greater at 575 K than at 750 $\mu$K. However, it soon became clear that the conventional picture of associative ionization, starting from the excited atomic states, could not be appropriate in the cold regime. Julienne[56] pointed out the essential problem with this picture. In the molasses cycle the optical field is only red-detuned by one line width, and the atoms must therefore be excited at very long range, near 1800 $a_0$. The collision travel time to the close internuclear separation where associative ionization takes place is long compared to the radiative lifetime, and most of the population decays to the ground state before reaching the autoionization zone. During the trap cycle, however, the excitation takes place at much closer internuclear distances due to a seventy-line-width red detuning and high-intensity field dressing. Therefore one might expect excitation survival to be better during the trap cycle than during the molasses cycle, and the NIST group, led by W. D. Phillips, set up an experiment to test the predicted cycle-dependence of the ion rate.

Lett *et al.*[57] performed a new experiment using the same hybrid trap. This time, however, the experiment measured ion rates and fluorescence separately as the hybrid trap oscillated between "trap" and "molasses" cycles. The results from this experiment are shown in Fig. 10.16. While

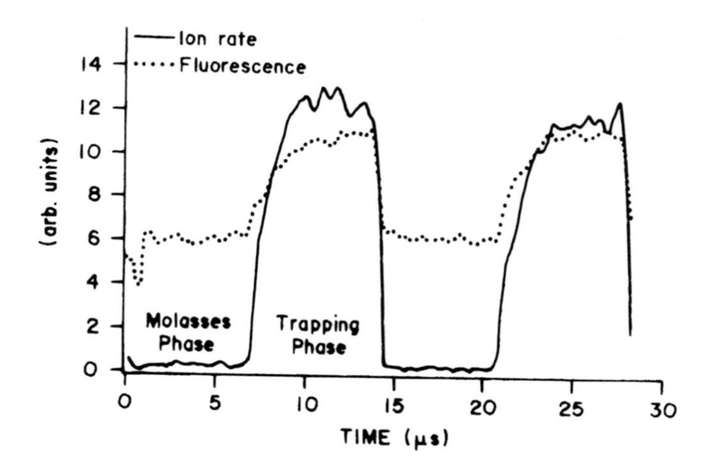


Fig. 10.16. Trap modulation experiment showing much greater depth of ion intensity modulation (by more than one order of magnitude) than fluorescence or atom number modulation, demonstrating that excited atoms are not the origin of the associative ionizing collisions, Lett *et al.*[57]

keeping the total number and density of atoms (excited atoms plus ground state atoms) essentially the same over the two cycles and while the excited state *fraction* changed only by about a factor of two, the ion rate increased in the trapping cycle by factors ranging from 20 to 200 with most observations falling between 40 and 100. This verified the predicted effect qualitatively even if the magnitude was smaller than the estimated $10^4$ factor of Julienne.[56] This modulation ratio is orders of magnitude more than would be expected if excited atoms were the origin of the associative ionization signal. Furthermore, by detuning the trapping lasers over 4 GHz to the red, Lett *et al.*[57] continued to measure ion production at rates comparable to those measured near the atomic resonance. At such large detunings reduction in atomic excited state population would have led to reductions in ion rate by over four orders of magnitude had the excited atoms been the origin of the collisional ionization. Not only did far off-resonance trap cycle detuning maintain the ion production rate,[57] evidence of peak structure was observed in the ion signal as the dipole trap cycle detuned to the red.

To interpret the experiment of Lett *et al.*[57] Julienne and Heather[45] proposed a mechanism which has become the standard picture for cold and ultracold photoassociative ionization. Fig. 10.17 details the model. Two colliding atoms approach on the molecular ground state potential. During the molasses cycle with the optical fields detuned only about one line width to the red of atomic resonance, the initial excitation occurs at very long range, around a Condon point at 1800 $a_0$. A second Condon point

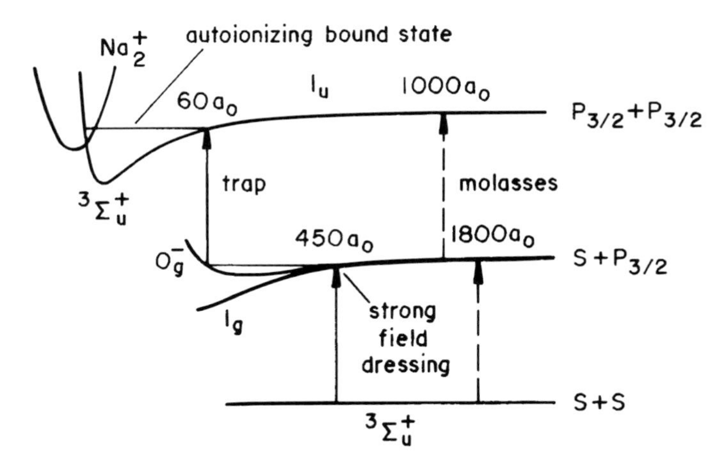


Fig. 10.17. Photoassociative ionization (PAI) in Na collisions, from Heather and Julienne.[58]

at 1000 $a_0$ takes the population to a $1_u$ doubly excited potential that, at shorter internuclear distance, joins adiabatically to a $^3\Sigma_u^+$ potential, thought to be the principal short-range entrance channel to associative ionization.[59,60] More recent calculations suggest other entrance channels are important as well.[61] The long-range optical coupling to excited potentials in regions with little curvature implies that spontaneous radiative relaxation will depopulate these channels before the approaching partners reach the region of small internuclear separation where associative ionization takes place. The overall probability for collisional ionization during the molasses cycle remains therefore quite low. In contrast, during the trap cycle the optical fields are detuned sixty line widths to the red of resonance, the first Condon point occurs at 450 $a_0$; and, if the trap cycle field couples to the $0_g^-$ long-range molecular state,[62] the second Condon point occurs at 60 $a_0$. Survival against radiative relaxation improves greatly because the optical coupling occurs at much shorter range where excited-state potential curvature accelerates the two atoms together. Julienne and Heather[45] calculated about a three-order-of-magnitude enhancement in the rate constant for collisional ionization during the trap cycle. The dashed and solid arrows in Fig. 10.17 indicate the molasses cycle and trap cycle pathways, respectively. The strong enhancement of the rate constant for collisional ionization in the trap cycle calculated by Julienne and Heather[45] is roughly consistent with the measurements of Lett *et al.*[57] although the calculated modulation ratio is somewhat greater than what was actually observed. Furthermore, Julienne and Heather calculate structure in the trap detuning spectrum. As the optical fields in the dipole trap tune to the red, a rather congested series of ion peaks appear which Julienne and Heather ascribed to free-bound association resonances corresponding to vibration-rotation bound levels in the $1_g$ or the $0_g^-$ molecular excited states. The density of peaks corresponded roughly to what Lett *et al.*[57] had observed; these two tentative findings together were the first evidences of a new photoassociation spectroscopy. In a subsequent full paper expanding on their earlier report, Heather and Julienne[58] introduced the term, "photoassociative ionization" to distinguish the two-step optical excitation of the quasimolecule from the conventional associative ionization collision between excited atomic states. In a very recent paper,[63] have developed a perturbative quantum approach to the theory of photoassociation which can be applied to the whole family of alkali homonuclear molecules. This study presents a useful table of photoassociation rates which reveals an important trend toward lower rates of molecule formation as the alkali mass increases and provides a helpful guide to experiments designed to detect ultracold molecule production.

## 10.6. Photoassociative Ionization in Atom Beams

Collisions in an atom beam can reveal polarization, orientation and alignment properties that are averaged out in "cell" experiments using MOTs.

Until the advent of cold, bright beams, no PAI spectra exhibiting alignment and orientation information had been measured in intra-beam collisions due to low beam density and relatively high divergence. However, DeGraffenreid *et al.*[64] have developed a dense, optically brightened beam specifically to meet the requirements of collision studies. In addition to low angular divergence (high brightness) combined with low velocity dispersion (high brilliance) the atom beam density of $(1 \pm 0.5) \times 10^{10}$ cm$^{-3}$ is comparable to that obtained in conventional MOTs. This arrangement produces an intra-beam PAI collision rate sufficient for high-resolution PAI spectroscopy.

In general, even if an atomic beam is very bright and brilliant, collisions within it will be subject to spatial averaging if the final scattering channel is coupled to a wide acceptance angle (in the molecular coordinate frame) of the reactant entrance channel. However, when the entrance channel angle is very narrow, the molecular collision axis and the atom beam laboratory axis will superpose; and fortunately PAI falls into this category. The range of possible molecular collision axes in PAI is restricted to a narrow acceptance angle determined by the long-range–short-range two-step mechanism expressed by Eqn. (10.63) and illustrated in Fig. (10.17). Therefore aligning the polarization axis of the laser light parallel or perpendicular to the bright atom beam axis is tantamount to aligning it parallel or perpendicular to the molecular collision axis.

Ramirez-Serrano *et al.*[49] used a cold, bright sodium beam to study PAI in intra-beam collisions at a collision temperature of $\simeq$4 mK. They recorded the one-color PAI rate as a function of red detuning from 0 to 45 GHz and as a function of the laser polarization alignment, parallel or perpendicular to the atom beam. Fig. 10.18 shows the sensitivity to polarization and a comparison with a typical MOT spectrum in a restricted range of detuning, within the first 5 GHz to the red of the atomic transition, $3\,^2S_{1/2}(F = 2) \rightarrow 3\,^2P_{3/2}(F = 3)$. The comparison in Fig. 10.18 illustrates three significant differences between MOT and beam spectroscopy. The beam measurements evidently show the expected marked polarization dependence in the peak amplitudes, but they also show a slightly improved spectral resolution over the MOT results, and new peaks within a detuning range from 0 to 750 MHz where the photoassociation (PA) laser beam destabilizes a MOT.

In the first 5 GHz detuning, the two-step PAI process consists of an initial excitation to a "pure long-range" state[62] of $0_g^-$ symmetry, followed

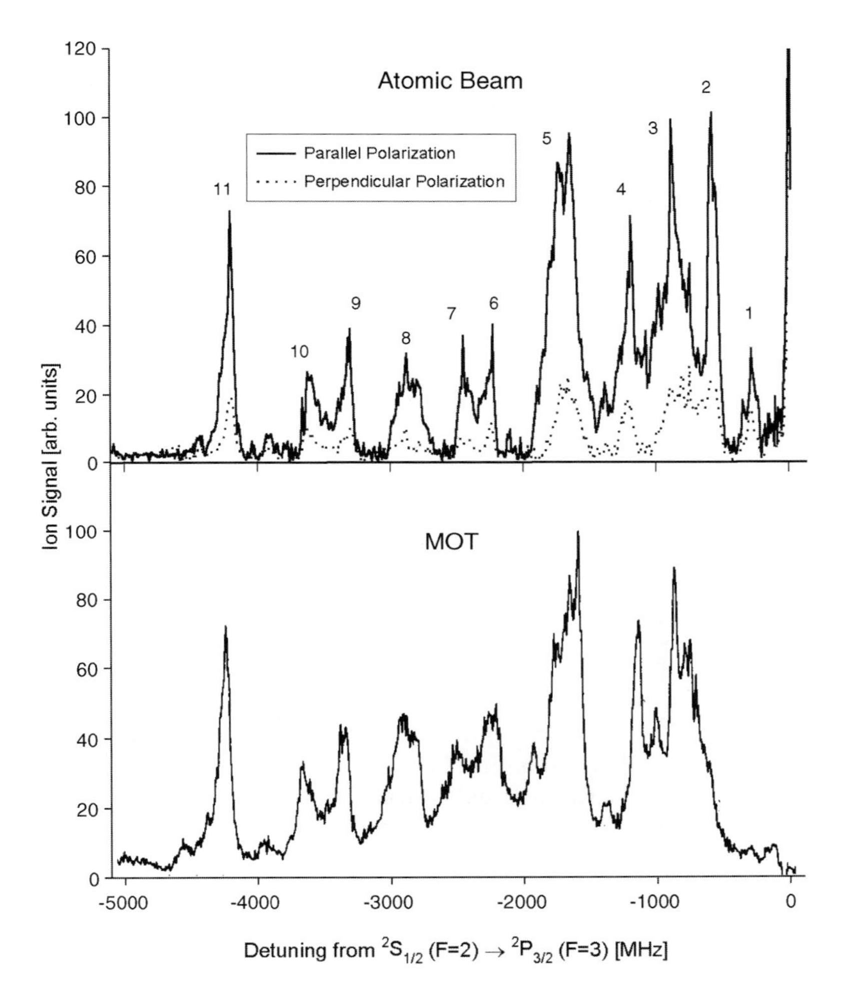


Fig. 10.18. Top panel: Bright beam PAI spectra within the first 5 GHz detuning to the red of resonance. Solid trace is parallel polarization. Dotted trace is perpendicular polarization. Peak numbered labels are for reference only. Bottom panel: PAI spectra obtained from a conventional MOT, from Fig. 5 of Ref. 65.

by a second step which transfers population to the doubly-excited state of either $1_u$ or $0_u^-$ symmetry.[65,66] The doubly-excited state subsequently autoionizes at short range. Since the first excitation step must pass through a $0_g^-$ state, dipole selection rules dictate that the starting ground molecular state must be of $1_u$ or $0_u^-$ symmetry, corresponding to components of the familiar $Na_2$ $^3\Sigma_u$ lowest triplet ("ground") state in Hund's case (b)

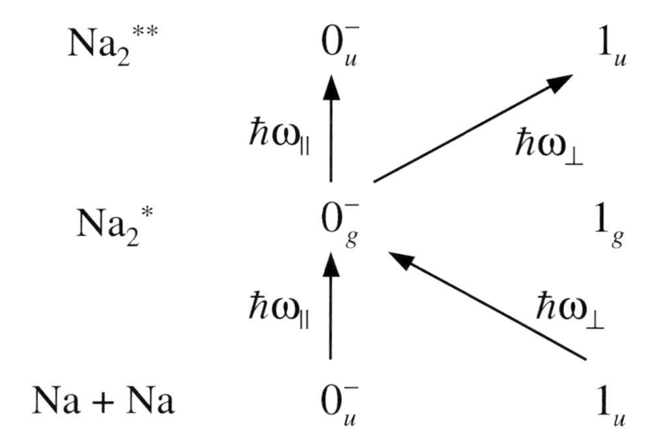


Fig. 10.19. Schematic diagram showing how two successive parallel or perpendicular transitions, passing through the $0_g^-$ intermediate state, selectively populate the $0_u^-$ or $1_u$ doubly excited states in the region 0 to 5 GHz detuning.

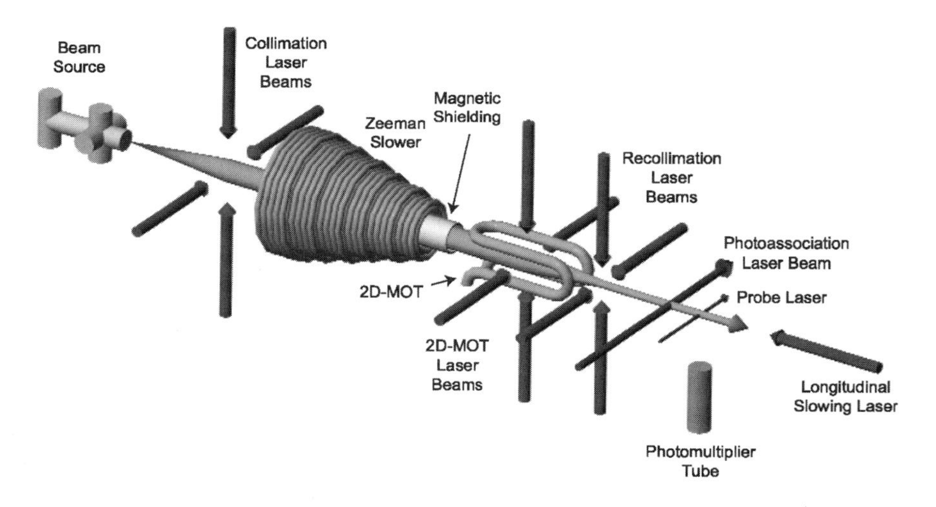


Fig. 10.20. Schematic of bright beam apparatus.

notation. Electric dipole transitions in which the quantum number for the total electronic angular momentum projection along the quantization axis is unchanged or differs by one unit, $\Delta\Omega = 0(\pm 1)$, are called parallel (perpendicular) transitions because the molecular states are coupled by the dipole transition moment parallel (perpendicular) to the quantization axis. Hence, PA light with parallel (perpendicular) polarization only couples the $0_u^-$ $(1_u)$ ground states to the intermediate $0_g^-$ state. Starting from the common $0_g^-$

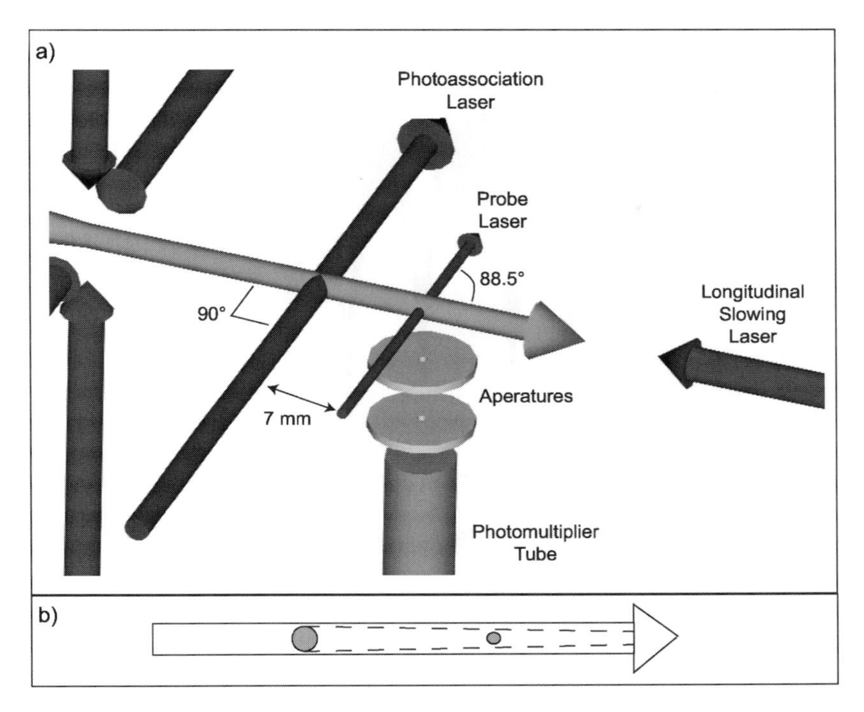


Fig. 10.21. Detail of the beam-loss interaction zone. (a) Physical disposition of the photoassociation and probe lasers. The probe laser is set at an angle of 88.5 degrees with respect to the atom beam axis and Doppler tuned onto resonance with the atoms in the beam. This "Doppler tuning" helps discriminate against background fluorescence from thermal atoms. (b) Diagram of how the fluorescence "hole" is created downstream.

intermediate state, parallel (perpendicular) polarized light populates the doubly-excited state $0_u^-(1_u)$ symmetry. Thus, as summarized in Fig. 10.19, light polarized parallel or perpendicular to the atom beam axis populates selectively the doubly excited state of either symmetry from the ground state. These excitation routes in Fig. 10.19 show that the second step should populate *either* the $0_u^-$ state (parallel polarization) *or* the $1_u$ state perpendicular polarization. Therefore, we would expect those peaks in Fig. 10.18 associated with $0_g^- \rightarrow 0_u^-$ to appear only with parallel polarization and the remaining peaks, associated with $0_g^- \rightarrow 1_u$ to appear only with perpendicular polarization. In fact very nearly the same set of peaks appears with excitation by both polarizations, but with quite different intensities. An analysis of the intensity ratios for the two polarizations indicates that the two doubly-excited states may not be well-characterized by the symmetry labels $0_u^-$ and $1_u$ and that the actual states may be about equal

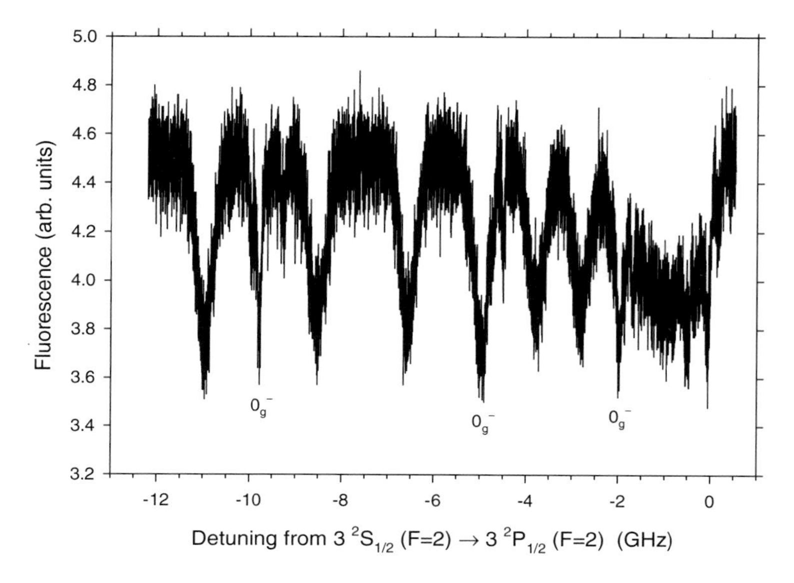


Fig. 10.22. Beam-loss spectrum for parallel polarization over the range of measurements from to 0 to 12 GHz detuning to the red of resonance for an unpolarized atomic beam.

mixtures of these two. The symmetry labels had been assigned from a spectroscopic analysis of the rotational branches of the rovibrational levels in the PAI spectra.[66] The polarization data appear to be at variance with this analysis, and the problem remains unresolved. In addition to photoassociative ionization spectroscopy, Ramirez-Serrano *et al.*[67] have developed a "beam-loss spectroscopy" in which photoassociation is induced in the cold beam flux by a scanning laser, and the resultant holes in the atom beam density are monitored by dips in the fluorescence intensity slightly downstream from the photoassociation zone. This technique is analogous to "trap-loss spectroscopy" reviewed in Weiner *et al.*[1] Fig. 10.20 shows a schematic of the experimental setup, and Fig. 10.21 a close-up of the interaction zone. This beam-loss spectroscopy also shows high-resolution photoassociation spectra, but the position and widths of the structures are modified by the high intensity (600 $\mathrm{W\,cm^{-2}}$) of the photoassociation laser. Fig. 10.22 shows a typical spectrum with some of the features identified as belonging to the pure long-range $0_g^-$ state of the photoassociated molecule. A zoom on the narrow feature around 9.8 GHz reveals the rotational state progression. When the colliding atoms are spin polarized they become identical bosons and all odd rotational states disappear, as shown in Fig. 10.23.

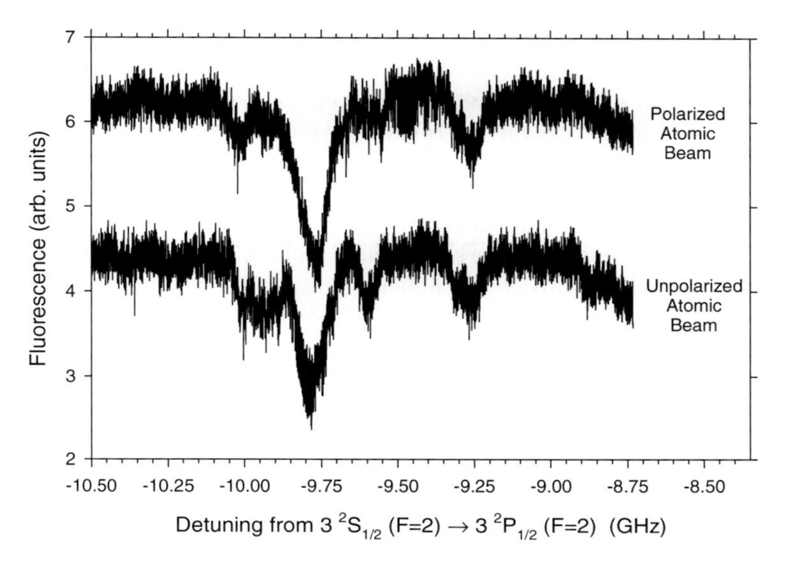


Fig. 10.23. A "zoom" on the spectrum shown in Fig. 10.22 around the sharp feature at ≃9.8 GHz. (a) Top panel shows rotational progression with spin polarized. (b) Bottom panel show rotational progression with spins unpolarized. Note that all odd rotational levels are missing from (a) consistent with the boson symmetry of two identical composite bosons. Note also that the unusually large peak at $J = 2$ indicates a "$d$-wave resonance"

## 10.7. Ground State Collisions

Much of this chapter has focused on collisions of cold, trapped atoms in a light field. Understanding such collisions is clearly a significant issue for atoms trapped by optical methods, and historically this subject has received much attention by the laser cooling community. However, there also is great interest in ground state collisions of cold neutral atoms in the absence of light. Most of the early interest in this area was in the context of the cryogenic hydrogen maser or the attempt to achieve Bose-Einstein condensation (BEC) of trapped doubly spin-polarized hydrogen. More recently the interest has turned to new areas such as pressure shifts in atomic clocks or the achievement of BEC in alkali systems. The actual realization of BEC in $^{87}$Rb,[68] $^{23}$Na,[69] $^{7}$Li,[70,71] $^{4}$He*[72,73] and H[74] has given a tremendous impetus to the study of collisions in the ultracold regime. Collisions are important to all aspects of condensates and condensate dynamics. The process of evaporative cooling which leads to condensate formation relies on elastic collisions to thermalize the atoms. The highly successful mean field theory of condensates depends on the sign and magnitude of the $s$-wave scattering length to parameterize the atom interaction energy that determines the mean field wave function. The ability of evaporative cooling to achieve the BEC transition, and the lifetime of the condensate, depend on having sufficiently small inelastic collision rates that remove trapped atoms through destructive processes. Therefore, ground state elastic and inelastic collision rates, and their dependence on magnetic or electromagnetic fields, is a subject of considerable importance.

This section will review work on ground state collisions of trapped atoms in the regime below 1 mK, with particular emphasis on the ultracold regime below 1 $\mu$K. The work on ground state collisions could easily be the subject of a major review article in its own right, so our review will be limited in scope. There is no claim to an exhaustive treatment. As for collisions in a light field, the subject is presented historically with emphasis on key concepts and measurements. The first section will review the early work in the field, including a brief survey of the work on hydrogen. A second section will discuss the role of collisions in BEC.

## 10.8. Early work

We noted in Sec. 10.3.2 that the quantum properties are quite well known for collisions where the de Broglie wavelength is long compared to the range of the potential. We confine our interest here to the special case of the collision of two neutral atoms at temperatures of less than 1 K. Interest in this subject was stimulated in the 1980's by two developments: the possibility of achieving Bose-Einstein condensation (BEC) with magnetically trapped spin-polarized hydrogen,[75–77] and the prospects of unparalleled frequency stability of the cryogenic hydrogen maser.[78–81] The ground state hydrogen atom has a $^2$S electronic state and a nuclear spin quantum number of 1/2. Coupling of the electron and nuclear spins gives rise to the well-known $F = 0$ and $F = 1$ hyperfine levels of the ground state. The transition between these two levels is the hydrogen maser transition, and the doubly spin polarized level, $F = 1$, $M = 1$, with both electron and nuclear spins having maximum projection along the same axis, is the one for which BEC is possible in a magnetic atom trap. Both the hydrogen maser and the phenomenon of BEC are strongly affected by atomic collisions of ground state hydrogen atoms. Collisions cause pressure-dependent frequency shifts in the maser transition frequency that must be understood and controlled,[82–85] and they cause destructive relaxation of the spin-polarized H atoms that can prevent the achievement of BEC.[86–89]

In the 1980s these developments stimulated theoretical calculations on the low temperature collision properties of atomic H and its isotopes.

Earlier work[90–92] had laid the groundwork for understanding inelastic spin exchange collisions by which two H atoms in the $F = 1$ state undergo a transition so that one or both of the atoms exit the collision in the $F = 0$ state. Reference 93 extended these calculations of the spin exchange cross section and collisional frequency shift of the hyperfine transition to the low temperature limit. These early calculations were based on extending the high temperature theory, based on knowing the phase shifts of the ground state molecular hydrogen $^1\Sigma_g$ and $^3\Sigma_u$ potentials alone without explicit inclusion of the hyperfine structure of the separated atoms. A proper quantum mechanical theory based on numerical solution of the coupled channel Schrödinger equation for the atoms with hyperfine structure, also known as the close coupling method, was introduced by the Eindhoven group, and applied to frequency shifts in hydrogen masers[84,85,94] and relaxation of doubly spin-polarized hydrogen in a magnetic trap.[88,95] The close coupling method is a powerful numerical tool and is the method of choice for quantitative calculations on ground state collisions. It is the best method currently available and has been applied to a variety of species, including mixed species. A recent discussion of the multichannel scattering theory for cold collisions has been given by Gao.[20]

Collisions of species other than hydrogen have been investigated. Uang and Stwalley[96] looked at collisions of hydrogen and deuterium in cold magnetic traps to assess the role of deuterium impurities in a cold spin-polarized hydrogen gas. Koelman *et al.*[97] and Koelman, Stoof *et al.*[98] calculated the lifetime of a spin-polarized deuterium gas, while Tiesinga *et al.*[99,100] examined frequency shifts in the cryogenic deuterium maser. More recently Jamieson *et al.*[101] have calculated collisional frequency shifts for the 1S-2S two-photon transition in hydrogen. Tiesinga *et al.*[99] calculated that the relaxation rate coefficient for doubly spin-polarized Na would be about ten times larger than for spin-polarized hydrogen. Tiesinga *et al.*[102] also calculated that the frequency shift in a cesium atomic fountain clock[103,104] might be large enough to limit the anticipated accuracy of such a clock. The use of neutral atoms in ultra-precise atomic clocks is discussed by Gibble and Chu[105,106] who measured large collisional frequency shifts in a Cs fountain. These have been verified by another experiment.[107] Verhaar *et al.*[108] argue that other cold collision properties can be deduced from these clock shift measurements, namely, that doubly spin-polarized $^{133}$Cs probably has a large negative scattering length; they criticize the opposite conclusion drawn from the same data by Gribakin and Flambaum[27] due to restrictive approximations used by the latter. Gibble and Verhaar[109] suggest that clock frequency shifts might be eliminated by using $^{137}$Cs. Kokkelmans *et al.*[110] use accurate collisional calculations for both isotopes of Rb to suggest that a Rb atomic clock will offer better performance than a

Cs one. Gibble *et al.*[111] have directly measured the $s$-wave scattering cross section and angular distribution in Cs atom collisions in an atomic fountain experiment at a temperature of $T = 0.89\,\mu$K.

## 10.9. Bose-Einstein Condensation

One of the major motivating factors in the study of collisions of cold ground state neutral atoms has been the quest to achieve Bose-Einstein condensation (BEC). This is a phase transition which occurs in a gas of identical bosons when the phase space density becomes large enough, namely, when there is about one particle per cubic thermal de Broglie wavelength. The specific criterion[112] for condensation is

$$n\left(\frac{2\pi\hbar^2}{mk_BT}\right)^{\frac{3}{2}} > 2.612, \tag{10.69}$$

where $n$ represents atom density. In the condensate, there is a macroscopic occupation by a large number of atoms of the single ground state of the many-body system, whereas in a normal thermal gas many momentum states are occupied with very small probability of occupying any given one. Achieving BEC means making the density large enough, or the temperature low enough, that Eqn. (10.69) is satisfied. The early work with spin-polarized hydrogen aimed at reaching high enough density using conventional refrigeration techniques. Unfortunately, this proved to be impossible due to the losses caused by collisions or surface recombination when the density was increased. An alternative approach was developed by using a magnetic trap without walls[113] and evaporative cooling[114] to reach much lower temperatures. The idea is to keep the density sufficiently low to prevent harmful collisions.

The success of laser cooling for alkali atoms gave an impetus to achieving BEC using alkali species. These are similar to hydrogen in that they have $^2$S ground electronic states with two hyperfine levels due to the nuclear spin (see Fig. 10.3). All alkali species have at least one stable isotope with odd nuclear spin, making the atom a composite boson. Unlike hydrogen, the alkali dimer $^3\Sigma_u$ state supports bound states; however, a metastable condensate is possible because of the long time scale required to make dimer bound states via three-body recombination at low condensate density. Ordinary laser cooling methods produce density and temperature conditions many orders of magnitude away from satisfying the phase space density criterion in Eqn. (10.69). The process of evaporative cooling was seen as a viable route to reaching BEC, and several groups set out to make it work. This approach was spectacularly successful, and within a few months of each other in 1995, three groups reached the regime of quantum degeneracy required for BEC. The first unambiguous demonstration of BEC in an evaporatively cooled atomic gas was reported by NIST/JILA group[68] for doubly spin polarized $^{87}$Rb, followed by evidence of BEC for doubly spin-polarized $^{7}$Li by a group at Rice University,[70] then demonstration of BEC at MIT[69] in the $F = 1, M = -1$ lower hyperfine component of $^{23}$Na. A much clearer demonstration of BEC for the $^{7}$Li system was given later.[71] By mid-1997, at least three additional groups used similar evaporative cooling and trap designs to achieve BEC in $^{87}$Rb and $^{23}$Na, and other attempts remain in progress elsewhere. Ten years later, BEC has been achieved by many different approaches by research groups all over the world. A large literature on the subject of BEC has already been generated. Some introductory articles on the subject are given by Refs. 115, 116 and 117.

Ground state collisions play a crucial role both in the formation of a condensate and in determining its properties. A crucial step in the formation of a condensate is the achievement of critical density and temperature by "evaporative cooling",[118] the process by which hot atoms are removed from the confined ensemble while the remaining gas thermalizes to a lower temperature. Elastic collisions are necessary for evaporative cooling to work, and the stability and properties of the condensate itself depend on the sign and magnitude of the elastic scattering length. Two- and three-body inelastic collisions cause destructive processes that determine the condensate lifetime. Therefore, these collisions have been of as much interest for alkali species as for hydrogen and have been the object of numerous experimental and theoretical studies.

First, the process of evaporation depends on elastic momentum transfer collisions to thermalize the gas of trapped atoms as the trapping potential is lowered. These elastic collisions represent "good" collisions, and they have cross sections orders of magnitude larger for alkali species than for hydrogen. During evaporation, the rate of inelastic collisions which destroy the trapped hyperfine level, the so-called "bad" collisions, must remain much less than the rate of elastic collisions. An excellent description of the role of these two types of collisions is given in the review on evaporative cooling by Ketterle and van Druten.[119] Long before alkali evaporative cooling was achieved, Tiesinga *et al.*[120] calculated that the ratio of the "good" to "bad" collisions appears to be very favorable ($\sim$1000) for both the $F = 4, M = 4$ and $F = 3, M = -3$ weak-field-seeking (i.e., capable of being trapped) states of $^{133}$Cs. Precise predictions were not possible due to uncertainties in the interatomic potentials. Monroe *et al.*[121] used time dependent relaxation of trapped atoms to measure the elastic cross section for $F = 3,\ M = -3$ ground state collisions near $30\,\mu$K. They measured a large value $(1.5(4) \times 10^{-12}\,\text{cm}^2)$ and found it to be independent of temperature between 30 and 250 $\mu$K. The measurements implied a scattering length magnitude near 46(12) $a_0$, if the cross section is assumed to be due to s-wave collisions. Newbury *et al.*[122] similarly measured the elastic cross section for the $F = 1,\ M = -1$ state of $^{87}$Rb to be $5.4(1.3) \times 10^{-12}$ cm$^2$, implying a scattering length magnitude of 88(21) $a_0$. Measurements on thermalization in a $F = 1,\ M = -1$ $^{23}$Na trap by Ref. 123 deduced a scattering length of 92(25) $a_0$ for this level. A very recent study[33] of thermalization of doubly spin-polarized $^{133}$Cs $F = 4, M = 4$ showed that the scattering length magnitude was greater than 260 $a_0$, and the elastic scattering cross section was near the upper bound given by the $s$-wave unitarity limit, $8\pi/k^2$, between 5 and 50 $\mu$K. All of these experimental studies measured total cross section only, and therefore were not able to determine the sign of the scattering length. Verhaar *et al.*,[108] however, calculated the sign to be negative for $^{133}$Cs $F = 4,\ M = 4$ .

The second way in which atomic interactions profoundly affect the condensate properties is through their effect on the energy. The effect of atom-atom interactions in the many-body Hamiltonian can be parameterized in the $T \to 0$ limit in terms of the two-body scattering length.[124] This use of the exact two-body T-matrix in an energy expression is actually a rigorous procedure, and can be fully justified as a valid approximation.[125] One simple theory which has been very successful in characterizing the basic properties of actual condensates is based on a mean-field, or Hartree-Fock, description of the condensate wave function, which is found from the equation[126]:

$$\left(-\frac{\hbar^2}{2m}\nabla^2 + V_{trap} + NU_0|\Psi|^2\right)\Psi = \mu\Psi. \tag{10.70}$$

Here $V_{trap}$ is the trapping potential which confines the condensate and

$$U_0 = \frac{4\pi\hbar^2}{m}A_0 \tag{10.71}$$

represents the atom-atom interaction energy, proportional to the $s$-wave scattering length for the trapped atomic state, and $\mu$ is the chemical potential, i.e., the energy needed to add one more particle to a condensate having $N$ atoms. The condensate wave function in this equation, called the Gross-Pitaevski equation or the nonlinear Schrödinger equation (NLSE), can be interpreted as the single ground state orbital occupied by each boson in the product many-body wave function

$$\Phi(1\ldots N) = \prod_{i=1}^{N}\Psi(i).$$

The wave function $\Psi$ could also be interpreted as the order parameter for the phase transition that produces the condensate.

The effect of atom interactions manifests itself in the NLSE through the mean-field term proportional to the local condensate density, $N|\Psi|^2$, and the coupling parameter $U_0$ proportional to the $s$-wave scattering length $A_0$. In an ideal gas, with no atom interactions, $A_0 = 0$, and this term vanishes. Equation (10.70) shows that the condensate wave function in such an ideal-gas case just becomes that for the zero point motion in the trap, that is, the ground state of the trapping potential. For typical alkali traps, which are harmonic to a good approximation, the zero point motion typically has a frequency on the order of 50 to 1000 Hz, and a range on the order of 100 $\mu$m to a few $\mu$m. For comparison, $k_BT/h = 20$ kHz at $T = 1\,\mu$K. In actuality, as atoms are added to the condensate the atom interaction term becomes the dominant term which affects the condensate wave function; and the shape of the condensate depends strongly on the size of the $U_0$ term, which is proportional to the product $NA_0$. The sign of the scattering length is crucial here. If $A_0$ is positive, the interaction energy increases as more atoms are added ($N$ increases) to the condensate. The condensate is stable and becomes larger in size as more atoms are added. In fact, a very simple approximation, called the Thomas-Fermi approximation, gives the condensate density $n$ by neglecting the kinetic energy term in Eqn. (10.70) in relation to the other terms:

$$n = N|\Psi|^2 = \frac{\mu - V_{trap}}{U_0}. \tag{10.72}$$

This equation is remarkably accurate except near the edge of the trap where $\mu - V_{trap}$ approaches 0 or becomes negative, and describes condensates of $^{87}$Rb $F = 2$, $M = 2$ and $F = 1$, $M = -1$ and $^{23}$Na $F = 1$, $M = -1$, all of which have positive scattering length. Condensates of such species can be made with more than $10^6$ atoms. On the contrary, if the scattering length is negative, as for $^7$Li, increasing the number of particles in the trap makes the interaction energy term in Eqn. (10.70) become more negative. The wave function contracts as more particles are added, and in fact, only about 1000 atoms can be added to a $^7$Li condensate before it becomes unstable and can hold no more atoms.[71,127,128] A condensate with negative scattering length is not possible in a uniform homogeneous gas, but in an atom trap the presence of zero-point motion does permit the existence of a very small condensate, as is the case for $^7$Li.

It is perhaps not obvious why a collisional property like a scattering length determines the energetics of the interacting particles. A simple heuristic argument to indicate why this is the case can be given in relation to Fig. 10.5 in Sec. 10.3.2. The long wavelength scattering wave has its phase shifted near $R = 0$ by the interaction potential. From the perspective of the asymptotic wave the effective origin of the oscillation near $R = 0$

is shifted by the presence of the potential, to $R = A_0 > 0$ for the case of positive $A_0$ and to $R = A_0 < 0$ for the case of negative $A_0$, as Fig. 10.5(b) shows. The kinetic energy associated with the long wavelength asymptotic wave is affected by this shift in effective origin of oscillation. If one thinks of the two-particle system in terms of a single reduced-mass particle-in-a-box, the left hand wall of the box at $R = 0$ is moved to larger or smaller R, depending on the sign of the scattering length. What is important for the energetics is whether the change in energy is positive or negative, relative to noninteracting atoms (the box with a wall at $R = 0$). Given that the energy of the ground state of a particle of reduced mass $m/2$ in a box of length $L$ is

$$E_{box} = \frac{\hbar^2}{m}\left(\frac{\pi}{L}\right)^2,$$

it is easy to work out that changing the length of the box from $L$ to $L + A_0$ changes the energy by an amount proportional to $A_0/m$, thus lowering the energy for the case of negative $A_0$ and raising it for the case of positive $A_0$. A rigorous analysis gives the coupling term in Eqn. (10.71).

The gas thermalization studies which measure the elastic scattering cross section cannot determine the sign of the scattering length. The sign can be determined from photoassociation spectroscopy, which is sensitive to the shape and phase of the ground state collisional wave function. Much tighter constraints have also been placed on the magnitudes of scattering lengths by photoassociation studies. Of course, observation of the condensate properties determines the sign as well. Even the magnitude of the scattering length can be found by measuring the expansion of the condensate when the trapping potential is removed, since the expansion rate depends on the strength of the interaction term in the initial condensate. This procedure has been carried out for both the $^{87}$Rb[129] and $^{23}$Na[130–132] condensates, confirming the magnitude of the scattering lengths determined more precisely by other means. Before the determinations by photoassociation spectroscopy, the sign and magnitude of the scattering lengths had also been estimated from the best available interatomic potentials based on spectroscopically derived potential wells and the long range van der Waals potentials. This was done for $^{7}$Li[133,134] and $^{23}$Na,[133,135] but the accuracy is not as good as for later determinations by photoassociation spectroscopy. Boeston *et al.*[136] have used spectroscopic data on K to calculate threshold scattering properties of $^{39}$K and $^{41}$K. Côté and Dalgarno[137] show that revised ground state potentials for the $K_2$ molecule need to be used, and these give different scattering lengths than the calculation of Ref. 136. The existence of photoassociation spectra for $K_2$ should permit determination of more accurate threshold scattering properties of both isotopes in the future.

The third way in which collisions are significant for BEC is the role of inelastic collisions which change the state of the trapped species. This process can produce strong-field-seeking states (i.e., states not capable of being trapped), as well as cause heating by releasing kinetic energy in the inelastic process. For example, if two doubly spin-polarized atoms collide and produce one or two atoms in the lower hyperfine state, then one or two units of the ground state hyperfine splitting is given to the atoms, to be shared equally among them. Since this splitting is typically many hundreds of mK, the atoms easily escape the shallow traps designed to hold atoms at a few $\mu$K or less. We have already discussed how these "bad" collisions can affect the process of evaporative cooling. Fortunately, the ratio of "good" to "bad" collisions is favorable in many cases, so that evaporation actually works well and results in BEC. One case where evaporation appears to be unlikely due to "bad" collisions is the case of doubly spin-polarized $^{133}$Cs. Experiments by Soding *et al.*[138] found that the inelastic rate coefficient for destruction of doubly spin-polarized $^{133}$Cs atoms is so large, about three orders of magnitude larger than predicted by,[120] that evaporative cooling of this level will be impossible. Leo *et al.*[34] explain this unusually large inelastic collision rate as being due to a second order spin-enhancement of the effective spin-dipolar coupling. Evaporative cooling of the lower hyperfine level, $^{133}$Cs $F = 3, M = -3$ may still be feasible, given the relatively low collisional destruction rate for this level[121]). However, quite recently the sign of the scattering length and its magnitude have been determined from Feshbach resonance spectroscopy.[139,140] These values are $a_S = (280 \pm 10)a_0$ and $a_T = (2400 \pm 100)a_0$ for the singlet and triplet scattering lengths, respectively. Earlier studies[141,142] had concluded that the triplet scattering length was negative, but this latest large positive value for $a_T$ means that a Cs BEC cannot be ruled out.

Measuring an inelastic collision rate in a condensate, compared to the corresponding collision rate in a thermal gas, provides a way to probe the quantum mechanical coherence properties of the condensate. Ref. 143 has shown how this can be done. Although the first-order coherence of a condensate can be measured by observing the interference pattern of two overlapping condensates,[144] collisions probe higher order coherence properties related to the different nature of the density fluctuations in a thermal gas and in a condensate. For example, for ordinary thermal fluctuations, the average of the square of the density is two times larger than the square of the average density, whereas for a condensate, these two quantities are equal. The second and third order coherence functions, $g^{(2)}(0)$ and $g^{(3)}(0)$ respectively, measure this effect, where the argument 0 implies that two or three particles are found at the same position; $g^{(2)}(0) = 2!$ and $g^{(3)}(0) = 3!$ for thermal gases, and both are 1 for a condensate. Thus atoms are bunched in a thermal source, but not in a condensate, analogous to photons in a thermal source and a laser. Kagan *et al.*[145] suggested that the three body recombination rate might provide a way to measure this property, since the rate coefficient for three-body recombination would be $3! = 6$ times smaller in a condensate than in a thermal gas. Burke *et al.*[143] in fact have done just this, measuring $g^{(3)}(0) = 7.4(2)$ by comparing the measured three body recombination rates for thermal and condensed $^{87}$Rb in the $F = 1, M = -1$ level.

Condensate coherence can also be probed with two-body inelastic collisions as well as three-body ones.[132] Stoof *et al.*[146] use a collision theory viewpoint to show the difference of collision rates in a thermal gas and a condensate, showing that the corresponding rate for a 2-body collision in a condensate is two times smaller than for the thermal gas. Reference 147 points out that a condensate will have a 2-body photoassociation spectrum, which will also show the factor of two decrease in rate coefficient relative to a thermal gas. In addition, a condensate photoassociation spectrum would probe the two-body part of the many-body wave function in much more detail than an overall collision rate, since it probes the wave function over a range of interparticle separations instead of just yielding $g^{(2)}(0)$, as an overall collision rate does. Photoassociation should be readily observable in a condensate, since at the frequencies of photoassociation lines, the light absorption rate due to two-body photoassociation at typical condensate densities will greatly exceed the light scattering rate by free atoms. An interesting question about photoassociation in a condensate is whether a three-body spectrum could be observed, in which excited triatomic molecules are formed from three nearby ground state atoms. If so, this could provide a means for a finer grained probing of three-body effects in a condensate.

Inelastic collisions, including three-body recombination, are important for condensates and condensate formation because these can lead to heating or removal of atoms from the system. We discussed above how "bad" collisions can affect evaporative cooling. The lifetime of a condensate itself will be determined by collision processes. If the vacuum is not sufficiently low, hot background gas species can collide with trapped or condensed atoms, thereby transferring momentum to the atoms. Since the background gas atoms typically have energies on the order of 300 K, the cold atoms receive enough momentum to be ejected from the trap. There is also the possibility that some glancing collisions may transfer a very slight amount of momentum, producing hot but still trapped atoms. This could be a source of heating processes of unknown origin that have been observed in magnetic atom traps.[121,130] Even if the vacuum is good enough, inelastic collisions among the trapped species themselves can limit the lifetime of the trapped gas or condensate. If the collision rate coefficient for the destructive process is $K_{in}^{(k)}$ for a k-body collision, the trap lifetime is $(K_{in}^{(k)} n^k)^{-1}$. For a nominal

$10^{14}$ cm$^{-3}$ atom density in a condensate, a one-second lifetime results from a two-body rate coefficient of $10^{-14}$ cm$^3$ s$^{-1}$ or a three-body rate coefficient of $10^{-28}$ cm$^3$ s$^{-1}$. These rate coefficients will be very dependent on the species and the particular hyperfine level which is trapped. The only hyperfine levels for which trapping and condensation are possible are those for which the inelastic rate coefficients are sufficiently low.

We noted above how two-body inelastic collision rate coefficients must be small in relation to elastic collision rate coefficients in order for evaporative cooling to work. This is true for the $F = 2,\ M = 2$ $^{87}$Rb and $^7$Li species and $F = 1,\ M = -1$ $^{23}$Na species that have been condensed. The inelastic rate is small for the doubly spin-polarized species for the reasons discussed in Sec. 10.7 above. It is small for $F = 1,\ M = -1$ collisions for basically the same reason. An inelastic collision requires a weak spin-dipolar mechanism, since the sum of atomic $M$ is not conserved, and additionally, the exit channel is a $d$-wave for an $s$-wave entrance channel. The small amplitude of the threshold $d$-wave leads to very small collisional destruction of $F = 1,\ M = -1$ for weak magnetic fields. The success of sympathetic cooling, and observation of a dual condensate of $^{87}$Rb $F = 1,\ M = -1$ and $F = 2,\ M = 2$,[148] raises the obvious theoretical question why the inelastic collision rate coefficient for the destructive collision of these two species was found to be so small, $2.8 \times 10^{-14}$ cm$^3$ s$^{-1}$. This inelastic process, which produces two $F = 1$ hot atoms, goes by the spin-exchange mechanism, and normally would be expected to be several orders of magnitude larger. Three theory groups immediately answered the puzzle by showing that this observation meant that $^{87}$Rb had a very special property, namely the scattering lengths of both the $^1\Sigma_g$ and $^3\Sigma_u$ states are nearly the same, and in fact, all scattering lengths between any two hyperfine levels of $^{87}$Rb are nearly the same.[149–151] The existence of nearly identical scattering lengths is a sufficient condition for the inelastic rate coefficient to be as small as it is. It is not a necessary condition, since a threshold scattering resonance could also lead to a low inelastic rate coefficient. Such a resonance does not exist for $^{87}$Rb, however. Julienne *et al.*[149] pointed out that reconciling the existing data on $^{87}$Rb required that the scattering lengths for collisions between any two pairs of $F = 2,\ M = 2$ or $F = 1,\ M = -1$ differ by no more than 4 $a_0$ and have a value of $103 \pm 5$ a$_0$. Both Julienne *et al.*[149] and Kokkelmans, *et al.*[150] calculated that the inelastic collisional destruction rate coefficient for collisions of $^{23}$Na $F = 2, M = 2$ and $F = 1, M = -1$ would be 3 to 4 orders of magnitude larger than the one measured for $^{87}$Rb. Consequently a dual species condensate would be impossible for $^{23}$Na. The issue of inelastic collision rate coefficients is crucial for the prospects of sympathetic cooling, which offers an attractive path for cooling species that cannot be cooled evaporatively, and needs to be investigated for mixed alkali species. For example, cooling is desirable for spin-polarized fermionic species such as $^6$Li, which may exhibit interesting Cooper pairing effects in the quantum degenerate regime.[125] Spin-polarized fermionic species cannot be cooled evaporatively, since only $p$-wave collisions are allowed, and these are strongly suppressed at low temperature (see Sec. 10.3.2).

Two-body rate coefficients for inelastic processes tend to be small for species that can be condensed, since otherwise the bad collisions will limit the trap density. But as density increases, three-body collisions will eventually provide a limit on trap density and lifetime. Three-body collisions produce a diatomic molecule and a free atom. These two products share the kinetic energy released due to the binding energy of the molecule. This is usually enough energy that the particles do not remain trapped; in any case, the molecule is unlikely to be trapped. Three-body collisions for spin-polarized hydrogen were studied by De Goey *et al.*[152,153] The collision rate coefficient is unusually small for this system, since the ground state triplet potential does not support any bound states in which to recombine, and making ground state singlet molecules requires a very weak spin-dipolar transition. The rate coefficient for alkali systems is orders of magnitude larger than for hydrogen, since the triplet potentials support several bound states with small binding energy. Moerdijk *et al.*[154] calculate the three-body rate coefficients for doubly spin-polarized $^7$Li, $^{23}$Na, and $^{87}$Rb to be 2.6, 2.0, and 0.04 $\times 10^{-28}\,\mathrm{cm}^3\,\mathrm{s}^{-1}$, respectively. Moerdijk and Verhaar[155] differ from the suggestion of Esry *et al.*[156] that the three-body rate coefficient may be strongly suppressed at low temperature. Reference 157 gives a simple formula based on the scattering length for the case when the last bound state in the potential is weakly enough bound. Reference 143 notes that this theory gives the magnitude of the measured three-body rate coefficient for $F = 1,\ M = -1$ $^{87}$Rb.

### 10.9.1. *Designer Condensates*

One of the more interesting prospects for tailoring the collisional properties of ground state species is to make use of an external field to modify the threshold collision dynamics and consequently change either the sign or magnitude of the scattering length or to modify the inelastic collision rates. This prospect was raised by Tiesinga, *et al.*[158] who proposed that threshold scattering properties for $^{133}$Cs could be changed by a magnetic field. This external control over the scattering length is possible because of the rapid variation in collision properties associated with a threshold scattering resonance. For example, a magnetic field can move a molecular bound state to be located at just the threshold energy for the collision energy of two levels of the lower hyperfine manifold. This sort of threshold phenomenon is called

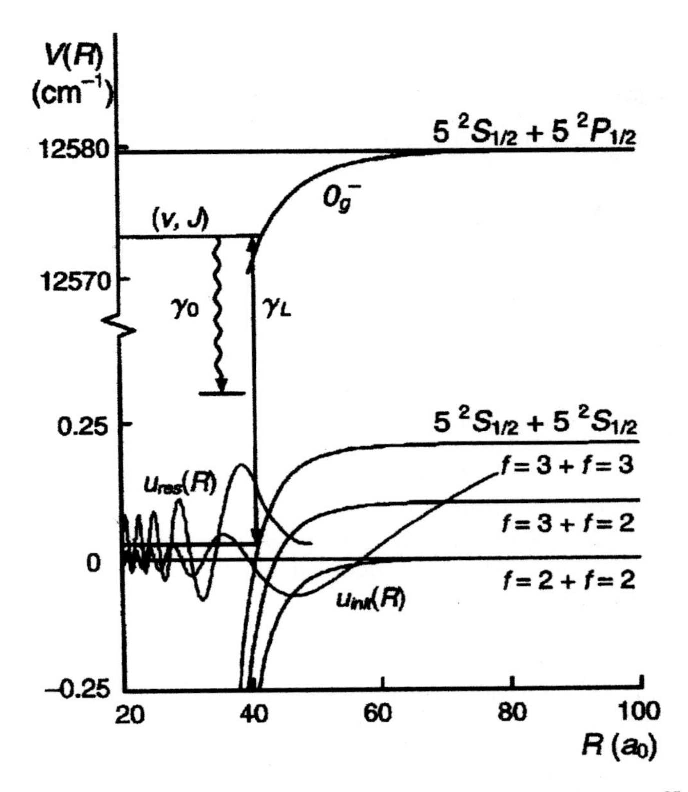


Fig. 10.24. Feshbach resonance in the case of ground state scattering from $^{85}$Rb atoms. The entrance channel wave function $u_{init}(R)$ couples to a quasibound sate with wave function $u_{res}(R)$. The energy level of the quasibound state is brought into a zero-energy resonance by tuning a magnetic field. Here the resonance is detected by enhancement of the photoassociation rate to a specific $v, J$ level of the $0_g^-$ excited state $\gamma_L$.

a *Feshbach resonance.* Fig. 10.24 shows the basic mechanism of a Feshbach resonance. An incoming $s$-wave scattering channel overlaps at zero collision energy with a quasibound state associated with a molecular asymptote of higher energy than the entrance channel. When the magnetic moments of the quasibound state and the $s$-wave are different, their Zeeman shifts will differ; and a magnetic field can be used to tune the quasibound level onto scattering resonance. Moerdijk *et al.*[159] discussed the role of resonances for $^6$Li, $^7$Li, and $^{23}$Na. An initial experimental attempt[122] to locate predicted resonances in $^{87}$Rb was unsuccessful but there was really no doubt that such scattering resonances existed. The question is whether they can be found in experimentally accessible regimes of magnetic fields. Using much more refined calculations of threshold scattering derived from photoassociation spectroscopy, Vogels *et al.*[160] made specific predictions that resonances in scattering of the $F = 2, M = -2$ lower hyperfine level of $^{85}$Rb would occur in experimentally accessible ranges of magnetic field. A Feshbach resonance was indeed found by Heinzen's group at the University of Texas[161] at a roughly measured magnetic field of about 167 Gauss, not too far from Vogels' prediction of 142 Gauss. Soon after the JILA group led by Wieman reported observing and measuring the same resonance with greatly improved precision[162] from which they were able to determine new and improved values for the triplet and singlet scattering lengths. Soon after they were able to report even newer and more improved values.[163] The JILA group has also used this tunable scattering length to study instabilities in a $^{85}$Rb BEC. By rapidly switching the scattering length from positive to negative values this group has been able to trigger a systematic and controlled collapse of the stable BEC.[164,165] Fig. 10.25 shows an image of an "exploded" BEC resulting from the sudden switching of the scattering length from an initial value $a_{init} = 0$ to negative values $a_{collapse}$ ranging from about $-2a_0$ to $-50a_0$. The detailed kinetics of this collapse is not fully understood. There appears to be an induction time, after the switch to $a_{collapse}$ but prior to the actual collapse, which decreases strongly with the absolute magnitude of $a_{collapse}$. Theory has predicted[166] that immediately after the switch to negative scattering length the BEC would begin to contract, raising the energy until an instability point is reached. The instability is thought to be caused by three-body collisions which release sufficient energy to expel some the atoms from the BEC but not to destroy the BEC entirely. However the BEC density at the point of collapse, inferred from imaging the post-collapse expansion of the residual BEC, does not appear to contract markedly, and the details of the collapsing mechanism is at this point uncertain. The MIT BEC group, led by W. Ketterle,[167] has also reported measurements of magnetically-induced Feshbach resonance effects on condensate mean-field energy and lifetime for Na $F = 1$, $M = +1$ confined in an optical trap. Reference 136 calculates that $^{41}$K may also offer good prospects for magnetic field tuning of the scattering length.

External fields other than magnetic ones can also change the scattering length. Reference 168 proposed that an optical field tuned near resonance with a photoassociation transition can be used to vary the ground state scattering length but not cause excessive inelastic scattering. Napolitano *et al.*[169] while investigating optical shielding, calculated that large changes in elastic scattering rates can be produced by light detuned by about 100 natural line widths to the blue of atomic resonance. This was partly because the optically induced mixing with the ground state and

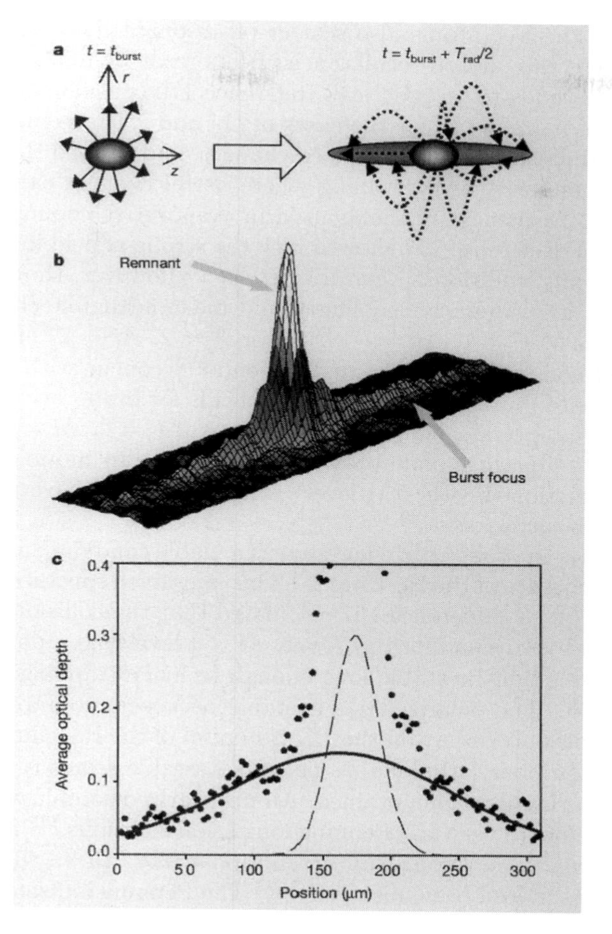


Fig. 10.25. This figure is from Ref. 165. Panel (a) shows a schematic of a "burst focus" in which the exploding BEC is expelled radially outward by inelastic collisions with energy sufficient to destabilize a fraction of the BEC but without enough kinetic energy to escape the trap. The atoms all fall back onto the radial axis after a time $T_{radial}/2$ where $T_{radial}$ is the period of the radial harmonic oscillation of the trap. The fact that all the atoms reach the radial axis at the same time produces the "focus". Panel (b) shows an image of the radial burst together with BEC remanent sharply peaked at the center. Panel (c) is a fit of the radial extent of the burst which is a measure of the kinetic energy.

an excited repulsive state with the $1/R^3$ long range form greatly changed the ground state collision by inducing contributions from angular momenta other than $s$-waves. The problem with using light to change ground state collision rates is that the light can also induce harmful inelastic processes.

Furthermore the free atoms also scatter off-resonant light and experience heating due to the photon recoil. Such effects must be minimized in order for optical methods to be practical. Reference 170 considered the proposal of Ref. 168 in more detail for the cases of $^7$Li and $^{87}$Rb, giving simple formulas for estimating the light-induced changes, and showed that there may be ranges of intensity and detuning where useful changes can be effected. Finally, radio frequency (rf) fields used in evaporative cooling can change collision rates. Reference 171 showed how the strong rf field in a microwave trap will modify collisions. Moerdijk, *et al.*[172] however, showed that the rf fields used in evaporative cooling would make negligible changes in the collision rates of $F = 1,\ M = -1$ $^{23}$Na or $F = 2,\ M = 2$ $^{87}$Rb in a magnetic trap. Suominen, Tiesinga, and Julienne[173] concur with this analysis for these species, but show that typical rf fields for evaporative cooling can cause enhanced inelastic collisional relaxation of $F = 2,\ M = 2$ $^{23}$Na. This is because rf-induced nonadiabatic transitions due to motion in the trap lead to production of other M-levels, which decay with very large spin-exchange rate coefficients.[149,150]

One species that has been suggested as a viable candidate for BEC is the metastable $^3$S$_1$state of the He atom.[174] This long-lived species can be cooled and trapped.[157,175] Reference 157 calculated that the collisional ionization rate coefficient is so small for the $J = 1,\ M = 1$ level that a polarized gas of such a species might be stable long enough to make trapping and condensation possible. The polarized gas is stable because a collision of two $j = 1, m = 1$ atoms only occurs on the $^5\Sigma_g$ potential of the He$_2$ dimer, for which Penning ionization is forbidden. A gas of metastable atoms is only stable if complete polarization is maintained. An unpolarized sample would rapidly destroy itself due to very fast Penning ionization collisions.[175] Since Shylapinikov's original suggestion, BEC in spin-polarized $^4$He in the metastable 2 $^3$S$_1$ state has indeed been observed.[72,73] The Penning ionization rate coefficient for other spin-polarized $j = 2$ metastable noble gases is not known. Experiments on cold, trapped $j = 2,\ m = 2$ Xe metastable atoms indicate that collisional ionization of the polarized gas is comparable to that for the unpolarized gas.[176]

## 10.10. Quantum-Information Collisions

The possibilities of operating on combinations of two-level quantum states (qubits) rather than classical binary states (bits) to implement computational tasks has excited the imaginations of theorists ever since the pioneering work of Feynman.[177] At present many ideas about how to best implement quantum information machines are swirling about the scientific community, and some of these ideas rely on quantum collisions between atoms trapped in optical lattices at very cold temperatures. DiVincenzo[178] has identified five requirements for quantum computation: (1) scalability with well-characterized qubits, (2) initialization of all qubits to a reference state, (3) coherence times much longer than gate operation time, (4) a "universal" set of quantum gates (interaction operators operating on the the qubits) that can be turned "on" and "off", and (5) readout of the quantum operation. Certain binary collisions between atoms satisfy these requirements. For example the Innsbruck theory group led by Zoller and Cirac have suggested the realization of qubit computations using entangled ground-state atoms in controlled collisions through traps.[179,180] The interaction potential through which the atom states are entangled is mediated by the $s$-wave scattering length,

$$U_{\alpha\beta}(\mathbf{x}, \mathbf{x}') = \frac{4\pi a_s^{\alpha\beta}\hbar^2}{m}\delta^3(\mathbf{x} - \mathbf{x}')$$

where $\alpha, \beta$ represent the internal states of the two atoms. Fig. 10.26 shows how the two-qubit gate would be implemented. Another proposal[181] for the implementation of a binary collision gate involves the use of an optical lattice to trap the atoms. Using a two-dimensional trap array fulfills DiVincenzo's scalability requirement, and the resonant dipole-dipole interaction should speed up the gate operation time compared to the decoherence time when the atoms are tightly confined and in the Lambe-Dicke regime.

Quantum computation and cryptography are in their infancy, and it is by no means clear that cold collisions will play a major role in developing

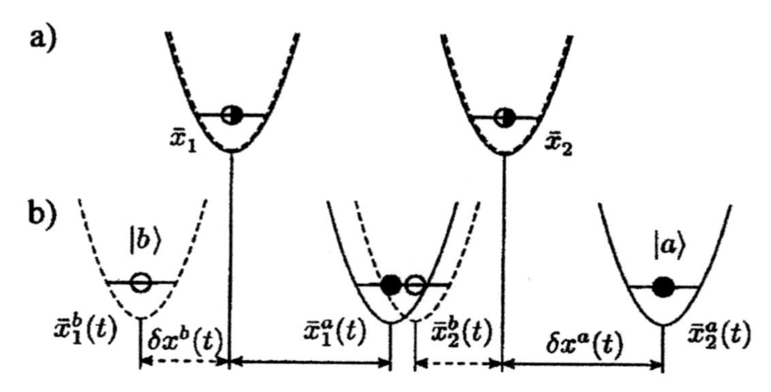


Fig. 10.26. From Ref. 179 — panel (a) shows the two positions of the two atoms before and after their interaction. The atoms are well-separated in their individual wells. Panel (b) shows the two traps brought physically together where the two atoms collide elastically through their $s$-wave scattering length. The labels $|a\rangle$ and $|b\rangle$ indicate the internal states of the two atoms. Notice that after entanglement both atoms have $|a\rangle$ and $|b\rangle$ character, denoted by the half-white, half-black circles in panel (a).

real quantum computers. The ability to "dial-an-$s$-wave" with Feshbach resonances and the relatively long intrinsic decoherence times associated with neutral atom interactions should at least provide a useful toolbox to investigate some of the physics aspects of quantum information.

## 10.11. Cold Molecules

Cold molecules hold fundamental interest for several reasons: they are the basis for an as yet unspecified "ultracold chemistry"[182] they can reversibly form composite bosons from fermion constituents in the degenerate quantum gas regime, they can be used to test fundamental symmetries at new levels of sensitivity,[183] they are arguably qubit candidates for the implementation of quantum gates[184,185]; and they are the object of a long-standing quest in chemical physics — control of chemical reactivity at the quantum level by means of well-defined and externally applied electric, magnetic and electromagnetic force fields. The state-of-the-art in the production and study of cold molecules has been recently reviewed.[186] We restrict discussion here to two principal areas of cold molecule research: cold molecules from ultracold gases and cold molecule formation from photoassociation.

### 10.11.1. *Cold Molecules From Ultracold Gases*

The use of the Feshbach resonance to alter the scattering length is not the only interesting aspect of this collisional phenomenon. As is evident from Fig. 10.27 the Feshbach resonance couples continuum atomic states to high-lying vibrational molecular states. By sweeping the magnetic field adiabatically through the resonance a significant fraction of cold molecules can be formed from an ensemble of cold atoms or even from an atomic BEC.[187] The detection of cold Cs molecules by photodissociation after applying a Feshbach resonance to an ensemble of cold Cs atoms in a far-detuned dipole trap has been reported by Chu's group.[188] Fifteen weakly coupled molecular states were detected. By using rate-equation modeling it was concluded that about $5 \times 10^5$ molecules co-existed with $10^8$ Cs atoms. This measurement implies a molecular conversion efficiency of about 0.5%. The atoms and molecules co-exist in a steady-state mixture since the Feshbach resonance back couples the molecules to atoms, and therefore the molecules can only be considered as quasibound.

The Chu experiment produced quasibound molecules from Raman-cooled Cs atoms, but other experiments have been able to form cold molecules directly from atomic Bose-Einstein condensates. The Ketterle group at MIT accomplished $Na_2$ formation from a Na BEC with an

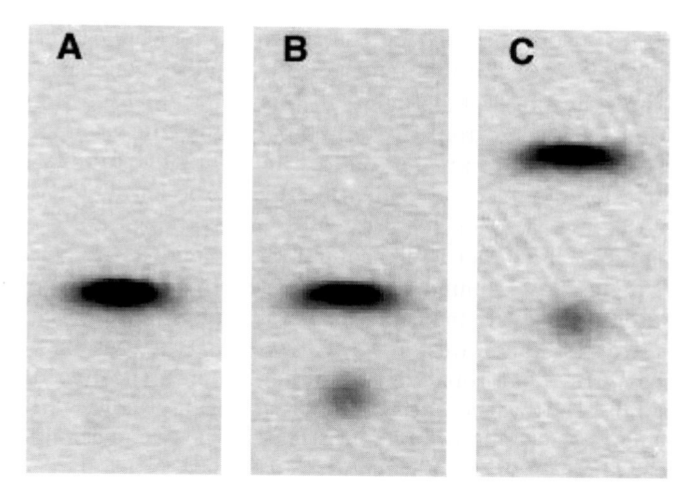


Fig. 10.27. Panel A shows the Cs BEC without the Feshbach resonance magnetic field sweep. Panel B shows the falling molecular cloud after the Feshbach sweep, and Panel C shows the molecular cloud levitated by a magnetic field gradient pointing vertically upward. The images of the molecular cloud are actually Cs atoms that have been reconverted from molecules after the spatial separation Ref. 187.

efficiency of 4%,[189] and the Rempe group in Garching produced cold molecules of $Rb_2$ from a condensate of atomic $^{87}$Rb atoms[190] and separated the atoms from the molecules by taking advantage of their differing magnetic moments. The Innsbruck group was also able to form $Cs_2$ molecules from a Cs BEC and to spatially separate the molecules from the atoms.[187] The efficiency of the conversion was reported to be about 24%. The kinetic temperatures of molecules formed from the BEC were very cold indeed — reaching a minimum of 19 nK in the vertical direction and 3 nK in the horizontal direction. These molecular temperatures are to be compared with the 5 $\mu$K molecules formed from Raman-cooled atoms in the Stanford experiment.[188]

In the preceding cases bosonic molecules were formed from clouds of cold bosonic atoms or from Bose-Einstein condensates. Another very interesting case occurs when molecular composite bosons are formed from fermionic atoms. The pairing of fermions is the key to unusual but very important physical states such as superfluidity and superconductivity. The Innsbruck group[191] Hulet's group at Rice University[192] and the group at the Ecole Normale in Paris headed by Christophe Salomon[193] have all produced $Li_2$ molecules from $^6$Li fermions. Similar experiments have been performed using $^{40}$K as the fermion atom.[194] An unexpected discovery from these experiments was that the efficiency for molecule formation and survival was very high (∼80% reported by Ref. 193) and molecular lifetimes ∼1 s. These conditions are favorable for the possible Bose condensation of the molecules since evaporative cooling can be applied to the cold molecular sample once it is formed from the fermionic atoms by sweeping through a Feshbach resonance. The explanation for the surprising stability was found to be suppression of molecular relaxation due to the Fermi statistics of the atoms.[195,196] The relaxation requires collisional stabilization of a dimer by a third atom, but the dimer itself requires two fermions to be close to each other ($\sim R_e$, the deeply bound dimer bond length) and in the presence of the third atom. Since the molecule formation takes place in a mixture of the lowest two $^6$Li spin states, at least two of the three atoms have to be identical fermions. The probability of finding them in such close proximity as $R_e$ is suppressed by the antisymmetry of Fermi statistics. The evaporative cooling of a mixture of $^6$Li spin states in the presence of a Feshbach resonance has indeed resulted in reports of the Bose-Einstein condensation of $^6Li_2$ molecules[191,197,198] and the emergence of a molecular BEC from a mixture of spin states in $^{40}$K.[199] This line of research is very important for the investigation of the BEC-BCS superfluid crossover region, but discussion of superfluidity is outside the scope of this chapter.

### 10.11.2. *Cold Molecules, Photoassociation and Coherent Control*

The molecules formed from ultracold quantum gases, *via* Feshbach resonance magnetic field sweeps, are always in very high vibrational states close to the dissociation limit and are very weakly bound. In the case of $^6Li_2$ the binding energy is only only $2\,\mu$K.[197] For $^{40}K_2$, the binding energy is only 60 nK,[199] and the size of the molecule (internuclear distance) is estimated to be 1650 $a_0$, where $a_0$ is the Bohr radius. From a physical chemistry perspective these "molecules" are not the familiar particles of conventional gas dynamics and molecular spectroscopy. One would like to find a way to transfer these cold molecules to the ground vibrational state and to more conventional molecular binding potentials.

A possible route to vibrationally and translationally cold molecules is through photoassociation of cold atoms in excited states with subsequent decay to chemically bound molecular electronic states. Fig. 10.28 shows how this process happens in the formation of $Cs_2$ molecules. Two MOT-cooled and trapped Cs atoms approach each other on the singlet or triplet ground state, undergo a vertical transition to the long-range $0_g^-$ excited molecular state from an excitation laser at $\lambda_1$. The transition takes place at the maximum Franck-Condon overlap near the outer turning point of the $0_g^-$ potential. At the inner turning point, the probability for

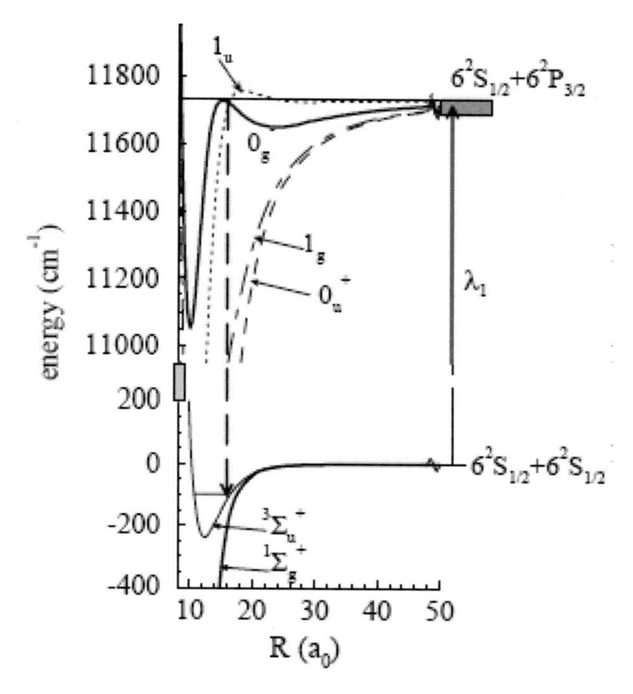


Fig. 10.28. Optical transitions and molecular states of $Cs_2$ that lead to cold molecules. Vertical transitions from the inner turning point of the long-range excited $0_g^-$ state leads to efficient production of translationally cold but vibrationally warm $Cs_2$ molecules.

spontaneous emission peaks and the $^3\Sigma_u^+$ state is populated. Although vibrationally excited levels of the $^1\Sigma_g^+$ single ground state are degenerate with the $^3\Sigma_u^+$ state, the *gerade-ungerade* selection rule greatly favors the triplet ground state. Similar experiments have been carried out for $Rb_2$ cold molecule production.[200] In addition to these homonuclear examples, heteronuclear cold molecules have also been formed by photoassociation. The interalkalis, RbCs,[201] KRb[202,203] and NaCs[204] have been identified by their spectroscopic signatures. This type of photoassociation does produce translationally cold molecules but vibrational states are still controlled essentially by a distribution of Franck-Condon transitions to the lower molecular electronic state. In order to achieve the goal of vibrational specificity and in particular to direct population to the lowest vibrational level it is necessary to actively intervene in the relaxation process. As a step toward this goal, chirped-pulse coherent control presents interesting possibilities.

There have been many proposals to apply coherent control to the photoassociation of cold atoms. A series of theoretical studies by the

Masnou-Seeuws group at Laboratoire Aimé Cotton in France has sought to optimize the transfer of population to the excited intermediate molecular state[205,206] and to direct the excited state population to specific vibrational levels of the triplet ground state, in particular to the $v'' = 14$ level of the ${}^3\Sigma_u^+$ ground triplet state in $Cs_2$.[207] They found that negative (blue to red) chirping of a broad-band "dump" pulse as well as negative chirping of the "pump" pulse could enhance overall probability for stable molecule formation by a factor of 24. The overall scheme is shown in Fig. 10.29. Furthermore by employing a longer dump pulse with duration of $\sim$3 ps they calculate that the population of the $v'' = 14$ vibrational level could be brought to about 12% of the total number of photoassociated molecules.

Experimental reports attempting to implement these theoretical proposals have, up to now, been relatively few. The Weidemüller group in Freiburg have applied shaped pulses with "evolution" in a feedback loop in an attempt to optimize photoassociation in cold Rb atoms[208] in MOT. A schematic of the experimental setup is shown in Fig. 10.30. Rather than finding enhanced $Rb_2$ production, however, the experiment showed that increasing the optical pulse energy actually decreased molecular signal due to photodissociation and photoionization of the formed molecules. Application of the evolutionary algorithm did result in an optimized pulse that excites molecules on the average 25% more efficiently than simple transform-limited pulses of the same energy. Nevertheless the net production rate, taking into account molecule formation and destruction

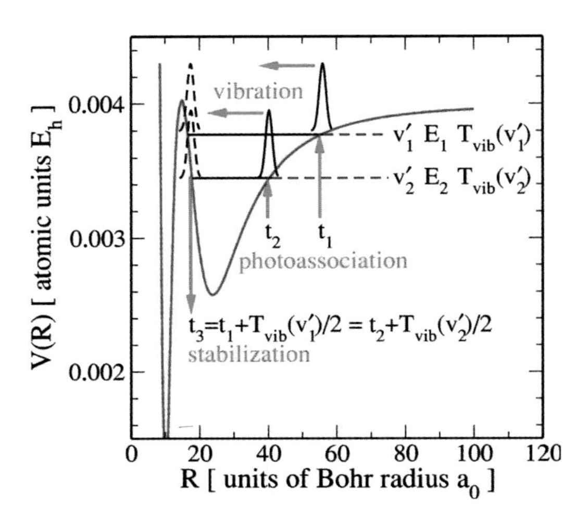


Fig. 10.29. The pump-dump scheme for specific vibrational level population in the ground triplet state of $Cs_2$ from Ref. 207.

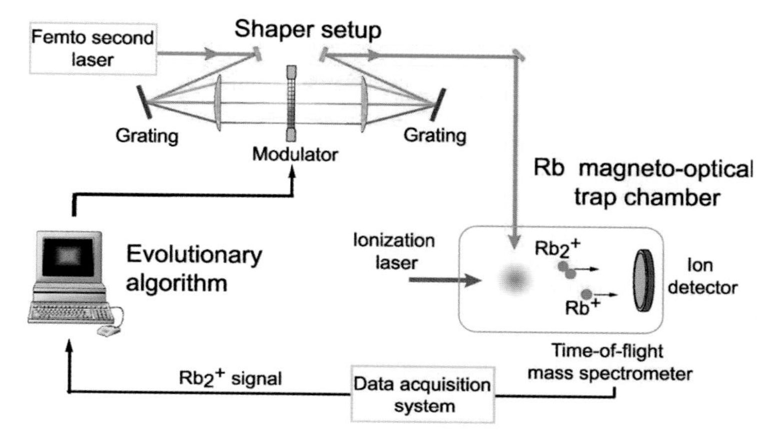


Fig. 10.30. Experimental setup for Freiburg experiment[208] searching for coherent control and optimization of photoassociation in $Rb_2$.

processes, showed a net decrease in $Rb_2$ production. Another experiment from the Walmsley group at Oxford, using chirped ∼100 femtosecond pulses, also investigated the coherent production of $Rb_2$ molecules from photoassociation of cold Rb atoms.[209] The experiment compared the effects of positive (red to blue) chirped pulses to nonchirped pulses under otherwise similar conditions on the production rate of $^{85}Rb_2$. The conclusion was that the chirped pulses contributed a net *quenching* effect rather than an enhancement effect to cold molecule production. Since the peak power in these experiments was on the order of $10^7$ W cm$^{-2}$ it is possible that weak-field potential curves cannot serve as a reliable guide to the collision dynamics in a strong-field regime. Both the Freiburg and Oxford experiments demonstrate that coherence effects can alter the photoassociation rate, but the physics of this alteration remains poorly understood and may require renewed theoretical effort to calculate collision interaction potentials in the strong-field regime. A very recent experiment from the Connecticut group[210] has demonstrated that the rate constant for collisional trap loss from a Rb MOT is markedly different for positive and negative chirped ∼40 ns pulses. For the negative chirp case, suppressed and enhanced trap loss rate constants in various detuning regimes are interpreted in terms of stimulated transfer between ground and excited molecular potential curves. Agreement with simple models is qualitative but suggestive. It appears therefore that coherence can indeed influence cold collisional processes, but it cannot be said at this time that the physics of "coherent control," marrying the ultrafast with the ultracold, is itself well understood or under control.

## References


1. Weiner J, Bagnato VS, Zilio S, Julienne PS. (1999) Experiments and theory in cold and ultracold collisions. *Rev. Mod. Phys.* **71**: 1–85.
2. Jones KM, Tiesinga E, Lett PD, Julienne PS. (2006) Ultracold photoassociation spectroscopy: Long-range molecules and atomic scattering. *Rev. Mod. Phys.* **78**: 483–535.
3. Phillips WD, Prodan JV, Metcalf HJ. (1985) Laser cooling and electromagnetic trapping of neutral atoms. *J. Opt. Soc. Am. B* **2**: 1751–1767.
4. Dalibard J, Cohen-Tannoudji C. (1985) Dressed-atom approach to atomic motion in lasr light: The dipole force revisted. *J. Opt. Soc. Am. B* **2**: 1701–1720.
5. Adams CS, Carnal O, Mlynek J. (1994) Atom interferometery. *Adv. At. Mol. Opt. Phys.* **34**: 1–33.
6. Adams CS, Sigel M, Mlynek J. (1994) Atom optics. *Phys. Rep.* **240**: 143–210.
7. Morinaga M, Yasuda M, Kishimoto T, Shimizu F. (1996) Holographic manipulation of a cold atomic beam. *Phys. Rev. Lett.* **77**: 802–805.
8. Jessen PS, Deutsch IH. (1996) Optical lattices. *Adv. At. Mol. Opt. Phys.* **37**: 95–138.
9. Dalfovo F, Giorgini S, Pitaevskii LP, Stringari S. (1999) Theory of bose-einstein condensation in trapped gases. *Rev. Mod. Phys.* **71**: 463–512.
10. Anglin JR, Ketterle W. (2002) Bose-einstein condensation of atomic gases. *Nature* **416**: 211–218.
11. Metcalf H, van der Straten P. (1994) Cooling and trapping of neutral atoms. *Phys. Rep.* **224**: 203–286.
12. Adams CS, Riis E. (1997) Laser cooling and trapping of neutral atoms. *Prog. Quantum Electron.* **21**: 1–79.
13. Suominen KA. (1996) Theories for cold atomic collisions in light fields. *J. Phys. B* **29**: 5981–6007.
14. Vigué J. (1986) Possibility of applying laser-cooling techniques to the observation of collective quantum effects. *Phys. Rev. A* **34**: 4476–4479.
15. Mies FH. (1973) Molecular theory of atomic collisions: Fine-structure transitions. *Phys. Rev. A* **7**: 942–957.
16. Mies FH. (1973) Molecular theory of atomic collisions: Calculated cross section for $H^+ + F(^2P)$. *Phys. Rev. A* **7**: 957–967.
17. Bayliss WE, Pascale J, Rossi F. (1987) Polarization and electronic excitation in nonreactive collisions: Basic formulation for quantum calculations of collisions between $^2P$-state alkali-metal atoms and $H_2$ or $D_2$. *Phys. Rev. A* **36**: 4212–4218.
18. Stoof HTC, Koelman JMVA, Verhaar BJ. (1988) Spin-exchange and dipole relaxation rates in atomic hydrogen: Rigorous and simplified calcualtions. *Phys. Rev. B* **38**: 4688–4697.
19. Herzberg G. (1950) *Molecular Spectra and Molecular Structure I, Spectra of Diatomic Molecules.* Vol. I, (van Nostrand, Princeton), 2nd edition.

20. Gao B. (1997) Theory of slow-atom collisions. *Phys. Rev. A* **54**: 2022–2039.
21. Bethe H. (1935) Theory of disintegration of nuclei by neutrons. *Phys. Rev.* **47**: 747–759.
22. Wigner EP. (1948) On the behavior of cross sections near thresholds. *Phys. Rev.* **73**: 1002–1009.
23. Delves H. (1958) Effective range expansions of the scattering matrix. *Nucl. Phys.* **8**: 358–373.
24. Julienne PS, Mies FH. (1989) Collisions of ultracold trapped atoms. *J. Opt. Soc. Am. B* **6**: 2257–2269.
25. Julienne PS, Smith AM, Burnett K. (1993) Theory of collisions between laser cooled atoms. *Adv. At. Mol. Opt. Phys.* **30**: 141–198.
26. Mott NF, Massey HSW. (1965) *The Theory of Atomic Collisions.* (Oxford, Clarendon), 3rd edition.
27. Gribakin GF, Flambaum VV. (1993) Calculation of the scattering length in atomic collisions using the semiclassical approximation. *Phys. Rev. A* **48**: 546–553.
28. Julienne PS. (1996) Cold binary atomic collisions in a light field. *J. Res. Natl. Inst. Stand. Technol.* **101**: 487–503.
29. Mies FH, Williams CJ, Julienne PS, Krauss M. (1996) Estimating bounds on collisional relaxation rates of spin-polarized $^{87}$Rb atoms at ultracold temperatures. *J. Res. Natl. Inst. Stand. Technol.* **101**: 521–535.
30. Boisseau C, Audouard E, Vigue J. (1998) Quantization of the highest levels in a molecular potential. *Europhys. Lett.* **41**: 349–354.
31. Trost J, Eltschka C, Friedrich H. (1998) Qunatization in molecular potentials. *J. Phys. B* **31**: 361–374.
32. LeRoy RJ, Bernstein RB. (1970) Dissociation energy and long-range ptoential of diatomic molecules from vibrational spacings of higher levels. *J. Ghem. Phys.* **52**: 3869–3879.
33. Arndt M, Dahan MB, Guery-Odelin D, Renolds MW, Dalibard J. (1997) Observation of a zero-energy resonance in Cs-Cs collisions. *Phys. Rev. Lett.* **79**: 625–628.
34. Leo P, Tiesinga E, Julienne PS, Walter DK, Kadlecek S, Walter TG. (1998) Elastic and inelastic collisions of cold spin-polarized $^{113}$Cs atoms. *Phys. Rev. Lett.* **81**: 1389–1392.
35. Joachan CJ. (1975) *Quantum Collision Theory.* (North-Holland, P.O. Box 103, 1000 AC Amserdam, The Netherlands).
36. Walker T, Feng P. (1994) Measurements of collisions between laser-cooled atoms. *Adv. At. Mol Opt. Phys.* **34**: 125–170.
37. Weiner J. (1995) Advances in ultracold collisions: Experimentation and Theory. *Adv. At. Mol. Opt. Phys.* **35**: 45–78.
38. Meath WJ. (1968) Retarded interaction energies between like atoms in different energy states. *J. Chem. Phys.* **48**: 227–235.
39. Scheingraber H, Vidal CR. (1977) Disctrete and continuous Franck-Condon factors of the $Mg_2$ $A^1\sum_u^+ - X^1\sum_g^+$ system and their J dependence. *J. Chem. Phys.* **66**: 3694–3704.

40. Inoue G, Ku JK, Setser DW. (1982) Photoassociative laser induced fluorescence of XeCl. *J. Chem. Phys.* **76**: 733–734.
41. Jones RB, Schloss JH, Eden JG. (1993) Excitation spectra for the photoassociation of Kr-F and Xe-I collision pairs in the ultraviolet (209–258) nm. *J. Chem. Phys.* **98**: 4317–4334.
42. Tellinghuisen J. (1985) *Photodissociation and Photoionization*. Vol. LX. *Advances in Chemical Physics* (Wiley, New York), pp. 299–369.
43. Thorsheim HR, Weiner J, Julienne PS. (1987) Laser-induced photoassociation of ultracold sodium atoms. *Phys. Rev. Lett.* **58**: 2420–2423.
44. Weiner J. (1989) Experiments in cold and ultracold collisions. *J. Opt. Soc. Am. B* **6**: 2270–2278.
45. Julienne PS, Heather R. (1991) Laser modification of ultracold atomic collisions: Theory. *Phys. Rev. Lett.* **67**: 2135–2138.
46. Leonhardt D, Weiner J. (1995) Direct two-color photoassociative ionization in a rubidium magneto-optic trap. *Phys. Rev. A* **52**: 1419–1429. see also erratum, *Phys. Rev. A* **53**: 2904(E), 1996.
47. Tsao C-C, Napolitano R, Wang Y, Weiner J. (1995) Ultracold photoassociative ionization collisions in an atomic beam: Optical field intensity and polarization dependence of the rate constant. *Phys. Rev. A* **51**: R18–R21.
48. Tsao C-C. (1996) *Photoassociative Ionization of Cold Sodium Atoms in an Atomic Beam*. PhD thesis, University of Maryland.
49. Ramirez-Serrano J, DeGraffenreid W, Weiner J. (2002) Polarization-dependent spectra in the photoassociative ionization of cold atoms in a bright sodium beam. *Phys. Rev. A* **65**: 052719-1–052719-8.
50. Weiner J, Masnou-Seeuws F, Giusti-Suzor A. (1989) Associative ionization: Experiments, potentials and dynamics. *Adv. At. Mol. Opt. Phys.* **26**: 209–296.
51. Thorsheim HR, Wang Y, Weiner J. (1990) Cold collisions in an atomic beam. *Phys. Rev. A* **41**: 2873–2876.
52. Gould PL, Lett PD, Julienne PS, Phillips WD, Thorsheim HR, Weiner J. (1988) Observation of associative ionization of ultracold laser-trapped sodium atoms. *Phys. Rev. Lett.* **60**: 788–791.
53. Bonanno R, Boulmer J, Weiner J. (1983) Determination of the absolute rate constant for associative ionization in crossed-beam collisions between Na $3^2P_{3/2}$ atoms. *Phys. Rev. A* **28**: 604–608.
54. Keller J, Baulmer J, Wang M-X, Weiner J. (1986) Strong velocity dependence of the atomic alignment effect in Na(3p) + Na(3p) associaitve ionization. *Phys. Rev. A* **34**: 4497–4500.
55. Wang M-X, Keller J, Boulmer J, Weiner J. (1987) Spin-selected velocity dependence of the associative ionization cross section in Na(3p) + Na(3p) associative ionization over the collision energy range from 2.4 to 290 mev. *Phys. Rev. A* **35**: 934–937.
56. Julienne PS. (1988) Laser modification of ultracold atomic collisions in optical traps. *Phys. Rev. Lett.* **61**: 698–701.

57. Lett PD, Jessen PS, Phillips WD, Rolston SL, Westbrook CI, Gould PL. (1991) Laser modification of ultracold collisions: Experiment. *Phys. Rev. Lett.* **67**: 2139–2142.
58. Heather RW, Julienne PS. (1993) Theory of laser-induced associative ionization of ultracold Na. *Phys. Rev. A* **47**: 1887–1906.
59. Dulieu OA, Giusti-Suzor A, Masnou-Seeuws F. (1991) Theoretical treatment of the associative ionization reaction between two laser-excited sodium atoms. Direct and indirect processes. *J. Phys. B* **24**: 4391–4408.
60. Henriet A, Masnou-Seeuws F, Dulieu O. (1991) Diabatic representation ofro the excited states of the $Na_2$ molecule: Application to the associative ionization reaction between two excited sodium atoms. *Z. Phys. D* **18**: 287–298.
61. Dulieu O, Magnier S, Masnou-Seeuws F. (1994) Doubly-excited states for the $Na_2$ molecule: Application to the dynamics of the associative ionization reaction. *Z. Phys. D* **32**: 229–240.
62. Stwalley WC, Uang Y-H, Pichler G. (1978) Pure long-range molecules. *Phys. Rev. Lett.* **41**: 1164–1166.
63. Pillet P, Crubeillier A, Bleton A, Dulieu O, Nosbaum P, Mourachko I, Masnou-Seeuws F. (1997) Photoassociation in a gas of cold alklai atoms: I. Perturbative quantum approach. *J. Phys. B* **30**: 2801–2820.
64. DeGraffenreid W, Ramirez-Serano J, Liu Y-M, Weiner J. (2000) Continuous, dense, highly collimated sodium beam. *Rev. Sci. Instr.* **71**: 3668–3676.
65. Amelink A, Jones KM, Lett PD, van der Straten P, Heideman HGM. (2000) Single-color photassociative ionization of ultracold sodium: The region from 0 to −5 GHz. *Phys. Rev. A* **62**: 013408-1–013408-8.
66. Amelink A, Jones KM, Lett PD, van der Straten P, Heideman H. (2000) Spectrsocopy of autoionizing doubly excited states in ultracold $Na_2$ molecules produced by photoassociation. *Phys. Rev. A* **61**: 042707-1–042707-9.
67. Ramirez-Serrano J, DeGraffenreid W, Weiner J. Beam-loss spectroscopy of cold atoms in a bright sodium beam. Manuscript in preparation; to be submitted to *Phys. Rev. A*.
68. Anderson MH, Ensher JR, Matthews MR, Wieman CE, Cornell EA. (1995) Observation of Bose-Einstein condensation a dilute atomic vapor. *Science* **269**: 198–201.
69. Davis KB, Mewes M-O, Andrews MR, van Druten NJ, Durfee DS, Kurn DM, Ketterle W. (1995) Bose-Einstein condensation in a gas of sodium atoms. *Phys. Rev. Lett.* **75**: 3969–3973.
70. Bradley CC, Sackett CA, Tollett JJ, Hulet RG. (1995) Evidence of Bose-Einstein condensation in an atomic gas with attractive interactions. *Phys. Rev. Lett.* **75**: 1687–1690.
71. Bradley CC, Sackett A, Hulet RG. (1997) Bose-Einstein condensation of lithium: Observation of limited condensate number. *Phys. Rev. Lett.* **78**: 985–988.

72. Pereira Dos Santo F, Wang J, Barrelet CJ, Perales F, Rasel E, Unnikrishnan CS, Leduc M, Cohen-Tannoudji C. (2001) Bose-Einstein condensation of metastable helium. *Phys. Rev. Lett.* **86**: 3459–3462.
73. Robert AB, Poupard AJ, Nowak S, Boiron D, Westbrook CI, Aspect A. (2001) A Bose-Einstein condensate of metastable atoms. *Science* **292**: 461–464.
74. Fried D, Killian TC, Willmann L, Landhuis D, Moss SC, Kleppner D, Greytak TJ. (1998) Bose-Einstein condensation of atomic hydrogen. *Phys. Rev. Lett.* **81**: 3811–3814.
75. Stwalley WC, Nosanow L. (1976) Possible new quantum systems. *Phys. Rev. Lett.* **36**: 910–913.
76. Silvera IF, Walraven JTM. (1980) Stabilization of atomic hydrogen at low temperature. *Phys. Rev. Lett.* **44**: 164–168.
77. Silvera IF, Walraven JTM. (1986) *Spin-polarized atomic H.* In Brewer DF (ed.) *Progress in Low Temperature Physics.* Vol. X. (Elsevier, Amsterdam), pp. 139–370.
78. Crampton SB, Phillips WD, Kleppner D. (1978) Proposed low temperature hydrogen maser. *Bull. Am. Phys. Soc.* **23**: 86.
79. Hess HF, Kochanski GP, Doyle JM, Greytak TJ, Kleppner D. (1986) Spin-polarized hydrogen maser. *Phys. Rev. A* **34**: 1602–1604.
80. Hurlimann MD, Hardy WN, Berlinski AJ, Cline RW. (1986) Recirculating cryogenic hydrogen maser. *Phys. Rev. A* **34**: 1605–1608.
81. Walsworth RL, Silvera IF, Gotfried HP, Agosta CC, Vessot RFC, Mattison EM. (1986) Hydrogen maser at temperatures below 1 K. *Phys. Rev. A* **34**: 2550–2553.
82. Kleppner D, Berg HC, Crapmton SB, Ramsey NF, Vessot RC, Peters HE, Vanier J. (1965) Hydrogen-maser principles and techniques. *Phys. Rev.* **138**: A972–A983.
83. Crampton SB, Wang HTM. (1975) Duration of hydrogen atom spin-exchange collisions. *Phys. Rev. A* **12**: 1305–1312.
84. Verhaar BJ, Koelman JMVA, Stoof HTC, Luiten OJ, Crampton SB. (1987) Hyperfme contribution to spin-exchange frequency shifts in the hydrogen maser. *Phys. Rev. A* **35**: 3825–3831.
85. Koelman JMVA, Crampton SB, Stoof HTC, Luiten OJ, Verhaar B. (1988) Spin polarized deuterium in magnetic traps. *Phys. Rev. A* **38**: 3535–3547.
86. Cline RW, Greytak TJ, Kleppner D. (1981) Nuclear polarization of spin-polarized atomic hydrogen. *Phys. Rev. Lett.* **47**: 1195–1198.
87. Sprik R, Walraven JTM, Yperen GH, Silvera IF. (1982) State-dependent recombination and suppressed nuclear relaxation in atomic hydrogen. *Phys. Rev. Lett.* **49**: 153–156.
88. Ahn RMC, van den Eijnde JPHW, Verhaar BJ. (1983) Calculation of nuclear spin relaxation rate for spin-polarized atomic hydrogen. *Phys. Rev. B* **27**: 5424–5432.
89. van Roijen R, Berkhout JJ, Jaakkola S, Walraven JTM. (1988) Experiments with atomic hydrogen in a magnetic trapping field. *Phys. Rev. Lett.* **61**: 931–934.

90. Dalgarno A. (1961) Spin-change cross sections. *Proc. R. Soc. London Ser. A* **262**: 132–135.
91. Allison A, Dalgarno A. (1969) Spin change in collisions of hydtrogen atoms. *Astrophys. J.* **158**: 423–425.
92. Allison AC. (1972) Spin-change frequency shifts in H-H collisions. *Phys. Rev. A* **5**: 2695–2696.
93. Berlinsky AJ, Shizgal B. (1980) Spin-exchange scattering cross sections for hydrogen atoms at low temperatures. *Can. J. Phys.* **58**: 881–885.
94. Maan AC, Stoof HTC, Verhaar BJ. (1990) Cryogenic H maser in a strong B field. *Phys. Rev. A* **41**: 2614–2620.
95. Lagendijk A, Silvera IF, Verhaar BJ. (1986) Spin-exchange and dipolar relazation rates in atomic hydrogen: Lifetimes in magnetic traps. *Phys. Rev. B* **33**: 626–628.
96. Uang Y-H, Stwalley WC. (1980) Close-coupling calculations of spin-polarized hydrogen-deuterium collisions. *Phys. Rev. Lett.* **45**: 627–530.
97. Koelman JMVA, Stoof HTC, Verhaar BJ, Walraven JTM. (1987) Spin polarized deuterium in magnetic traps. *Phys. Rev. Lett.* **59**: 676–679.
98. Koelman JMVA, Stoof HTC, Verhaar BJ, Walraven JTM. (1988) Lifetime of magnetically trapped ultracold atomic deuterium gas. *Phys. Rev. B* **38**: 9319–9322.
99. Tiesinga E, Kuppens SJM, Verhaar BJ, Stoof HTC. (1991) Collisions between cold grond state Na atoms. *Phys. Rev. A* **43**: R5188–R5191.
100. Tiesinga E, Crampton SB, Verhaar BJ, Stoof HTC. (1993) Collisional frequency shifts and line broadening in the cryogenic deuterium maser. *Phys. Rev. A* **47**: 4342–4347.
101. Jamieson MJ, Dalgarno A, Doyle JM. (1996) Scattering lengths for collisions of ground state and metastable state hydrogen atoms. *Mol. Phys.* **87**: 817–826.
102. Tiesinga E, Verhaar BJ, Stoof HTC, van Bragt D. (1992) Spin-exchange frequency shift in cesium atomic fountain. *Phys. Rev. A* **45**: 2671–2674.
103. Kasevich MA, Riis E, Chu S, DeVoe RG. (1989) Rf spectroscopy in an atomic fountain. *Phys. Rev. Lett.* **63**: 612–615.
104. Clairon A, Salomon C, Guellati S, Phillips WD. (1991) Ramsey resonance in a Zacharias fountain. *Europhys. Lett.* **16**: 165–170.
105. Gibble K, Chu S. (1992) Future slow-atom frequncy standards. *Metrologia* **29**: 201–212.
106. Gibble K, Chu S. (1993) Laser-cooled Cs frequency standard and a measurement of the frequency shift due to ultracold collisions. *Phys. Rev. Lett.* **70**: 1771–1774.
107. Ghezali S, Laurent P, Lea SN, Clairon A. (1996) An experimental study of the spin-exchange frequency shift in a laser-cooled cesium fountain frequency standard. *Europhys. Lett.* **36**: 25–30.
108. Verhaar BJ, Gibble K, Chu S. (1993) Cold-collision properties derived from frequency shifts in a Cs fountain. *Phys. Rev. A* **48**: R3429–R3432.
109. Gibble K, Verhaar BJ. (1995) Eliminating cold-collision frequency shifts. *Phys. Rev. A* **52**: 3370–3373.

110. Kokkelmans SJJMF, Verhaar BJ, Gibble K, Heinzen DJ. (1997) Predictions for laser-cooled Rb clocks. *Phys. Rev. A* **56**: R4389–R4392.
111. Gibble K, Chang S, Legere R. (1995) Direct observation of $s$-wave atomic collisions. *Phys. Rev. Lett.* **75**: 2666–2669.
112. Lee TD, Yang CN. (1958) Low-temperature behavior of dilute Bose system of hard spheres i. Equilibrium properties. *Phys. Rev.* **112**: 1419–1429.
113. Hess HF, Kochanski GP, Doyle JM, Mashuhara N, Kleppner D, Greytak TJ. (1987) Magnetic trapping of spin-polarized atomic hydrogen. *Phys. Rev. Lett.* **59**: 672–675.
114. Masuhara N, Doyle JM, Sandberg JC, Kleppner D. (1988) Greytak TJ, Evaporative cooling of spin-polarized atomic hydrogen. *Phys. Rev. Lett.* **61**: 935–938.
115. Burnett K. (1996) Bose-Einstein condensation with evaporatively cooled atoms. *Contemp. Phys.* **37**: 1–14.
116. Cornell E. (1996) Very cold indeed: The nanokelvin physics of Bose-Einstein condensation. *J. Res. Natl. Inst. Stand. Technol.* **101**: 419–434, and other articles in this special issue on Bose-Einstein condensation.
117. Townsend C, Ketterle W, Stringari S. (1997) Bose-Einstein condensation. *Phys. World* **10**: 29–34.
118. Hess HF. (1986) Evaporative cooling of magnetically trapped and compressed spin-polarized hydrogen. *Phys. Rev. B* **34**: 3476–3479.
119. Ketterle W, Druten NJV. (1996) Evaporative cooling of trapped atoms. *Adv. At. Mol. Opt. Phys.* **37**: 181–236.
120. Tiesinga E, Moerdijk AJ, Verhaar B, Stoof HTC. (1992) Conditions for Bose-Einstein condensation in magnetically trapped atomic cesium. *Phys. Rev. A* **46**: R1167–R1170.
121. Monroe CR, Cornell EA, Sackett CA, Myatt CJ, Wieman CE. (1993) Measurement of Cs-Cs elastic scattering at $T = 30\,\mu K$. *Phys. Rev. Lett.* **70**: 414–417.
122. Newbury NR, Myatt CJ, Wieman CE. (1995) $s$-wave elastic collisions between cold ground state $^{87}$Rb atoms. *Phys. Rev. A* **51**: R2680–R2683.
123. Davis KB, Mewes M, Joffe MA, Andrews MR, Ketterle W. (1995) Evaporative cooling of sodium atoms. *Phys. Rev. Lett.* **74**: 5202–5205.
124. Huang K, Yang CN. (1957) Quantum mechanical many-body problem with hard-sphere interactions. *Phys. Rev.* **105**: 767–775.
125. Stoof HTC, Bijlsma M, Houbiers M. (1996) Theory of interacting quantum gases. *J. Res. Natl. Inst. Stand. Technol.* **101**: 443–457.
126. Edwards M, Burnett K. (1995) Numerical solution of the nonlinear Schroedinger equation for small samples of trapped neutral atoms. *Phys. Rev. A* **51**: 1382–1386.
127. Dodd R, Edwards M, Williams CJ, Clark CW, Holland MJ, Ruprecht PA, Burnett K. (1996) Role of attractive interactions on Bose-Einstein condensation. *Phys. Rev. A* **54**: 661–664.
128. Bergmann T. (1997) Hartree-Fock calculations of Bose-Einstein condensation of $^7$Li atoms in a harmonic trap for $T > 0$. *Phys. Rev. A* **55**: 3658–3669.

129. Holland MJ, Jin DS, Chiofalo ML, Cooper J. (1997) Emergence of interaction effects in Bose-Einstein condensation. *Phys. Rev. Lett.* **78**: 3801–3805.
130. Mewes M-O, Andrews MR, van Druten NJ, Kurn DM, Dur-fee DS, Ketterle W. (1996) Bose-Einstein condensation in a tightly confining dc magnetic trap. *Phys. Rev. Lett.* **77**: 416–419.
131. Castin Y, Dum R. (1996) Bose-Einstein condensation in time-dependent traps. *Phys. Rev. Lett.* **77**: 5315–5319.
132. Ketterle W, Miesner H-J. (1997) Coherence properties of Bose condensates and atom lasers. *Phys. Rev. A* **56**: 3291–3293.
133. Moerdijk AJ, Verhaar BJ. (1994) Prospects for Bose-Einstein condensation in atomic $^{7}$Li and $^{23}$Na. *Phys. Rev. Lett.* **73**: 518–521.
134. Moerdijk AJ, Stwalley WC, Hulet RG, Verhaar BJ. (1994) Negative scattering length of ultracold $^{7}$Li gas. *Phys. Rev. Lett.* **72**: 40–43.
135. Moerdijk AJ, Verhaar BJ. (1995) Laser cooling and the highest bound states of the Na diatom system. *Phys. Rev. A* **51**: R4333–R4336.
136. Boesten HMJM, Vogels JM, Tempelaars JGC, Verhaar BJ. (1996) Properties of cold collisions of $^{39}$K atoms and of $^{41}$K atoms in relation to Bose-Einstein condensation. *Phys. Rev. A* **54**: R3726–R3729.
137. Côté R, Dalgarno A. (1997) Mechanism for the production of vibrationally excited ultracold molecules of $^{7}Li_2$. *Chem. Phys. Lett.* **279**: 50–54.
138. Soding J, Guery-Odelin D, Desbiolles P, Ferrari G, Dahan MB, Dalibard J. (1998) Giant spin relaxation of an ultracold cesium gas. *Phys. Rev. Lett.* **80**: 1869–1872.
139. Cheng C, Vuleti/'c V, Kerman AJ, Chu S. (2000) High resolution Feshbach spectroscopy of cesium. *Phys. Rev. Lett.* **85**: 2717–2720.
140. Leo PJ, Williams CJ, Julienne PS. (2000) Collision properties of ultra-cold $^{133}$Cs atoms. *Phys. Rev. Lett.* **85**: 2721–2724.
141. Kokkelmans SJJMF, Verhaar BJ, Gibble K. (1998) Prospects for Bose-Einstein Condensation in Cesium. *Phys. Rev. Lett.* **81**: 951–954.
142. Drag C, Tolra BL, T'Jampens B, Comparat D, Allegrini M, Crubellier A, Pillet P. (2000) Photoassociative spectroscopy as a self-sufficient tool for the determination of the Cs triplet scattering length. *Phys. Rev. Lett.* **85**: 1408–1411.
143. Burt EA, Christ RW, Myatt CJ, Holland MJ, Cornell EA, Wieman CE. (1997) Coherence, correlations and collisions: What one learns about Bose-Einstien condensates from their decay. *Phys. Rev. Lett.* **79**: 337–340.
144. Andrews MR, Townsend CG, Miesner H-J, Durfee DS, Kurn DM, Ketterle W. (1997) Observation of interference between two Bose-Einstein condensates. *Science* **275**: 637–641.
145. Kagan Y, Svistunov BV, Shlyapnikov GV. (1985) Effect of Bose-Einstein condensation on inelastic processes in gases. *Pis'ma Zh. Eksp Tear. Fiz.* **42**: 169–172 (JETP Lett. 42, 209–212).
146. Stoof HTC, Janssen AML, Koelman JMVA, Verhaar BJ. (1989) Decay of spin-polarized atomic hydrogen in the presence of a Bose condensate. *Phys. Rev. A* **39**: 3157–3169.

147. Burnett K, Julienne PS, Suominen K-A. (1996) Laser driven collisions between atoms in a Bose-Einstein condensed gas. *Phys. Rev. Lett.* **77**: 1416-1419.
148. Myatt CJ, Burt EA, Christ RW, Cornell EA, Wieman CE. (1997) Production of two overlapping Bose-Einstein condensates by sympathetic cooling. *Phys. Rev. Lett.* **78**: 586–589.
149. Julienne PS, Mies FH, Tiesinga E, Williams CJ. (1997) Collisional stability of double Bose condensates. *Phys. Rev. Lett.* **78**: 1880–1883.
150. Kokkelmans SJJMF, Boesten HMJM, Verhaar BJ. (1997) Role of collisions in creation of overlapping Bose condensates. *Phys. Rev. A* **55**: R1589–R1592.
151. Burke JP, Bohn JL, Esry BD, Greene CH. (1997) Impact of the $^{87}$Rb singlet scattering length on suppressing inelastic collisions. *Phys. Rev. A* **55**: R2511–R2514.
152. De Goey LPH, der Berg THMV, Mulders N, Stoof HTC, Verhaar BJ. (1986) Three-body recombination in spin-polarized atomic hydrogen. *Phys. Rev. B* **34**: 6183–6191.
153. De Goey LPH, Stoof HTC, Verhaar BJ, Glockle W. (1988) Role of three-body correlations in recombination of spin-polarized atomic hydrogen. *Phys. Rev. B* **38**: 646–658.
154. Moerdijk AJ, Boesten HMJM, Verhaar BJ. (1996) Decay of trapped ultracold alkali atoms by recombination. *Phys. Rev. A* **53**: 916–920.
155. Moerdijk AJ, Verhaar BJ. (1996) Collisional two- and three-body decay rates of dilute quantum gases at ultralow temperatures. *Phys. Rev. A* **53**: R19–R22.
156. Esry B, Greene CH, Zhou Y, Lin CD. (1996) Role of the scattering length in three-boson dynamics and Bose-Einstein condensation. *J. Phys. B* **29**: L51–L57.
157. Fedichev PO, Reynolds MW, Rahmanov UM, Shlyapnikov GV. (1996) Inelastic decay processes in a gas of spin-polarized triplet helium. *Phys. Rev. A* **53**: 1447–1453.
158. Tiesinga E, Verhaar BJ, Stoof HTC. (1993) Threshold and resonance phenomena in ultracold ground-state collisions. *Phys. Rev.* **47**: 4114–4122.
159. Moerdijk AJ, Verhaar BJ, Axelsson A. (1995) Resonance in ultracold collisions of $^{6}$Li, $^{7}$Li, $^{23}$Na. *Phys. Rev. A* **51**: 4852–4861.
160. Vogels JM, Tsai CC, Freeland RS, Kokkelmans SJJMF, Verhaar BJ, Heinzen DJ. (1997) Prediction of Feshbach resonances in collisions of ultracold rubidium atoms. *Phys. Rev. A* **56**: R1067–R1070.
161. Courteille P, Freeland RS, Heinzen DJ, van Abeelen FA, Verhaar BJ. (1998) Observation of a Feshbach Resonance in Cold Atom Scattering. *Phys. Rev. Lett.* **81**: 69–72.
162. Roberts JL, Claussen NR, Burke JP, Greene CH, Cornell EA, Wieman CE. (1998) Resonant Magnetic Field Control of Elastic Scattering in Cold $^{85}$Rb. *Phys. Rev. Lett.* **81**: 5109–5112.
163. Roberts JL, Burke JP, Claussen NR, Cornish SL, Donley EA, Wieman CE. (2001) Improved characterization of elastic scattering near a Feshbach resonance in $^{85}$Rb. *Phys. Rev. A* **64**: 024702-1–024702-3.

164. Roberts JL, Claussen NR, Cornish SL, Donley EA, Cornell EA, Wieman CE. (2001) Controlled Collapse of a Bose-Einstein Condensate. *Phys. Rev. Lett.* **86**: 4211–4214.
165. Donley EA, Claussen NR, Cornish SL, Roberts JL, Cornell EA, Wieman CE. (2001) Dynamics of collapsing and expliding Bose-Einstein condensates. *Nature* **412**: 295–299.
166. Kagan Y, Muryshev AE, Shlyapnikov GV. (1998) Collapse and Bose-Einstein Condensation in a Trapped Bose Gas with Negative Scattering Length. *Phys. Rev. Lett.* **81**: 933–937.
167. Inouye S, Andrews MR, Stneger J, Miesner H-J, Stamper-Kurn DM, Ketterle W. (1998) Observation of Feshbach resonances in a Bose-Einstein condensate. *Nature* **392**: 151–154.
168. Fedichev PO, Kagan Y, Shlyapnikov GV, Walraven JTM. (1996) Influence of nearly resonant light on the scattering length in low-temperature atomic gases. *Phys. Rev. Lett.* **77**: 2913–2916.
169. Napolitano R, Weiner J, Julienne PS. (1997) Theory of optical suppression of ultracold-collision rates by polarized light. *Phys. Rev. A* **55**: 1191–1207.
170. Bohn J, Julienne PS. (1997) Prospects for influencing scattering lengths with far-off-resonant light. *Phys. Rev. A* **56**: 1486–1491.
171. Agosta C, Silvera IF, Stoof HTC, Verhaar BJ. (1989) Trapping of neutral atoms with resonant microwave radiation. *Phys. Rev. Lett.* **62**: 2361–2364.
172. Moerdijk AJ, Verhaar BJ, Nagtegaal TM. (1996) Collisions of dressed ground-state atoms. *Phys. Rev. A* **53**: 4343–4351.
173. Suominen K-A, Tiesinga E, Julienne PS, (1998) Nonadiabatic dynamics in evaporative cooling of trapped atoms by a radio frequency field. *Phys. Rev. A* **58**: 3983–3992.
174. Shlyapnikov GV, Walraven JTM, Rahmanov UM, Reynolds MW. (1994) Decay kinetics and Bose condensation in a gas of spin-polarized triplet helium. *Phys. Rev. Lett.* **73**: 3247–3250.
175. Bardou F, Emile O, Courty J-M, Westbrook CI, Aspect A. (1992) Magneto-optical trapping of metastable helium: Collisions in the presence of resonant light. *Europhys. Lett.* **20**: 681–686.
176. Rolston S. (1997) NIST. private communication.
177. Feynman RP. (1996) In Hey AJG, Allen R (eds.), *Feynman Lectures on Computation*, Perseus Press.
178. DiVincenzo DP. (2000) Physical implementation of quantum computation. *Fortschr. Phys.* **48**: 771–783.
179. Jaksch D, Briegel H-J, Cirac JI, Gardiner CW, Zoller P. (1999) Entaglement of Atoms via Cold Controlled Collisions. *Phys. Rev. Lett.* **82**: 1975–1978.
180. Calarco T, Hinds EA, Jaksch D, Schmiedmayer J, Cirac JI, Zoller P. (2000) Quantum gates with neutral atoms: Controlling collisional interaction in time-dependent traps. *Phys. Rev. A* **61**: 022304-1–02304-11.
181. Deutsch IH, Brennen GK. (2000) Quantum computing with neutral atoms in an optical lattice. *Fortschr. Phys.* **48**: 925–943.

182. Heinzen DJ, Wynar R, Drummond PD, Kheruntsyan KV. (2000) Superchemistry: Dynamics of coupled atomic and molecular Bose-Einstein condensates. *Phys. Rev. Lett.* **84**: 5029–5033.
183. Hudson JJ, Sauer BE, Tarbutt MR, Hinds EA. (2002) Measurement of the electron electric dipole moment using YbF molecules. *Phys. Rev. Lett.* **89**: 023003–1-4.
184. DeMille D. (2002) Quantum computation with trapped polar molecules. *Phys. Rev. Lett.* **88**: 067901–1-4.
185. Tesch CM, de Vivie-Riedle R. (2002) Quantum computation with vibrationally excited molecules. *Phys. Rev. Lett.* **89**: 157901–1-4.
186. Dulieu O, Raoult M, Tiemann E. (2006) Cold molecules. *J. Phys. B* **39**: (19). special issue.
187. Herbig J, Kraemer T, Mark M, Weber T, Chin C, Nagerl H-C, Grimm R. (2003) Preparation of a pure molecular quantum gas. *Science* **301**: 1510–1513.
188. Chin C, Kerman J, Vuletić V, Chu S. (2003) Sensitive detection of cold cesium molecules formed on Feshbach resonances. *Phys. Rev. Lett.* **90**: 033201–1-4.
189. Xu K, Mukaiyama T, Abo-Shaeer JR, Chin JK, Miller D, Ketterle W. (2003) Formation of quantum-degenerate sodium molecules. *Phys. Rev. Lett.* **91**: 210402–1-4.
190. Diirr S, Volz T, Marte A, Rempe G. (2004) Observation of molecules produced from a Bose-Einstein condensate. *Phys. Rev. Lett.* **92**: 020406–1-4.
191. Jochim S, Bartenstein M, Altmeyer A, Hendl G, Reidl S, Chin C, Denschlag JM, Grimm R. (2003) Bose-Einstein condensation of molecules. *Science* **302**: 2101–2103.
192. Strecker KE, Partridge GB, Hulet RG. (2003) Conversion of an atomic Fermi gas to a long-lived molecular Bose gas. *Phys. Rev. Lett.* **91**: 080406–1-4.
193. Cubizolle J, Bourdels T, Kokkelmans SJJMF, Shlyapnikov GV, Salomon C. Production of long-lived ultracold $Li_2$ molecules from a Fermi gas. *Phys. Rev. Lett.* **91**: 240401–1-4, (2003).
194. Regal CA, Ticknor C, Bohn JL, Jin DS. (2003) Creation of ultracold molecules from a Fermi gas of atoms. *Nature* **424**: 47–50.
195. Petrov DS, Salomon C, Shlyapnikov GV. (2004) Weakly bound dimers of fermionic atoms. *Phys. Rev. Lett.* **93**: 090404–1-4.
196. Petrov DS, Salomon C, Shlyapnikov GV. (2005) Scattering properties of weakly bound dimers of fermionic atoms. *Phys. Rev. A* **71**: 012708.
197. Zwierlein MW, Stan CA, Schunck CH, Raupach SMF, Gupta S, Hadzibabic Z, Ketterle W. (2003) Observation of Bose-Einstein condensation of molecules. *Phys. Rev. Lett.* **91**: 250401–1-4.
198. Bourdel T, Khaykovich L, Cubizolles J, Zhang J, Chevy F, Teich-rnann T, Tarruell L, Kokkelmans SJJMF, Salomon C. Experimental study of the BEC-BCS crossover region in lithium 6. *Phys. Rev. Lett.* **93**: 050401–1-4, (2004).
199. Greiner M, Regal CA, Jin DS. Emergence of a molecular Bose-Einstein condensate from a Fermi gas. *Nature* **426**: 537–540, (2003).

200. Fioretti A, Fazzi M, Mazzoni M, Ban T, Gabbanini C. (2004) Ultra-cold molecules. *Physica Scripta* **T112**: 13–19.
201. Kerman AJ, Sage JM, Sainis S, Bergeman T, DeMille D. (2004) Production of Ultracold, Polar RbCs* Molecules via Photoassociation. *Phys. Rev. Lett.* **92**: 033004–1-4.
202. Mancini MW, Telles GD, Caires ARL, Bagnato VS, Marcassa LG. (2004) Observation of ultracold ground-state heteronuclear molecules. *Phys. Rev. Lett.* **92**: 133203–1-4.
203. Wang D, Qui J, Stone MF, Nikolayeva O, Wang H, Hattaway B, Gensemer SD, Gould PL, Eyler EE, Stwalley WC. (2004) Photoassociative production and trapping of ultracold KRb molecules. *Phys. Rev. Lett.* **93**: 243005–1-4.
204. Haimberger C, Kleinert J, Bhattacharya M, Bigelow NP. (2004) Formation and detection of ultracold ground-state polar molecules. *Phys. Rev. A* **70**: 021402–1-4.
205. Vala J, Dulieu O, Masnou-Seeuws F, Fillet P, Kosloff R. (2000) Coher-ent control of cold-molecule formation through photoassociation using a chirped-pulsed-laser field. *Phys. Rev. A* **63**: 013412–1-12.
206. Luc-Koenig E, Kosloff R, Masnou-Seeuws F, Vatasescu M, Coherent control of cold-molecule formation through photoassociation using a chirped-pulsed-laser field. *Phys. Rev. A* **70**: 033414–1-19. (2004)
207. Koch CP, Luc-Koenig E, Masnou-Seeuws F. (2006) Making ultra-cold molecules in a two-color pump-dump photoassociation scheme using chirped pulses. *Phys. Rev. A* **73**: 033408–1-13.
208. Salzman W, Poschinger U, Wester R, Weidemiiller M, Merli A, Weber SM, Sauer F, Plewicki M, Weise F, Esparza AM, Woste L, Lindinger A. (2006) Coherent control with shaped femtosecond laser pulses applied to ultracold molecules. *Phys. Rev. A* **73**: 023414–1-5.
209. Brown BL, Dicks AJ, Walmsley IA, (2006) Coherent control of ultracold molecule dynamics in a magneto-optical trap by use of chirped femtosecond laser pulses. *Phys. Rev. Lett.* **96**: 173002–1-4.
210. Wright MJ, Pechkis JA, Carini JL, S. Kallush, R. Kosloff, Gould PL, (2007) Coherent control of ultracold collisions with chirped light: Direction matters. *Phys. Rev. A* **75**: 051401–1-4.

# Index